KU-767-263

AUSTRALIAN MUSEUM / REED NEW HOLLAND

# THE MAMMALS OF AUSTRALIA

**The National Photographic Index of Australian Wildlife**

**EDITED BY RONALD STRAHAN**

Reed New Holland

AUSTRALIAN
MUSEUM

A Reed New Holland book
Published 1998 by
New Holland Publishers Pty Ltd
Sydney • Melbourne • London • Cape Town

Produced and published in Australia by
New Holland Publishers Pty Ltd
3/2 Aquatic Drive, Frenchs Forest NSW 2086
167 Drummond St, Carlton VIC 3053

24 Nutford Place
London W1H 6DQ
United Kingdom

80 McKenzie Street
Cape Town 8001
South Africa

First published in 1983 as The Australian Museum
Complete Book of Australian Mammals by Angus and
Robertson Publishers.
Second edition published in 1995 by Reed Books.

© Text: The Australian Museum Trust, 1995
© Photography: Australian Museum and photographers as
credited, 1995

Library of Australia

ISBN 1 87633 401 0

Publisher: **Clare Coney**
Managing Editor: **Mary Halbmeyer**
Design: **Linda Maclean**
Printed and bound: **Imago Productions, Singapore**

All rights reserved. No part of this publication may be
reproduced, stored in a retrieval system or transmitted, in
any form or by any means, electronic, mechanical, photo-
copying, recording or otherwise, without the prior written
permission of the publishers and copyright holders.

## EDITOR

R. Strahan

## CONSULTANTS

J.E. King, Seals
G.C. Richards, Bats
C.H.S. Watts, Rodents
P.A. Woolley, Dasyurids

## PHOTO RESEARCHER

D. Greig

## CONTRIBUTORS

F.R. Allison, Maryborough, Qld

H.J. Aslin, Department of Environment and Planning, Adelaide, SA

R.G. Atherton, Cawongla via Kyogle, NSW

M.L. Augee, University of New South Wales, Sydney, NSW

P.R. Baverstock, Southern Cross University, Lismore, NSW

R.J. Begg, Conservation Commission of the Northern Territory, Darwin, NT

C. Belcher, Ecosystems Environmental Consultants, Timboon, Vic

A.R. Bentley, Australian Deer Reseaarch Foundation, Croydon, Vic

D. Berman, Conservation Commission of the Northern Territory, Alice Springs, NT

A.J. Bradley, University of Queensland, Brisbane, Qld

R.W. Braithwaite, CSIRO Division of Wildlife & Ecology, Darwin, NT

W.G. Breed, University of Adelaide, SA

L.S. Broome, NSW National Parks & WIldlife Service, Queanbeyan, NSW

M.M. Bryden, University of Sydney, NSW

A.A. Burbidge, Western Australian Wildlife Research Centre, Perth, WA

C.J. Burwell, University of Queensland, Brisbane, QLD

J.H. Calaby, CSIRO Division of Wildlife and Ecology, Canberra, ACT

F.N. Carrick, University of Queensland, Brisbane, QLD

D. Choquenet, Agricultural Research and Veterinary Centre, Orange, NSW

P. Christensen, Forests Department, Manjimup, WA

R.L. Close, University of Western Sydney, Macarthur, NSW

A. Cockburn, Australian National University, Canberra, ACT

R.J. Cole, Alice Springs, NT

B. Coman, Vernox Pest Management, Strathfieldsaye, Vic

P. Conder, Healesville, Vic

L.K. Corbett, CSIRO Division of Wildlife and Ecology, Darwin, NT

J. Covacevich, Queensland Museum, Brisbane, Qld

M.C. Crawley, Foundation for Research, Science and Technology, Wellington, New Zealand

A.J. Dennis, Malanda, Qld

C.R. Dickman, University of Sydney, NSW

J.M. Dixon, Museum of Victoria, Melbourne, Vic

B. Dorges, Lucklum, Germany

J.N. Dunlop, Murdoch University, Murdoch, WA

P.D. Dwyer, University of Queensland, Brisbane, Qld

R. Edgar, Armidale, NSW

M.D.B. Eldridge, Macquarie University, Sydney, NSW

M. Evans, Queensland Department of Environment and Heritage, Brisbane, Qld

T.F. Flannery, Australian Museum, Sydney, NSW

M.R. Fleming, Conservation Commission of Northern Territory, Alice Springs, NT

B.J. Fox, University of New South Wales, Sydney, NSW

G.R. Friend, WA Wildlife Research Centre, Perth, WA

J.A. Friend, WA Wildlife Research Centre, Perth, WA

P.J. Fuller, WA Wildlife Research Centre, Perth, WA

J. Godsell, Australian National University, Canberra, ACT

S.D. Goldsworthy, University of Tasmania, Hobart, Tas

G. Gordon, Queensland National Parks and Wildlife Service, Bribane, Qld

N.J. Goudberg, Natural Resource Assessments, Cairns, Qld

R.H. Green, Queen Victoria Museum, Launceston, Tas

A.T. Haffenden, Queensland National Parks and Wildlife Service, Brisbane, Qld

L.S. Hall, University of Queensland, Brisbane, Qld

S. Hand, University of New South Wales, Sydney, NSW

K.A. Handasyde, University of Melbourne, Vic

M. Happold, Australian National University, Canberra, ACT

D.C.D. Happold, Australian National University, Canberra, ACT

G. Heinsohn, James Cook University of North Queensland, Townsville, Qld

R.P. Henzell, Animal and Plant Control Commission, Adelaide, SA

J. Heucke, Lucklum, Germany

L.A. Hinds, CSIRO Division of Wildlife and Ecology, Canberra, ACT

R.A. How, Western Australia Museum, Perth, WA

G.A. Hoye, Singleton, NSW

S. Ingleby, Taronga Zoo, Sydney, NSW

P. Jarman, University of New England, Armidale, NSW

C.N. Johnson, James Cook University of North Queensland, Townsville, Qld

K.A. Johnson, Arid Zone Research Institute, Alice Springs, NT

P.M. Johnson, Queensland Department of Environment and Heritage, Townsville, Qld

P.G. Johnston, Macquarie University, Sydney, NSW

S. Jolly, Landcare Research New Zealand, Christchurch, New Zealand

B. Jones, Melville, WA

M. Jones, University of Tasmania, Hobart, Tas

C. Kemper, South Australian Museum, Adelaide, SA

J.A. Kerle, Biological Consultant, Alice Springs, NT

J.E. King, Springwood, NSW

T.H. Kirkpatrick, Queensland National Parks and Wildlife Service, Warwick, Qld (retired)

D.J. Kitchener, Western Australian Museum, Perth, WA

D.G. Langford, Arid Zone Research Centre, Alice Springs, NT

B.S. Law, State Forests of NSW, Sydney, NSW

L.K-P. Leung, CSIRO Division of Wildlife and Ecology, Canberra, ACT

L. Lim, Cremorne, NSW

J.K. Ling, Clare, SA

D. Lunney, National Parks and Wildlife Service, Sydney, NSW

T.H. Maddock, University of Adelaide, SA

M. Maguire, Territory Wildlife Park, Palmerston, NT

B.J. Marlow, Springwood, NSW (deceased)

R.W. Martin, Monash University, Melbourne, Vic

G.M. Maynes, Population Assessment Unit, Canberra, ACT

M. McCoy, Biological Consultant, Alice Springs, NT

J.C. McIlroy, CSIRO Division of Wildlife and Ecology, Canberra, ACT

G.M. McKay, Macquarie University, Sydney, NSW

J.L. McKean, Conservation Commission of the Northern Territory, Darwin, NT

N.L. McKenzie, WA Wildlife Research Centre, Perth, WA

J.C. Merchant, CSIRO Division of Wildlife and Ecology, Canberra, ACT

N. Mooney, Tasmania Parks & Wildlife Service, Hobart, Tas

L.A. Moore, CSIRO Division of WIldlife and Ecology, Atherton, Qld

K.D. Morris, WA Wildlife Research Centre, Perth, WA

S.R. Morton, CSIRO Division of Wildlife and Ecology, Canberra, ACT

K. Myers, Jewells, NSW

A.E. Newsome, CSIRO Division of Wildlife and Ecology, Canberra, ACT

P.D. Olsen, Sutton, NSW

P. Ong, Monash University, Melbourne, Vic

H. Parnaby, Australian Museum, Sydney, NSW

C.R. Pavey, University of Queensland, Brisbane, Qld

P. Pavlov, Pavlov Ecology, Cape Tribulation, Qld

D.J. Pearson, WA Wildlife Research Centre, Perth, WA

W. Phillips, Australian Nature Conservation Agency, Canberra, ACT

W.E. Poole, CSIRO Division of WIldlife and Ecology, Canberra, ACT

R.I.T. Prince, WA Wildlife Research Centre, Perth, WA

D.G. Read, University of New South Wales, Sydney, NSW

T.B. Reardon, South Australian Museum, Adelaide, SA

T.D. Redhead, CSIRO Division of Wildlife and Ecology, Canberra, ACT

J.R.W. Reid, CSIRO Division of Wildlife and Ecology, Alice Springs, NT

M.B. Renfree, University of Melbourne, Vic

G.C. Richards, CSIRO Division of Wildlife and Ecology, Canberra, ACT

A.C. Robinson, Department of Environment and Natural Resources, SA

R. Rose, University of Tasmania, Hobart, Tas

D.E. Rounsevell, Tasmanian Parks and Wildlife Service, Hobart, Tas

R. Russell, Mossman, Qld

G.D. Sanson, Monash University, Melbourne, Vic

J.H. Seebeck, Arthur Rylah Institute, Melbourne, Vic

M.S. Serena, Species Survival Australia, Melbourne, Vic

G.B. Sharman, Evandale, Tas

P.D. Shaughnessy, CSIRO Division of Wildlife and Ecology, Canberra, ACT

G. Singleton, CSIRO Division of Wildlife and Ecology, Canberra, ACT

A.P. Smith, University of New England, Armidale, NSW

M.J. Smith, South Australian Museum, Adelaide, SA

T. Soderquist, Monash University, Melbourne, Vic

H.J. Spencer, Tropical Research Centre, Cape Tribulation, Qld

A.N. Start, Department of Conservation and Land Management, Perth, WA

E. Stodart, Curtin, ACT

R. Strahan, Australian Museum, Sydney, NSW

G.M. Suckling, Department of Conservation and Natural Resources, Melbourne, Vic

R.A. Tedman, University of Adelaide, SA

B. Thomson, Conservation Commission of NT, Alice Springs, NT

C. Tidemann, Australian National University, Canberra, ACT

M. Trenerry, Queensland Department of Environmental Health and Heritage, Cairns, Qld

V. Turner, Monash University, Melbourne, Vic

A. Valente, Kyabram, Vic

S.M. Van Dyck, Queensland Museum, Brisbane, Qld

K.A. Vernes, James Cook University, Townsville, Qld

J.W. Wainer, University of Melbourne, Melbourne, Vic

S.J. Ward, University of Melbourne, Vic

R.M. Warneke, Warneke Marine Mammal Services, Yolla, Tas

C.H.S. Watts, South Australian Museum, Adelaide, SA

R. Wells, Flinders University, Adelaide, SA

L. Whisson, Department of Conservation and Land Management, Perth, WA

D. Whitford, South Perth, WA

C.K. Williams, CSIRO Division of Wildlife and Ecology, Perth, WA

B.A. Wilson, Deakin University, Geelong, Vic

J.W. Winter, Ravenshoe, Qld

J.C.Z. Woinarski, Conservation Commission of the NT, Darwin, NT

D.P. Woodside, Taronga Zoo, Sydney, NSW

P.A. Woolley, La Trobe University, Melbourne, Vic

D. Wurst, Wildlife Research Centre, Alice Springs, NT

R.A. Young, University of Southern Queensland, Toowoomba, Qld

# *Preface to the Second Edition*

This book is a radical revision, with a new publisher, of *The Australian Museum Complete Book of Australian Mammals* (1983). Its new title is a more succinct description of its contents and deliberately mimics John Gould's *The Mammals of Australia*, published in the first half of the nineteenth century. Using the unique resources of the Australian Museum's National Photographic Index of Australian Wildlife, it has been possible to produce a comparable work but, whereas it was possible for Gould to be familiar with all that was known about the natural history of every Australian mammal known in his time, the repository of knowledge is now so great that I have had to call upon the expertise of more than 150 individuals to encompass the field.

The stimulus for the first edition was a fine collection of mammal photographs. Since then, the collection has grown much larger and, thanks to a well-targeted search for seldom-photographed species, by the Collection Manager, Denise Grieg, most of the illustrations in this edition are new and better. In 1994 the National Photographic Index went out of existence, but its collections are now in the hands of the Museum's commercial photo-library. Very few Australian mammalogists have not contributed to the Index and it is appropriate here to give them final thanks.

There being considerably less resources of staff and funds for the production of the second edition, I have relied heavily upon voluntary assistance and it is with deep gratitude that I acknowledge the dedication with which my Editorial Assistants, first Suzi O'Rear, then Jan Noble, undertook the boring, repetitive clerical operations involved in dealing with a multitude of contributors. I am also pleased to mention the patient persistence of a Clerical Officer, Peter Klobe, who saw four successive drafts of the manuscript through his computer.

Finally, and with due humility, I must express my appreciation of the confidence with which many authors of this work have entrusted their findings into my hands and—not always with unalloyed joy—seen their deathless prose transformed into a form that I preferred. I hope that the end has justified my means.

*Amendments to the Second Printing of the Second Edition.*

This revised printing has been produced by New Holland Publishers. It takes into account various errors or omissions but does not attempt to update the text.

*Ronald Strahan*
*Australian Museum*
*October 1997*

# Contents

Australia lies between latitudes 10°S and 44°S and longitudes 113°E and 154°E. It comprises two major landmasses—the mainland, and the island of Tasmania—and numerous small islands. Total land area is 7 682 300 square kilometres. East–west across the mainland the longest distance is about 4000 kilometres; and north–south it is 3180 kilometres from Cape York (Queensland) to Wilsons Promontory (Victoria). Nearly two-thirds of the continent is arid or semi-arid, where inland rivers flow only after rare rainfall, and drainage into basins such as Lake Eyre occurs only a few times in a century.

ARAFURA SEA

COBOURG
PENINSUL

TIMOR

Melville Is.

Bathurst Is.

SEA

DARWIN · Alligator
Rivers

JOSEPH
BONAPARTE
GULF

Daly R.

ARNHE
LAN

Katherine

Roper R

Wyndham ·

KIMBERLEY

Ord R.

Victoria R.

BARKL

KING LEOPOLD
RANGES

Derby ·

Fitzroy R.

· Broome

Tennant Creek ·

TANAMI
DESERT

NORTHERN

TERRITORY

GREAT SANDY DESERT

Port Hedland ·

De Grey R.

WESTERN AUSTRALIA

Alice
Springs

MACDONNELL RANGES

EXMOUTH GULF

Fortescue R.

HAMERSLEY RANGE

Ashburton R.

PILBARA

GIBSON DESERT

INDIAN

WARBURTON RANGE

· Carnarvon

SHARK BAY

MUSGRAVE RANGES

SOUTH

AUSTRALIA

GREAT VICTORIA DESERT

Coober Pedy ·

Murchison R.

Houtman
Abrolhos · Geraldton

NULLARBOR

PLAIN

Lak
Gairdne

· Kalgoorlie

Eucla ·

PERTH

Swan R.

Esperance ·

GREAT

AUSTRALIAN

BIGHT

OCEAN

Blackwood R.

Recherche
Archipelago

· Albany

0        200        500                1000 kilometres

0              200        400            620 miles

The northernmost regions are affected by tropical weather patterns, notably a season of heavy rains from November to April in the Kimberley, Arnhem Land and far north Queensland. Sites on the Great Dividing Range inland from Cairns record an average yearly rainfall of more than 4 metres.

At the southern end of the Great Dividing Range are the Snowy Mountains, with Mount Kosciusko the highest peak (2228 metres). West of the Range, streams flow into the Murray–Darling river system, traversing landscapes dominated by agriculture. Just over 5 per cent of Australia is incorporated in national parks and reserves.

TORRES STRAIT

Wessel Islands

GULF
OF
CARPENTARIA

Groote Eylandt

Sir Edward
Pellew Group

Mornington Is.

CAPE
YORK
PENINSULA

GREAT BARRIER REEF

• Cooktown

Mitchell R.

• Cairns

BLELAND

Gregory R.

Norman R.

Burdekin R.

• Townsville

•Mount Isa

Georgina R.

QUEENSLAND

GREAT DIVIDING RANGE

MPSON
ESERT

Diamantina R.

CHANNEL
COUNTRY

• Rockhampton

TROPIC OF CAPRICORN

PACIFIC

•Birdsville

Cooper Creek

Bulloo R.

• Charleville

Condamine R.

Fraser Is.

ke R.

Lake Eyre

• BRISBANE

Lake
rrens

RANGE

Darling R.

FLINDERS RANGES

BARRIER

Macquarie R.

Armidale

• Coffs Harbour

OCEAN

•Broken
Hill

NEW SOUTH WALES

Hunter R.

Port Macquarie

yalla•

Lachlan R.

GREAT DIVIDING RANGE

Newcastle

• SYDNEY

NINSULA

ER GULF

•Mildura

Murrumbidgee R.

• ADELAIDE

Murray R.

Wagga Wagga
Albury

• CANBERRA

A.C.T.

Kangaroo
Is.

VICTORIA

•Bendigo

Snowy R.

▲ MT KOSCIUSKO

Mount
Gambier

•Ballarat

MELBOURNE

Geelong •

WILSONS PROMONTORY

King Is.

BASS STRAIT

Flinders Is.

TASMANIA

• Launceston

• HOBART

# Introduction

DAVE WATTS

This work, to the best of my knowledge, provides an account of every species of native mammal known to have existed in Australia since European settlement and every introduced species now living in a wild state. In this context, Australia is regarded as the continent, Tasmania and offshore islands; more distant areas under Australian administration are not considered. It is a little difficult to define just which exotic species exist in a feral state and this point is discussed further on page 722. Accounts are given of several species of seals that are only occasional visitors to Australian shores and the Dugong—which does not come ashore—is included because it is confined to shallow coastal waters and may enter estuaries. Whales and dolphins may also enter estuaries, but they are left out of consideration because those recorded from around the Australian coast are primarily inhabitants of the open ocean.

The naming of animal species is a discipline with a multitude of procedural rules, one of which is that a scientific name is not regarded as valid until it has been published together with a formal description of the animal to which it applies. Several species included here are referred to only by suggested common names.

*The snout of the mature male Southern Elephant-seal is developed into a trunk-like structure.*

To be more precise, then, it should be said that the native mammals treated here are those species, currently regarded as valid, that are believed to have been extant for at least some time since 1788 and had been formally described by mid-1994 (plus a few that have yet to be described).

In the accounts of each species we have tried to set down where it lives, what it eats, what it does, how it reproduces and grows and the various factors that lead to death: in summary, we attempt to evaluate its present status.

It is not easy to define where a species lives. The blank area on a distribution map, showing where a species is *not* found, is more informative than the blocked-in areas that indicate where it *may* be found. The distribution of a species within its range is usually quite discontinuous, being restricted to particular habitats which provide its specific requirements of food, water and shelter. We have attempted to define habitats in terms of local climate, vegetation and soil, but the reader will find that the treatment is variable because, in some instances, all this information is not available.

In respect of diet, information is also variable: only in a few instances have we been able to be quite specific. In order to utilise certain foods, animals require very specific adaptations for capture, grazing, chewing or digestion and mention is made of these anatomical or physiological specialisations, either in the account of a particular species or in the remarks that introduce an animal group. The extent to which a species depends upon access to drinking water is usually mentioned and the reader will find that quite a large number of Australian mammals either do not drink or can survive for long periods without drinking.

A reasonably complete account of the behaviour of any one species could fill a book, but such an embarrassment of riches exists for less than half a dozen Australian mammals. What we have attempted is to describe briefly the daily and annual patterns of activity, including feeding, grooming, sheltering, attack and defence, social interactions, courtship, mating and care of the young, but the lives of some species are still insufficiently understood to enable all of these activities to be covered.

In most cases, we have been able to specify the mating season(s) of a species, the length of gestation, the litter size and the period of infancy. Information on parental care of the young is more fragmentary and there are few instances where we can give a reasonable account of the life (and mortality) of animals in the period between weaning and adulthood.

It seems that all Australian mammals have been affected by European settlement. A few large grazing kangaroos have benefited from the clearing of forests and the extension of grasslands and some native rodents may have increased in numbers in response to the introduction of sugar cane in north-eastern Australia, but most species appear to have decreased in range and numbers; some to the point of extinction. Where possible, we have indicated the factors leading to the decline of a species and what should be done to arrest this.

Although this is a book about individual species, it must be recognised that each one exists in a web of life in which it eats other species and is itself eaten in a cycle which, if undisturbed, remains relatively stable for tens of thousands of years. Yet changes do occur, sometimes forced by variation in climate, sometimes by shifts in balance as new species arise in the course of evolution. This is not a book about ecology or evolution but, in order to make some sense of the variety of native mammals, some mention is made of their ecological and evolutionary relationships.

The hierarchy of categories used here in the classification of Australian species of the class Mammalia extends through subclass, order, suborder and superfamily to family (the category given greatest prominence), but where a family includes a large number of Australian species, it may be treated at the level of subfamilies or tribes. In the text, we have used the conventional suffixes -oid, -id and -ine respectively to refer to members of a superfamily, family or subfamily. For example, every member of the superfamily Dasyuroidea is a dasyuroid, every member of the family Dasyuridae is a dasyurid and every member of the subfamily Dasyurinae is a dasyurine.

In the first edition, I attempted to place related species close together, but this proved confusing. Here, to emphasise that relationships should not be inferred from proximity, species are arranged alphabetically within their genera (which themselves are usually arranged alphabetically within their families or subfamilies).

The information summarised in these pages is based on the researches of many hundreds of biologists recorded in thousands of publications. It would have been unwieldy and of no interest to the general reader to cite each of these sources, so the text has been written as a continuous (dogmatic) narrative. Individuals are mentioned only when they are directly quoted or in respect of their historical interest. However, the account of each species is followed, where possible, by references to a few publications that can lead the reader into the relevant literature.

Although the species accounts are designed to be accessible to the general reader, this book is used by students, field naturalists and zoologists. A certain amount of technical information is therefore provided in condensed form.

# The Species Accounts

## Name

In the first edition, I included only the common and scientific names of each species. For the convenience of professionals, I here include citation of the author and date of the original description. Where the author's name is in brackets, this means that the describer placed the species in a genus different from the one to which it is here assigned.

The system of biological nomenclature was developed two centuries ago when every educated westerner could read Latin and had some knowledge of Greek. Descriptive names constructed from Latin and Greek roots were widely understood and served as reminders of the features of a genus or species. At present, when hardly anybody has much knowledge of classical languages, I believe it worthwhile to translate these names. Moreover, once one recognises the roots that have been used in a name, the proper pronunciation of that name becomes more apparent.

I have therefore provided, where possible, a brief translation of each name and an indication of a preferred pronunciation. I need hardly add that I expect to have little influence on the established generation of Australian mammalogists (those who refer to themselves as mam-ol'-oh-jists and to *Pseudomys* as sude-oh'-meez) but I hope to act as a subversive influence on their students.

## Size

This is usually expressed (in respect of adult animals) as the combined length of the head and body, the length of the tail and the weight. In the case of bats, the forearm length (a useful diagnostic parameter in this group) is also given. Where the sexes are known to differ significantly in size, their measurements are separated. Measurements are usually expressed as a range from the smallest to the largest, followed (in brackets) by the mean, or average, of the measurements available to the author. Sometimes the available data are insufficient to permit this, in which case they are presented either as a range or as estimates.

## Identification

No attempt has been made to provide a full or diagnostic description of every species. The information given is designed merely to *differentiate it from those of similar size and appearance that may be found in the same general area*. In most cases, the information given enables the species to be identified from its size, colour and shape. Where identification depends upon details of the teeth, skull, chromosomes, or blood proteins, we mention the fact and go no further.

## RECENT SYNONYMS

There are few species that have not been known in the past by several scientific names (synonyms). These have been well summarised in the *Zoological Catalogue of Australia: 5, Mammals* (1988). Under the heading of 'Recent Synonyms', we include only those that have had some currency in the past 50 years or so.

In some instances, a synonym is followed by the qualification '(part)'. This means that the species under discussion is not the complete equivalent of those animals referred to by the synonym. Such a situation arises when an assemblage of animals once regarded as constituting a single species is subsequently judged to consist of two or more.

## COMMON NAMES

Ignorance and local loyalties will always lead to confusion over common names. Those used in this book conform largely to the 1980 recommendations of the Australian Mammal Society. Under the heading 'Other Common Names', we include only those which we believe still to have some usage in some part of Australia. For common names used in the past, the reader is referred to my *A Dictionary of Australian Mammal Names* (1981).

## STATUS

When a species is widespread and common, it is easy to designate it as such. It is much harder to determine the status of a species that is seldom seen or trapped. It may always have existed in small numbers and, indeed, owe its continued existence to being a minor element of an ecosystem. It may be an opportunistic breeder which exists in small numbers in restricted refuge areas for many years, erupting only when unusual rainfall provides particularly favourable conditions. It may be prevalent but evasive and trap-shy. There may be insufficient biologists interested enough to devote time to searching for it in appropriate areas, or the areas in which it lives may be inaccessible. Such situations have led to revision of the status of some species previously regarded as endangered or extinct and these revisions serve as salutary warnings against premature pessimism. Nevertheless, there is much well-documented evidence of the drastic decline of many species that were once widespread and in many cases it is possible to attribute the decline to alienation of habitat. Here one can be dogmatic: if the habitat disappears, so will the species. In evaluating the status of each species, we have attempted to be as objective as possible, basing judgement on all the evidence available and it may be noted that, even during the three years during which this book was being written, new information has led us to upgrade the status of several species. Nevertheless the prognosis for many remains poor. These questions are addressed at greater length in the following section on 'Conservation of Australian Mammals'.

## SUBSPECIES

Biologists differ in their definitions of groups—subspecies, races, forms—below the species level but recognise that, in different parts of its range, populations of a species may differ in size, shape, colour, behaviour, chromosomes or biochemical characteristics. Where such variation appears to be continuous (from warm to cold or wet to dry climate) it is usually referred to as a cline; discrete groups, often separated by some geographical feature, are usually regarded as subspecies. There are numerous cases where present knowledge is insufficient to distinguish between these conditions but, where subspecies, geographical races, or forms are recognised, these are mentioned.

## EXTRALIMITAL DISTRIBUTION

Some species native to Australia are also found elsewhere: in New Guinea, South-East Asia, or even further afield. In such cases, their distribution outside the geographical limits of Australia is briefly indicated.

## DISTRIBUTION MAPS

Each map has been drawn by the author of the species account and checked by the section editor or myself. Peter Menkhorst provided detailed information on the distribution of species found in Victoria, while Andrew Burbidge and Norman McKenzie checked the maps of species occurring in Western Australia. Nevertheless, no map provides more than a crude indication of where a species may be found and each must be subject to continual revision in the light of further research.

Where the range of a species is known to have contracted since European settlement of Australia, its former distribution is indicated by a lighter tinted area.

At a quite late stage in finalisation of the maps for this edition Andrew Burbidge drew my attention to the possibility of adding a third component to the maps of at least some species, particularly from the western half of the continent: this is their recent fossil distribution. Thanks largely to data provided by Alex Baynes, but with the assistance of Burbidge and McKenzie, this information is indicated in hatched areas, *where known*. The absence of either lighter tinted or hatched areas from a map does not imply that the species did not once have a wider distribution; it usually reflects a lack of information.

Similarly, the known distribution of marine animals is provided, and where the species sometimes strays into an area outside the usual range, a stipple indicates this.

# The Australian Mammal Fauna

Native Australian mammal species comprise only 5 to 6 per cent of the estimated world total, but this is appropriate to a continent that constitutes only 5.7 per cent of the land surface of the earth (excluding Antarctica). However, the diversity is markedly less than in other major landmasses for, of the 17 orders of living terrestrial mammals, only four (monotremes, marsupials, rodents and bats) include species that are native to Australia, whereas seven to ten orders are represented in each of the other continents. All three groups of marine mammals (cetaceans, seals and dugongs) occur in Australian coastal waters.

There is a marked disproportion in the representation of each order. Monotremes occur only in Australia and New Guinea; about two-thirds of the living marsupials occur in Australia and New Guinea; but Australian rodents comprise only about 4 per cent of the world total. Species of bats found in Australia make up about 7 per cent of the world total but many of these are also found outside Australia: only about 4 per cent are endemic.

## ORIGINS

The peculiar composition of the Australian mammal fauna is to some extent explicable in terms of the origin of the Australian continent. It was once part of Gondwanaland, a southern super-continent that also included the landmasses that we now recognise as South America, Africa, Madagascar, India and New Zealand. Africa became separated some 160 million years ago; moving northwards to abut against southern Europe. Then, about 125 million years ago, India broke away, drifted northwards and collided with Asia about 45 million years ago; the Himalayas are a result of the collision. New Zealand started moving eastwards into the Pacific about 80 million years ago, but Australia, Antarctica and South America remained connected until about 55 million years ago (at which time about one-third of the polar continent was north of the Antarctic Circle). The Australian landmass (including New Guinea) began to move across the Indian Ocean to Asia and the northern edge of its continental shelf met that of South-East Asia about 15 million years ago, leading to a buckling that created the New Guinean highlands.

Australia is still on the move at the rate of about 6 centimetres a year but it remains an isolated continent.

The native mammals of Australia are derived from two sources. The oldest are descendants of animals (monotremes and marsupials) that were on the continent when it broke its moorings from Antarctica and remained as passengers while Australia was drifting across the Indian Ocean: they may be said to have had the run of the ship. Rodents began to come aboard—probably first into New Guinea—as the continent approached the Indonesian Archipelago. The Asian and Australian faunas are now so close that, at their nearest, they are separated only by a narrow channel between the islands of Bali and Lombok, but this still constitutes a significant barrier to most mammals. Some bats undoubtedly flew across when the marine barriers were much wider: rodents presumably drifted from island to island on floating vegetation.

In 1992, the discovery in Patagonia of a 54-million-year-old monotreme tooth established the fact that monotremes originally had at least a partially Gondwanan distribution. We do not know when monotremes became extinct in South America, nor whether they moved from Australia across Antarctica to South America or vice versa. The Patagonian tooth is similar to that of the oldest fossil platypus teeth from Australia but we cannot necessarily assume that the most primitive monotreme was platypus-like. It seems much more likely that it was a very less specialised mammal. Perhaps the most significant lesson learnt from this discovery is that we should avoid making evolutionary generalisations based on the *lack* of evidence.

Fossil marsupials about 100 million years old are known from North America. It seems that marsupials moved thence into Europe, where they were not particularly successful and into South America where they underwent an extensive evolutionary radiation while becoming extinct in the northern continent. South America is still the home of about one-third of the living species of marsupials, mostly members of the opossum family, Didelphidae (one species of which has reinvaded North America).

These largely arboreal and omnivorous marsupials are similar to Australian possums in general appearance (which is why the Australian animals were so named). They also resemble Australian carnivorous marsupials in respect of their dentition. However, current opinion is that the living marsupials of the world fall into two very distinct groups: the cohorts Ameridelphia and Australidelphia, which not only have very distinct distributions but are only distantly related. Surprisingly, it is not easy to define the differences between the groups on the basis of a few easily recognised characters. Perhaps the most fundamental difference is that the sperms of ameridelphians are paired (heads joined, tails free), while those of australidelphians are single, as in most other mammals.

In the first edition, I followed a very old tradition of linking the Australian carnivorous marsupials and bandicoots with the American opossums on the grounds that all three shared a polyprotodont dentition (many incisors in the upper and lower jaws). However, since this is probably an ancient, relatively unspecialised condition, present in the common ancestor of the ameridelphians and australidelphians, the coincidence is of little evolutionary significance.

Although a few fossil marsupials have been found in Antarctica (not very far from Patagonia), these are not intermediate between the Ameridelphia and Australidelphia.

In the course of some 40 million years during which Australia was adrift in the Indian Ocean, its marsupials underwent a considerable adaptive radiation. The carnivorous dasyurids probably remain closest to the original stock, while the kangaroos are among the most specialised of a herbivorous group that includes the possums. The relationships of the bandicoots remain obscure.

The fact that New Guinea has a significant marsupial fauna is often overlooked. Although lacking some of the more specialised families, it has a good representation of dasyurids, bandicoots, ringtail and brushtail possums and kangaroos. Several groups, notably the cuscuses, are so successful that they were able to move from New Guinea into northern Australia prior to the formation of Torres Strait. A few marsupials have also spread westwards with human assistance to Timor and eastwards to the Solomons.

As remarked earlier, the diversity of the Australian bats is a little larger than might be expected in terms of the area of Australia. The first bats probably entered the continent 50–55 million years ago; some Australian fossil remains are of that age. It is unclear how they came into Australia. For some species that arrived earlier, it is highly likely that entry was primarily through north-western Australia and Cape York—on a number of occasions. There has been some evolution of bats within Australia and New Guinea but there has been no great radiation. Only two genera are endemic to Australia.

Entry to Australia across marine barriers was more difficult for rodents. Present evidence indicates that a few species managed to migrate from eastern Indonesia before 3 million years ago and to found a diverse rodent fauna in the New Guinea region. Some 4 million years ago, some members of this group entered Australia and embarked upon a further evolutionary radiation, resulting in a large number of species known as the 'old endemics'. Much more recently, not more than a million years ago, some 'true' rats entered Australia from New Guinea and evolved into a small number of species known as the 'new endemics'.

Old endemics occupy habitats ranging from rainforest to desert. They include aquatic and arboreal species and, among the more numerous terrestrial forms, there is a genus of hopping-mice. None of the eight species of new endemics varies much in appearance from typical Asian rats and few are adapted to arid habitats. The number of species of Australian bats and rodents is almost equal, but the latter have undergone a much greater evolutionary radiation.

At present, some of the most widespread and successful species are those that have been introduced by humans. Although the Dingo is often regarded as a member of the indigenous fauna, it may have been introduced as recently as 5000–4000 years ago. The feral Cat and House Mouse, which came to Australia with European settlers, have now spread over almost the whole of the continent; the Rabbit and Fox each occupy about two-thirds of the land. Large herbivores such as the Horse, Donkey, Camel, Goat, Pig, feral cattle and deer are also so well established that they, too, must be regarded, with whatever degree of regret, as members of Australia's mammal fauna.

## CLASSIFICATION OF MAMMALS

In order to come to grips with the immense numbers and variety of animal species, they are named and classified according to international conventions. Names used in classification often run parallel to those used in ordinary language, but zoologists deal with many species that are unknown to the general public and therefore have no common names. Classification—the grouping of species into a hierarchy of categories—may also reflect common usage but, because zoologists attempt to arrange such hierarchies to reflect probable evolutionary relationships, they are subject to revision in the light of new information coming to hand. The elements of the classification of mammals are set out below. The treatment serves mainly to introduce a number of names applied to the subclasses, orders, suborders, superfamilies and families which comprise the Australian mammal fauna.

Evidence that mammals evolved from reptiles is provided by fossils 250–180 million years old, which include a large number of species sharing so many reptilian and mammalian features that they cannot be assigned definitely to either group. Mammals are, in effect, 'improved reptiles' (as are birds) and it seems that a number of reptilian groups independently achieved a mammalian level of organisation. Some of these groups are extinct but three have survived to the present time. These are the egg-laying monotremes; the marsupials, which give birth to small 'embryonic' young; and the eutherian mammals, which nourish their embryos by means of a placenta and give birth to young that are fully formed, or almost so. All living mammals are readily distinguishable from living reptiles by being warm-blooded, feeding their young on milk and possessing hair.

## Monotremes: Subclass Prototheria

This group, consisting of the Platypus and the echidnas, includes only one order, the Monotremata. Female monotremes lay soft-shelled eggs and suckle their young on milk secreted through numerous ducts opening onto the abdomen, not via teats. The name Monotremata ('one hole') refers to the single opening through which both faeces and urine are voided from the body. The eggs of females are laid through this common aperture but the semen of males is discharged through a penis.

On the ankle of males is a hollow spur which, in the Platypus, is connected to a venom gland but which appears to have no function in the echidna. Monotremes lack vibrissae ('whiskers') but this, like the absence of teeth and of external ears, may be the result of loss in the course of their evolution.

Living monotremes are grouped into two families: **Ornithorhynchidae**, including the Platypus; and **Tachyglossidae**, including the echidnas. Each of these subgroups probably warrants a much higher status.

## Marsupials: Subclass Marsupialia

The most striking feature of marsupials is that they are born in an embryonic condition: tiny, naked and with the hindlimbs and tail still incomplete. Contrary to many textbook statements, some marsupials (e.g. bandicoots and the Koala) have relatively well-developed placentas, but the presence of a placenta does not lead to better development of the neonate.

A newborn marsupial clambers from its mother's cloacal opening to a teat and attaches itself to this until it is relatively well developed. Teats are usually enclosed in a pouch but this is not always the case, particularly in some small carnivorous marsupials, where the young simply dangle from the mother's teats and are dragged along with her.

## Cohort Australidelphia: Australian and New Guinean marsupials

As remarked earlier, this group is difficult to define. It comprises four orders. The Dasyuromorphia are carnivorous or insectivorous. The Peramelemorphia are omnivorous. The Diprotodontia are primarily herbivorous but include secondarily insectivorous and nectarivorous species. The Notoryctemorphia comprises the Marsupial Mole.

**ORDER DASYUROMORPHIA: CARNIVOROUS MARSUPIALS** Except for the Numbat, members of this group have four pairs of needle-like incisors in the upper jaw and three similar pairs in the lower jaw; the canine teeth are well developed; the premolars and molars are sharply serrated. The snout is usually elongate and the legs of similar length. The first digit of the hindfoot is never strongly developed and is lacking in some species. No digits of the hindfoot are fused. The tail is never prehensile.

The order comprises three families. The **Dasyuridae** includes a large number of rather unspecialised predators (quolls, antechinuses, dunnarts, etc. and the mainly scavenging Tasmanian Devil. The **Thylacinidae** has only one species, the Thylacine ('Tasmanian Tiger'), which has been extinct for more than 50 years. The Numbat, sole member of the family **Myrmecobiidae**, is a specialised termite-eater, with degenerate dentition.

**ORDER PERAMELEMORPHIA: BANDICOOTS AND BILBIES** Members of this order have four or five pairs of blunt incisors in the upper jaw and three similar pairs in the lower jaw. The snout is long and tapered; the head gives the impression of being an extension of the body. The second, third and fourth toes of the forefoot bear stout, flattened claws, used in digging; the other digits are reduced or absent. The hindlimbs are much larger than the forelimbs and contribute to a bounding gait. The fourth toe of the hindfoot is longer and much more powerful than the others. The second and third toes are small and joined together except for their claws. The tail is not prehensile.

The order is currently divided into two families. The **Peroryctidae** includes many New Guinean species but the only Australian member is the Rufous Spiny Bandicoot. All other Australian species (bandicoots and bilbies) are placed in the family **Peramelidae**.

**ORDER DIPROTODONTIA: KOALA, WOMBATS, POSSUMS AND MACROPODS** This large and diverse order of the cohort Australidelphia is characterised by the presence of only one pair of *functional* incisors in the lower jaw: there may be one to three pairs in the upper jaw. The second and third toes of the hindfoot are joined together, except for their claws. The order comprises two suborders.

The suborder Vombatiformes comprises two distantly related families. The **Phascolarctidae** contains only the Koala. The **Vombatidae** includes the wombats.

The suborder Phalangerida is a large group divided into five superfamilies, covering eight families.

The Phalangeroidea has only one family, the **Phalangeridae**, comprising the brush-tail possums, cuscuses, and the Scaly-tailed Possum.

The Burramyoidea has only one family, the **Burramyidae**, comprising the pygmy-possums.

The Petauroidea includes two families. The **Petauridae** comprises the wrist-winged gliders, Leadbeater's Possum and the peculiar Striped Possum. The **Pseudocheiridae** comprises the ringtail possums and the Greater Glider.

The Tarsipedoidea includes at least one family, the **Tarsipedidae**, which has one species, the Honey Possum. It has been suggested that it should also include the family **Acrobatidae**, which also has only one species, the Feathertail Glider. If this is not accepted, the Acrobatidae can perhaps be placed within the Burramyoidea.

The Macropodoidea includes two families. The **Potoroidae** comprises the potoroos, bettongs and rat-kangaroos. The **Macropodidae** comprises the wallabies, kangaroos and tree-kangaroos.

**ORDER NOTORYCTEMORPHIA: MARSUPIAL MOLE** This highly specialised marsupial, which normally spends its entire life underground, is the only member of this order and its single family **Notoryctidae**.

## Eutherian or Placental Mammals: Subclass Eutheria

This large group, the most successful of the three subclasses, is most simply defined as comprising those mammals that are neither monotremes nor marsupials. Living species fall into 16 or 17 orders, of which three form part of the terrestrial or partly terrestrial native Australian fauna. These are the orders Chiroptera (bats), Rodentia (rodents) and Carnivora (Dingo, Fox, Cat and seals).

Human activities have led to the introduction of representatives of three other orders: Lagomorpha (Rabbit and Hare); Perissodactyla (Horse and Donkey); and Artiodactyla (Pig, Camel, Sheep, Goat, deer and cattle).

**ORDER CHIROPTERA: BATS**  Bats divide into two very distinct suborders.

The Megachiroptera, comprise the flying-foxes, fruit-bats and blossom-bats. All Australian megachiropteran species are included in the family **Pteropodidae**.

The suborder Microchiroptera, is a diverse group of mainly insectivorous bats. The Australian microchiropterans include representatives of six families.

The **Emballonuridae** includes seven species of sheathtail-bats.

The **Megadermatidae** has only one representative, the Ghost Bat.

The **Molossidae** has five species of freetail-bats.

The **Rhinolophidae** has three species of horseshoe-bats and the very similar **Hipposideridae** has five species of leafnosed-bats.

The **Vespertilionidae** has 11 genera and more than 30 species.

**ORDER RODENTIA: RODENTS**  All the Australian rodents are members of the family **Muridae**, which also includes the introduced mouse and rats.

**ORDER SIRENIA: DUGONG AND MANATEES**  The family **Dugongidae** has only one living species; the manatees are related.

**ORDER CARNIVORA: DINGO, DOG, FOX, CAT AND SEALS**  Australian carnivores fall into four families: the **Canidae** (Dingo and introduced Fox), **Felidae** (introduced Cat); **Otariidae** (fur-seals and sea-lions); and **Phocidae** ('true' seals).

**ORDER LAGOMORPHA: RABBIT AND HARE**  The Rabbit and Hare are members of the family **Leporidae**.

**ORDER PERISSODACTYLA: HORSE AND DONKEY**  The Horse and Donkey, both introduced, are members of the **Equidae**.

**ORDER ARTIODACTYLA: PIG, CAMEL, GOAT, SHEEP, DEER AND CATTLE**  Australian artiodactyls include representatives of four families: **Bovidae** (cattle and Goat); **Camelidae** (Camel); **Cervidae** (deer); and **Suidae** (Pig).

# Conservation of Australian Mammals

by A.A. Burbidge

Australia's mammals have not fared well since the arrival of Europeans. Over the past two centuries, 17 species have become extinct, more than in any other continent during this period. Nine species that once occurred over much of the mainland now survive only on offshore islands and many others are threatened with extinction. Three categories are used to define the degree of threat.

### * Presumed Extinct
Species not definitely located in the wild during the past 50 years or have not been found in recent years despite thorough searching.

### * Endangered
Species in danger of extinction and whose survival is unlikely if the causal factors continue operating. (Causal factors are all those contributing to reduction in numbers and/or habitat to a critical level.) Also included in this category are those species that may be extinct but which have been seen in the wild in the past 50 years although not subject to a thorough search that could settle the question of their continued survival.

### * Vulnerable
Species believed likely to move into the 'endangered' category in the near future if the causal factors continue to operate, or for which there are significant potential threats to their survival.

## Lists of Threatened Species

The Endangered Fauna Network, a body of scientists from all Australian government conservation agencies, has produced an official list of threatened Australian mammals. This is open to variation in the light of changing circumstances or increased information and is likely to be extended by inclusion of the recently rediscovered Mahogany Glider as an endangered species. If the Lord Howe Island Long-eared Bat (known only from a single skull) can be shown to have been alive less than 200 years ago, it must be added to the 'presumed extinct' category. On the other hand, the Percy Island Flying-fox (also known only from a single specimen) will need to be deleted if, as is possible, it is found not to be a valid species.

## Patterns of Decline

Extinctions and significant declines among terrestrial Australian mammals are virtually confined to non-flying species weighing between 35 and 5000 grams, referred to as the Critical Weight Range (CWR). Mammals within the CWR have declined most in arid and semi-arid areas, less so in the wetter south-east, south-west and moderately wet tropical areas, though hardly at all in the wet tropics. No forest-dwelling species has become extinct, but Leadbeater's Possum is threatened by a felling cycle that may not permit the retention of sufficient trees old enough to have holes suitable for its nests. The Mahogany Glider will be similarly disadvantaged by the replacement of eucalypt forest by softwood plantations.

## Introduced Herbivores

Grazing of cattle and Sheep is the major land-use of most of the arid zone but many other introduced herbivores, such as the Rabbit, Goat, Horse, Donkey, Camel and Pig, extended beyond leased areas into conservation reserves and unoccupied deserts. These exotic mammals have had an immense impact on native vegetation and, consequently, upon the indigenous mammals that depend upon that vegetation for food and shelter. Overgrazing in times of drought has destroyed habitats and, even outside Sheep country, the small areas of high-quality habitat that once provided refuges for many native species in times of drought have been severely degraded by other introduced herbivores, especially the Rabbit.

## Introduced Predators

The Dingo, introduced 8000–3000 years ago, must have had an impact on the pre-existing fauna (almost certainly being responsible for elimination of the Thylacine from the mainland), but, by the time of European settlement, the Dingo had become, in effect, a member of the indigenous fauna.

The Cat may have become established prior to European settlement, perhaps from shipwrecks on the west coast in the seventeenth century, or even earlier by Asian fishermen. Desert Aborigines report that it has 'always been present', or that it 'came from the west'. Explorers of the western desert in the 1890s found the Cat to be a common food item of Aborigines. There is, however, no firm evidence of extinctions caused by the Cat. It appears to have eliminated populations of CWR mammals from some offshore islands but not others. There is no doubt that it preys on mammals as large as the Rufous Hare-wallaby (up to 1960 grams) and Burrowing Bettong (1500 grams), but extinct species such as the Desert Rat-kangaroo (900 grams), Central Hare-wallaby (1500 grams) and Pig-footed Bandicoot (200 grams) persisted until at least the 1930s, probably into the 1950s, suggesting that the Cat was not the only causal factor in their extinction. Recent research has shown that predation by Cats on Rufous Hare-wallabies and Burrowing Bettongs is a major impediment to experimental reintroductions of these species into parts of their former range.

The Red Fox did not become established until the 1860s but colonised most of the continent (except the wet tropics) by the 1930s. It is clear that, in some parts of Australia, native mammals had declined or disappeared before the Fox arrived.

Nevertheless, the Fox is known to have eliminated remnant CWR populations on the mainland.

Control measures in parts of south-western Western Australia have led to a resurgence of such mammals as the Numbat, Western Quoll, Brush-tailed Bettong, Black-footed Rock-wallaby, Tammar Wallaby and Common Brush-tailed Possum. On Dolphin island, off Dampier, control of the Fox has led to an enormous increase in Rothschild's Rock-wallaby (5250 grams). Evidence such as this suggests that the Fox has been the primary cause of the decline and/or extinction of many CWR species.

**Fire Regimes**  For many thousands of years Aborigines burnt the country as part of their hunting-gathering way of life. Many small fires throughout the year created a mosaic of vegetation at different stages of regeneration, providing shelter and a variety of plant and insect food. Displacement of Aborigines by European pastoralists and farmers led to infrequent, large summer fires that have a homogenising effect on the vegetation. It is significant that the disappearance of many mammals from the deserts coincided with depopulation of the Aborigines and the subsequent change in fire regime.

**Interactions**  It is clear that the interaction of several factors has been responsible for the high rate of extinction of Australian mammals since European settlement.

Chief among these are habitat changes brought about by introduced herbivores, predation by introduced carnivores and changed fire regimes. Clearing from agriculture or urbanisation is against the interests of *most* species but some, such as the large kangaroos appear to have benefited from the extension of grasslands and watering points.

## Management of Threatened Species

Two main approaches are being taken to the study and management of threatened species. One is to treat a species on its own, according to a Recovery Plan that documents the threats to it and prescribes actions to counteract these: the Western Quoll and Greater Stick-nest Rat are among species to be so treated. The other approach is to concentrate on threatening processes: recent efforts to control the Fox in south-western Australia have had beneficial effect on many native mammal species.

Where direct causes of decline can be identified, management of species or communities can be addressed with some confidence. Sadly, our knowledge of the biology of many threatened mammals is so slight that we do not yet have the information necessary for rational programs of management.

| | Marsupials | Bats | Rodents |
|---|---|---|---|
| **PRESUMED EXTINCT** | Thylacine | Percy Island Flying Fox | White-footed Rabbit-rat |
| | Desert Bandicoot | | Gould's Mouse |
| | Pig-footed Bandicoot | | Lesser Stick-nest rat |
| | Lesser Bilby | | Big-eared Hopping-mouse |
| | Broad-faced Potoroo | | Long-tailed Hopping-mouse |
| | Desert Rat-kangaroo | | Short-tailed Hopping-mouse |
| | Eastern Hare-wallaby | | Darling Downs Hopping-mouse |
| | Central Hare-wallaby | | |
| | Crescent Nailtail Wallaby | | |
| | Toolache Wallaby | | |
| **ENDANGERED** | Western Quoll | | Central Rock-rat |
| | Kowari | | Shark Bay Mouse |
| | Dibbler | | Heath Rat |
| | Red-tailed Phascogale | | Greater Stick-nest Rat |
| | Julia Creek Dunnart | | Dusky Hopping-mouse |
| | Numbat | | |
| | Golden Bandicoot | | |
| | Western Barred Bandicoot | | |
| | Northern Hairy-nosed Wombat | | |
| | Leadbeater's Possum | | |
| | Long-footed Potoroo | | |
| | Brush-tailed Bettong | | |
| | Northern Bettong | | |
| | Burrowing Bettong | | |
| | Rufous Hare-wallaby | | |
| | Banded Hare-wallaby | | |
| | Bridled Nailtail Wallaby | | |
| **VULNERABLE** | Eastern Quoll | Ghost Bat | False Water-rat |
| | Mulgara | | Golden-backed Tree-rat |
| | Sandhill Dunnart | | Plains Rat |
| | Eastern Barred Bandicoot | | Western Mouse |
| | Bilby | | Pilbara Pebble-mound Mouse |
| | Western Ringtail | | Pilliga Mouse |
| | Mountain Pygmy-possum | | Northern Hopping-mouse |
| | Black-footed Rock-wallaby | | |
| | Brush-tailed Rock-wallaby | | |
| | Proserpine Rock-wallaby | | |

# Subclass Prototheria
## *Monotremes*

### ORDER MONOTREMATA
#### *Platypus and Echidnas*

# ORDER MONOTREMATA

*Platypus and Echidnas*

*M*onotremes differ from other mammals in laying eggs and in lacking teats. Some aspects of their internal anatomy are also unique. Textbooks written in the northern hemisphere often state or imply that there has been an evolutionary progression from reptiles, through monotremes and marsupials to the eutherians (a name meaning 'good' or 'perfect' mammals) that are characteristic of Eurasia, North America and Africa. This simple idea does not fit the facts. Given what we know about how anatomical structures can change in the course of evolution—and what changes are impossible—there is no conceivable scenario in which monotremes could be the ancestors of marsupials or that either of these groups could have given rise to the more familiar eutherians.

The anatomy and physiology of all living mammals testify in some degree to their origin from reptiles. The egg-laying habit of the monotremes is a notable reptilian feature that has been retained in this group of mammals. So is the presence of an interclavicle and a pair of epicoracoids in the shoulder girdle. However, if by some quirk of history, the classification of animals had originated in Sydney rather than in Athens two millennia ago, egg-laying would have been regarded simply as one of the ways in which mammals produce young and what is cited as a 'reptilian' shoulder girdle would have been seen as a structure common to some mammals and reptiles.

Categories such as 'mammal' and 'reptile' were originally constructed in terms of recognisably different assemblages of *living* animals but, even on these criteria, monotremes are no less mammalian in most aspects of their structure and function than marsupials or eutherians. Palaeontology provides evidence of what might be termed a number of 'experiments' on the part of various reptile groups involving a simplification of the lower jaw, improvements in the middle ear and more efficient suspension of the body from the four limbs. One such group was the ancestor of all living mammals but we have not yet identified it. Indeed, even if we had a complete fossil record of all the mammal-like reptiles and early mammals, we would recognise that the reptile–mammal boundary is a human construct.

*The leathery beak of the Platypus is a device for sifting small invertebrates from the bottom. It is richly supplied with touch receptors and with electro-receptors that are able to detect weak currents emitted by the muscles of its prey.*
E. BEATON/TERRA AUSTRALIS

Previous page: *Echidnas vary considerably in the lengths of hairs between the spines, the fur being longer in cooler regions. This individual from Victoria has long hairs; those from northern Australia may have almost no visible fur.*
J. LOCHMAN

Inset: *The female Echidna lays a single egg into a pouch on its belly. The hatchling is quite embryonic, lacking hindlimbs.* E. SLATER

Nevertheless, if we work within such constructs, it makes sense to hypothesise that the monotremes are survivors of an early branch from the mammal evolutionary tree and that, somewhat later, the trunk of the tree split into what we call marsupials and eutherians.

Among the features contributing to the success of mammals is the ability to maintain a relatively constant warm body temperature. Monotremes were long regarded as 'inferior' and 'quasi-reptilian' in this respect, mostly due to inadequate knowledge. Their usual body temperature of 32–33°C is lower than that of most eutherian mammals but, if this were pertinent, mammals would have to be regarded as inferior to birds. The interesting situation is that the Platypus, which has to cope with cold water, maintains a remarkably constant body temperature. The Short-beaked Echidna maintains a relatively constant body temperature when active but 'turns off' its regulatory system when inactive in cold conditions. So do many small eutherians and birds.

Another feature of mammals is the nourishment of young on milk secreted by the mother. In marsupials and eutherians, this is delivered by teats but, in monotremes, milk is secreted from numerous pores on the belly. This is demonstrably less efficient than in other mammals but is not a 'reptilian' feature.

Recent discoveries of fossil teeth show that monotremes once occurred in South America. These teeth are similar to those of fossil platypuses from Australia, leading to the hypothesis that the ancestral monotremes were platypus-like. However, there is not enough evidence to justify this view and it seems safer to postulate that platypuses and echidnas are specialised survivors of ancestral monotremes that had a more generalised anatomy. Extrapolating from the anatomical features common to the surviving forms, we arrive at a rather 'ordinary-looking' mammal, possessing a pouch, laying eggs, producing milk but lacking teats, probably without vibrissae and possibly specialised for burrowing.

It seems likely that any such ancestral form would have constituted the base of an adaptive radiation that included a wide variety of forms less improbable than the echidnas and platypuses but these have yet to be found and, if found, recognised.

Meanwhile, it seems that—in competition with marsupials and eutherians—the Platypus and echidnas (particularly the short-beaked species) have survived because they occupy ecological niches in which there is little or no competition. As a feeder on freshwater benthic invertebrates, the Platypus faces no competition except to a slight extent from the Water Rat. The Short-beaked Echidna is the only Australian mammal to feed predominantly on ants. The Long-beaked Echidna (relatives of which were once present in Australia) is in competition with many other vertebrates for food resources in the forest floor.

# FAMILY ORNITHORHYNCHIDAE
## *Platypus*

*The Platypus propels itself under water by its webbed forefeet; the hindfeet are employed in steering and braking.*

Ornithorhynchids are known in Australia as fossils up to about 130 million years old. The only surviving species lacks teeth in the adult: molars that develop in the embryo are resorbed before the young complete their development. The transient teeth make it possible to link the living Platypus with fossil forms, including species that lived in Australia at a time when dinosaurs were still the dominant terrestrial vertebrates. The discovery in 1992 of fossil 'platypus' teeth from Patagonia, dated to about 60 million years ago, is fascinating evidence of a Gondwanan distribution of monotremes.

We have no evidence of the anatomy of the animals from which these early teeth were derived. There is a natural tendency to assume that they may have resembled platypuses but it is quite possible that their possessors were generalised terrestrial insectivores.

Most of which can be usefully said about the living member of the family is included in the account of the species.

# Platypus

## *Ornithorhynchus anatinus*

### (Shaw, 1799)

or'-nith-oh-rink'-us  ah'-nah-teen'-us:
'duck-like bird-snout'

Since no living mammal remotely resembles the Platypus, a description would be superfluous. In the wild, it is most frequently seen in the mid-distance, often in poor light. Even when it is at rest on the surface, only the tip of its bill and the top of its head, back and tail can be seen. However, its smooth swimming action, with characteristic bow-wave, low silhouette, absence of visible ears and rolling dive, readily distinguish it from the dog-paddle swimming style of the Water Rat, *Hydromys*, the only other aquatic mammal in Australian fresh water.

The underfur is extremely fine and dense. Long, flattened guard hairs, which project through the underfur, undergo a moulting cycle throughout the year while the underfur remains intact: even after long periods of immersion it remains dry. Grooming sometimes occurs in the water, but more frequently when an animal rests on a log or rock.

The Platypus is distributed from the winter snows and high altitudes of Tasmania and the Australian alps, to the tropical rainforest lowlands and plateaus of far northern Queensland. The western limits of its range are poorly known. In some localities it appears not to extend west of the Great Dividing Range but it once occurred in much of the Condamine River in Queensland and in the Murray River and its

*Ornithorhynchus anatinus*

*A Platypus may bask in the sun for periods during the day but is usually found either in its burrow or in water.*

tributaries in South Australia. It now seems to be extinct in that state, except for an introduced population on Kangaroo Island.

The Platypus feeds mainly on a wide variety of adult and larval aquatic invertebrates: small fishes and amphibians may also be eaten. Smaller prey items are sifted from the substrate by the complex bill apparatus, but larger prey can be snapped up individually. Apart from its superficial appearance and a comparable sifting function, the bill of the Platypus is nothing like

that of a duck. It is pliable—like soft moist rubber—and very sensitive. Infants have milk teeth, but these are shed without replacement and the adult has no calcified teeth. Horny grinding plates and sharp, shearing ridges provide a functional, if unconventional, 'dentition'. Food obtained during a dive is stored in large cheek-pouches, opening from the back of the bill, until it can be thoroughly masticated while the animal floats on the surface.

The Platypus is essentially solitary but substantial numbers may share the use of a relatively small body of water. It has a range of vocalisations, most commonly an annoyed growling sound when disturbed.

---

### *Ornithorhynchus anatinus*

**Size** Size varies with location but does not appear to reflect a north–south clinal variation. Length is measured from tip of bill to tip of tail ($\pm$ standard deviation where available).

**REGION**

**North Queensland**

| | |
|---|---|
| *Length* | 44.1 ($\pm$ 3.1) mm (male) |
| | 41.0 ($\pm$ 1.8) mm (female) |
| *Weight* | 1018 ($\pm$ 208) g (male) |
| | 704 ($\pm$ 49) g (female) |

**South-east Queensland**

| | |
|---|---|
| *Length* | 49.3 ($\pm$ 2.7) mm (male) |
| | 43.8 ($\pm$ 1.6) mm (female) |
| *Weight* | 1556 ($\pm$ 194) g (male) |
| | 1222 $\pm$ 94 g (female) |

**New South Wales**
**East of divide**

| | |
|---|---|
| *Length* | 50.5 ($\pm$ 2.4) mm (male) |
| | 41.5 $\pm$ 2.0 mm (female) |
| *Weight* | 1434 ($\pm$ 218) g (male) |
| | 857 $\pm$ 107 g (female) |

**New South Wales**
**On divide**

| | |
|---|---|
| *Length* | 47.4 ($\pm$ 3.5) mm (male) |
| | 40.3 $\pm$ 2.0 mm (female) |
| *Weight* | 1379 ($\pm$ 132) g (male) |
| | 888 $\pm$ 92 g (female) |

**New South Wales**
**West of divide**

| | |
|---|---|
| *Length* | 54.9 ($\pm$ 29) mm (male) |
| | 47.0 mm (female) |
| *Weight* | 2215 ($\pm$ 364) g (male) |
| | 2000 g (female) |

**Tasmania**

| | |
|---|---|
| *Length* | 53.2 mm (male) |
| | 53.5 mm (female) |
| *Weight* | 1900 g (male) |
| | 1500 g (female) |

**Identification** Unmistakable.

**Synomyns** None.

**Other Common Names** None.

**Status** Common but vulnerable.

**Subspecies** None.

**References**

Augee, M.L. (ed.) (1978). *Monotreme Biology*. Roy. Zool. Soc. NSW, Sydney.

Grant, T. (1984). *The Platypus*. Univ. NSW Press, Sydney.

Griffiths, M.E. (1978). *The Biology of Monotremes*. Academic Press, New York.

When not in the water, it resides in a burrow: usually a short, simple construction in the river bank, just above water level and often under a tangle of tree roots. A Platypus burrow is frequently distinguishable from other holes in the bank by its horizontally oval cross-section; it is sometimes open-ended. When incubating eggs or suckling young, a female constructs a more elaborate nursing burrow which may be 20 metres long, plugged with earth at intervals and terminating in a nest chamber containing a bed of damp herbage brought in by the mother.

The Platypus is usually active around dawn and dusk, but activity patterns are determined in a complex manner by such variables as locality, human activity, day-length, air and water temperatures and stream productivity. It may be crepuscular, nocturnal or diurnal.

C. ANDREW HENLEY/LARUS

*When at the surface, a Platypus opens its nostrils and its eyes; both are closed underwater.*

The adult male has a strong, sharp spur (about 12 millimetres long) on each ankle. The canal running through it and opening near its tip is connected to a large venom gland in the thigh. This gland, which shows greatest activity in the breeding season, produces a venom capable of causing excruciating pain and incapacity in humans when injected. Its function is probably involved with competition between males in the breeding season. In juvenile males, the spur develops within a horny sheath. In juvenile females, a rudimentary spur sheath is present, but this is lost before sexual maturity is attained. Development of the spur is an approximate indication of age.

The breeding season varies with latitude. Mating takes place (in the water) during September in much of New South Wales and Victoria; a month or so earlier in Queensland; and somewhat later in Tasmania. Intrauterine gestation probably lasts about two weeks. Two eggs are usually laid (sometimes three or one). These are about 17 millimetres in diameter, have a parchment-like shell, usually stick together and contain a partly developed embryo.

The female incubates the eggs by holding them, pressed to her belly with her tail, while she lays curled up in the nest chamber of the burrow. Hatching takes place about one to two weeks after the eggs are laid. The young then feed by sucking milk from their mother's abdominal surface, where her mammary ducts open. During the four to five month period of lactation, the young remain in the burrow, but the female leaves to forage.

Predation on adults is probably insignificant. The Platypus is distributed along the seaboard of the eastern States and through Tasmania, but in a very discontinuous manner. It occurs in some bodies of fresh water that are under considerable and increasing pressure. There is only a limited potential for recolonisation should a calamity eradicate an isolated population. Alienation of habitat may present a significant hazard leading to a reduction in range. Pollution, inappropriate fishing practices and stream and river bank disturbance for agricultural, industrial or dam construction purposes are probably responsible for local extinctions, as has happened in South Australia. Although the Platypus is still common over much of its range, it should be regarded as vulnerable.

F.N. CARRICK

# FAMILY TACHYGLOSSIDAE
## *Echidnas*

*The feet of the Echidna have powerful toes and very strong claws, enabling an individual to dig into an ant nest or—using all four limbs—to sink horizontally into the soil and disappear in a few minutes.* K. GRIFFITHS

There are only two living monotypic genera in this family. The Short-beaked Echidna is a very variable species that occurs in almost all Australian environments. Several genera of much larger long-beaked echidnas were widespread in Australia and New Guinea until the late Pleistocene but there is now only one species, rare and restricted to the highlands of New Guinea. Both species feed by means of a long, sticky tongue, but, whereas the Short-beaked Echidna feeds almost exclusively on ants, the Long-beaked Echidna forages in forest litter, eating earthworms and larger solitary insects.

*A juvenile Echidna, removed from its nest.*

D. WATTS

# Short-beaked Echidna

## *Tachyglossus aculeatus*
### (Shaw, 1792)

tak'-ee-glos'-us ah-kue'-lay-ah'-tus:
'spiny swift-tongue'

*The Echidna is a surprisingly good swimmer.*

The species that is usually referred to in Australia as the Echidna is more properly referred to as the Short-beaked Echidna, distinguishing it from the larger Long-beaked Echidna of New Guinea. It is readily recognised by its covering of long spines. Fur is present between the spines and, in the Tasmanian form may be so long as to obscure these. It is distributed over most of Australia from regions of winter snow to the deserts and has no particular habitat requirements other than a supply of the ants and termites on which it feeds.

A toothless and highly specialised feeder, it breaches an ant or termite nest with its forepaws or snout and extends its long tongue into the galleries. Insects adhering to the copious sticky saliva with which the tongue is covered are drawn into the mouth and masticated between a horny pad at the back of the tongue and a similar structure on the palate. A considerable amount of soil and nest material is ingested and this forms the bulk of the distinctive cylindrical droppings.

It is solitary, occupying overlapping home ranges which lack fixed nest sites. The Short-

K. ATKINSON

*Tachyglossus aculeatus*

Inset: *When disturbed, an Echidna curls into a ball almost completely covered with spines. Nevertheless, the belly remains vulnerable and experienced dogs are able to attack this weak spot.* K. ATKINSON

beaked Echidna usually seeks shelter under thick bushes, in hollow logs, under piles of rabbit and wombat debris, or occasionally in a rabbit or wombat burrow. Self-constructed burrows are used by females incubating and suckling young, from which the mother often departs for several days to forage.

In arid regions, it avoids temperature extremes by sheltering in caves or crevices, with activity restricted to the night. In more temperate climates the pattern of activity

*The Echidna feeds on ants using its powerful forefeet to dig into the nest.*

depends upon air temperature: it tends to forage around dawn and dusk but, in winter, particularly in southern latitudes, it may be active in the middle of the day. Echidnas have been shown to hibernate in several regions of eastern Australia, including the Snowy Mountains.

The male has a spur on the ankle of its hind-leg similar to that of the male Platypus but lacks a functional venom gland. Mating occurs in July

J. MCCANN

### Tachyglossus aculeatus

**Size**
*Head and Body Length*
30–45 cm
*Weight*
2–7 kg

**Identification** Dorsal surface of body and rudimentary tail covered with spines. Fur usually present between spines (in Tasmania fur may obscure the spines). Long tubular snout.
**Recent Synonyms** None.
**Other Common Names** Spiny Anteater, Porcupine.
**Status** Common.
**Subspecies** *Tachyglossus aculeatus acanthion*, Western Australia, Northern Territory and arid zones of all mainland states. *Tachyglossus aculeatus aculeatus*, coastal regions and dividing range slopes of southern Queensland, New South Wales, Victoria and South Australia. *Tachyglossus aculeatus lawesii*, Papua New Guinea. *Tachyglossus aculeatus multiaculeatus*, Kangaroo Island. *Tachyglossus aculeatus setosus*, Tasmania.
**References** Griffiths, M. (1989). Tachyglossidae (in) D.W. Walton and B.J. Richardson (eds) *Fauna of Australia*, Vol. 1B. Aust. Govt. Publ. Serv., Canberra. pp. 407–435

Augee, M. & B. Gooden (1993). *Echidnas of Australia and New Guinea*. Univ. NSW Press, Sydney.

and August, during which time a single female has been observed to be followed by as many as ten males. About two weeks after copulation, a single soft-shelled egg is laid—probably directly into the pouch on the belly of the female. It hatches after about ten days and sucks up milk exuded from the numerous pores of the paired mammary glands. It is not known how long the young remains in the pouch or whether occupation is discontinuous. It was once thought that the young remained in the mother's pouch, in the burrow, until spines erupted at about three

months of age, but recent studies show that females with pouch-young forage outside the burrow and begin to leave young behind in the burrow well before three months old. The period of suckling is unknown but juveniles tend to be first seen from September to November, when about one year old and weighing 1–2 kilograms.

The spiny coat provides an excellent defence. When suddenly disturbed in the open on hard ground, an Echidna curls into a ball of radiating spines; if on soil, it may dig itself below the surface while remaining horizontal, disappearing like a sinking ship. By extending its spines and limbs, it can wedge itself securely in a rock crevice or hollow log.

Adults have no significant predators but they are occasionally eaten by the Dingo. Young animals are perhaps eaten by large goannas. Distribution is sparse, particularly in the arid parts of the range, but the species remains ubiquitous and in no apparent danger.

M.L. AUGEE

# Subclass Marsupialia
## *Marsupials*

**ORDER DASYUROMORPHIA**
*Carnivorous Marsupials*

**ORDER PERAMELEMORPHIA**
*Bandicoots and Bilbies*

**ORDER DIPROTODONTIA**
*Koala, Wombats, Possums and
Macropods*

**ORDER NOTORYCTEMORPHIA**
*Marsupial Mole*

*A* marsupial is born in a very incomplete state: minute, blind, hairless and with hindlimbs only partially formed. The forelimbs, however, are precociously developed and the toes are armed with sharp, curved claws. As soon as it is born and free of the embryonic membranes, it uses the forelimbs, in a manner reminiscent of an overarm swimming stroke, to pull itself over the mother's abdomen, seeking a teat, to which it firmly attaches itself by its mouth. It does not release itself from this attachment until it has become a miniature adult. In many marsupials the teats are covered by an extensive fold of skin, forming a pouch that provides a protective environment for the young during their second phase of development. The name of the group is derived from the Latin *marsupium*, meaning 'pouch' and the young inside this structure are aptly called 'pouch-embryos'.

The nature of marsupial development was recognised by European biologists long before they knew anything about Australian animals. Opossums from central and southern America had been studied since early in the sixteenth century and the pouch of these animals was regarded as a second, external womb. The generic name of the commonest opossums, *Didelphis*, from the Greek roots *di-*, two; *delphys*, womb, reflects this interpretation although, as it happens, marsupials also have two uteri or 'true' wombs. We now know that many of the smaller species of opossums do not have pouches and the same is true of a number of small carnivorous Australian marsupials: in some species the pouch is a temporary structure which develops only in the breeding season. Thus, possession of a pouch does not define a (female) marsupial. Nor is this structure exclusive to marsupials: the egg of an echidna is carried in a temporary pouch on its mother's abdomen and the hatchling remains there until its development is nearly complete.

It is usually stated that marsupials are born in an embryonic state because they lack a placenta, an efficient device (well developed in those mammals that are neither monotremes nor marsupials) whereby the blood supply of the mother and embryo are brought into extensive close contact in the uterus, permitting the embryo to be a perfect parasite. It obtains food and oxygen from the mother's blood and passes its carbon dioxide and other waste products into the mother's blood to be excreted or processed by her.

Possession of a placenta certainly enables some mammals, such as horses and elephants, to be born large and sufficiently mature to run with their parents after a few hours. However, to say that marsupials are born small and incomplete *because* they do not have a placenta suggests that they have somehow failed; that they would be much better off with such a structure. In fact, bandicoots and the Koala have a well-developed placenta but their young are no more advanced at birth than those of other marsupials.

The fault in the argument lies in the facile assumption that because the placental mammals are the most successful of the three major living groups, they are superior *in*

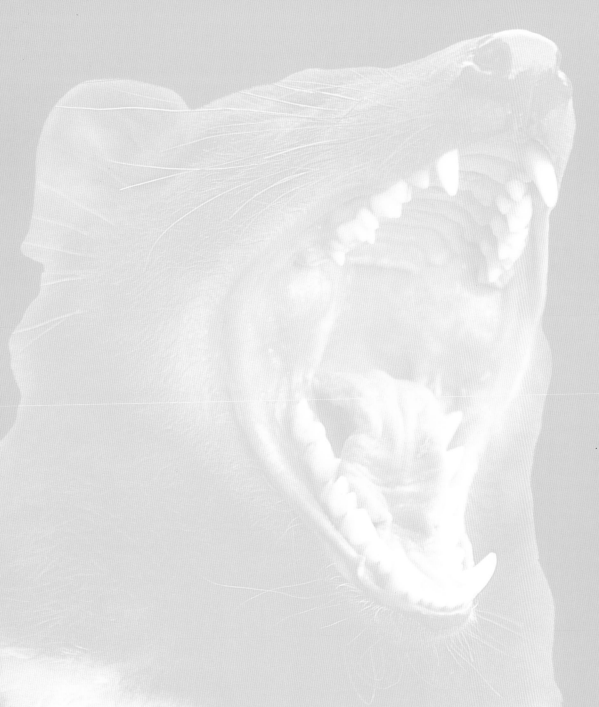

*all respects*. Yet a female marsupial is hardly more incommoded by the young in her pouch than a pregnant placental mammal is by the foetuses in her uterus. The marsupial method of reproduction works: it just happens to be different from that of the placental mammals. According to the needs of the particular species, a marsupial may produce large numbers of young almost continuously, a few, or only one a year. A number of marsupial species have diminished in numbers or become extinct in competition with placental mammals but there is no indication in any of these cases, that this was due to the *reproductive* superiority of the latter.

Marsupials have a slightly lower body temperature and metabolic rate than placental mammals. This too, has been cited as an indication of a lower grade of organisation but such an argument would imply that birds are superior to all mammals. Again, the point is not that marsupials are second-rate energy converters but that they expend less energy than placental mammals of similar size and activity. Where food is scarce, this could be an advantage.

The senses of marsupials appear to span much the same range as in placental mammals (except those bats and cetaceans which specialise in echolocation) but the brain of a marsupial is usually smaller than that of a placental of equivalent size. The cerebral hemisphere—centres of information storage, association and what may loosely be termed intelligence—are markedly less well developed. In this respect, marsupials are different *and* inferior and it is hard to avoid the conclusion that, in competition with placental mammals, marsupials have often been outwitted.

Most mammals that climb in trees do so, like cats, with the aid of their claws. In the course of evolution, only two groups of mammals, the primates and the marsupials, have developed a device for gripping a branch. Monkeys and most of the primates have an opposable first digit on the hand (thumb or pollex) and on the foot (big toe or hallux); marsupials usually have an opposable hallux on the hindfoot. Because this acts by pressure and friction rather than by digging into bark, it does not need a claw and this structure appears to have been lost early in the course of marsupial evolution. The hallux is well developed in the opossums of South America—the least specialised of the living marsupials—and also in the possums of Australia. Its presence indicates that marsupials were originally arboreal. In addition to gripping with the hindfoot, the Koala and most possums are also able to exert a pincer-like grip between the first two and last three digits of the forefoot. Such adaptations are of no use to those that live on the ground: the hallux has been lost in all but one species of the kangaroo group, is absent or very weakly developed in bandicoots and is small or absent in dasyurids.

Several species glide by means of a membrane between the fore- and hindlimbs. A number are terrestrial and some of these construct burrows; one spends almost its entire life 'swimming' beneath the desert sand. One South American opossum is aquatic but no Australian marsupial is adapted for life in the water although many, including kangaroos and the Koala, can swim quite well when necessary.

The first marsupials were almost certainly insectivorous. Living forms include general insectivores, a termite-eater, predatory carnivores, a carnivorous scavenger, omnivores, fruit-eaters, leaf-eaters, nectar-eaters, browsers and grazers. All authorities recognise that the Australian marsupials fall into three major groups: the insectivorous–carnivorous quolls, dunnarts, etc.; the insectivorous–omnivorous bandicoots; and the basically herbivorous possums and kangaroos. However, the evolutionary relationships between these groups remains unclear and the Marsupial Mole is an enigma. While it is generally agreed that the possum–kangaroo assemblage is the

most specialised and most recently evolved and majority opinion regards the carnivorous marsupials (dasyuroids) as derived from the earliest Australian marsupials, there is a strong argument that bandicoot-like forms may have been the stem group.

Relationships between the Australian and American marsupials are also undetermined. A current view is that those marsupials on the western side of the Pacific form a monophyletic group, the Australidelphia, while those on the eastern side (with the possible exception of one small family) comprise the very distinct Ameridelphia. While such a division appears to be rational, the two supposed groups still evade convincing definition—apart from the strange phenomenon of pairing of the sperms of ameridelphians, so that they are propelled by two flagella.

# ORDER DASYUROMORPHIA

## *Carnivorous Marsupials*

*T*his taxonomic category includes exactly the same families as the superfamily Dasyuroidea. Its function is to indicate the considerable evolutionary distance between the carnivorous marsupials and the other three major divisions of the cohort Australidelphia: the orders Peramelemorphia, Diprotodontia and Notoryctemorphia.

*Thylacine. Illustration in John Gould's* The Mammals of Australia.

# Superfamily Dasyuroidea

The superfamily Dasyuroidea includes the basically carnivorous Australian marsupials (as distinct from the omnivorous bandicoots and the basically herbivorous possum–kangaroo group). They are characterised by the possession of three pairs of approximately equal-sized lower incisors and an absence of fusion between the second and third toes of the hindfoot.

The vast majority of the living dasyuroids are members of the family Dasyuridae, ranging in size from the Tasmanian Devil (up to 8 kilograms) to the Narrow-nosed Planigale (5–9 grams), one of the smallest of the known mammals. Their diet ranges from mammals and birds to insects and other invertebrates.

Fairly closely allied to the dasyurids is the family Thylacinidae (page 163), the last member of which became extinct in the twentieth century. It was a predator and, in adaptation to its way of life, had a head remarkably like that of a wolf.

The termite-eating Numbat is specialised for its exclusive diet and its teeth are reduced to peg-like structures. Otherwise, this single member of the family Myrmecobiidae resembles the dasyurids.

Some authorities place the enigmatic Marsupial Mole within the Dasyuroidea. To emphasise its uniqueness, it is treated here as the sole member of a separate order Notoryctemorphia (page 409).

# FAMILY DASYURIDAE
## *Dasyurids*

*Juvenile Eastern Quolls just outside their nest.* C. ANDREW HENLEY/LARUS

The diversity of body forms appropriate to herbivorous placental mammals is illustrated by such vastly different animals as giraffes, elephants, rhinoceroses, horses, pigs and mice. Much less variety is possible among those that hunt, catch and eat other animals: stripped of their skins, a lion, wolf, weasel or shrew differ only slightly in shape. A similar situation exists in the marsupials: herbivores range in form from possums and gliders to kangaroos and wombats, but the exclusively carnivorous species such as quolls, antechinuses and dunnarts, appear to be minor anatomical variations upon a common theme. Moreover, the constraints placed upon the body of a terrestrial predatory mammal are such that there is even considerable resemblance between placentals and marsupials that share this habit. The similarity of quolls to European carnivores was noted by early settlers, who referred to these animals as 'native polecats' (a polecat being a weasel-like animal) but, with the passage of time, this name was contracted to the quite inappropriate 'native cat'. The predatory nature of the smaller carnivorous marsupials was not immediately recognised and, in reference to their size, shape and fur, they laboured until recently under the name of 'marsupial mice': 'marsupial shrews' would have been a far better name.

Dasyurids are characterised by a biting, cutting dentition with four pairs of pointed upper incisors and three lower pairs; well developed upper and lower canines; two or three pairs of upper and lower blade-like premolars and four pairs of upper and lower molars with sharp, shearing cusps.

Except for the Kultarr, which has very long hindlimbs, the legs of dasyurids are unspecialised. The forefeet have five clawed toes radiating from a more or less circular palm which is in contact with the ground when the animal runs. In dasyurids that climb, the forefoot provides a simple grip by movement of all the toes in towards the palm. In arboreal species the hindfoot is broad and has a small, mobile hallux. The hindfoot is somewhat longer in those that combine terrestrial with arboreal life and the hallux is less well developed in these species. In completely terrestrial species, the hindfoot is long and the hallux is absent or reduced to a useless nubbin.

Depending on the species, the number of teats on female dasyurids ranges from 4 to 12, but individuals may vary from the number typical of a species. Many species lack a pouch, this condition being seen at its starkest in the phascogales, where eight teats are situated on a circular patch of skin on the abdomen. In others, low lateral folds develop at the sides of the mammary area of a lactating female. These folds do not constitute a pouch in the strict sense and provide very little protection for the young which, as they develop, dangle from the mother's belly between her fore- and hindlegs. Although such exposure appears to be extremely hazardous, most young survive to weaning and mortality is highest after the young have ceased suckling and begin to disperse. A permanent pouch is present in most ningauis—the Kultarr, Spotted-tail Quoll and Tasmanian Devil. After vacating the pouch, young dasyurids may be left in a nest while the mother forages, or may cling to the fur of the mother's body as she forages.

As is the case in most mammals, smaller dasyurids tend to have more young in a litter than the larger ones. It has been observed that some species bear more young than can be accommodated on the teats but it is difficult to determine the extent of this phenomenon: no information is available from wild-caught animals and, even in captivity, small supernumerary young may easily be lost in the debris at the bottom of a cage. Such overproduction is probably common in the smaller dasyurids.

There being little variation in the body shape of dasyurids, speciation has been mainly in terms of size, behaviour and physiology. Probably the most significant physiological adaptations are those related to life in arid environments and it is interesting that a number of species fall into closely related pairs, one adapted to a relatively damp environment and the other to dry or desert conditions. The most extreme variation from the basic body plan of dasyurids is found in the desert-adapted Kultarr, once known as the 'Jerboa-marsupial'. With elongated hindlegs and a long,

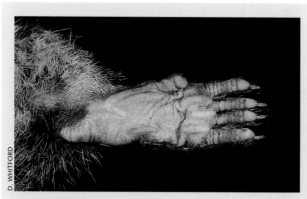

*Dasyurids have relatively long hindfeet. The first toe is either short or absent: its presence in a species is indicative of aboreal habits.*

tufted tail, it resembles the jerboas of northern Africa and the kangaroo-mice of America and it was assumed, on the basis of this similarity to be saltatorial. However, although it can leap, its fast gait is a gallop.

On the basis of their anatomy, serum proteins and chromosomes, the Australian Dasyuridae is currently divided into four subfamilies. The quolls, Tasmanian Devil, Kowari, Mulgara, Kaluta, dibblers, pseudantechinuses and parantechinuses comprise the Dasyurinae; phascogales and other antechinuses comprise the Phascogalinae; dunnarts and the Kultarr comprise the Sminthopsinae; while the planigales and ningauis comprise the Planigalinae.

Of the six dasyurine species, three are now rare. The Eastern Quoll, which once spread from southern Queensland through New South Wales and Victoria to eastern South Australia is known certainly to exist only in Tasmania. Its decline has been attributed to a variety of causes but it is probably more than coincidental that it is the most terrestrial of the quolls and that there are no foxes in Tasmania. The range of the Western Quoll has contracted from the greater part of non-arid Australia to the south-western corner of Western Australia. The reasons for its decline are unknown. The desert-dwelling Mulgara and Kowari have also diminished in range, again for no obvious reason.

Among the phascogalines, the range of the Red-tailed Phascogale has been reduced from at least four widely separated areas in semi-arid Australia to a region in south-western Western Australia. The Southern Dibbler, is also now restricted to a small portion of this region and appears to be both rare and endangered.

The White-tailed, Sandhill, Hairy-footed, Long-tailed and Julia Creek Dunnarts of the Sminthopsinae are rare and possibly were so prior to European settlement.

## Subfamily Dasyurinae

This group, defined largely on biochemical criteria, includes the quolls; dibblers, pseudantechinuses, parantechinuses and Kaluta (all of which were once regarded as antechinuses); and the Kowari, Mulgara and Tasmanian Devil.

K. ATKINSON

# Mulgara

## *Dasycercus cristicauda*

### (Krefft, 1867)

daz'-ee-ser'-kus  kris'-tee-kaw'-dah:
'crest-tailed hairy-tail'

The Mulgara inhabits the arid sandy regions of Australia, living in burrows which it digs on the flats between low sand-dunes (central Australia) or the slopes of high dunes (eastern edge of the Simpson Desert). The complexity of the burrow varies; those in central Australia usually have only one entrance with two or three side tunnels and pop-holes, while those in Queensland have more than one entrance, deeper branching tunnels and numerous pop-holes. When a burrow is excavated it is usual to find only a single animal in residence, suggesting that the tolerance displayed in captivity to other individuals may not be exhibited under natural conditions.

The most striking feature of these small, robustly built animals is the crest of black hairs on the tail, which is short and fattened at the base. The Mulgara probably hunts at night for its

*The Mulgara is a rather generalised predator, mostly upon large insects, spiders and scorpions. These two are feeding on a cricket.*

food—insects, other arthropods and small vertebrates—but like many of its relatives it is not strictly nocturnal. It may sun-bathe at the entrance to its burrow and, in captivity, it will emerge from cover during the day to stretch out under an infra-red lamp. If two or more are housed together they often lie with their heads on each other's backs.

The Mulgara appears to be well adapted to life in the desert. Its ability to produce urine

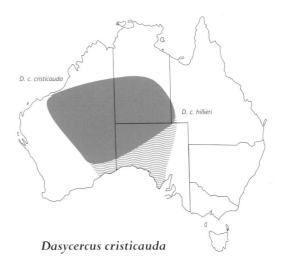

*Dasycercus cristicauda*

55

BABS & BERT WELLS

*The crest of black hairs on the tail is a distinguishing feature.*

with a high concentration of urea permits it to excrete the large amounts of nitrogenous waste arising from its carnivorous diet in a relatively small volume of urine and thus to survive on the water it obtains from its food.

Little is known about breeding in the wild but females with up to eight young in the pouch have been captured between June and December. Among captive animals, mating has been observed from mid-May to mid-June and young have been born in late June, July and

August after a gestation period of five to six weeks. The young suckle for three to four months and become reproductively mature when 10–11 months old. Individuals of both sexes have been known to come into breeding condition each year for six years, suggesting that they are fairly long-lived animals. If they continue to grow throughout life, as appears to be the case from examination of their bones, the variation seen in the size of adult specimens may be, at least in part, a reflection of their age.

Mulgaras appear to undergo fluctuations in numbers, at least in some parts of their wide range. For example, a population of the subspecies *Dasycercus cristicauda hillieri* found in Queensland in 1967 could not be found in the same locality in 1971 and 1979 but was present in 1990 and 1991.

P.A. WOOLLEY

## *Dasycercus cristicauda*

**Size**
*Head and Body Length*
125–220 mm (males)
125–170 mm (females)
*Tail length*
75–125 mm (males)
75–100 mm (females)
*Weight*
75–170 g (males)
60–95 g (females)

**Identification** Light sandy brown above, greyish-white below. Hairs reddish on base of tail, black on distal two-thirds. Black hairs increase in length towards tip and form a dorsal crest. Ears short and rounded. Five toes on fore and hindfeet.
**Recent Synonyms** *Dasycercus blythi*.
**Other Common Names** Crest-tailed Marsupial Mouse.

**Status** Common in Northern Territory. Rare, possibly underestimated in Western Australia, South Australia and Queensland.
**Subspecies** *Dasycercus cristicauda cristicauda*, greater part of range from Western Australia, Northern Territory and South Australia.
*Dasycercus cristicauda hillieri*, south-western Queensland and north-eastern South Australia.
**References**
Jones, F.W. (1949). The study of a generalised marsupial (*Dasycercus cristicauda* Krefft). *Trans. Zool. Soc. Lond.* 26: 409–501.
Woolley, P. (1971). Maintenance and breeding of laboratory colonies of *Dasyuroides byrnei* and *Dasycercus cristicauda*. *Int. Zoo. Yb.* 11: 351–354.
Woolley, P.A. (1990). Mulgaras, *Dasycercus cristicauda* (Marsupialia: Dasyuridae); their burrows and records of attempts to collect live animals between 1966 and 1979. *Aust. Mammal.* 13: 61–64.

P.A. WOOLLEY & D. WALSH

*The Little Red Kaluta (once known as the Little Red Antechinus) inhabits dense spinifex thickets.*

# Little Red Kaluta

## *Dasykaluta rosamondae*

### (Ride, 1964)

daz'-ee-kah-lue'-tah   roz'-ah-mon'-dee:
'Rosamond's hairy-kaluta'

This rufous dasyurid, described in 1964 from a number of specimens collected on Woodstock Station and nearby localities in the Pilbara region of Western Australia, is often found among maze-like tussocks of Woolly Spinifex (*Triodia lanigera*). Its specific names make a whimsical allusion to the red-haired Rosamond, mistress of Henry II, who was hidden away in the royal manor of Woodstock behind an elaborate maze. Woolly Spinifex, which is fire resistant and relatively unpalatable to sheep, may help to safeguard areas of suitable habitat within the restricted range of the Little Red Kaluta.

It is an inquisitive animal and has the curious habit of flicking its thick, tapering tail in the air as it investigates its surroundings. Mainly nocturnal, it feeds voraciously on insects and small vertebrates such as lizards.

*Dasykaluta rosamondae*

## Dasykaluta rosamondae

**Size**

*Head and Body Length*
95–110 mm (males)
90–100 mm (females)
*Tail Length*
55–70 mm (males)
60–65 mm (females)
*Weight*
25–40 g (males)
20–30 g (females)

**Identification** Russet brown to coppery above and below; rather rough fur. Head and ears short. Forepaws strong and well haired on back. General form similar to *Dasycercus* but easily distinguished by small size, colour and lack of black hair on tail.

**Recent Synonyms** None.
**Other Common Names** Russet Antechinus, Spinifex Antechinus, Little Red Antechinus.
**Status** Common, limited.
**Subspecies** None.
**References**

Ride W.D.L. (1964). *Antechinus rosamondae*, a new species of dasyurid marsupial from the Pilbara district of Western Australia; with remarks on the classification of *Antechinus. West. Aust. Nat.* 9: 58–65.

Woolley, P.A. (1991). Reproduction in *Dasykaluta rosamondae* (Marsupialia: Dasyuridae): field and laboratory observations. *Aust. J. Zool.* 39: 549–568.

The Little Red Kaluta is one of ten species of dasyurid marsupials in which the males are known to die shortly after their first brief mating season and, as in three species that have been well studied, this mortality is probably induced by stress. Mating occurs in September and the young are born about seven weeks later in November. The female has eight teats, all of which may be occupied by young. Laboratory-reared young are weaned when three-and-one-half to four months old, in February and March and juveniles appear in the field population at this time. The young are ready to breed by the following September. Although males can breed in only one season, some females may breed in at least two. No more than one litter is produced in each breeding season.

P.A. WOOLLEY

BABS & BERT WELLS

*The Little Red Kaluta is a predator on large arthropods and small vertebrates such as lizards.*

K. JOHNSON

# Kowari

## *Dasyuroides byrnei*

### Spencer, 1896

daz'-ee-yue-roy'-dayz birn'-ee:
'Byrne's *Dasyurus*-like (animal)'

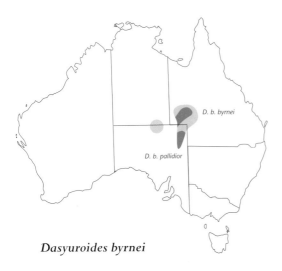

D. b. byrnei

D. b. pallidior

*Dasyuroides byrnei*

*The Kowari is a compact-bodied dasyurid, distinguished by a dense black brush on the distal half of the tail.*

The sparsely vegetated gibber deserts of south-western Queensland and north-eastern South Australia are the home of this predatory marsupial. It is an active digger which either constructs its own burrows or modifies those of other mammals, such as the Long-haired Rat and Bilby. Many freshly dug burrows have been observed in sand islands, mounds and under saltbushes following a wet period. Remains have been found under airport markers in the Channel Country and early reports suggest that it entered station homesteads.

In parts of the Channel Country, it has frequently been seen bounding along station roads at night. Overlapping home ranges of several square kilometres have been observed in both males and females. All individuals use several burrows, some with multiple entrances. Although captive animals bask in the sun, often

HANS & JUDY BESTE

*This Kowari is devouring a locust. The photograph displays the unspecialised structure of the hindfoot.*

### Dasyuroides byrnei

**Size**

*Head and Body Length*
140–180 (165) mm (males)
135–160 (150) mm (females)
*Tail Length*
110–140 (120) mm (males)
110–130 (115) mm (females)
*Weight*
85–140 (120) g (males)
70–105 (100) g (females)

**Identification** Distinguished from *Dasycercus* by brush of black hairs completely encircling terminal half of tail (crest in *Dasycercus*). Four toes on hindfoot (five in *Phascogale*).

**Recent Synonyms** None.

**Other Common Names** Brush-tailed Marsupial Rat, Byrne's Crest-tailed Marsupial Rat, Bushy-tailed Marsupial Rat.

**Status** Possibly extinct west of Lake Eyre; rare and scattered between Simpson Desert and Cooper Creek.

**Subspecies** Two subspecies of doubtful validity and distribution have been described:
*Dasyuroides byrnei byrnei*, north-eastern part of the range.
*Dasyuroides byrnei pallidior*, south-western part of the range.

**References**

Aslin, H.J. (1974). The behaviour of *Dasyuroides byrnei* (Marsupialia) in captivity. *Z. Tierpsychol.* 35: 187–208.

Hutson G. (1975). Sequences of prey-catching behaviour in the Brush-tailed Marsupial Rat (*Dasyuroides byrnei*). *Z. Tierpsychol.* 39: 39–60.

Lim, L. (1992). *The recovery plan for the Kowari.* Aust. Nat. Parks and Wildl. Serv. and S. Aust. Dept. of Env. and Planning, Adelaide.

stretching out at full length, the species appears to be nocturnal in the wild. During cold weather, captive animals show signs of torpor and, if disturbed in the nest, may shiver visibly for some minutes while the body temperature is returning to normal.

The diet in the wild probably consists of insects, small vertebrates and carrion of larger animals. The stomach contents of one road-killed animal showed that it had been eating a Long-haired Rat, *Pseudomys villosissimus*. In captivity, it readily accepts a variety of insects and has no difficulty in killing House Mice and small chickens. Meat, eggs and small quantities of vegetable foods are also eaten. Free water is not required if the diet is moist.

A variety of sounds are produced, including an open-mouthed hissing and a loud, staccato chattering, both made in response to threats from predators or other Kowaris; dependent young make a 'grating' call. Vigorous tail-switching, reminiscent of an angry cat, is used as a threat display. Smell is an important medium of communication and urine and faeces are used to mark home ranges and burrow sites. Both sexes have a sternal gland which is rubbed on objects for scent-marking. Apart from associations between mother and young, the Kowari appears to be solitary; males and females only coming together briefly to mate.

Breeding occurs between May and December, the onset of oestrus probably being determined by seasonal changes. Mating is prolonged and pairs may copulate for periods of up to three hours, repeated over one to three days. Young are born 30–35 days after mating, by which time the female's pouch area, normally an inconspicuous depression, has become hairless and reddened. The female normally has six, (sometimes seven) teats and if more than this number of young are born, some fail to survive. Newly born young are about 3 mm long.

For the first 30 days, the young are partially enclosed by skin-folds forming the pouch.

Subsequently they hang below the female's body, becoming detached when about 55 days old. They may then be left in the nest, or cling to the mother's back. They continue to suckle until about 110 days old and, from the age of about 95 days, engage in play. Both sexes are capable of breeding at about nine months of age. In the wild, most adult females appear to produce a second litter in November. This is probably the stimulus for the dispersal of the surviving young from the first litter of the year.

H.J. ASLIN AND L. LIM

*The Kowari is a very competent predator. This individual is eating a House Mouse.*

HANS & JUDY BESTE

# Western Quoll

## *Dasyurus geoffroii*

### Gould, 1841

daz'-ee-yue'-rus jef-roy'-ee: 'Geoffroy's hairy-tail'

When first described by John Gould in 1841, the Western Quoll (or Chuditch) occupied a diverse array of forest, shrub and desert habitats across most of mainland Australia apart from the south-eastern coast and 'Top End'. Following European settlement this range contracted dramatically: museum specimens were last collected in New South Wales in 1841, in Victoria in 1857 and in Queensland no later than 1907. The most recent reports of the Western Quoll in the central deserts date from the mid-1950s.

It currently survives only in south-western Western Australia, in areas dominated by sclerophyll forest or drier woodland and mallee shrubland.

Desert populations would den in earth burrows, hollow logs or tree limbs and hollows in termite mounds. Most diurnal resting sites in sclerophyll forest consist of hollow logs or earth burrows, although bandicoot nests and hollow tree bases may be used occasionally. The roofs of suburban residences around Perth commonly served as nesting sites until at least the 1930s.

While the Western Quoll forages primarily on the ground and at night, it is sometimes active during the day, especially at the height of the breeding season or when cold, wet weather restricts nocturnal foraging. The diet of desert animals included mammals (up to at least the size of a Rabbit), lizards, frogs and invertebrates. In forest habitats, it eats insects (ranging in size from

---

### *Dasyurus geoffroii*

**Size**

*Head and Body Length*
310–400 (360) mm (males)
260–360 (310) mm (females)
*Tail Length*
250–350 (305) mm (males)
210–310 (275) mm (females)
*Weight*
710–2185 (1310) g (males)
615–1130 (890) g (females)

**Identification** Brown above, with conspicuous white spots; creamy-white below. Tail bushy, black on distal half, occasionally with one to two spots proximally. Five toes on hindfoot, pads granular.
**Recent Synonyms** *Dasyurinus geoffroii.*
**Other Common Names** Western Native-Cat, Chuditch.

**Status** Rare and endangered, scattered in south-western Australia; extinct elsewhere.
**Subspecies** *Dasyurus geoffroii geoffroii*, inland Australia.
*Dasyurus geoffroii fortis*, south-western Western Australia.
**References**
Serena, M. and T.R. Soderquist (1989). Spatial organisation of a riparian population of the carnivorous marsupial *Dasyurus geoffroii. J. Zool. Lond.* 219: 273–283.
Soderquist, T.R. and M. Serena (1990). Occurrence and outcome of polyoestry in wild Western Quolls, *Dasyurus geoffroii* (Marsupialia: Dasyuridae). *Aust. Mammal.* 13: 205–208.

*The Western Quoll, which is quite aboreal, has five toes on the hindfoot.* HANS & JUDY BESTE

winged termites to large cockroaches and beetles), freshwater crustaceans and a variety of terrestrial reptiles, mammals and birds (up to the size of bandicoots and parrots). Western Quolls occasionally climb small trees to obtain prey or escape from predators, clasping the trunk between the long hindfeet and gripping with the forepaws.

Reflecting its carnivorous nature, the Western Quoll occurs at low densities even in high-quality habitats. Along the Murray River valley in Western Australia, each adult female dens within a stable core area of 55–120 hectares, while male dens are distributed over an area of 400 hectares or more. The core areas of neighbouring females typically show little or no overlap, suggesting that they are actively defended. In contrast, male core areas overlap extensively with those of other males as well as females.

*D. g. geoffroii*

*D. g. fortis*

**Dasyurus geoffroii**

J. LOCHMAN

*The Western Quoll usually gives birth to six young.*
*After quitting the pouch, they are suckled in a nest.*
*Siblings can vary in coat colour.*

Litters are born from May to September, most appearing in June and July. Up to six newborn young are accommodated in the moist, enclosed pouch. Juveniles begin to protude through the pouch opening at the age of six to seven weeks. By nine weeks they have outgrown the pouch entirely and are left in a den while their mother forages. At this age they can barely crawl, are very poorly insulated by fur and are not yet able to shiver. By 16 weeks, juveniles are well furred and beginning to eat solid food. They are weaned by 22–24 weeks and typically disperse shortly thereafter, in summer. Males and females are both capable of breeding at the age of one year. The majority (about 80 per cent) of weaned juveniles are the offspring of first-year mothers. This partly reflects the fact that first-year females comprise slightly more than half the breeding female population. First-year females also tend to have larger litters than older females and are about three times more successful in raising young to weaning age. While Western Quolls have been known to live for more than five years in captivity, wild individuals seldom survive for more than three years. Monitoring of radio-tagged individuals indicates that many factors may contribute to mortality: being hit by motor vehicles; illegal shooting; predation by both raptors and Foxes; disease and natural accidents such as drowning; and injury in leg-hold traps set for Foxes or Rabbits or around poultry sheds.

The total population of the Western Quoll is probably less than 6000 and it is currently classified as endangered. Its long-term survival will depend on implementing sympathetic silvicultural practices and prescribed burning regimes, protecting remaining natural habitats from being cleared and progressively reintroducing captive-bred animals to parts of the original range.

M. SERENA AND T. SODERQUIST

# Northern Quoll

## *Dasyurus hallucatus*

### Gould, 1842

hall'-ue-kah'-tus:
'hairy-tail with notable first digit
(on hindfoot)'

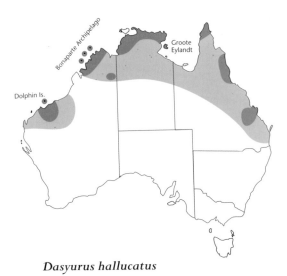

*Dasyurus hallucatus*

The Northern Quoll formerly occurred across northern Australia from the Pilbara region of Western Australia to south-eastern Queensland. It now appears to be restricted to six areas: the Hamersley Range; the North Kimberley; northern and western Top End; Cape York tip; Atherton Tableland; and Carnarvon Range. It is found within most treed habitats within its current range, sometimes living in or around human dwellings and camping grounds. It is most abundant in broken, rocky country and in open eucalypt forest within 150 kilometres of the coast. It is the smallest, most arboreal and probably most aggressive of the four quolls. Like the other three, it has undergone a substantial reduction in range since European settlement, but much

later. In North Queensland and the Kimberley, its disappearances from areas are a recent memory. The severity of the problem is difficult to ascertain due to the paucity of survey work in northern Australia.

Although it is widely distributed through a range of habitats, the most successful breeding occurs near creeklines. Those individuals near small waterholes regularly drink when water is available. It usually dens in hollow tree trunks, particularly favouring eucalypts with reddish bark, such as the Darwin Woolybutt. Accumulation of faeces indicates that it also spends considerable time on the tops of large termite mounds.

It is particularly aggressive, its 'pugnacious disposition' being noted by early collectors. The diet is varied and may include mammals such as the Large Rock-rat, Common Rock-rat and Sandstone Antechinus, as well as reptiles,

*The Northern Quoll has five toes on the hindfoot and is both aboreal and terrestrial. Smallest of the quolls, it feeds on large insects, small vertebrates and soft fruits.*

HANS & JUDY BESTE

worms, ants, termites, grasshoppers, beetles and their larvae, moths, honey and a variety of figs and other soft fruits. Its faeces and body smell strongly. An anthropological zoologist, Donald Thomson, reported that Aboriginal people did not much relish it because of the odour of its flesh. It also appears not to be eaten much by large predators such as the Dingo and Rufous Owl.

In the denser populations of the rocky country, both sexes often live for two years, sometimes three and attain greater size than in the savanna, where few males survive the breeding season and none are known to participate in two breeding seasons. There is a resemblance here to the male die-off in species of *Antechinus*. However, physiological studies show that females are also stressed in the breeding season and, with near-annual fires, both sexes may approximate to an annual life history in the savanna.

Females lack a pouch, but, in May, the area surrounding the six to eight teats becomes enlarged and partially surrounded by a flap of skin. Mating occurs in late June and one to eight young are born in July. Less than a quarter of all females carry an infant on each available teat, the average number being six. Young are carried by the mother for eight to ten weeks, some litters ceasing attachment to the teats in August, others in September. At this stage the eyes are still closed, but the back and flanks are covered with short, grey-brown hair, with cream spots visible. As many as one-third of the young are lost by September. The remainder appear to be suckled in the nest until about five months old, by which time the mother's teats may be cut and even suppurating, due to injury from her infants' sharp teeth. By this time they may even be feeding from a stretched teat while clinging to the back of the hard-pressed mother.

R.W. BRAITHWAITE AND R.J. BEGG

## Dasyurus hallucatus

### Size
*Head and Body Length*
123–310 mm (males)
125–300 mm (females)
*Tail Length*
127–308 mm (males)
200–300 mm (females)
*Weight*
400–900 g (males)
300–500 g (females)

**Identification** Grey-brown to brown above with large white spots; cream to white below. Smallest Australian member of the genus. Distinguished from other Australian quolls by combination of unspotted tail and striated pads on five-toed hindfoot.

**Recent Synonyms** *Satanellus hallucatus*.

**Other Common Names** Little Northern Native Cat, Satanellus, North Australian Native Cat. Indigenous name: Njanmak (Mayali).

**Status** Locally common but absent from much of former range.

**Subspecies** None.

**References**

Begg, R.J. (1981). The small mammals of Little Nourlangie Rock, N.T. 3. Ecology of the Northern Quoll, *Dasyurus hallucatus* (Marsupialia: Dasyuridae). *Aust. Wildl. Res.* 8: 73–85.

Schmitt, L.H., D.J. Kitchener, W.F. Humphreys and R.A. How (1989). Ecology and physiology of the northern quoll, *Dasyurus hallucatus* (Marsupialia, Dasyuridae), at Mitchell Plateau Kimberley, Western Australia. *J. Zool. Lond.* 217: 539–558.

Braithwaite, R.W. and A.D. Griffiths (1994). Demographic variation and range contraction in the Northern Quoll *Dasyurus hallucatus* (Marsupialia: Dasyuridae). *Wildl. Res.* 21: 203–17.

*The Spotted-tailed Quoll is an agile climber but spends most of its time on the forest floor.*

# Spotted-tailed Quoll

## *Dasyurus maculatus*

### (Kerr, 1792)

mak'-yue-lah'-tus: 'spotted hairy-tail'

The Spotted-tailed Quoll is the largest marsupial carnivore on the Australian mainland: its spotted tail and size readily distinguish it from the other quolls. It has been described as 'savage' and 'very ferocious', which it can be when cornered or handled, but its characteristic gaping jaw and large 'sabre-like' canines belie its frequent indifference to human presence.

It is usually nocturnal, although it may bask in the sun and sometimes hunt during daylight hours, perhaps in response to increased nutritional requirements while rearing young. Partly arboreal, it has well developed ridges or striations on the pads of its feet. It has a distinctive bound-

ing gate and a call 'like a blast from a circular saw'. Defecating localities or 'latrines' may define territories and indicate reproductive status.

It is an efficient predator, taking prey ranging from small wallabies to insects. Medium-sized mammals comprise about two-thirds of the diet in south-eastern Australia: birds, small animals and carrion from Dingo or wild Dog kills are also important components of the diet.

Both sexes become mature when about one year old and mating takes place from April to July. During this period, unmated females come on heat for about three days every three weeks.

*Dasyurus maculatus*

Copulation lasts up to eight hours, with the male grasping or licking the female's neck while she remains in a low crouch with eyes half-closed and head lowered. The pouch develops from an area with marginal folds into a fairly deep receptacle opening anteriorly and containing six teats arranged in two crescentic rows. The gestation period is 21 days and the average litter size is five. The young become free of the teats when about seven weeks old.

Social play is well developed at 13 weeks and juveniles are fully independent at 18 weeks. Both parents defend the nest site but although

*As its common name implies, the Spotted-tailed Quoll is distinguished by having spots on its body and tail. No other quoll has this feature.* D. WATTS

the male sometimes brings food to the female while her young are dependent, it has little contact with the offspring. Maturity is attained at the age of one year.

The species is recorded from a wide range of habitats, including rainforest, open forest, woodland, coastal heathland and inland riparian forest. It occurs from the coast to the snowline and inland to the western plains in Red Gum forest along the Murray River. Den sites have been recorded in caves, rock crevices and hollow logs.

The Spotted-tailed Quoll was formerly widely distributed on either side of the Great Dividing Range from southern Queensland to South Australia and Tasmania. A small subspecies is found in northern Queensland. Numbers in Tasmania appear to have recovered from the dramatic decline earlier this century (attributed to a supposed epidemic disease). It is probably extinct in South Australia and uncommon to rare in Victoria, New South Wales and southern Queensland. Its distribution is now disjunct over much of its present range. Loss of habitat through land clearing for agriculture and forestry, poisoning and trapping are also implicated in the decline. The Spotted-tailed Quoll's taste for domestic poultry has also resulted in its demise in many rural areas abutting areas of suitable habitat. Current threats to the species include competition with introduced predators such as the Fox and feral Cat, continued alienation of suitable habitat through logging, and primary and secondary poisoning from wild Dog, Fox and Rabbit baiting and trapping.

Except in Tasmania, where numbers appear to have increased, the Spotted-tailed Quoll now exists mostly in isolated areas that may be too small to support long-term viable populations while current land management practices continue and introduced predator numbers remain uncontrolled.

R. EDGAR AND C. BELCHER

## *Dasyurus maculatus*

**Size**

*Head and Body Length*
380–759 mm (males)
350–450 mm (females)

*Tail Length*
370–550 mm (males)
340–420 mm (females)

*Weight*
up to 7 kg (males)
up to 4 kg (females)

**Identification** Rich rufous brown to dark brown above, with white spots of varying size; pale below. Distinguishable from all other quolls by spotted tail.

**Recent Synonyms** *Dasyurops maculatus.*

**Other Common Names** Tiger Cat, Tiger Quoll, Spotted-tailed Native Cat, Spotted-tail Dasyure.

**Status** Common to sparse.

**Subspecies** *Dasyurus maculatus maculatus*, southern Queensland to Tasmania.
*Dasyurus maculatus gracilis*, northern Queensland.

**References**

Fleay, D.F. (1940). Breeding of the Tiger Cat. *Victorian Nat.* 56: 158–163.

Settle, G.A. (1978). The quality of Quolls. *Aust. Nat. Hist.* 19: 164–169.

Green, R.H. and Scarborough, T.J. (1990). The Spotted-tailed Quoll *Dasyurus maculatus* (Dasyuridae, Marsupialia) in Tasmania. *Tasmanian Nat.* 100: 1–15.

# Eastern Quoll

## *Dasyurus viverrinus*

### (Shaw, 1800)

viv'-e-ree'-nus: 'ferret-like hairy-tail'

The Eastern Quoll with its striking white-spotted black or fawn fur is distinguishable from the larger chocolate-coloured Spotted-tailed Quoll by the absence of spots on its tail. This slightly built animal with its large, sensitive ears, moist nose and agile movements is a more graceful predator than its larger relative.

Once ranging over most of south-eastern Australia, it suffered a drastic decline around the beginning of the twentieth century, possibly as the result of an epidemic. It is extinct in South Australia and no animals have been caught elsewhere on the mainland in the past decade. Recent claims of sightings in Victoria and New South Wales contribute to the hope that populations may still exist in these States, but the species is common only in Tasmania.

A variety of habitats including dry sclerophyll forest, scrub, heathland and cultivated land are utilised by the Eastern Quoll. In Tasmania, the highest densities occur where eucalypt forest and pastures are interdispersed. It is an opportunistic carnivore with insects as its most important prey and agricultural pests such as corbie grubs comprising a large portion of the diet. Ground-nesting birds and small mammals such as bandicoots, rabbits and rats are frequently eaten and the carcasses of larger animals such as wallabies, possums and sheep are scavenged when available. Grasses are eaten regularly and fruit, such as blackberry, is seasonally popular.

In Tasmania, mating occurs from mid-May until early June. After a gestation of about three weeks, a female may give birth to as many as 30 minute young but, since she has only six teats, survival is limited to the first six young to attach themselves to these. Six millimetres long at

birth, they remain attached to the teat until mid-August when they each weigh about 20 grams and are too bulky for the pouch. The female then deposits them in a grass-lined den, carrying them on her back if she moves from one den to another. Weaning is complete by the end of October, when the juveniles become independent. The female provides no protection for them when they emerge from the den nor does she assist them to obtain food. Juveniles engage in vigorous play and may be seen at night in pairs or groups chasing each other's tails. They become sexually mature by the next breeding season. Individuals with either black or fawn

---

### *Dasyurus viverrinus*

**Size**
*Head and Body Length*
320–450 (370) mm (males)
280–400 (340) mm (females)
*Tail Length*
200–280 (240) mm (males)
170–240 (220) mm (females)
*Weight*
900–2000 (1300) g (males)
700–1100 (880) g (females)

**Identification** Distinguished from *Dasyurus maculatus* by smaller size, presence of only four toes on hindfoot and absence of spots on tail.
**Recent Synonyms** None.
**Other Common Names** Eastern Native Cat, Quoll.
**Status** Tasmania, common; mainland Australia, rare, possibly extinct.
**Subspecies** None.
**References**
Blackhall, S. (1980). Diet of the Eastern Native-cat, *Dasyurus viverrinus* (Shaw), in Southern Tasmania. *Aust. Wildl. Res.* 7: 191–197.
Godsell, J. (1982). The population ecology of the Eastern Quoll, *Dasyurus viverrinus*, in southern Tasmania (in) M. Archer (ed.) *Carnivorous Marsupials.* Royal Zoological Society of NSW, Sydney. pp. 199–207.

D. WATTS

coat colour occur in the same litter, independent of their sex or the colour of the parents.

Mortality of young while in the pouch and den is low, so large numbers of juveniles enter the population in November. Death and dispersal of juveniles and adults over the summer and autumn reduce the population by the following breeding season. Although both males and females can breed for several years, most breeding adults are those of the previous season.

Although the Eastern Quoll is a solitary feeder, the home ranges of individuals overlap considerably. Males may travel over a kilometre in a night but females restrict their movements to a few hundred metres around their dens. Dens, which can consist of several chambers, may be made in a diversity of structures including underground burrows, hollow logs, rock piles and hay sheds. A female may use one or several dens and, except when nursing her young, may share these with males or other females. Males occupy many dens, frequently changing these nightly but rarely sharing them with other males. Despite overlaps in home ranges and occasional

*Two colour phases occur in the Eastern Quoll, one fawn, the other black. Individuals of different colour often occur in the same litter.*

instances of den sharing, adults usually avoid one another. Social interactions increase during the short breeding season when fights between males become more frequent.

J. GODSELL

***Dasyurus viverrinus***

DICK WHITFORD

DICK WHITFORD

## *Parantechinus apicalis*

**Size**

*Head and Body Length*

145 mm (males)

140 mm (females)

*Tail Length*

105–115 (males)

95 mm (females)

*Weight*

60–100 g (males)

40–75 g (females)

**Identification** Brownish-grey above, freckled with white; greyish-white tinged with yellow below. Readily distinguished by tapering hairy tail, white ring around eye and the freckled appearance of the rather coarse fur.

**Recent Synonyms** *Antechinus apicalis.*

**Other Common Names** Freckled Antechinus, Speckled Marsupial Mouse.

**Status** Rare, limited.

**Subspecies** None.

**References**

Woolley, P. (1971). 'Observations on the reproductive biology of the Dibbler, *Antechinus apicalis* (Marsupialia: Dasyuridae).' *J. Roy. Soc. West. Aust.* 54: 99–102.

Woolley, P.A. (1991). 'Reproductive pattern of captive Boullanger Island Dibblers, *Parantechinus apicalis* (Marsupialia: Dasyuridae).' *Wildl. Res.* 18: 157–163.

# Southern Dibbler

## *Parantechinus apicalis*

### (Gray, 1842)

pa-ran'-tek-ine'-us ah'-pik-ah'-lis: 'pointed antechinus-like (animal)'

The Southern Dibbler (originally referred to simply as the Dibbler) was seen for the first time in 83 years when a pair was collected by chance in 1967 at Cheyne Beach, on the south coast of Western Australia. Since then, despite intensive efforts, only small numbers of individuals have been captured at each of a few localities on the mainland, from the Fitzgerald River National Park to the east of Cheyne Beach and Torndirrup National Park to the west.

In the early days of European settlement in Western Australia it was much more widespread, being recorded from the Moore River region to King George Sound. Clearing of land for farming is probably responsible for reducing its distribution but the fact that geologically recent fossil remains are found from Shark Bay

*The Southern Dibbler (long known simply as the Dibbler) is restricted to a few mainland locations and offshore islands in south-western Australia. It feeds on large insects and small reptiles.*

to Bremer Bay indicates that its range was contracting prior to the arrival of Europeans.

Some were found in 1985 on Boullanger and Whitlock Islands off the west coast near Jurien Bay, to the north of the Moore River. The existence of apparently thriving populations on these small islands suggests that isolation has afforded them protection. Individuals from the islands are smaller than those on the mainland.

Breeding occurs annually in March and females carry as many as eight young in the shallow pouch. The young remain dependent for three to four months and are ready to breed when 10–11 months old. A single copulation may continue for several hours and a pair may copulate more than once during the mating period. Captive individuals, both male and female, from mainland and island populations may come into

breeding condition in at least two successive seasons. In the wild, in some years at least, all males in the island population are reported to die after their first breeding season, as do those of several other species of small dasyurids. This difference in behaviour between captive and wild island males requires further investigation.

P.A. WOOLLEY

**Parantechinus apicalis**

# Northern Dibbler

## *Parantechinus bilarni*

(Johnson, 1954)

bil-ar'-nee: 'Bill Harney's antechinus-like (animal)'

---

### *Parantechinus bilarni*

**Size**
*Head and Body Length*
90–115 mm (males)
90–115 mm (females)
*Tail Length*
90–125 mm (males)
90–120 mm (females)
*Weight*
20–40 g (males)
15–35 g (females)

**Identification** Fur greyish-brown above, pale grey below. Patch of darker hairs on forehead and chestnut patches behind ears.
**Recent Synonyms** *Antechinus bilarni*, has been confused with *Pseudantechinus macdonnellensis*.
**Other Common Names** Harney's Antechinus, Sandstone Antechinus.
**Status** Common, limited.
**Subspecies** None.
**References**
Begg, R.J. (1981). The small mammals of Little Nourlangie Rock, NT. 2. Ecology of the Sandstone Antechinus, *Antechinus bilarni* (Marsupialia: Dasyuridae). *Aust. Wildl. Res.* 8: 57–72.
Calaby, J.H. and J.M. Taylor (1981). Reproduction in two marsupial-mice *Antechinus bellus* and *Antechinus bilarni* (Dasyuridae), of tropical Australia. *J. Mammal.* 62: 329–341.

---

*The Northern Dibbler (once known as the Sandstone Antechinus) is not well known. It mostly inhabits rocky slopes or screes.*

The Sandstone Antechinus has been known only since 1948 when it was collected on the American–Australian expedition to Arnhem Land. Its specific name was derived from the local Aboriginal pronunciation of the name of an Australian naturalist and author of books on folklore, Bill Harney, who accompanied the expedition.

It is largely restricted to the Western Arnhem Land escarpment where it is found on rugged, dissected quartz sandstone. Within this habitat it occurs in all types of vegetation associations, but most commonly on scree slopes of large boulders covered by open eucalypt forest with perennial grasses up to 3 metres in height for much of the year. The region is subject to

D.A. MATTHEWS

*The long tail of the Northern Dibbler may contribute to its agility in its rocky habitat.*

monsoonal rains from November to March, with a dry season from May to September, when there may be a shift of populations to patches of deciduous vine thicket. Here the more complete canopy and dense litter layer provide humid conditions where insects, which constitute a major part of its diet, may proliferate.

Breeding occurs once a year, from late June to early July when the scrota of males are maximally developed. Subsequently, the males undergo a temporary decrease in body weight

and the size of the scrotum decreases until shortly before the next breeding season. Many males die after mating, but about a quarter of the individuals of each sex survive to breed in a second season. Some females even survive to produce a third litter. Throughout their lives, males are heavier than females.

The female lacks a pouch. She usually has six teats, but seldom carries more than four or five young, which hang from these as she moves about. Perhaps because of their exposed position, up to 15 per cent of the young are lost within the first month or two of life. Females are more active during this period, presumably having to hunt more to produce sufficient milk for the growing young. Maternal mortality is also high at this time, perhaps as a result of increased predation on the heavily burdened females. The young are weaned late in the year and reach sexual maturity by the following June. About three-quarters of the young that survive to maturity move away from the area in which they were born. Captive animals at La Trobe University have been observed to mate in July and give birth to young in August, the interval between mating and parturition being 38 days.

P.A. Woolley and R.J. Begg

*Parantechinus bilarni*

B.G. THOMSON

*When prey is plentiful the Fat-tailed Pseudantechinus stores fat in its tail: this provides a reserve of energy for when food is scarce.*

wide range. Insects form the main component of its diet and the tail, which is shorter than the body, becomes very fat when food is plentiful. Although predominantly nocturnal, individuals may emerge from shelter among the rocks to sunbathe and Aborigines take advantage of this habit to catch them.

Breeding occurs in winter and spring but each female produces only a single litter of up

# Fat-tailed Pseudantechinus

## *Pseudantechinus macdonnellensis*

### (Spencer, 1895)

sude'-an-tek-ine'-us mak-don'-el-en'-sis:
'MacDonnell Ranges false-antechinus'

The Fat-tailed Pseudantechinus was first collected in the MacDonnell Ranges, near Alice Springs. It is found mainly on rocky hills and breakaways but also lives in termite mounds in some parts of its

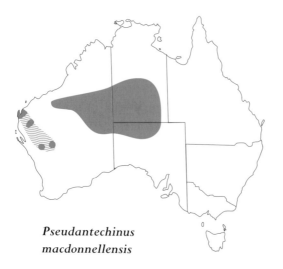

*Pseudantechinus macdonnellensis*

---

**Pseudantechinus macdonnellensis**

**Size**
*Head and Body Length*
95–105 mm (males)
95–105 mm (females)
*Tail Length*
75–80 mm (males)
75–85 mm (females)
*Weight*
25–45 g (males)
20–40 g (females)

**Identification** Greyish-brown above with chestnut patches behind ears; greyish-white below. Tapering tail usually very fat at base. Hindfoot broad.
**Recent Synonyms** *Antechinus macdonnellensis.*
**Other Common Names** Red-eared Antechinus, Fat-tailed Marsupial Mouse.
**Status** Common.
**Subspecies** None.
**References**
Spencer, W. (1896). *Report on the Work of the Horn Scientific Expedition to Central Australia, Part 2— Zoology.* London and Melbourne. pp. 1–52.
Woolley, P.A. (1991). Reproduction in *Pseudantechinus macdonnellensis* (Marsupialia: Dasyuridae): field and laboratory observations. *Wildl. Res.* 18: 13–25.

to six young in a year. These are born between late July and early September in Central Australia and in October in the more westerly part of their range. In captivity, the interval between mating and parturition is from 45 to 55 days and the young suckle for about 14 weeks. Sexual maturity is reached in the first year of life and both sexes may survive to breed in more than one year. The male has an appendage on the penis, the function of which is not known.

In a study in progress, S. Gilfillan has found that animals living in the MacDonnell Ranges near Ormiston Gorge can be trapped most readily in the more sparsely vegetated parts of their rocky habitat and that some degree of site fidelity is displayed, especially by females.

P.A. WOOLLEY

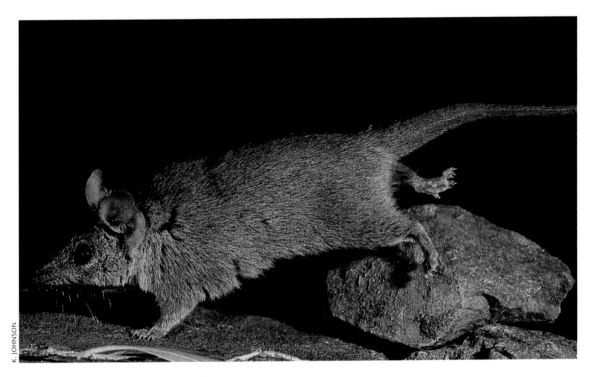

K. JOHNSON

# Carpentarian Pseudantechinus

## *Pseudantechinus mimulus*

### (Thomas, 1906)

mim'-yue-lus: 'little mimic of false–antechinus'

*The Carpentarian Pseudantechinus has not been collected on the Australian mainland since 1905 but has been found on islands in the Gulf of Carpentaria since 1967.*

The type specimen of the Carpentarian Pseudantechinus was collected in 1905 by W. Stalker on Alexandria Station in the Northern Territory. Stalker had been employed by the Hon. John Forrest of Brisbane and Sir William Ingram to obtain zoological specimens from the region including the adjoining Alroy Downs Station. *Pseudantechinus mimulus* was named by Oldfield Thomas in 1906 on the basis of a single female. Its location was given as 'Alexandria' and the general location as 19°S and 137°E, but the station at that time was much larger than its present 8390 square kilometres and a more precise location may never be known. The name was suppressed by Ride in 1971 when he placed it in synonymy with *P. macdonnellensis* but it was returned to full species status by Kitchener in 1991.

No additional specimens have been collected from mainland Australia since 1905 despite an appreciable amount of mammal survey work in apparently suitable habitat in the region. Three specimens were collected by John Calaby in 1967 from North Island in the Sir Edward Pellew Group in the Gulf of Carpentaria, Northern Territory. Additional collections were made in 1988 from North Island as well as Centre and South West Islands in the same island group.

There are no habitat notes to indicate where Stalker found the species on the mainland but, on the Sir Edward Pellew Islands, it occurs on gently sloping, rocky sandstone hills with a pebbly to rocky surface. The tree overstorey is scattered and contains mostly *Eucalyptus tetradonta* with a relatively dense shrubby understorey. The ground layer consists mostly of spinifex, *Plectrachne pungens*.

## *Pseudantechinus mimulus*

### Size
*Head and Body Length*
80–90 (84) mm (males)
77–91 (86) mm (females)
*Tail Length*
66–70 (68) mm (males)
59–75 (70) mm (females)
*Weight*
14–18 (16) mm (males)
14–25 (18) mm (females)

**Identification** Buff brown above with slight greyish tinge showing from base of hairs. Greyish white below; rufous patch behind the ears. Distinguished from congeners by smaller size.
**Recent Synonyms** *Pseudantechinus macdonnellensis.*
**Other Common Names** None.
**Status** Rare on mainland. Common on three of the Sir Edward Pellew Islands.
**Subspecies** None.
**Reference**
Kitchener, D.J. (1991). *Pseudantechinus mimulus* (Thomas, 1906) (Marsupialia, Dasyuridae): rediscovery and description. *Rec. West. Aust. Mus.* 1: 191–202.

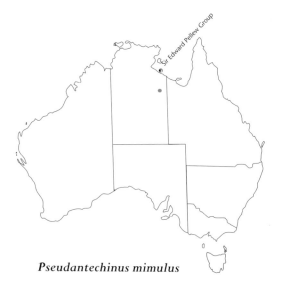

*Pseudantechinus mimulus*

Little is known of the biology of the species in the wild but, of the seven females caught in July–August, 1988, none had pouch young. In captivity, the Carpentarian Pseudantechinus is active at night and moves with great agility among rocks. It feeds voraciously on insects and the tail can become fattened. It hides by day among rocks and does not appear to build a nest of any sort.

K.A. JOHNSON AND D.G. LANGFORD

P.A. WOOLLEY & D. WALSH

# Ningbing Pseudantechinus

## *Pseudantechinus ningbing*

### Kitchener, 1988

ning'-bing: 'Ningbing false-Antechinus'

*Little is known of the biology of the Ningbing Pseudantechinus, which was not described until 1988.*

appendage to the penis. The penis does, however, have accessory erectile tissue, the form of which allies it to some extent with the Fat-tailed Pseudantechinus.

The Ningbing Pseudantechinus, only recently recognised as a distinct species, was first collected by Harry Butler on the abandoned Ningbing Station in the Kimberley region of Western Australia. It was at first considered to be a long-tailed form of the Fat-tailed Pseudantechinus but it differs from this in a number of ways, notably in lacking an

*Pseudantechinus ningbing*

79

The Ningbing Pseudantechinus has not been bred in captivity but observations made on wild-caught and captive animals indicate that breeding occurs only once a year. Mating occurs in June and the young are born in late July and early August. At an age of 16 weeks they are independent and at 11 months are capable of breeding. Both males and females may survive to breed in a second year.

P.A. WOOLLEY

### Pseudantechinus ningbing

**Size**
*Head and Body Length*
90–100 mm (males)
80–90 mm (females)
*Tail Length*
80–90 mm (males)
75–90 mm (females)
*Weight*
20–25 g (males)
15–20 g (females)

**Identification** Light greyish-brown above and below: chestnut patches behind ears. Similar in form to *Pseudantechinus macdonnellensis* but tail longer, with long hairs covering base. Remainder of tail sparsely haired and slightly scaly in appearance.
**Recent Synonyms** *Antechinus macdonnellensis*.
**Other Common Names** None.
**Status** Sparse, possibly underestimated.
**Subspecies** None.
**References**
Woolley, P.A. (1982). Phallic morphology of the Australian species of *Antechinus* (Marsupialia: Dasyuridae): A new taxonomic tool? (in) M. Archer (ed.) *Carnivorous Marsupials*. Royal Zoological Society of NSW, Sydney, pp. 767–781.
Woolley, P.A. (1988). Reproduction in the Ningbing Antechinus (Marsupalia: Dasyuridae): field and laboratory observations. *Aust. Wildl. Res.* 15: 149–156.

# Woolley's Pseudantechinus

## *Pseudantechinus woolleyae*

### Kitchener and Caputi, 1988

wool'-ee-ee: 'Woolley's false-Antechinus'

Very similar in appearance to the Fat-tailed Pseudantechinus, this species was not recognised until recently. It differs in some details of skull anatomy and lacks an appendage to the penis. However, the penis has accessory erectile tissue similar to that of the Ningbing Pseudantechinus.

It has been found in a variety of habitats ranging from rugged, stony country (where most specimens have been collected) to sand plains. It is less widely distributed than the Fat-tailed Antechinus but occurs with it in parts of its range in Western Australia.

Little is known of its habits in the wild. A female with pouch young estimated to be less than one week old, was captured in early October 1987. She reared her litter of six (four females and two males) in captivity and the

*Pseudantechinus woolleyae*

BABS & BERT WELLS

### Pseudantechinus woolleyae

*Woolley's Antechinus is named after Patricia Woolley, an authority on dasyurid marsupials.*

**Size**

*Head and Body Length*
107–108 mm (males)
100–108 mm (females)

*Tail Length*
75 mm (males)
75–85 mm (females)

*Weight*
35–50 g (males)
30–45 g (females)

**Identification** Fur rich brown above, buff below. Chestnut patches behind ears. Tail fattened at base.

**Recent Synonyms** None.

**Status** Unknown.

**Subspecies** None.

**References**

Kitchener, D.J. and N. Caputi (1988). A new species of false antechinus (Marsupialia: Dasyuridae) from Western Australia, with remarks on the generic classification within the Parantechini. *Rec. West. Aust. Mus.* 14: 43–59.

How, R.A., J. Dell and N.K. Cooper (1991). Vertebrate Fauna (in) Ecological Survey of Abydos-Woodstock Reserve, Western Australia. *Rec. West. Aust. Mus.* Supplement 37. pp. 78–123.

female and four of her offspring were maintained at La Trobe University. The young reached sexual maturity when they were about ten months old. Both males and females came into breeding condition during the short annual season in as many as four successive years. A litter of four resulted from a mating between the wild-caught female and her son in the breeding season of their second year in captivity.

P.A. WOOLLEY

D. WATTS

# Tasmanian Devil

## *Sarcophilus harrisii*

(Boitard, 1841)

sar-kof'-il-us ha'-ris-ee-ee:
'Harris's flesh-lover'

Largest of the living marsupial carnivores, the Tasmanian Devil resembles a small, robustly built dog with very powerful jaws. It is primarily a nocturnal predator on vertebrates but despite

### *Sarcophilus harrisii*

**Size**

*Head and Body Length*

652 mm (males)

570 mm (females)

*Tail Length*

258 mm (males)

244 mm (females)

*Weight*

9 kg (males)

7 kg (females)

**Identification**  Black all over; white marks usually on chest, sometimes on rump. Size of small dog.

**Recent Synonyms**  None.

**Other Common Names**  None.

**Status**  Common.

**Subspecies**  None.

**References**

Guiler, E.R. (1970). Observations on the Tasmanian Devil, *Sarcophilus harrisii* (Marsupialia: Dasyuridae), Parts 1 and 2. *Aust. J. Zool.* 18: 49–70.

Pemberton, D and D. Renouf (1993). A field study of communication and social behaviour of the Tasmanian Devil at feeding sites. *Aust. J. Zool.* 41: 507–526.

its name, appearance and reputation, it is a shy creature, easily killed by a dog. It is now restricted to and abundant in, Tasmania but, like the Thylacine, it was once widespread over the Australian mainland. It probably became extinct there due to increasing aridity and the spread of the Dingo, which was prevented by Bass Strait from entering Tasmania. (Absence of the Dingo also favoured the Thylacine.) Sub-fossil remains dated to 430 ybp have been found in south-western Western Australia.

Populations in Tasmania have fluctuated widely in the recent past, being rare in the first decade of the twentieth century and again in the 1940s, possibly as a result of a distemper-like disease that is also thought to have ravaged quoll and Thylacine populations. Perhaps facilitated by partial land clearance and subsequent increase in food availability, it is now widespread and abundant in suitable habitats, reaching highest densities in the north-east of the island. Barring further epidemics, or establishment of wild dog populations, its status appears to be secure.

Although occurring in all major habitat types in Tasmania, including outer city suburbs, it is most abundant in dry sclerophyll forest and coastal woodland, especially where this is inter-spersed with open grassland. It forages mainly at night, but may be active at dawn and dusk; newly independent young are sometimes sighted during daylight. Devils usually den in burrows, concealed in wooded terrain and each may use several dens in its home range, sometimes shared by two or more individuals. Devils some-

*The Tasmanian Devil, largest surviving dasyurid, has distinctive white patches on its black fur.*

times rest in logs, crevices underneath buildings, wombat burrows and thick patches of scrub.

*Sarcophilus harrisii* is sexually dimorphic: males grow larger and have a broader head and thicker neck than females. Mating is promiscuous. The species is not territorial, each individual occupying a home range of 8–20 square kilometres that may overlap extensively with others. Although usually foraging singly, several individuals may feed simultaneously on a large carcass: as many as 22 have been observed feeding on a dead cow. This gives rise to much squabbling, with champing of jaws, shouldering, chasing and close-range displays of teeth. Most of the time this does not result in physical contact but bites can be substantial and older animals are very scarred. Interactions are accompanied by

a wide array of vocalisations from snorts and soft barks to monotone growlings that develop into blood-curdling screeches. Communication by scent and smell is also important. Devils have been observed to drag the cloacal region on the ground and, in some areas, they defecate in latrines which may function as 'noticeboards' of information on individual movements, diet and reproductive status.

Moving in a loping gait, an individual can cover eight kilometres in a night. Intensive foraging occurs in dense vegetation, while tracks and forest verges are sometimes used for more direct travel. All Devils are able to climb trees, the young particularly so. A wide variety of food is eaten, including insects, beached fish and birds, but small mammals—ranging from possums and wallabies to wombats—are dominant in

*Female Tasmanian Devil with juveniles.*

D. WATTS

D. WATTS

*Although normally solitary, Tasmanian Devils frequently aggregate around a carcass, competing vociferously.*

the diet. In captivity, small prey is killed with a pounce, using the front paws and a bite. Devils hunt pademelons by following them at a persistent lope and they have been seen to attack wombats. It is unlikely that they are able to kill larger animals such as sheep and wallabies unless they are already moribund. The Devil's reputation as a sheep-killer is based on sightings of animals feeding on the carcasses of sheep that died from other causes.

Devils are efficient scavengers that rapidly locate dead animals in the bush and forage on the tide-lines of beaches and along verges for road-kills. The massive jaw muscles and strong teeth are well adapted to scavenging, enabling every part of a carcass (including hide and skull) to be consumed: an individual can eat 40 per cent of its body weight in a night. The proportion of pre-dation to scavenging is not known and, indeed, may well be determined by circumstances.

Breeding is highly synchronised. Mating occurs in March and births occur in April. Devils are polyovular; females produce an average of 40 eggs although there are only four teats. Up to four young (average 2.8) are carried in the shallow, backward-opening pouch until August,

when the fully-furred young are left in a den. They start to roam from the den in November and become independent in February when 40 weeks old. Mortality is high in the first year of independent life. Most females breed when two years old and both sexes grow to adult size by two to three years of age. Males, (but not females) disperse from the natal area. Longevity in the wild is about six years.

M. JONES

recent fossils only

***Sarcophilus harrisii***

## Subfamily Phascogalinae

This group, defined largely on biochemical criteria, includes the phascogales and antechinuses.

The Fawn Antechinus, palest and one of the largest of its genus, occasionally climbs trees.

# Fawn Antechinus

## *Antechinus bellus*

### (Thomas, 1904)

ant'-ek-ine'-us bel'-us:
'beautiful hedgehog-equivalent'

The Fawn Antechinus, an isolated tropical relative of the yellow-footed species, is the palest member of the genus and one of the largest. It was long regarded as very rare but the great increase in faunal surveys of tropical Australia over the past 15 years has shown it to be locally common.

It occupies woodland and open forest in the hot monsoonal part of the Northern Territory, north of about 14 degrees latitude. These communities are dominated by eucalypts, the majority of which have hollow trunks and limbs

*Antechinus bellus*

**Size**
*Head and Body Length*
121–148 (134) mm (males)
110–130 (117) mm (females)
*Tail Length*
105–126 (114) mm (males)
93–110 (100) mm (females)
*Weight*
42–66 (55) g (males)
26–41 (34) g (females)

**Identification** Pale to medium grey above, sometimes with a fawn or brownish tinge; cream to very pale grey below; white chin, hands and feet. Distinguished from *Parantechinus bilarni* by larger size, paler colour and absence of reddish patches behind ears; from *Sminthopsis virginiae* by pale colour and absence of reddish cheeks.
**Recent Synonyms** None.
**Other Common Names** Fawn Marsupial Mouse.
**Status** Common.
**Subspecies** None.
**References**
Calaby J.H. and J.M. Taylor (1981). Reproduction in two marsupial-mice, *Antechinus bellus* and *A. bilarni* (Dasyuridae), of tropical Australia. *J. Mammal.* 62: 329–41.
Friend, G.R. (1985). Ecological studies of a population of *Antechinus bellus* (Marsupialia: Dasyuridae) in tropical northern Australia. *Aust. Wildl. Res.* 12: 151–62.

as a result of termite attack and there is a grassy or open, shrubby understorey. Much of the habitat is burnt during the dry season. It is the only antechinus in this habitat but the Northern

HANS & JUDY BESTE

*Antechinus bellus*

Dibbler, which is confined to rocky habitat such as boulder piles and scree slopes, is found in the same general area.

The Fawn Antechinus shelters in hollows in standing or fallen trees and logs. It is active at night and may be seen foraging on the ground or running up and around the trunks of trees. Disturbed individuals have been observed to run up stringybarks and enter holes more than 10 metres from the ground. The few stomach samples examined have contained only insect remains.

The breeding pattern is generally similar to that of the better known Brown Antechinus. During the period in which they are sexually active, males display a sternal gland, usually bare of hair, with surrounding fur stained yellow to brown. Testes of males contain sperm from June to August and all mating takes place over a two week period in late August. Males die at the end of the mating period. Litters are born from the latter part of September into October and up to ten young have been observed on the ten teats. However, since as many as 16 embryos have been found in pregnant females, there is a loss of young at, or prior to, birth. Young remain attached to the teats for four to five weeks and are suckled in the nest until weaned in early January.

J.H. CALABY

# Yellow-footed Antechinus

## *Antechinus flavipes*

### (Waterhouse, 1838)

flah'-vee-pez: 'yellow-footed hedgehog-equivalent'

The Yellow-footed Antechinus is one of the few small nocturnal marsupials still seen around gardens and houses in suburban areas. Its nervous and cheeky disposition usually makes it a welcome and amusing visitor but its tendencies to pilfer from the kitchen and to build nests inside television sets and lounge chairs sometimes make it a nuisance.

The most widespread of antechinuses, it occurs from north-eastern Queensland to south-

---

### *Antechinus flavipes*

**Size**
*Head and Body Length*
93–165 (121) mm (males)
86–127 (105) mm (females)
*Tail Length*
70–151 (100) mm (males)
65–107 (86) mm (females)
*Weight*
26–79 (56) g (males)
21–52 (34) g (females)

**Identification** Distinguished by the distinct change in fur colour from the slate-grey head to the warm rufous rump, feet, belly and sides. Prominent light eye-rings and a black tip to tail.
**Recent Synonyms** *Antechinus stuartii* (part).
**Other Common Names** Yellow-footed Marsupial Mouse, Mardo.
**Status** Abundant.

*After quitting the rudimentry pouch, young Yellow-footed Antechinuses are suckled in a nest. The young individuals illustrated are in the vicinity of the nest.*

HANS & JUDY BESTE

*In the absence of a complete pouch, infant Yellow-footed Antechinuses dangle from their mother's teats.*

around Australia, the largest, most superbly red individuals occurring in northern Queensland and more drably coloured, white-bellied individuals in south-western Western Australia.

The darting movements of the Yellow-footed Antechinus make it a difficult animal to observe. It can run upside-down along branches and rock surfaces with as much speed as when it

western Western Australia in a broad spectrum of habitats from tropical vine-forests through swamps to dry mulga country. Its diet consists mostly of insects but may include almost anything from flowers and nectar to small birds and house mice. Many farmers and bird fanciers are familiar with signs of its activity: as it devours vertebrate prey, its victims such as mice or birds are neatly turned inside-out and the everted skin is left as a testimony to the feast. (Other dasyurids have similar feeding behaviour but few leave their traces where humans can find them.)

It is one of the most colourful of small marsupials and is distinguished from other antechinuses by the definite change in fur colour from its slate-grey head to the warm, orange-brown sides, belly, rump and feet. It has white to light grey eye-rings and a prominent black tip to the tail. Fur colour varies greatly

**Subspecies** *Antechinus flavipes flavipes*, south-eastern Queensland, New South Wales, Victoria, South Australia.
*Antechinus flavipes leucogaster*, Western Australia.
*Antechinus flavipes rubeculus*, north-eastern Queensland.

### References

Van Dyck, S. (1982). The relationship of *Antechinus stuartii* and *A. flavipes* (Dasyuridae: Marsupialia), with special reference to Queensland (in) M. Archer (ed.) *Carnivorous Marsupials.* Roy. Zool. Soc. NSW and Surrey Beatty and Sons, Sydney. pp. 723–766.

Smith, G.C. (1984). The Biology of the Yellow-footed Antechinus, *Antechinus flavipes* (Marsupialia: Dasyuridae), in a swamp forest on Kinaba Island, Cooloola, Queensland. *Aust. Wild. Res.* 11: 465–80.

M. TRENERRY

*The Yellow-footed Antechinus has yellow-brown feet, but these are noticeably paler in southern populations.*

scampers along the forest floor, pouncing on insects and 'bulldozing' amongst leaf litter in search of prey. Nevertheless, its erratic, quick movements do not always enable it to escape predators such as owls and feral Cats.

Like most other Australian antechinuses, it breeds only once a year. Mating takes place either in late winter or spring, depending on the locality. Copulation may last up to 12 hours, the male ensuring the female's cooperation by tightly gripping the scruff of her neck in his jaws and bear-hugging her abdomen from behind with its forelegs. Shortly after mating, all males die, leaving pregnant females and young-to-be in an environment free from their competition. After about a month's gestation, the female gives birth to as many as 12 young which are carried in the pouch for up to five

weeks and weaned after about three months. The young share a leafy nest until the following winter when they begin to become territorial and more intolerant of each other's company.

S.M. VAN DYCK

*Antechinus flavipes*

# Atherton Antechinus

## *Antechinus godmani*

(Thomas, 1923)

god'-man-ee: 'Godman's hedgehog-equivalent'

*Antechinus godmani*

For many years the Atherton Antechinus was confused with the common and more widely distributed Yellow-footed Antechinus. For most of 60 years since its discovery in 1923 it was thought to occur at only a handful of localities

*The Atherton Antechinus has small eyes and seems to depend more on hearing and smell to locate its prey.*

D.WHITFORD

near Ravenshoe on the southern edge of the Atherton Tableland. However, more intensive research in recent years has shown it to be present at nearly 30 localities, from the summit of Bellenden Ker in the north, to near Cardwell in

## Antechinus godmani

**Size**

*Head and Body Length*
120–160 (143) mm (males)
90–135 (122) mm (females)
*Tail Length*
105–145 (126) mm (males)
85–115 (104) mm (females)
*Weight*
85–125 (95) g (males)
53–73 (58) g (females)

**Identification** Recognisable by its large size, dull brown body fur, bright orange to ginger cheeks and almost naked tail.

**Recent Synonyms** *Antechinus flavipes godmani.*

**Other Common Names** Godman's Marsupial Mouse, Godman's Antechinus.

**Status** Rare, limited.

**Subspecies** None.

**References**

Van Dyck, S. (1982). The status and relationships of the Atherton Antechinus, *Antechinus godmani* (Marsupialia: Dasyuridae). *Aust. Mammal.* 5: 195–210.

McDonald, K.R. (1991). New distribution records for *Antechinus godmani* (Thomas), a restricted rainforest endemic. *Mem. Qld. Mus.* 30: 487–491.

Laurance, W.F. (1993). The pre-European and present distributions of *Antechinus godmani* (Marsupialia: Dasyuridae), a restricted rainforest endemic. *Aust. Mammal.* 16: 23–27.

the south. This, nevertheless, represents a narrow latitudinal range of around 130 kilometres and an area of approximately 2000 square kilometres. Its refuge here is in mountain-top rainforest above 600 metres. With an average rainfall of over 2700 millimetres, this area is always damp and misty.

The Atherton Antechinus is predominantly nocturnal, although an individual may spend some time foraging during the early morning or late afternoon. Most of its active hours are occupied with rummaging through fallen vegetation, rotting logs, or climbing through the forest canopy. Although it is known to feed on insects, arachnids, earthworms, frogs and lizards, the size and strength of an adult male gives the impression that it would be capable of dealing smartly with small rats and other vertebrates nearly as big as itself. The few researchers who have handled this species in the wild all testify to its unfailing capacity to deal swiftly and uncompromisingly with slow human fingers.

It has very small eyes and appears to depend more on hearing and smell than on vision. In one locality, males blinded by diseased eyes were observed moving about with ease on the rainforest floor, catching insects and even entering the live traps set to catch them for study. Disease appears to characterise old males, most of which probably die when they are about 12 months old, immediately after the mating season in July.

Females become particularly shy after the mating season and are rarely observed outside their leafy nests built in tree hollows or epiphytes. Births occur in August and the females are capable of carrying up to six young. These are suckled continuously for the first five weeks, by which time they hang from the mother's teats like grapes. For the remainder of the suckling period, the female leaves the young in her nest while she forages for food.

The Atherton Antechinus is not common within its restricted and isolated range and for this reason has been little studied. Although its range falls within the boundaries of protection from logging offered by World Heritage listing of Wet Tropics rainforest, much more information is required on the biology and habitat requirements of this rare and secretive species.

S.M. Van Dyck

C.ANDREW HENLEY/LARUS

# Cinnamon Antechinus

## *Antechinus leo*

### Van Dyck, 1980

lee'-oh: 'lion hedgehog-equivalent'

The Cinnamon Antechinus is found only in the semi-deciduous mesophyll rainforest between the Iron and McIlwraith Ranges on the east coast of Cape York Peninsula. It nests in a hollow in a tree or log during the day and forages at dusk and into the night for a variety of invertebrates including ants, beetles, caterpillars, centipedes, roaches, spiders and worms. It is an excellent climber, darting up and down trees with great speed and agility. On the forest floor, it tends to run silently along logs and buttresses rather than scamper noisily through leaf litter.

*The Cinnamon Antechinus, described in 1980, is an agile climber in tropical rainforest.*

Its life history is remarkably synchronised. Mating begins around mid-September when males become very active, increasing their home ranges and extending their activity into the day, presumably to search for females. Encountering one, the larger male violently subdues her by

*Antechinus leo*

C. ANDREW HENLEY/LARUS

biting the back of her neck, clasping her body and attempting intromission. Copulation was recorded in the field for up to 12 minutes and possibly lasts much longer. Occasionally the female may free herself before genital lock is achieved and hide by clinging motionlessly on a branch, but the male

*The Cinnamon Antechinus also feeds on the forest floor.*

pursues her relentlessly; mated females often have patches of hair missing from the back. Towards the end of the mating season, males rapidly lose weight and condition; some individuals have been seen lying on the forest floor, too weak to run away. By about mid-October all males die. Females give birth from late October to early November and carry up to ten young in a pouch. By about mid-December the young have grown too large for the mother to carry and she deposits them in the nest. In late January the young venture out with the mother. Male juveniles soon move away, but daughters remain in their mother's area and inherit her resources (e.g. home range and nest sites) once she dies. The onset of mating in *A. leo* is timed so that the lactation and weaning of young coincide with the peak of food abundance in the wet season. Some females live for more than two years to rear a second litter but very few survive to breed a third time.

Snakes and owls are probably the main predators of this small marsupial. The Black Butcherbird, *Cracticus quoyi* and Grey Goshawk, *Accipiter novaehollandiae* prey on it in the mating season when it is active during the day. The Cinnamon Antechinus is common within its restricted and isolated range and is at no immediate risk if its habitat remains intact.

L.K-P. LEUNG

---

### Antechinus leo

**Size**

*Head and Body Length*
142–160 (150) mm (males)
109–136 (127) mm (females)
*Tail Length*
127–140 (134) mm (males)
81–125 (110) mm (females)
*Weight*
67–124 (94) g (males)
32–74 (54) g (females)

**Identification**  Characterised by uniform, rich cinnamon fur and a darker mid-dorsal head stripe.
**Recent Synonyms**  Previously confused with *Antechinus flavipes rubeculus* and *A. godmani*.
**Other Common Names**  Iron Range Antechinus, Cape York Antechinus.
**Status**  Common in suitable habitat; habitat limited.
**Subspecies**  None.
**Reference**
Van Dyck, S. (1980). The Cinnamon Antechinus, *Antechinus leo* (Marsupialia: Dasyuridae), a new species from the vine forest of Cape York Peninsula. *Aust. Mammal.* 3: 3–17.

D. WATTS

*The Swamp Antechinus is restricted to cool, dense, damp vegetation, where it digs in the soil and leaf litter for small invertebrates.*

# Swamp Antechinus

## *Antechinus minimus*

### (Geoffroy, 1803)

min'-im-us: 'smallest hedgehog-equivalent' (species was first assigned to *Dasyurus*)

The Swamp Antechinus lives mainly in cool to cold, wet closed heath and dense wet tussock grassland and sedgeland. On the mainland it is predominantly coastal but it is distributed widely in Tasmania, mostly in buttongrass sedgelands up to an elevation of 1000 metres, also in rainforest and wet-forested gullies; it avoids very open vegetation. Unlike the Brown Antechinus, which is partly arboreal and nests in trees, the Swamp Antechinus is terrestrial, making a grass-lined nest at ground level, usually at the base of a tussock. It forages like a bandicoot, using the long foreclaws to dig for insects and other invertebrates in leaf litter and soil. Strictly nocturnal, it is most active in the first few hours after dark.

## *Antechinus minimus*

**Size**
*Head and Body Length*
103–140 (120) mm (males)
98–117 (110) mm (females)
*Tail Length*
65–100 (80)mm (males)
67–85 (75) mm (females)
*Weight*
30–103 (65) g (males)
24–65 (42) g (females)

**Identification** Leaden grey on head and shoulders grading into rich yellowish-brown on rump and flanks; greyish-yellow or buff below. Tail short-haired, grizzled dark brown above, lighter below. Fur coarse, grizzled. Foreclaws long. The tail and ears short; eyes and ears small.
**Recent Synonyms** None.
**Other Common Names** Little Tasmanian Marsupial Mouse.
**Status** Rare, limited on mainland; sparse in Tasmania.
**Subspecies** *Antechinus minimus minimus*, Tasmania and Bass Strait Islands.
*Antechinus minimus maritimus*, mainland and Glennie Island.
**References**
Wainer, J.W. (1976). Studies of an island population of *Antechinus minimus* (Marsupialia, Dasyuridae). *Aust. J. Zool.* 19: 1–7.
Wilson, B.A. (1986). Reproduction in the female dasyurid *Antechinus minimus maritimus* (Marsupialia, Dasyuridae). *Aust. J. Zool.* 34: 189–97.

Mainland females have eight teats; those in Tasmania have six. A rudimentary pouch (a fold of skin around the mammary area) develops at the onset of breeding, which occurs annually between May and August, the time varying quite consistently with locality. Births occur about a month after mating; young are attached to the teats for about two months and begin to wander

# Brown Antechinus

## *Antechinus stuartii*

Macleay, 1841

stue'-art-ee-ee: 'Stuart's hedgehog-equivalent'

A.m. maritimus

Great Glennie Is.

Flinders Is.

King Is.

A.m. minimus

**Antechinus minimus**

from the next when about three months old. The litter size at weaning usually equals the number of teats. All adult males die shortly after mating but some females survive to breed in a second year.

Females maintain small home ranges. Adult males range widely, presumably in defence of territory and in search of females. Population densities vary from a minimum of one lactating female per hectare to 14 per hectare on the mainland and 19 per hectare on Glennie Island, off Wilson's Promontory. Since its preferred habitat is limited, the Swamp Antechinus is patchily distributed. It is considered to be rare in South Australia and Victoria, less so in Tasmania. It appears to prefer late successional vegetation and it is significant that populations in the eastern Otway Ranges, Victoria, that were eliminated by the 1983 'Ash Wednesday' bush-fires, have not become re-established after ten years.

The scientific name of the species suggests that it is the smallest antechinus. In fact, it is one of the largest. The specific name was applied in 1803, when this marsupial was regarded as a member of the genus *Dasyurus* (making it the smallest of the quolls).

Very recently, what had long been known as the Brown Antechinus was discovered to consist of two genetically distinct populations, the more southern of which is a new species, the Agile Antechinus, yet to be formally described. Consequently, the range of the Brown Antechinus (in the strict sense) is now less than before. Although very similar in appearance, the two species show great electrophoretic differences and overlap in distribution only in a small area of southern New South Wales around Kioloa. The two forms also appear to be reproductively isolated, ovulation being stimulated by different rates of increase in photoperiod. The Brown Antechinus (in the strict sense) has a disjunct distribution. One subspecies, *A. stuartii stuartii*, occurs from southern Queensland to southern NSW. The other, *A. s. adustus*, is restricted to tropical Queensland.

The Brown Antechinus is widespread in a variety of forested habitats in eastern Australia and occurs in highest density where there is thick ground cover and abundant logs. In Queensland, its distribution is restricted to the wettest, densest forests, leaving the more open forests and heathlands to the Yellow-footed Antechinus. It is usually terrestrial but may be quite arboreal in dry forests that have little ground cover and in

J.W. WAINER AND B.A. WILSON

Beetles, spiders, amphipods and cockroaches make up the greater part of the diet of this opportunistic insectivore, which may forage on trees and on the ground. Most hunting takes place at night but it may be active during the day, particularly during winter when food is scarce. Another means of coping with the scarce winter food supply is for an individual to slip into torpor for a few hours at a time, thus reducing its energy requirements.

Males, which are 20–100 per cent heavier than females, become increasingly aggressive as winter progresses: their aggressive staccato bursts of 'chee' vocalisations are heard more frequently as the males sort out their relationships before the mating season (August in southern Australia, September in southern Queensland). This local synchrony of the breeding season is triggered by a particular rate of increase in daylength after the winter solstice. Both sexes nest communally until late May and then become more solitary in the period leading up to the mating season. When male and female get together for mating, copulation is prolonged, generally occurring for about six hours, with a few of the males mating many times.

At the end of the two-week mating season, not a single male is left alive. Death results from the stress associated with the social demands of the mating season, a time when the males stop

---

### Antechinus stuartii

**Size**

*Head and Body Size*
77–140 (100) mm (males)
74–120 (92) mm (females)

*Tail Length*
73–110 (95) mm (males)
65–100 (85) mm (females)

*Weight*
29–71 (35) g (males)
17–40 (20) g (females)

**Identification** Uniform greyish-brown above, paler below. Head broad and flat, coming to a point at the nose. Eyes moderately prominent but without surrounding ring of pale fur. Moderately hairy, thin tail almost as long as head and body.

**Recent Synonyms** *Antechinus flavipes burrelli*, *A. flavipes adustus*.

**Other Common Names** Macleay's Marsupial Mouse, Stuart's Antechinus. Indigenous name: Berruth (Woi Wurrong).

**Status** Abundant.

**Subspecies** *Antechinus stuartii stuartii*, south-eastern Australia.

*Antechinus stuartii adustus*, vicinity of Cairns, Queensland.

**References**

Braithwaite, R.W. (1979). Social dominance and habitat utilisation in *Antechinus stuartii* (Marsupialia). *Aust. J. Zool.* 27: 517–528.

Dickman, C.R., D.H. King, M. Adams and P.R. Baverstock (1988). Electrophoretic identification of a new species of *Antechinus* (Marsupialia: Dasyuridae) in south-eastern Australia. *Aust. J. Zool.* 36: 455–463.

---

wetter forests in New South Wales, where the Dusky Antechinus occupies the forest floor. In the rainforests of North Queensland it frequently nests in epiphytic ferns.

A.s. adamsi

A.s. stuartii

*Antechinus stuartii*

ESTHER BEATON

*The Brown Antechinus, one of the more common members of the genus, is terrestrial in its wetter habitats but quite aboreal everywhere else.*

C.A. HENLEY

*The Brown Antechinus lacks the pale eye-ring of the Agile Antechinus.*

feeding, live on their reserves and seek all opportunities to mate. As in the Agile Antechinus, it seems that stress hormones reduce the effectiveness of the immune system, allowing the males to succumb to parasites of the blood and intestine and to bacterial infections of the liver. Females avoid this extreme syndrome and may breed in a second season but the success of reproduction in such subsequent years is very low.

Females lack a pouch. There are usually eight teats but the number varies from six to ten, being constant for a locality. Gestation takes 27 days and sufficient young are usually born to occupy every teat. Although the young are dragged awkwardly across the ground until they are five weeks old, their survival rate is high. From an age of about five weeks they are left in a spherical nest of dry plant material hidden in a hollow log or tree trunk, from which they later venture out in company with the mother. The association between a mother and her offspring decreases shortly after weaning at about 90 days.

The Brown Antechinus is less well known than the Agile Antechinus but it is likely that the two species differ in only minor aspects of their biology.

R. W. BRAITHWAITE

*Dusky Antechinus feeding.*  P. WHITFORD

# Dusky Antechinus

## *Antechinus swainsonii*

### (Waterhouse, 1840)

swane'-sun-ee-ee: 'Swainson's hedgehog-equivalent'

This small, ground-dwelling marsupial is found only in Tasmania and on the east coast of mainland Australia. The densest populations occur in mountainous areas, such as Kosciusko National Park and the Brindabella Range, where the annual rainfall exceeds 1000 millimetres, the preferred habitats in these areas being alpine heath or tall open forest with a dense understorey of fern or shrub. Although individuals less than six months old sometimes climb in the lower limbs of trees, activity is mostly restricted

*A.s. mimetes*

*A.s. insulanus*

*A.s. swainsonii*

**Antechinus swainsonii**

to the rich and friable topsoils typical of these habitats. Here the Dusky Antechinus forages—as much by day as by night—using its long claws and powerful limbs to dig out a wide variety of soil invertebrates, occasionally supplementing this diet with fruits such as blackberry. It locates prey principally by smell and to less extent, by sight and hearing. Food is held in the forepaws and passed voraciously to the mouth; water is lapped up by the short tongue. Both males and females appear alert and inquisitive, an impression that is enhanced by their unusual and characteristically jerky movements.

During the winter, females excavate nests in creek banks or just below the soil surface, often under the cover of decaying logs or grass. The nest chamber, roughly spherical and about 10 centimetres in diameter, is loosely lined with dried grass and leaves and has a single opening, less than 1 metre long, leading to the outside. Tunnels are sometimes constructed along the topsoil–vegetation interface but more commonly the larger tunnels of bush rats are used. Wombat pathways may be used as trails.

Seasonal fluctuations in population size arise in part from the highly synchronised life history, with mating restricted to a short period in winter. Copulation is violent and is not preceded by evident courtship: the larger male attempts to subdue the female immediately by biting the fur of her neck and by tightly clasping her chest region with his forelimbs before

C. ANDREW HENLEY/LARUS

*Dusky Antechinus sniffing heath flowers, possibly in search of insects.*

achieving intromission, which lasts for up to six hours. All males die within three weeks of mating. Females give birth a month after mating and carry their six to ten young in an open pouch for up to eight weeks. Young are thereafter left in the nest until able to venture out on their own at about three months of age. The influx of juveniles into the population in spring coincides with a peak in the availability of their invertebrate food, but mortality is nevertheless high at this time, probably due to predation on the inquisitive but inexperienced young.

The Dusky Antechinus has a relatively wide range of vocalisations, some of which are related to age or sex. When threatened, nestlings utter a 'siss' cry, which increases in frequency and intensity throughout nest life. Males and females 'chit' (occasionally 'whitter') when encountering an unfamiliar object. Fighting males utter shrill, open-mouthed 'siss' cries, 'wheezes' and 'cackle laughs' and smear objects with a secretion from a chest gland.

Although the Dusky Antechinus is not threatened as a species, local populations throughout the range have been reduced by controlled burning and the replacement of

C. ANDREW HENLEY/LARUS

*Female Dusky Antechinus with eleven-week-old young in underground nest.*

# Agile Antechinus

## *Antechinus* sp.

During the course of a study of genetic variation in the Brown Antechinus in 1980, evidence of a second, related species, was obtained at a single locality on the south coast of New South Wales. Further surveys showed that the second species occurs widely in south-eastern Australia, in all forest and heathland habitats from the coast to an altitude of about 2000 metres. Although yet to receive a scientific name, the swift movements and acrobatic climbing ability of this species have given it the unofficial common name of the Agile Antechinus. Formerly confused with the Brown Antechinus, the Agile Antechinus is distinguished by its relatively small size, grey body fur, certain skull characters and distinctive tissue proteins.

Invertebrates comprise the bulk of the diet, with a preference for large beetles, spiders and cockroaches; small lizards and soft berries are ingested occasionally. Prey are detected by sight

---

### *Antechinus swainsonii*

**Size**

*Head and Body Length*
103–188 (128) mm (males)
89–140 (116) mm (females)

*Tail Length*
80–121 (107) mm (males)
75–100 (92) mm (females)

*Weight*
43–178 (65) g (males)
37–100 (41) g (females)

**Identification** Characterised by deep brown-black or grizzled, grey-brown, soft upper fur; small eyes and ears; long, curved claws.

**Recent Synonyms** None

**Other Common Names** Swainson's Antechinus, Dusky Marsupial Mouse.

**Status** Abundant in suitable habitat; habitat common.

**Subspecies** *Antechinus swainsonii swainsonii*, Tasmania.

*Antechinus swainsonii mimetes*, south-eastern Queensland, eastern New South Wales, southern Victoria, except Grampians range.

*Antechinus swainsonii insulanus*, Grampians range, western Victoria.

**References**

Green, R.J. (1972). The murids and small dasyurids in Tasmania, Part 5. *Antechinus swainsonii* (Waterhouse, 1840). *Rec. Queen Victoria Mus.* Launceston 46: 1–13.

Dickman, C.R. (1986). An experimental study of competition between two species of dasyurid marsupials. *Ecol. Monogr.* 56: 221–241.

---

native forests by pine plantation. These practices remove complex understorey vegetation and litter and, with it, much of the invertebrate food. Farming has a smaller influence, but feral Cats and Foxes may be very effective in locally reducing numbers.

C.R. DICKMAN

*Antechinus* sp.

C. ANDREW HENLEY/LARUS

*Copulation in the Agile Antechinus is very vigorous. Shortly after mating all males die (as in other antechinuses).*

## *Antechinus* sp.

**Size**

*Head and Body Length*
80–110 (98) mm (males)
75–95 (90) mm (females)

*Tail Length*
80–110 (92) mm (males)
75–100 (80) mm (females)

*Weight*
20–40 (30) g (males)
16–25 (18) g (females)

**Identification** Medium grey or greyish brown above, paler below. Eyes prominent, with no ring of lighter fur surrounding them. Thin tail, almost as long as the head and body length. Distinguished from *Antechinus stuartii* by smaller size, greyish fur and skull characters, including more massive upper incisors, presence of lingual cusp on premolars 1–3 and relatively broader inter-orbital area.
**Recent Synonyms** *Antechinus stuartii* (part).
**Other Common Names** None, but has been confused with Brown Antechinus.
**Status** Abundant.
**Subspecies** Non.

or sound on the ground surface or on the trunks and limbs of trees and are often captured after active pursuit. Foraging is mostly nocturnal, but occurs also by day if food is scarce. Animals may enter torpor in response to food shortages, especially in winter.

The life cycle is highly synchronised. The mating period occupies two to three weeks in winter each year and is terminated by the death of all males. Ovulation is facilitated by social contact and occurs at about the time of male deaths (when the day length is increasing at a rate of 127–137 seconds per day). The number of teats tends to vary with locality: females with six teats occur in the wettest areas; those with ten in the driest and highest localities.

Females give birth after a gestation of 27 days and carry a littler of six to ten young in the pouch for five weeks. Young are then left in a leaf-lined nest, usually above ground in a tree-hollow, until they are weaned at about three months of age. Most males disperse shortly after weaning in summer, but females remain in their natal area. Population size is highest at this time of year, but declines rapidly until autumn due to high mortality among adult females and juveniles. Except during lactation, when females are solitary, groups of up to 20 unrelated animals may nest together.

Males forage over an area of about a hectare, almost three times the area used by females, but both sexes use much larger areas for social interactions. Soft 'chit' cries are produced by lactating females near their young.

*Female Agile Antechinus with eight-week-old young.*

**References**

Dickman, C.R. (1986). An experimental study of competition between two species of dasyurid marsupials. *Ecol. Monogr.* 56: 221–241.

Dickman, C.R., D.H. King, M. Adams and P.R. Baverstock (1988). Electrophoretic identification of a new species of Antechinus (Marsupialia: Dasyuridae) in south-eastern Australia. *Aust. J. Zool.* 36: 455–463.

*The pouch of antechinuses is rudimentary. The individuals illustrated below are Agile Antechinus, about three weeks old.*

Weak 'siss' cries are uttered by nestlings, whereas more intense and prolonged 'siss' cries are produced by adults during aggressive encounters.

The Agile Antechinus is abundant over much of its range, but has suffered local reductions in numbers due to clearing, planting of exotic pines, harvesting operations and control-burning. Although animals fall prey to introduced predators such as the feral Cat and Fox and native predators such as owls, there is no indication that the species is of current conservation concern.

C.R. DICKMAN

C. ANDREW HENLEY/LARUS

# Red-tailed Phascogale

### *Phascogale calura*

## Gould, 1844

fas'-koh-gah'-lay kah-lue'-rah:
'beautiful-tailed pouched-weasel'

J. LOCHMAN

Notable among dasyurid marsupials because of the rusty red colour of the fur on the upper proximal half of its tail, the Red-tailed Phascogale now has a quite restricted distribution in those parts of the south-western Australian wheatbelt that receive an annual rainfall of 300–600 millimetres. Within this region it is confined largely to isolated reserves that exceed 450 hectares but it also occurs in some small patches of forest that have not been disturbed by farming activities.

Its preferred habitat appears to be the denser and taller climax vegetation communities within Wandoo (*Eucalyptus wandoo*) and Rock Oak (*Casuarina huegeliana*) alliances that include species of *Gastrolobium* and *Oxylobium*; plants that produce monosodium fluoroacetate, which can be lethal to sheep and cattle. Native herbivores

have evolved a tolerance to fluoroacetate, as have the native carnivores that prey upon them. However, the bodies of native herbivores can contain sufficient fluoroacetate to poison introduced carnivores—particularly the Fox—that eat them. The disappearance of the Red-tailed Phascogale from most of southern and central Australia and its survival in the Avon and Stirling vegetation subprovinces of Western Australia may be due to the protection provided by these toxic plants.

The Red-tailed Phascogale is arboreal, able to make leaps of up to 2 metres and thus to cover considerable distances while remaining within the forest canopy. It also feeds extens-

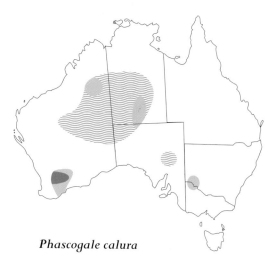

*Phascogale calura*

*The Red-tailed Phascogale is a strongly aboreal dasyurid, distinguished by reddish fur on the proximal half of its tail.*

ively on the ground. Although mainly nocturnal, it has been seen to emerge during the day to investigate potential food sources. Curiously, where *Casuarina huegeliana* is found in dense patches, the commonly occurring abraded bark caused by the rubbing together of branches reveals a colour identical to the rusty red of the tail. It is in this vegetation type, where occasional senescent *Eucalyptus wandoo* trees provide nest sites, that the species reaches its highest density.

It is an opportunistic feeder, taking a wide range of insects and spiders (with a preference for those less than 10 millimetres in length); small birds and small mammals, particularly the House Mouse. It does not need to drink.

Predators include the Cat and the Fox: as early as 1843 John Gilbert reported that the original specimen was captured by a domestic Cat while he was at the Williams River in south-western Western Australia.

A feature of the life history (shared with the Brush-tailed Phascogale antechinuses and the Kaluta), is the annual post-mating mortality of all males. Mating occurs during a three-week period in July. After 28–30 days gestation, females give birth to up to 13 young, no more than eight of which can be accommodated on the teats: the average litter size is 7.5. Males move greater distances around mid-June and during the later stages of the mating period than at other times of the year.

In the wild, males live only 11.5 months but some females survive to reproduce in a second or even third season. Captive males, held in isolation but allowed to mate in their first year, survive for up to three years.

Males possess a sternal scent gland and paracloacal glands, both of which become secretory just before and during the mating period. Social behaviour appears to change during the life his-

tory, as evidenced by pregnant females becoming quite aggressive toward males.

Most of the reserves in which the species is more commonly found have not been burnt for many years and, as a consequence, carry a climax vegetation community which provides the animal with potential nest sites and with sufficiently dense foliage for protection and foraging. Recognition of this requirement seems crucial for the conservation of the species.

A.J. BRADLEY

## *Phascogale calura*

**Size**

*Head and Body Length*
105–122 (113) mm (males)
93–105 (101) mm (females)
*Tail Length*
134–145 (141) mm (males)
119–144 (132) mm (females)
*Weight*
39–68 (60) g (males)
38–48 (43) g (females)

**Identification** Ashy grey above, cream to white below; blackish patch in front of eye. Ears and upper proximal part of tail reddish, distal half of tail with brush of long black hairs. Distinguished from *Phascogale tapoatafa* by reddish base of tail.

**Recent Synonyms** None.

**Other Common Names** Red-tailed Wambenger.

**Status** Common, limited. At risk.

**Subspecies** None.

**Reference**

Kitchener, D.J. (1981). Breeding, diet and habitat preference of *Phascogale calura* (Gould 1844) (Marsupialia: Dasyuridae) in the southern wheatbelt, Western Australia. *Rec. West. Aust. Mus.* 9: 173–86.

# Brush-tailed Phascogale

## *Phascogale tapoatafa*

### (Meyer, 1793)

tah'-poh-ah-tah'-fah: an indigenous name for this species

Known to the earliest settlers in Sydney, the Brush-tailed Phascogale has, in the past, been referred to as a 'vampire marsupial' and 'blood-thirsty killer'. In fact, it feeds mainly on cockroaches, beetles, centipedes, spiders and even bull ants, but it is an occasional predator of small vertebrates and even of penned poultry. One of the most arboreal of the dasyurids, it seldom feeds on the ground, preferring to forage in large trees, especially on dead branches. It is an agile hunter, spiralling up tree trunks, running along or underneath major branches and leaping up to 2 metres between trees. The claws are long and sharp and the hindfoot can grip either by inward folding of the edges of the sole or by flexion of the first toe. The hindfoot can be rotated through about 180° at the ankle, enabling the phascogale to climb upwards or downwards with equal ease.

Hidden prey is captured by tearing away bark with the long front incisors and using the dexterous forepaws to probe into crevices. Large insects that fall are often hastily pursued to the ground. Nectar forms part of the diet and the Brush-tailed Phascogale may spend much of a night foraging in a single heavily flowering eucalypt, dashing from cluster to cluster. It is active only between dusk and dawn but, on long winter nights, individuals often do not emerge from their nests until after midnight. When alarmed, the phascogale repeatedly taps its forefeet against bark.

Mating occurs over a three-week period between mid-May to early July, varying with locality. Prior to breeding, males engage in short pursuits of females, without contact: females permit several such chases. Females display an interest in scent from the chest gland of males, perhaps identifying future mates. During the breeding season, a male may move long distances, sometimes beyond its home range. A female can repel an approaching male by a vocal threat. A female in oestrus may attract the attention of a male by approaching him, making quick movements around him and uttering a chirping sound. Mating usually occurs in a tree hollow. The only mating to have been directly observed lasted one hour, after which the pair separated and each moved off in a different direction. Gestation lasts about 30 days and more young are born than can be accommodated on the eight teats. The usual (initial) litter size is seven to eight but it may be as low as three. As pregnancy advances, the exposed mammary area deepens and a fleshy rim develops, completely enclosing the newborn young. Despite rapid growth, the young remain fully protected until they are five weeks old, when they barely protrude from the pouch. At about the age of seven weeks they are deposited in the maternal nest. Hairless and weighing only 4

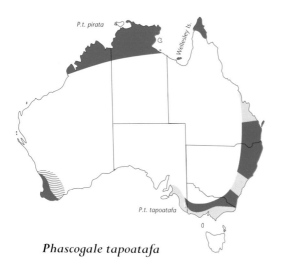

*Phascogale tapoatafa*

grams, they lose heat rapidly on cold nights and become torpid within an hour of the mother's departure. The female returns from foraging every few hours to warm and nurse her litter. Nest attendance gradually decreases until, when the young are about 20 weeks old, the mother does not return until dawn and weaning is complete. It is common for the size of the litter to decrease during the course of lactation and there is significant mortality of newly weaned young.

Play among juveniles is limited to brief chases; they explore and learn to hunt alone. Dispersal occurs in mid-summer with juvenile males moving many kilometres while females settle in nearby vacant home ranges or occasionally share the maternal home range. Phascogales rarely nest together after dispersal. Distribution is usually very sparse. Female home ranges span 20–70 hectares and do not overlap with those of unrelated females. Male home ranges are more than twice as large, overlap both with those of females and other males and expand greatly during the breeding season. During the day, it sleeps in a nest. An individual may use more than 20 nests in a single year, including hollow tree limbs, rotted stumps and even globular bird nests. Large tree cavities with small, secure entrances are preferred by lactating females, which build spherical nests of bark strips, feathers and fur. The female is industrious, dragging bark strips up to one-third of her weight and constructing a nest over 30 times her own size. Nests often contain an accumulation of pungent, black faeces, which may serve as a territorial marker or perhaps

*The Brush-tailed Phascogale is a strongly aboreal dasyurid, distinguished by a black tail with a large, dark brush on the distal two-thirds of the tail.*

J. LOCHMAN

discourage nest-site competitors such as Sugar Gliders. Feral honeybees may usurp nest cavities, providing serious competition where suitable nursery hollows are rare. As in the Red-tailed Phascogale, antechinuses and the Kaluta, all male Brush-tailed Phascogales die after mating but, unlike antechinuses, they show little physical decline during breeding, continuing to forage and groom while roaming several kilometres each night in search of mates. Weakened by stress-induced illnesses such as gastric ulcers, they often fall prey to owls, Foxes and Cats. Females are more vulnerable when lactating; juveniles are sometimes taken from the nest by goannas or kookaburras. Females can live three years in the wild, by which time their canines are blunt and their front incisors worn nearly to the gum.

The species was formerly distributed throughout the dry sclerophyll forests and woodlands of temperate and tropical Australia. It prefers open forest with sparse ground cover. Since European settlement, much of this habitat has been cleared for agriculture, considerably reducing its range. Residual habitat in temperate areas is often fragmented and limited to soils too poor for agriculture, thus isolating small populations and impeding genetic interchange. The status of the tropical subspecies is uncertain: records indicate it to be widespread but rare. Low population densities and annual male die-off make the Brush-tailed Phascogale very vulnerable to localised extinction. Surveys are urgently needed to define current distribution.

T. SODERQUIST

## Phascogale tapoatafa

### Size

*Phascogale tapoatafa tapoatafa* from Victoria*
*Head and Body Length*
160–261 (199) mm (males)
148–223 (181) mm (females)
*Tail Length*
175–234 (207) mm (males)
160–226 (194) mm (females)
*Weight*
175–311 (231) g (males)
106–212 (156) g (females)

*Weights of *P. t. tapoatafa* from northern New South Wales and *P. t. pirata* are approximately 80 per cent and 70 per cent respectively of those of *P. t. tapoatafa* from Victoria.

**Identification**  Uniform grizzled grey above, cream to white below. Large naked ears. Conspicuous black 'bottle-brush' tail with hairs up to 55 millimetres long.

**Recent Synonyms**  *Phascogale penicillata*.

**Other Common Names**  Tuan (Victoria), Common Wambenger (Western Australia), Black-tailed Phascogale.

**Status**  Widespread, populations localised, uncommon to rare.

**Subspecies**  *Phascogale tapoatafa tapoatafa*, southern Australia.
*Phascogale tapoatafa pirata*, northern Australia.

### References

Cuttle, P. (1982). Life history Strategy of the Dasyurid Marsupial *Phascogale tapoatafa*, pp. 13–22 (and) A Preliminary Report on Aspects of the Behaviour of the Dasyurid Marsupial *Phascogale tapoatafa*, pp. 325–332 (in) M. Archer (ed.) *Carnivorous Marsupials, Vol.1*. Roy. Zool. Soc. NSW, Sydney.

Soderquist, T.R. (1993). Maternal strategies of *Phascogale tapoatafa* (Marsupialia: Dasyuridae). I. Breeding Seasonality and Maternal Investment, (and) II. Juvenile Thermoregulation and Maternal Attendance. *Aust. J. Zool.* 41: 549–576.

## Subfamily Planigalinae

This group, defined largely on biochemical criteria, includes the planigales and ningauis. These are very small dasyurids with somewhat compressed skulls.

# Giles' Planigale

## *Planigale gilesi*

### Aitken, 1972

plan'-ee-gah'-lay jile'-zee: 'Giles' flat-weasel'

Named after Ernest Giles, who explored much of the Australian deserts, this Planigale is also 'an accomplished survivor' in the arid and semi-arid areas eastward from the Lake Eyre region to vicinity of Moree, New South Wales. From Bedourie, Queensland, its distribution extends south to Victoria along the Murray River.

When the species was described in 1972 it was known from only a dozen specimens but subsequent trapping surveys have found it in

C. ANDREW HENLEY/LARUS

*Giles' Planigale has short legs which are extended outward when moving over the ground.*

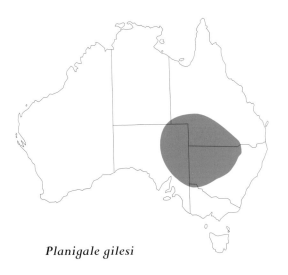

*Planigale gilesi*

many additional localities and have extended its known distribution. It is found in habitats with deep cracking clay soils: floodplains of creeks and rivers, away from creeks on grassy plains and in the clay interdune areas among sandhills. The presence of soil cracks is a significant habitat requirement, providing essential shelter from heat and cold. This species is frequently found sympatrically with the Narrow-nosed Planigale, although the former is perhaps less common where soil cracks are shallow.

Like other planigales, it has a flattened triangular head, beady eyes and a pointed nose. When walking, the hindfeet are moved in an outwardly extended arc, not in line under the body. Running is a scurrying motion, often with short leaps and bounds.

The short, relatively broad feet are well adapted for climbing within soil cracks. Above ground, it climbs easily among grass and shrubs or slithers through the litter, hunting for beetles, locusts, spiders or other arthropods; even the occasional small lizard or mammal.

Unlike other planigales, which have three premolars on the upper and lower jaws, Giles' Planigale has only two, but this does not make it a less efficient killer. Small prey are killed with quick bites, but it is bulldog-like when attacking larger prey; biting, hanging on and chewing. In good seasons, with abundant prey, the tail becomes swollen and carrot-shaped with stored fat; when conditions are poor, it is thin and bony.

Individuals (except lactating females with dependent young in nests) have shifting home ranges: some travel more than a kilometre in a few days. Generally, Giles' Planigale spends more time above ground than the Narrow-nosed Planigale, especially in the crepuscular period. During summer and winter its daily activity has a broad bimodal pattern, most activity occurring in the three hours after sunset and the hour before sunrise. There is more diurnal activity in winter, particularly during the period before sundown and after sunrise. On average, individuals are active for about five hours each day. During winter they may bask in the sun for warmth or become torpid for short periods to conserve energy. When animals are active at air temperatures below the normal body temperature of about 34°C, the metabolic rate increases

## *Planigale gilesi*

**Size**
*Head and Body Length*
60–80 (76) mm (males)
60–70 (68) mm (females)
*Tail Length*
55–70 (64) mm (males)
55–65 (60) mm (females)
*Weight*
9.5–16.0 (11.5) g (males)
5.0–9.0 (6.9) g (females)

**Identification**
Brindled cinnamon-grey above, olive buff below. Flattened body and pointed snout. Ears small, rounded and lying against the head. Two premolar teeth. Claws black. Legs short in comparison with body length; moves low to ground.
**Recent Synonyms** None.
**Other Common Names** Paucident Planigale.
**Status** Sparse.
**Subspecies** None.
**References**

Read, D.G. (1984). Movements and home ranges of three sympatric dasyurids, *Sminthopsis crassicaudata*, *Planigale gilesi* and *P. tenuirostris* (Marsupialia), in semi-arid western New South Wales. *Aust. Wildl. Res.* 11: 223–234.

Read, D.G. (1987). Habitat use by *Sminthopsis crassicaudata*, *Planigale gilesi* and *P. tenuirostris* (Marsupialia: Dasyuridae) in semi-arid New South Wales. *Aust. Wildl. Res.* 14: 385–395.

Read, D.G. (1989). Microhabitat separation and diel activity patterns of *Planigale gilesi* and *P. tenuirostris* (Marsupialia: Dasyuridae). *Aust. Mammal.* 12: 45–53.

D. WHITFORD

*Suckling Giles' Planigales, 54 days old. The flattened head is well illustrated.*

to offset the cold. At air temperatures of 14°C to 20°C this increase may be tenfold.

Vocalisations are loud and include a 'chh, chh' or sharp 'ca, ca', associated with aggressive or defensive behaviour—as when chasing another individual or defending a refuge. A high-pitched twittering, like that of a small bird, is made when frightened. In the breeding season, females in oestrus and males call to each other with high-pitched clicking sounds.

Breeding is from mid-July to late January, with one or two litters per female in a season.

A single mating can last over two hours. Gestation is 16 days. The oestrous cycle is 21 days and oestrus lasts for three days. The pouch, which opens to the rear, encloses 12 teats. At 65 days the young are independent and weigh about 4 grams. The usual litter is six to eight.

Population densities are generally low and fluctuate markedly from season to season and from year to year. It is possible that, in the wild, fewer than 20 per cent of individuals survive for more than two years. In captivity, some have attained the age of five years.

D.G. READ

109

*The Long-tailed Planigale has a particularly flattened head, enabling it to pursue prey and to shelter in the cracks of drying soils.*

# Long-tailed Planigale

## *Planigale ingrami*

(Thomas, 1906)

in'-gram-ee: 'Ingram's flat-weasel'

The flattened head, characteristic of all plani-gales, is particularly marked in the Long-tailed Planigale in which the height of the skull (about 3 millimetres) is only about one-fifth of the length. The spatulate head appears to have evolved as a means of exploiting narrow crevices in cracked soils, a feature of the seasonally flooded grass-lands and savanna woodlands which comprise the habitat of this species. During the dry season it occupies grass tussocks, spaces under rocks and crevices in the moist, contracting soils of drying swamps and watercourses. With the onset of summer floods, it seeks refuge on higher ground.

The Long-tailed Planigale is the smallest of the marsupials and, probably, of the world's mammals. Notorious for its rapacious appetite, it hunts throughout the night, wedging its way

*Planigale ingrami*

**Size**
*Head and Body Length*
55–63 (59) mm (males)
57–65 (59) mm (females)
*Tail Length*
57–60 (59) mm (males)
44–57 (52) mm (females)
*Weight*
3.9–4.5 (4.2) g (males)
4.2–4.5 (4.3) g (females)

**Identification** Smallest of the planigales. Permanently thin tail, usually longer than head and body. Very flat head; three upper and lower premolars.
**Recent Synonyms** *Planigale subtilissima*.
**Other Common Names** Ingram's Planigale, Northern Planigale.
**Status** Rare, scattered. Possibly underestimated.
**Subspecies** The latest revision of the genus does not recognise subspecies but designates three forms:
*Planigale ingrami* forma *ingrami*, Northern Territory to Townsville, Queensland.
*Planigale ingrami* forma *subtilissima*, Kimberley, Western Australia.
*Planigale ingrami* forma *brunnea*, Richmond area, Queensland.

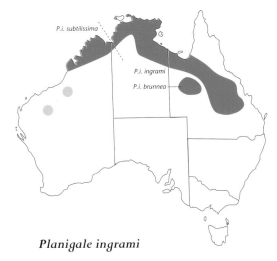

*Planigale ingrami*

through narrow spaces in search of insects and their larvae and even lizards and young mammals almost as large as itself. Larger insects such as grasshoppers, are pounced upon and eventually killed or subdued by persistent biting. The head, thorax and wings are usually discarded and only the soft parts eaten.

Males are slightly larger than females. The female has a well developed pouch that opens to the rear and encloses eight to ten teats. Births occur throughout the year but mostly from December to March. Litters vary from four to eight. Young become detached from the teats when about six weeks old and are left in a grassy nest under fallen bark, leaf litter or a tussock, while the mother forages. They become independent at three months of age.

Because of its minute size and secretive nocturnal habits, the Long-tailed Planigale is extremely difficult to study under natural conditions and no accurate assessment has been made of its numbers.

S.M. VAN DYCK

### References

Archer, M. (1976). Revision of the marsupial genus *Planigale* Troughton (Dasyuridae). *Mem. Qld. Mus.* 17: 341–365.

Troughton, E. le G. (1928). A new genus, species and subspecies of marsupial mice (Family Dasyuridae). *Rec. Aust. Mus.* 16: 281–288.

G.E. SCHMIDA

*Common Planigale eating a King Cricket.*

# Common Planigale

## *Planigale maculata*

### (Gould, 1851)

mak'-yue-lah-tah: 'spotted flat-weasel'

The contrast between the audacious ferocity and tiny size of this small marsupial arouses respect in anyone who observes it. A predator of a wide range of insects—some not much smaller than itself—it occupies a range of habitats from rainforest, sclerophyll forest and grasslands to marshlands and rocky areas. It even manages to exist in low density in the outer suburbs of cities such as Brisbane, where it falls prey to domestic cats. In captivity it eats insects, eggs, meat and honey.

Under laboratory conditions, females build small saucer-shaped nests of grass and bark and it is reasonable to assume that they do the same in

K. ATKINSON

*The Common Planigale is widely distributed and occurs in some suburbs.*

exposed to view. After they leave the pouch, this aperture begins to regress and eventually becomes no more than a shallow, almost hairless region with the hairs stained a yellowish-brown. In females that have not borne a litter, this area is hairy and unpigmented.

T.D. REDHEAD

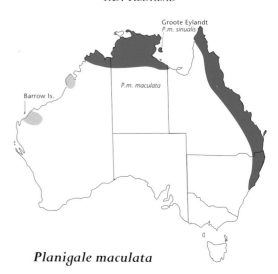

**Planigale maculata**

the wild. In coastal Arnhem Land, breeding probably occurs throughout the year but it may be restricted to late spring and summer in Queensland. Following a gestation period of 19–20 days, a litter of 4–12 (average eight) is born. At the time that these enter the pouch, its aperture is small but, as the young increase in size, it becomes more open and the young are

---

## Planigale maculata

### Size

#### Planigale maculata sinualis

*Head and Body Length*
70–100 (81) mm (males)
71–87 (79) mm (females)
*Tail Length*
61–95 (79) mm (males)
64–82 (76) mm (females)
*Weight*
6–22 (12) g (males)
7–15 (10) g (females)

**Identification**  Grey to cinnamon above, grizzled, occasionally with tiny white spots; paler below, white under chin. Head flattened. Distinguished from *Planigale ingrami* and *P. tenuirostris* by greater size; from *P. gilesi* in

possessing three premolars above and below (two in *P. gilesi*).
**Recent Synonyms**  *Antechinus maculatus.*
**Other Common Names**
Pygmy Marsupial Mouse, Coastal Planigale.
**Status**  Common.
**Subspecies**  *Planigale maculata maculata,* mainland.
*Planigale maculata sinualis,* Groote Eylandt.
**References**
Aslin, H. (1975). Reproduction in *Antechinus maculatus* Gould. *Aust. Wildl. Res.* 2: 77–80.
Taylor, J.M., J.H. Calaby and T.D. Redhead (1982). Breeding in wild populations of the marsupial-mouse *Planigale maculata sinualis* (Dasyuridae) (in) M. Archer (ed.) *Carnivorous Marsupials.* Roy. Zool. Soc. of NSW and Surrey Beatty and Sons, Sydney. pp. 83–87.

WHITFORD

# Narrow-nosed Planigale

## *Planigale tenuirostris*

### Troughton, 1928

ten'-yue-ee-rost'-ris: 'slender-snouted flat-weasel'

This diminutive marsupial is found throughout the eastern interior of Australia in a variety of habitats that have cracking clay soils, including open grassy areas, mallee scrubs and densely vegetated flats beside creeks. In some soils, the cracks may be only 30 centimetres deep, in oth-

*The Narrow-nosed Planigale basks in the morning sun to conserve its body heat.*

ers, more than 2 metres. These provide essential shelter from heat in summer and cold in winter. Winter night-time temperatures above ground can be below zero but in the cracks below ground, it can be relatively mild at 10–13°C.

An agile climber, the Narrow-nosed Planigale hunts beetles, grasshoppers, crickets, spiders and moths among soil cracks, grass and shrubs. The narrow, flattened head is pushed like a wedge into the litter, prizing apart grass stems and turning over leaves in search of hidden prey. Aided by soft, silky fur and a loose skin, it can wriggle its lissom little body into very narrow nooks and crannies. Fearless, pugnacious and with quick, deft use of the forepaws, it will attack a grasshopper or centipede larger than

*113*

## *Planigale tenuirostris*

### Size

*Head and Body Length*
50–75 (65) mm (males)
50–70 (62) mm (females)
*Tail Length*
52–65 (56) mm (males)
50–60 (55) mm (females)
*Weight*
4.5–9.0 (6.8) g (males)
4.0–7.0 (5.3) g (females)

**Identification**  Russet brown above; off white below. Flattened body; flat triangular head. Silky fur. Thin tail almost as long as head and body. Claws caramel. Three premolar teeth.

**Recent Synonyms**  None.

**Other Common Names**  None.

**Status**  Sparse.

**Subspecies**  None.

**References**

Read, D.G. (1984). Reproduction and breeding season of *Planigale gilesi* and *P. tenuirostris* (Marsupialia: Dasyuridae). *Aust. Mammal.* 7: 161–173.

Read, D.G. (1985). Development and growth of *Planigale tenuirostris* (Marsupialia: Dasyuridae) in the laboratory. *Aust. Mammal.* 8: 69–78.

Read, D.G. (1987). Diets of sympatric *Planigale gilesi* and *P. tenuirostris* (Marsupialia: Dasyuridae): relationships of season and body size. *Aust. Mammal.* 10: 11–21.

winter it is most active in the three hours after sunset and the four hours preceding dawn. Activity occurs mostly in short bursts of less than five minutes, followed by a rest for a minute or so. More than 70 per cent of the nocturnal activity is below ground, where it scurries over the vertical sides of the cracks; its short legs moving alongside the body. The hind legs have remarkable flexibility and rotation at the hip and ankle joints. Inquisitive by nature, the Narrow-nosed Planigale continually investigates its surroundings, sniffing for anything new or different. It may even stand on its hindlegs to get a better sniff.

On average, an individual rests or sleeps for more than 21 hours each day, alone or in a group. Especially in winter, several huddle together in a nest made from dried grass and littered with discarded bits of prey such as wings and other hard insect fragments. Huddling reduces the loss of body heat and helps to conserve energy. On a daily basis, they also reduce energy use by going into torpor, during which metabolism is reduced by 30–50 per cent. Energy loss during winter is reduced by basking in the sun, perched near the top of a crack, always alert and ready to scuttle to safety below if danger threatens.

Individuals are not restricted to a territory or definite home range. The home range shifts

itself, subduing and killing it with rapid bites to the head and body. While eating, it sometimes sits on its hind legs and uses both forepaws to hold larger pieces of food. The diet of insects and other arthropods provides sufficient water and the Narrow-nosed Planigale does not need to drink.

It is primarily nocturnal. In summer, it can be active at any time during the night, but in

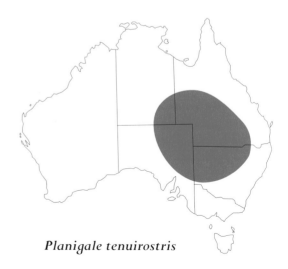

*Planigale tenuirostris*

D. WHITFORD

*The Narrow-nosed Planigale has long facial vibrissae that must be groomed.*

continuously throughout the year, perhaps as a response to seasonal and yearly fluctuations in prey abundance. Population densities are low and vary considerably from year to year and during the course of a year. Even in the best habitats, densities are probably less than one individual per hectare.

Breeding begins in late July or early August and continues until mid-January. Females are polyoestrous, having an oestrous cycle of 33 days and an oestrus of one day. During the breeding season, males make short 'tsst, tsst' calls, presumably to contact potential mates. Females make similar calls, but only during oestrus. Litters are born 19 days after mating. Although females have 12 teats in the pouch, the average litter size is six. The pink, hairless young are 3 millimetres long at birth and remain continuously attached to the teats until about 40 days old, when they can detach themselves. At this time, the mother builds a nest in which the young are left while she forages for food. The eyes open at 51 days and young are suckled until development is completed at about 95 days and they become independent. It is possible for a female to raise two litters in a season but it seems unlikely that many do so in the wild. Individuals continue to grow throughout life and, in any age group, males are larger than females. In captivity, some individuals have reached three years of age but it is unlikely that more than 15 per cent of the population survives, in the wild, to an age of two years or more.

D.G. READ

# Wongai Ningaui

## *Ningaui ridei*

### Archer, 1975

nin-gow'-ee ride'-ee: 'Ride's ningaui (mythical Aboriginal creature)'

### *Ningaui ridei*

**Size**
(WA and NT specimens)
*Head and Body Length*
57–75 mm
*Tail Length*
59–71 mm
*Weight*
6.5–10.5 g

**Identification** Distinguished from *Sminthopsis* by smaller size and relatively broader hindfoot. Distinguished from *Planigale* by relatively narrower hindfoot. Most readily distinguished from *Ningaui timealeyi* and *N. yvonneae* by the shape and size of the auditory bulla and almost entirely allopatric ranges.

**Recent Synonyms** None.

**Other Common Names** Inland Ningaui.

**Status** Common in suitable habitat.

**Subspecies** None.

**References**

Johnson, K.A. and A.D. Roff (1980). Discovery of Ningauis, (*Ningaui* sp: Dasyuridae: Marsupialia) in the Northern Territory, Australia. *Aust. Mammal.* 3: 127–129.

McKenzie, N.L. and N.J. Hall (1992). The biological survey of the Eastern Goldfields of Western Australia: Kurnalpi-Kalgoorlie study area. *Rec. West. Aust. Mus.* Suppl. No. 41.

Kitchener, D.J., J. Stoddart and J. Henry, (1983). A taxonomic appraisal of the genus *Ningaui* Archer (Marsupialia: Dasyuridae), including description of a new species. *Aust. J. Zool.* 31: 361–379.

This minute, nocturnal marsupial kills its prey by biting around the head in a swift and ferocious attack. Although known to eat beetles, grasshoppers, spiders, moths and cockroaches, it prefers small invertebrates less than 10 millimetres long.

The Wongai Ningaui was described in 1975 from two subadults collected among spinifex on red sandplains near Laverton, Western Australia. Since then, it has been found on sandy surfaces throughout the arid districts of Western Australia, Northern Territory, South Australia and Queensland. Its range extends eastwards to the 500 millimetre isohyet in Queensland; elsewhere it is confined to areas receiving less than 350 millimetre average annual rainfall. It is not known from districts that are subject to tropical monsoons, such as the Pilbara and the northern parts of the Great Sandy Desert.

Inland desert specimens have been collected only on sandy surfaces (dunes, sandplains and buckshot plains) supporting spinifex hummock grasslands of *Triodia* or *Plectrachne*, shrubs such as *Thryptomene maisonneuvei*, *Acacia ligulata* or *Grevillea eriostachya*, mallee such as *Eucalyptus youngiana* and/or such trees as Desert Oak, Corkwood, Cypress Pine, Gidgee, Mulga or Bara Gum.

Although it favours sandy surfaces in the eastern goldfields of Western Australia and the Channel Country of Queensland, subadults have also been recorded on gibbers and heavy alluvial

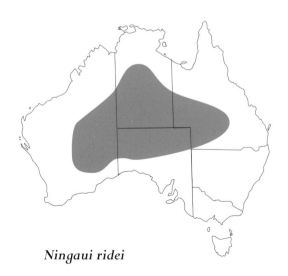

*Ningaui ridei*

*The Wongai Ningaui is a tiny dasyurid which preys upon arthropods. Copulation is vigorous.*

*The suckling young of the Wongai Ningaui are large in comparison with their mother's size.*

earths during late summer. These soils support communities dominated by trees, mallee, or shrubs over tussock or spinifex hummock grasses.

The Wongai Ningaui is nocturnal and shelters during the day in a low bush, hollow log or a small burrow. It seldom digs its own burrows, preferring to use those abandoned by small lizards or large spiders. Radio-tracking in the Simpson Desert suggests that individuals do not have fixed home ranges and are capable of moving at least 1.5 kilometres from point of capture within three days. Captive specimens have been active at night, climbing among spinifex leaves and using the tail in a semi-prehensile manner.

Males and females are similar in size and appearance. Females have six to seven teats and litters of five to seven young are known. Births occur from October to January in both the eastern goldfields of Western Australia and Simpson Desert of Queensland. A female from South Australia was captured with pouch young 6 mm long. Three weeks later these had ceased to be attached to the teats and either remained in the nest or clung to the mother's back as she foraged. Seven weeks after the female had been captured, the young fed independently.

The Wongai Ningaui is common in suitable habitat throughout its range, most of which is unsuitable for agricultural development. Secure populations occur in the Wanjarri and Gibson Desert Nature Reserves of Western Australia, Uluru National Park in the Northern Territory and Simpson Desert National Park in Queensland.

N.L. McKenzie and C.R. Dickman

# Pilbara Ningaui

## *Ningaui timealeyi*

### Archer, 1975

tim-eel'-ee-ee: 'Tim Ealey's ningaui'

BABS & BERT WELLS

*The Pilbara Ningaui inhabits slightly damper areas in the semi-arid grasslands of the Hammersley Plateau.*

The Pilbara Ningaui is a minute, bristly furred, predatory marsupial which inhabits the semi-arid grasslands of the Hamersley Plateau, Western Australia. It is most common on outwash plains near ridges and outcrops along drainage lines where moisture encourages the growth of large, dense hummocks of spinifex, scattered emergent shrubs, mallees and trees. By day, the hummocks provide it with shelter and refuge and at night it hunts on the ground in open spaces and over the branches and stems of shrubs. Animals maintained in a laboratory have alternating periods of activity and inactivity throughout the day and night but it is probable that, in nature, daylight activity is restricted to grooming and defecation within, or close to, the refuge. There is no evidence of periods of torpor.

The prey of the Pilbara Ningaui includes desert centipedes and cockroaches, animals which may be much larger than itself and must be subdued by a struggle. In captivity it eats cockroaches, centipedes, grasshoppers, crickets and small skinks, but ignores beetles and millipedes. Captive animals will drink water but, in the wild, most water is probably obtained in its food and, possibly, by licking dew.

Following good seasons, sexual maturity is attained in late winter and females can be found with young in the pouch from September to March, a period encompassing most of the annual rainfall in the region. In poorer years the breeding season is shorter, with pouch development restricted to November, December and January. Females have four to six teats arranged in a concentric ring within a pouch which is a simple unfurred depression on the belly. Usually five to six young are carried in the pouch to the stage of weaning. They become independent when about 2 grams in weight. During the breeding period males become aggressive

*Ningaui timealeyi*

## Ningaui timealeyi

**Size**

*Head and Body Length*
46–57 mm

*Tail Length*
59–79 mm

*Weight*
2.0–9.4 g

**Identification** Differs from planigales and dunnarts by being bristly furred. Distinguishable from *N. ridei* in having a relatively larger outer edge to the supratragus of the ear.

**Recent Synonyms** None.

**Other Common Names** Ealey's Ningaui.

**Status** Common.

**Subspecies** None.

**Reference**

Archer, M. (1975). *Ningaui*, a new genus of tiny dasyurids (Marsupialia) and two new species, *N. timealeyi* and *N. ridei* from arid Western Australia. *Mem. Qld. Mus.* 17: 237–49.

towards each other and females with pouch young drive away other adults.

By March in most years the population consists predominantly of the now independent young. Few, if any, individuals of either sex survive into a second breeding season. In dry years, when reproductive success is low, the Pilbara Ningaui may survive only in pockets of moister habitat, later recolonising from these refuges. At present, it is common over much of the broad, wild Hamersley Plateau and is widespread in the Hamersley Range National Park.

J.N. DUNLOP

*The Southern Ningaui preys upon arthropods. The individual illustrated is eating a mouse spider.*

# Southern Ningaui
## *Ningaui yvonneae*

## Kitchener, Stoddart and Henry, 1983

ee-von'-ee: 'Yvonne (Kitchener)'s ningaui'

This ningaui is widely distributed in sand plains of the semi-arid regions of southern Australia. In Western Australia, it is found only where spinifex (*Triodia* spp.) is present. Highest densities occur on deep red sand plains vegetated with dense spinifex and occasional mallee emergents. It also occurs on yellow sands with sparse spinifex interspersed with moderately dense proteaceous and myrtaceous heaths. In Victoria, it inhabits diverse heath communities with scattered *Eucalyptus incrassata* and *E. foecunda* in dune and interdune situations. In South Australia and New South Wales, it inhabits mallee scrub over

C. ANDREW HENLEY/LARUS

spinifex. It is sympatric with *Ningaui ridei* at Bungalbin Hill and the goldfields of Western Australia, where both have been collected from the same vegetation association.

Little is known of its natural history. During the day it shelters in low bushes, hollow logs, small burrows and, probably, under spinifex.

*The Southern Ningaui is mainly a creature of spinifex grasslands but inhabits heathlands in the cooler parts of its range.*

### Ningaui yvonneae

**Size**

*Head and Body Length*
48–74 mm
*Tail Length*
57–71 mm
*Weight*
4–10 g

**Identification** Distinguished from *Ningaui timealeyi* by longer hindfoot with pads on sole of foot less clearly demarcated, tail shorter, ear longer, alisphenoid tympanic wing more developed and foramen rotundum further from anterior edge of alisphenoid bulla. Differs from *Ningaui ridei* in hallux which does not reach interdigital pads of foot, alisphenoid tympanic wings less well developed and foramen rotundum much further from anterior edge of alisphenoid bulla.

**Recent Synonyms** None.

**Other Common Names** None.

**Status** Usually rare, but may be locally common.

**Subspecies** None.

**References**

Kitchener, D.J., Stoddart J. and Henry, J. (1983). A taxonomic appraisal of the genus *Ningaui* Archer (Marsupialia: Dasyuridae), including description of a new species. *Aust. J. Zool.* 31: 361–79.

Kitchener, D.J., Cooper, N. and Bradley, A. (1986). Reproduction in the male *Ningaui* (Marsupialia: Dasyuridae). *Aust. Wildl. Res.* 13: 13–25.

The breeding season begins in late August or early September. Apparently independent juveniles have been collected in January and March. Males appear to mature sexually at about eight months of age; they are not able to reproduce in the season of their birth. There is no suggestion that the spermatogenic cycle of males is disrupted immediately following the onset of breeding. The testes continue to be active until some time after January, when they regress. It is not clear whether or not adult males survive to mate in second or subsequent breeding seasons.

D.J. KITCHENER

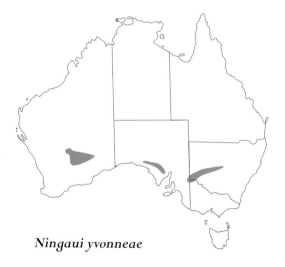

*Ningaui yvonneae*

## Subfamily Sminthopsinae

This group, largely defined on biochemical criteria, includes the dunnarts and the Kultarr.

# Kultarr

## *Antechinomys laniger*

(Gould, 1856)

an'-tek-ine'-oh-mis lah'-ni-jer: 'woolly antechinus-mouse'

BABS & BERT WELLS

*The Kultarr resembles a dunnart in most of its features, except its long, brush-tipped tail and very long hindlegs.*

Once considered to be the marsupial equivalent of hopping-mice, the Kultarr was long known as the Jerboa-marsupial. The great length of the hindlegs supported this view but studies of its locomotion have shown that it is consistently quadrupedal, bounding rapidly from hindlegs to forelegs. This gait gives it great manoeuvrability, permitting it to change direction rapidly by pivoting on its forefeet to evade a predator or perhaps to avoid being bitten by potentially dangerous prey such as spiders or centipedes.

Gould provided the first description of a Kultarr in 1856 but incorrectly depicted it on the branch of a tree. A terrestrial animal, predominantly adapted to life in open country, it inhabits desert plains, stony and sandy land, where grasses and small bushes constitute the principal vegetation; and *Acacia* scrubland. The eastern Australian subspecies shows a preference for sparsely vegetated claypans among *Acacia* woodland, while the central and Western Australian form prefers stony, granite plains dominated by *Acacia eremophila* and *Cassia* scrubland. It has been found sheltering in logs or stumps, beneath saltbush and spinifex tussocks and in deep cracks in the soil at the base of *Acacia*

and *Eremophila* trees. It occurs in the burrows of other animals such as trapdoor spiders and hopping-mice, goannas and agamids, but it is not known whether it digs its own burrow in the wild. Captive animals have been observed to dig shallow burrows and to cover the entrance with grass. It is nocturnal and probably spends much of the night foraging for insects such as spiders, crickets and cockroaches.

The Kultarr has a long breeding season and both sexes are potentially capable of breeding in more than one season. The timing of breeding is known to be different in geographically distant populations. Captive unmated females from south-western Queensland pass through successive oestrous cycles from July to February. Oestrous cycles occurring from August to January are typical of unmated Western Australian females. Field evidence suggests that females become pregnant during the first

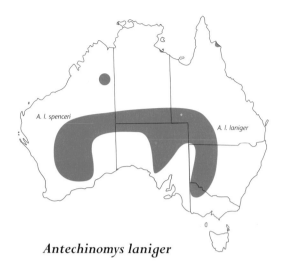

*A. l. spenceri*

*A. l. laniger*

### Antechinomys laniger

oestrous cycle of the season and that most females have a full complement of pouch young. Females have either six or eight teats. Young have been reported in the pouch from August to November.

The pouch, a crescent-like fold of skin covering the anterior part of the mammary area, develops during the breeding season and subsequently regresses. It provides protection for the young during the initial stages of suckling. After about 30 days, when they are about 25 millimetres long, the young can relinquish the teat and may be left in the nest. Later they ride on the mother's back. Exchanges of calls between the mother and young are used for mutual location and are important in stimulating the retrieval of young that have strayed from the nest or become dislodged from her back. Weaning takes place at about three months.

The Kultarr is uncommon over most of its range and populations appear to fluctuate seasonally. It is not directly affected by human activity but its security may be reduced by changed or intensified land use. Overall, the species appears to be neither endangered nor vulnerable but some populations, such as those at Cedar Bay, Queensland and in southern New South Wales (where no specimens have been recorded since 1900) may have disappeared.

A. VALENTE

### Antechinomys laniger

**Size**
(varies with locality)
*Head and Body Length*
80–100 (85) mm (males)
70–95 (85) mm (females)
*Tail Length*
100–150 (130) mm (males)
100–140 (120) mm (females)
*Weight*
circa 30 g (males)
circa 20 g (females)

**Identification**  Grizzled fawn-grey to sandy brown above, white on chest and belly; mid-line of face, crown of head and eye-ring darker. Very large ears; large protruding eyes. Long thin tail with prominent pencil of dark brown to black hairs. Hindfoot greatly elongated with only four toes.

**Recent Synonyms**  *Antechinomys spenceri*, *Sminthopsis laniger*.

**Other Common Names**  Jerboa-marsupial, Jerboa Pouched-mouse, Jerboa Marsupial Mouse, Wuhl-wuhl, Pitchi-pitchi (last two names referring to *Antechinomys laniger spenceri*).

**Status**  Rare, scattered. Presumed extinct at Cedar Bay, Queensland and southern New South Wales.

**Subspecies**  *Antechinomys laniger laniger*, eastern Australia.
*Antechinomys laniger spenceri*, Central and Western Australia.

**References**

Happold, M. (1972). Maternal and juvenile behaviour in the marsupial jerboa *Antechinomys spenceri* (Dasyuridae). *Aust. Mammal.* 1: 27–37.

Woolley, P.A. (1984). Reproduction in *Antechinomys laniger* ('spenceri' form) (Marsupialia: Dasyuridae): field and laboratory observation. *Aust. Wildl. Res.* 11: 481–489.

P. DEMPSEY

# Kangaroo Island Dunnart

## *Sminthopsis aitkeni*

### Kitchener, Stoddart and Henry, 1984

Sminth-op'-sis ate'-ken-ee: 'Aitken's mouse-like (animal)'

In April and again in June, 1969, a dog caught a male dunnart that was escaping from the bases of recently felled yaccas in mallee heath on the north-central portion of Kangaroo Island. These were placed in the South Australian Museum where the Curator of Mammals, Peter Aitken, described them as *Sminthopsis murina* and noted that they represented a new record for Kangaroo Island.

*Sminthopsis murina*, as understood at that time, had an extremely wide geographic distribution from Western Australia, across the southern part of the continent and north to north-eastern Queensland. In 1984, electrophoretic studies of the *S. murina* complex revealed that a number of distinct forms were involved. These were formal-

*Described in 1984, the Kangaroo Island Dunnart is restricted to that island, where it is rare.*

ly described and all the Kangaroo Island specimens examined were assigned to a new species, *Sminthopsis aitkeni*.

Since its original discovery, *S. aitkeni* has proved to be rather elusive on Kangaroo Island. There have been a number of attempts to learn more of its distribution and habitat preference, but only seven specimens are known from four locations. It has been found in a variety of mallee heath vegetation types on the lateritic soils of the Kangaroo Island plateau, a habitat shared with the Bush Rat and the Western Pygmy-possum. Some indication of the rarity of (or perhaps the difficulty in

Kangaroo Is.

*Sminthopsis aitkeni*

123

## Sminthopsis aitkeni

**Size**

*Head and Body Length*

84–93 mm (males)

77–82 mm (females)

*Tail Length*

90–106 mm (males)

91–94 mm (females)

*Weight*

20–25 g (males)

**Identification** Dark sooty above, light grey below. A slender, pointed muzzle and a tail always longer than the body and never incrassated. Distinguished from *Sminthopsis murina* and *S. dolichura* by the blackish rather than brown dorsal colour, a grey rather than white belly and a longer tail.

**Recent Synonyms** *Sminthopsis murina*.

**Other Common Names** Sooty Dunnart.

**Status** Rare, limited.

**Subspecies** None.

**References**

Aitken, P.F. (1972). *Sminthopsis murina* (Waterhouse) 1838. A new record from Kangaroo Island, South Australia. *S. Aust. Nat.* 46: 36–37.

Baverstock, P.R., M. Adams and M. Archer (1984). Electrophoretic resolution of species boundaries in the *Sminthopsis murina* complex (Dasyuridae). *Aust. J. Zool.* 32: 823–832.

Kitchener, D.J., J. Stoddart and J. Henry (1984). A taxonomic revision of the *Sminthopsis murina* complex (Marsupialia: Dasyuridae) in Australia, including descriptions of four new species. *Rec. West. Aust. Mus.* 11: 201–248.

capturing) this species is indicated by a detailed biological survey of Kangaroo Island in 1990, where a total of over 3000 pitfall and 8000 cage trap-nights resulted in the capture of only one *Sminthopsis aitkeni*. It seems that Kangaroo Island's only endemic mammal may be very rare and perhaps restricted to a particular type of habitat that we do not yet recognise.

A.C. ROBINSON

# Chestnut Dunnart

## *Sminthopsis archeri*

### Van Dyck, 1986

arch'-er-ee: 'Archer's mouse-like (animal)'

Most of what is known about the Chestnut Dunnart is history, not as the result of extinction, but because it comes from vast, northern, monsoonal woodlands rarely visited by zoologists. To date, it is known from only 20 museum specimens, 13 from southern Papua New Guinea and seven from Cape York Peninsula.

During a 1973 mammal survey of the Trans-Fly Plains, south-western Papua New Guinea, John Waitham and Harry Parnaby inspected a series of pot-holes dug during the course of road constructions. From these pits, over a period of three months, they collected 17 tan-brown dunnarts, distinguished by their 'roman noses' and prominent eye-rings. This sparked an investigation which resulted in the rediscovery of the species in northern Australia:

*Sminthopsis archeri*

D. WHITFORD

*Described in 1986, the little-known Chestnut Dunnart occurs in southern Papua New Guinea and Cape York.*

it had, in fact, been collected there as early as 1898 by A.S. Meek, who sent two specimens to the British Museum of Natural History in London. In 1933, the anthropologist and naturalist, Donald Thomson, collected four specimens (one from a pandanus palm) from the lower Archer River on Cape York Peninsula. In 1980, Rob Atherton and John Winter trapped a single specimen in the dry, grassy woodlands near Mapoon on the north-west coast of Cape York Peninsula, where the larger and more colourful Red-cheeked Dunnart was abundant. Acting on this find, Ann Kerle and Dick Whitford trapped at a nearby locality in 1981,

where they caught two more Chestnut Dunnarts. In 1993, a male was trapped near Iron Range by Luke Leung. This specimen was to be the first record of the species from the eastern side of Cape York Peninsula.

Both the New Guinea and Cape York Peninsula series of dunnarts shared dental and foot pad features which distinguish them from the Carpentarian Dunnart, *Sminthopsis butleri*,

## Kakadu Dunnart

### *Sminthopsis bindi*

### Van Dyck, Woinarski and Press, 1994

bin'-dee: 'small dasyurid (Jamoyn Aborig.) mouse-like (animal)'

The first live specimen of this small dunnart was recorded in 1980 during a CSIRO faunal survey of Kakadu National Park. Its taxonomic position remained uncertain until 12 further specimens were collected during the survey of the newly

---

### *Sminthopsis archeri*

**Size**

*Head and Body Length*
98–107 (102) mm (males)
83–85 (84) mm (females)
*Tail Length*
92–105 mm (males)
82–87 (84) mm (females)
*Weight*
No records, *circa* 16 g

**Identification** A thin-tailed, medium sized dunnart with light grey to tan-brown fur, prominent black eye-rings and a distinctive roman-nose.
**Recent Synonyms** None.
**Other Common Names** None.
**Status** Unknown, probably limited.
**Subspecies** None, but has been confused with the Carpentarian Dunnart.
**Reference**
Van Dyck, S. (1986). The Chestnut Dunnart, *Sminthopsis archeri* (Marsupialia: Dasyuridae) a new species from the savannas of Papuan New Guinea and Cape York Peninsula, Australia. *Aust. Mammal.* 9: 111–124.

---

### *Sminthopsis bindi*

**Size**

*Head and Body Length*
58–84 (75) mm (males)
52–80 (72) mm (females)
*Tail Length*
62–105 (87) mm (male)
61–97 (81) mm (females)
*Weight*
*circa* 12–14 g

**Identification** Pale grey above, with dark eye-ring, pale area above the eye and sometimes with a poorly defined, darker head stripe. Thin tail. Small size. Unfused interdigital pads on narrow hindfeet with large striate apical granules.
**Recent Synonyms** None.
**Other Common Names** None.
**Status** Uncertain but possibly reasonably abundant in a limited range.
**Subspecies** None.
**Reference**
Van Dyke, S., J.C.Z. Woinarski and A.J. Press (1994). The Kakadu Dunnart, *Sminthopsis bindi* (Marsupialia: Dasyuridae), a new species from the stony woodlands of the Northern Territory. *Mem. Qld. Mus.* 37: 311–323.

---

with which the Australian examples had previously been confused.

Virtually nothing is known about the ecology or habitat requirements of the Chestnut Dunnart. In Australia, it is known from tall stringybark woodlands on red earth soils of the laterite–bauxite plateau, where the canopy species include *Erythrophleum chorostachys* and *Eucalyptus nesophyla* with an understorey of *Parinari nonda*, *Planchonia careyi*, *Grevillea parallela* and *Acacia rothia*.

The number of teats varies from six to eight. New Guinean records indicate births occurring from July to October.

S. VAN DYCK

*The Kakadu Dunnart has an unusually slender tail.*

added Stage III of Kakadu from 1988 to 1990. These confirmed its specific identity and close relationship to *Sminthopsis archeri* from Cape York. It has been found subsequently at six other localities in the Top End. Although two of these records were roadkills and one was a mangled leftover under a Ghost Bat roost, the increase in its recording has largely been due to the recent widespread use of pitfall trapping for wildlife surveys in this region.

One characteristic common to all sites where this dunnart has been found is extensive surface cover of gravel, the significance of which is unknown. Most specimens have been from hilly areas with woodlands dominated by Salmon Gum, *Eucalyptus tintinnans*; Red-barked Bloodwood, *E. dichromophloia*; and Ironwood *Erythrophleum chlorostachys*.

Very little is known about the ecology of this species. There is a record of a Kakadu Dunnart sheltering in a small burrow. The breeding season has not been defined but juveniles have been reported in February, March and November and pouch young in October. The species does not appear to develop the swollen-based tail typical of many other dunnarts. The diet is unknown but captive individuals have eaten a wide range of invertebrates, including large moths, crickets and caterpillars.

**Sminthopsis bindi**

J. WOINARSKI AND S. VAN DYCK

# Butler's Dunnart

## *Sminthopsis butleri*

### Archer, 1971

but'-ler-ee: 'Butler's mouse-like (animal)'

*Sminthopsis butleri*

This dunnart was named after the naturalist, Harry Butler, who found it at Kalumburu, near the mouth of the King Edward River, in the Kimberley region of Western Australia. On 14 December 1965, he collected a female with seven pouch young that were subsequently reared in captivity. According to his field notes, Butler found it while spotlighting at the bottom of the airstrip where the blacksoil country

meets the sand plain. The area was well vegetated with eucalypts and grass. Another very small individual, weighing about 4 grams and perhaps not yet independent, was found in flood debris near the Mission on 20 January 1966. To the Aborigines at Kalumburu this dunnart and other small-sized mammals, are known as 'munjols'.

There has been little collecting activity specifically directed towards finding Butler's Dunnart. On a visit to Kalumburu in June 1991, I found that the airstrip had been extended at the bottom end and that the surrounding area had been very recently burnt. Trapping in the vicinity was unsuccessful and the only mammal seen when spotlighting around the perimeter of the airstrip was the Cat. Further searches should be made in other areas on the mainland where there is similar habitat, less affected by humans, Cats and Dogs, or by frequent burning.

Very recently it has been found on Bathurst Island and a previously misidentified specimen collected in 1913 on Melville Island was found in a Museum collection.

Dunnarts collected on Cape York and previously thought to be *S. butleri* are now recognised as a distinct species, *S. archeri*.

P.A. WOOLLEY

---

### *Sminthopsis butleri*

**Size**

*Head and Body Length*
88 mm
*Tail Length*
90 mm

**Identification** Fur soft, greyish above and white below. Vague head stripe. Tail thin and sparsely furred. Distinguished from the sympatric *S. virginiae* by differences in colouration.

**Recent Synonyms** None.

**Other Common Names** Carpentarian Dunnart, Munjol.

**Status** Very rare.

**Subspecies** None.

**Reference**

Archer, M. (1979). Two new species of *Sminthopsis* Thomas (Dasyuridae: Marsupialia) from northern Australia, *S. butleri* and *S. douglasi*. *Aust. Zool.* 20: 327–345.

# Fat-tailed Dunnart

## *Sminthopsis crassicaudata*

## (Gould, 1844)

kras'-ee-kaw-dah'-tah: 'Fat-tailed mouse-like (animal)'

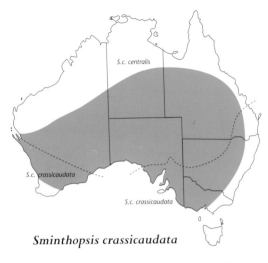

*Sminthopsis crassicaudata*

This species can be immediately recognised by its large ears and eyes and by a swollen tail which is much shorter than that of the Stripe-faced Dunnart. It occurs widely in southern Australia, in a variety of vegetation habitats including open woodland, low shrublands of saltbush and bluebush, tussock grasslands on clay or sandy soils, gibber plain and, in the southern part of its range, farmlands. The range extends from relatively moist regions near the southern coast through the arid inland and into the inhospitable plains of the Lake Eyre basin. The solitary males and females occupy large, drifting, home ranges.

The Fat-tailed Dunnart is completely nocturnal, emerging after dark to hunt for invertebrates on the surface of the ground.

*The Fat-tailed Dunnart stores fat in its tail, which can be carrot-shaped in well-nourished individuals. Some other dunnarts do the same.*

C. ANDREW HENLEY/LARUS

The Fat-tailed Dunnart is a typical member of its genus, displaying large ears and prominent eye-rings. Grooming is an elaborate ritual. D. WHITFORD

## Sminthopsis crassicaudata

**Size**
*Head and Body Length*
60–90 (75) mm
*Tail Length*
40–70 (55) mm
*Weight*
10–20 (15) g

**Identification** Tail slightly shorter than head and body in northern populations; markedly less than head and body in southern populations. Interdigital pads of hindfoot with uniformly finely granulated pads, tending towards development of mid-line rows of slightly enlarged granules in northern populations.

**Recent Synonyms** None.

**Other Common Names** Fat-tailed Marsupial Mouse.

**Status** Common.

**Subspecies** *Sminthopsis crassicaudata crassicaudata*, south-eastern mainland.
*Sminthopsis crassicaudata centralis*, most of temperate Australia.

**Reference**
Morton, S.R. (1978). An ecological study of *Sminthopsis crassicaudata* (Marsupialia: Dasyuridae), Parts I, II and III. *Aust. Wild. Res.* 5: 151–211.

Unusually for a small mammal, it forages on bare open areas, but populations are probably denser in areas of thick vegetation. There do not appear to be especially preferred types of insects or other invertebrates in the diet and vertebrates are rarely taken in the wild. Because enough water is obtained from its food, the Fat-tailed Dunnart does not need to drink. It avoids temperature extremes by sheltering in nests of grass or other dried plant material constructed beneath logs or rocks or, in arid regions, within deep cracks in the soil. During autumn and winter, groups of individuals frequently huddle together in communal nests throughout the day, conserving energy by sharing body heat. When food becomes very hard to find, individuals may enter torpor. Fat stores in the tail function as an extra energy reserve, which tends to be depleted during winter.

Breeding takes place from July to February. The female has a well-developed pouch with eight to ten teats and sufficient young are usually born to occupy all of these, although the average number to survive to weaning, at the age of about ten weeks, is only five. No long-term bonds are formed either between parents or

between mother and young. Animals reared in the laboratory may attain sexual maturity at about five months of age but there is no evidence that they breed at this age in the wild.

Populations in the moister parts of the range are relatively stable but, in arid regions, the population density fluctuates according to rainfall and the consequent supply of invertebrate prey. In contrast to most other dasyurids, the Fat-tailed Dunnart has probably increased in range following clearing of land by European settlers. Because of its preference for open grasslands and low shrublands, it has been able to inhabit farmlands derived from forests and scrubs in south-eastern and south-western Australia.

S.R. MORTON

# Little Long-tailed Dunnart

## *Sminthopsis dolichura*

### Kitchener, Stoddart and Henry, 1984

dol'-ik-yue'-rah: 'long-tailed
mouse-like (animal)'

This dunnart was described as recently as 1984, having previously been considered part of the Common Dunnart, *S. murina*, complex. The external similarity of *S. dolichura* to other species of *Sminthopsis* makes field identification difficult, particularly when dealing with subadults. Indeed, there is a need for further research on the taxonomy of the complex in order to clearly define the various species and their distributions.

The Little Long-tailed Dunnart occurs in a variety of different habitats in semi-arid and arid country in south-western Western Australia and South Australia and has been recorded in numerous conservation reserves and national parks. Preferred habitats include eucalypt woodlands, woodlands dominated by *Acacia* and *Casuarina* species, shrublands, heaths of myrtaceous and proteaceous species and hummock grasslands with an overstorey of low trees or mallees.

It is an active nocturnal hunter, subduing its prey of beetles, crickets, spiders and geckoes with an onslaught of rapid bites. Individuals captured in pitfall traps together with House Mice will readily kill these and devour the head and/or hindquarters: it is not known whether such predation occurs in the wild. During the day it shelters in a nest of dry grass and leaves constructed within a hollow log, a grass tussock or grass-tree (*Xanthorrhoea* spp.). In hummock grasslands, individuals may shelter in the abandoned burrows of hopping-mice.

## *Sminthopsis dolichura*

**Size**
*Head and Body Length*
63–99 (74.5) mm (males)
63–92 (73.7) mm (females)
*Tail Length*
88–109 (94.6) mm (males)
84–97 (89.6) mm (females)
*Weight*
11–20 (14.8) g (males)
10–21 (12.4) g (females)

**Identification** Dorsal fur pale to dark grey; head pale grey with a thin, black eye-ring; face, cheeks and patches behind the ears brownish; ventral surfaces white. Ears long and bare. Tail thin, about equal to or slightly longer than head and body, the dorsal surface light grey and the ventral white. Differs from *S. hirtipes* and *S. granulipes* in lacking granular terminal pads or hair on the interdigital pads on the hindtoes. The lack of prominent, dark head stripe or patch and a non-incrassated tail distinguishes it from *S. crassicaudata* and *S. macroura*. Adults differ from *S. ooldea* by their larger size and lack of an incrassated tail. Differs from *S. murina* by its longer tail and the dorsal fur being grey rather than brownish. Distinguished from *S. gilberti* by its longer tail and shorter ears and feet. Differs from *S. griseoventer* by its longer tail and white rather than grey ventral fur.

**Recent Synonyms** *Sminthopsis murina fuliginosa,* part of south-western Western Australia. *Sminthopsis murina murina*, South Australia.

**Other Common Names** None.

**Status** Common in suitable habitat.

**Subspecies** None.

**Reference**

Kitchener, D.J., J. Stoddart and J. Henry, (1984). A taxonomic revision of the *Sminthopsis murina* complex (Marsupialia, Dasyuridae) in Australia, including descriptions of four new species. *Rec. West. Aust. Mus.* 11: 201–248.

Throughout its range, the Little Long-tailed Dunnart is often common in areas in early stages of regeneration following fire. It becomes particularly abundant three to four years after fire and may temporarily displace other species of dunnarts in such areas. The breeding season extends from August to March, most females giving birth to one litter of up to eight young. There is limited evidence that some females may rear two litters in a season, but this is yet to be confirmed in the field. The oestrous cycle, gestation length and developmental rates are not known, since no laboratory studies of breeding have yet been carried out.

The young become independent of the mother when they weigh about 5 grams and disperse widely into a range of habitats. Females may commence breeding when eight to nine months old and live up to two years. Males are capable of breeding when four to five months old; the longest time between captures for a male in the field is only 14 months.

*The Little Long-tailed Dunnart, described in 1984, occurs in a variety of habitats from eucalypt woodland to heath.*

***Sminthopsis dolichura***

The Fox and feral Cat probably prey on the species, particularly during spring and autumn when large numbers of juvenile animals are dispersing. Despite this, the Little Long-tailed Dunnart appears to be secure and under no immediate threat, although populations may fluctuate greatly in response to different seasonal conditions.

G.R. FRIEND AND D.J. PEARSON

BABS & BERT WELLS

P.A. WOOLLEY & D. WALSH

# Julia Creek Dunnart

## *Sminthopsis douglasi*

### Archer, 1979

dug'-las-ee: 'Douglas' mouse-like (animal)'

Until recently, this dunnart was known only from four specimens, lodged in museums between 1911 and 1972. These were obtained from three properties lying between the towns of Julia Creek and Richmond, Queensland. In 1992, it was found in another eight localities in the downs country of Queensland. It appears

*Apparently the largest member of its genus, the Julia Creek Dunnart is rare and restricted to the Downs region of Queensland.*

to be the largest species of *Sminthopsis*.

It is nocturnal and probably rests in the maze of underground cavities provided by the grass-covered, cracking brown soils of the region, which are known also to provide shelter for the Long-haired Rat, *Rattus villosissimus*; Forrest's Mouse, *Leggadina forresti*; Stripe-faced Dunnart, *S. macroura*; Fat-tailed Dunnart, *S. crassicaudata*; and the Long-tailed Planigale, *Planigale ingrami*. Like the Stripe-faced Dunnart, the Julia Creek Dunnart has a prominent facial stripe and, when in good condition, a long tail that is fattened at the base. However, it is distinguished by dark hairs on the tip of the tail, in a ring around the eyes and on the outer mesial edge of the pinna. When fully grown, it is larger than the Stripe-faced Dunnart and even recently weaned young are readily distinguished by the length of their hindfeet (more than 20 millimetres).

Studies in progress on reproduction in captive animals suggest that breeding may occur throughout the year. The females have eight

*Sminthopsis douglasi*

### Sminthopsis douglasi

**Size**

*Head and Body Length*
130–135 mm (males)
*Tail Length*
120–130 mm (males)
*Weight*
50–70 g (males)
40–60 g (females)

**Identification** Brown, speckled with grey above; buffy white below. Distinct face-stripe, dark hairs in ring around eyes and on outer mesial edge of pinna. Rufous hairs on cheeks and at base of ears. Fattened, tapering tail slightly shorter than head and body; dark hairs towards tip. Largest species of *Sminthopsis*.
**Recent Synonyms** None.
**Other Common Names** None.
**Status** Endangered.
**Subspecies** None.
**References**
Woolley, P.A. (1992). New records of the Julia Creek Dunnart, *Sminthopsis douglasi* (Marsupialia: Dasyuridae). *Wildl. Res.* 19: 779–783.
Archer, M. (1979). Two new species of *Sminthopsis* Thomas (Dasyuridae: Marsupialia) from northern Australia, *S. butleri* and *S. douglasi*. *Aust. Zool.* 20: 327-345.

teats and are able to rear as many young in a single litter. Young reared in captivity reach sexual maturity when 17–27 weeks old (females) and 28–31 weeks (males), females always maturing earlier than their male littermates. Should this also occur in the field it may provide a mechanism that reduces inbreeding in the wild population.

P.A. WOOLLEY

BABS & BERT WELLS

# Gilbert's Dunnart
## *Sminthopsis gilberti*

### Kitchener, Stoddart and Henry, 1984

gil'-bert-ee: 'Gilbert's mouse-like (animal)'

Gilbert's Dunnart was described as a new species in 1984 following a revision of the *Sminthopsis murina* complex. It is endemic to Western Australia and was named after John Gilbert, who collected fauna extensively for John Gould in south-western Western Australia between 1843 and 1844.

Its range extends in a band from the Darling scarp near Perth, across the central and southern wheat belt. Outlying populations occur on the Roe Plain at the southern margin of the Nullarbor. There are no records from the South-west Interzone. *S. gilberti* overlaps with the southern distribution of *S. dolichura* in the jarrah forest

*Sminthopsis gilberti*

**Size**
*Head and Body Length*
81–92 mm
*Tail Length*
75–92 mm
*Weight*
14–25 g

**Identification**  Until recently, this species was included in the *Sminthopsis murina* species complex. Externally it is very similar to two other species from this complex that occur in Western Australia, *S. dolichura* and *S. griseoventer*, but is larger. Light grey above (similar to *Sminthopsis dolichura*), but hind foot longer than *S. dolichura* and *S. griseoventer* (17.5–19.0 mm against 16.8–17.5 mm). Ears longer than *S. dolichura* and *S. griseoventer* (21.1 mm against 17.7–19.3 mm). Whiter ventral fur than *S. griseoventer*. Tail length less than head-vent length, except for Roe Plain populations.

*Like other members of its genus, Gilbert's Dunnart does not have a pouch large enough to contain its litter.*

near Collie and is allopatric with *S. griseoventer* which occurs to the west and south.

In the Darling Range it favours myrtaceous heaths dominated by bottlebrush, grass-trees and zamias on shallow granitic soils and also mixed woodlands of Wandoo, Marri and Jarrah on sandy loams. In the central and southern wheat belt it occurs in casuarina or mallee-heaths on gravelly soils. Further east it is known from mallee shrublands on sand plain (Frank Hann National Park, Lake Cronin and Roe Plain) and in open woodlands of Salmon Gum or *Eucalyptus oleosa* on loams (Lake Cronin and Roe Plain respectively).

Throughout its range, breeding occurs over the spring to early summer period. Pouch young are present in October and November and juveniles are present in the population in February.

**Sminthopsis gilberti**

Females have eight teats and one captured in October on the Roe Plain had seven hairless young attached. Adult males are present after mating has occurred and do not appear to exhibit the male die-off phenomenon seen in some other small dasyurids. Like other temperate-zone *Sminthopsis* species, it is nocturnal and insectivorous and probably nests above the ground in logs or vegetation.

Secure populations occur in several nature conservation reserves including Tuttanning, Dragon Rocks and Nuytsland Nature Reserves and in the Western Australian State Forest.

K.D. MORRIS AND N.L. MCKENZIE

*S. gilberti* has distinctive patches of white fur behind the ears.
**Recent Synonyms** *Sminthopsis murina* (part).
**Other Common Names** None.
**Status** Common, limited distribution.
**Subspecies** None.
**References**

How, R.A., J. Dell and B.G. Muir (1988). Vertebrate fauna (in) The biological survey of the eastern goldfields of Western Australia: Part 4. Lake Johnston-Hyden study area. *Rec. West. Aust. Mus.* Suppl. 30: 44-83.

Kitchener, D.J., J. Stoddart and J. Henry, (1984). A taxonomic revision of the *Sminthopsis murina* complex (Marsupialia, Dasyuridae) in Australia, including descriptions of four new species. *Rec. West. Aust. Mus.* 11: 202–248.

McKenzie, N.L. and A.C. Robinson (1987). *A biological survey of the Nullarbor Region: South and Western Australia in 1984.* Department of Environment and Planning, Adelaide.

*Described in 1984, Gilbert's Dunnart inhabits semi-arid areas of southern Western Australia.*

BABS & BERT WELLS

BABS & BERT WELLS

*The White-tailed Dunnart inhabits low shrubland. Its hindfeet are more or less uniformly granulated.*

# White-tailed Dunnart

## *Sminthopsis granulipes*

Troughton, 1932

gran-yue'-lee-pez: 'granular-footed mouse-like (animal)'

Although regarded as rare for a century after its discovery at King George Sound, Western Australia, the White-tailed Dunnart has been trapped on a number of occasions over the past decade in the course of faunal surveys in Western Australia. Its present distribution is now fairly well defined. At several localities it occurs in the same habitat as the Common and Fat-tailed Dunnarts, which are similar to it in size.

The White-tailed Dunnart appears to prefer a low shrubland vegetation. Adequate habitat records are available for 34 captures involving 19 different sites. Two-thirds of the captures were in shrublands of mixed species, almost all of which were lower than 1 metre and had a sparse, medium-dense canopy. Most other captures were in sparse mallee overlying a low, to medium-dense to dense shrub layer. There was no apparent reference for particular plant species. Most captures were on sand (including red dunes in goldfields) or sandy loams, but several were on gravelly loam.

*Sminthopsis granulipes*

138

## *Sminthopsis granulipes*

**Size**

*Head and Body Length*
69–88 (83) mm
*Tail Length*
56–66 (60) mm
*Weight*
18–37 (25) g

**Identification** Light fawn above, mottled by dark brown tipping of hairs, basal four-fifths of hairs blue-grey; cream below, basal parts of hairs blue-grey; tail white with thin dorsal line of dark brown hairs; paws white. Hind margin of ears notched. Pads on palm and sole finely and evenly granulated, without enlarged granules, smooth areas or hairs.

**Recent Synonyms** None.

**Other Common Names** Granular-footed Marsupial Mouse, Ash-grey Dunnart.

**Status** Common, limited.

**Subspecies** None.

**Reference**
Kitchener, D.J. and A. Chapman (1978). Mammals of Dongolocking Nature Reserve. *Rec. West. Aust. Mus.* Suppl. No. 6: 53–58.

Stomach contents of eight specimens indicate that it feeds on arthropods. It appears to be an opportunistic feeder, eating moths, bugs, ants, wasps, sawfly larvae, scarab beetles, weevils, termites, cockroaches, horseflies, grasshopper eggs, purse-web, web- and wolf-spiders and centipedes.

No pregnant females have been collected, but the condition of the pouch, mammary glands and teats of six females collected between late September and late October suggests that they had recently weaned their young. Obvious juveniles, weighing less than 10 grams, have been captured only in November, lending strong support to the inference that births occur in winter.

D.J. KITCHENER

# Grey-bellied Dunnart

## *Sminthopsis griseoventer*

### Kitchener, Stodart and Henry, 1984

griz'-ay-oh-vent'-er: 'grey-bellied mouse-like (animal)'

The Grey-bellied Dunnart is one of a complex of related species occurring in south-western Western Australia. It is restricted to coastal plain and lateritic ranges in a narrow arc from Mount Peron to Point Dempster, with one distinctive population occurring on Boullanger Island in Jurien Bay. Within this distribution it occurs in a wide range of habitats from open woodland of *Eucalyptus* and *Banksia*, to dense heath and seasonal swampland. The densest populations occur in coastal heath on sandy soils, usually at least ten years after fire.

Invertebrates form the bulk of the diet, but young mice, lizards and soft fruits are taken occasionally. It searches for prey under leaf litter, locating it by a combination of all senses, especially smell. Prey larger than 3 centimetres long is preferred and is held in the forepaws

*Sminthopsis griseoventer*

before being killed; small prey items are taken directly in the jaws. It is strictly nocturnal, most activity occurring in the early and late parts of the night.

Breeding is restricted to winter and spring. In the population on Boullanger Island, both sexes mate promiscuously in July and females carry litters of up to eight young in August. Pouch life lasts four to five weeks. Young are thereafter deposited in a leaf-lined nest just under the soil surface and first emerge in late October at ten weeks of age. Young females remain near their area of birth, but males disperse up to several hundred metres within two months of weaning; survival of both sexes up to this time is less than 50 per cent. Females give birth to only one litter a year and, exceptionally, may produce young in two consecutive

seasons. Males and females become sexually mature at about one year and live to a maximum age of two-and-one-half years. Population density ranges from three individuals per hectare in spring to a peak of nine per hectare in summer, when young become independent. Life history events appear to be similar in mainland populations, but delayed by up to two months. Population densities do not exceed one per hectare.

Despite urban development, changes in the fire regime and clearing for agriculture over parts of its range, this species appears to be secure. It is recorded regularly from non-reserved land and is represented in National Parks throughout its geographical range.

C.R. DICKMAN

---

## Sminthopsis griseoventer

**Size**

*Head and Body Length*
68–94 (85) mm (males)
65–88 (83) mm (females)
*Tail Length*
66–98 (83) mm (males)
65–95 (81) mm (females)
*Weight*
15–24 (19) g (males)
14–20 (16) g (females)

**Identification** Dorsal fur dark at base, tipped with light grey anteriorly and becoming slightly darker on back and flanks; fur long and fluffy in appearance. Ventral fur olive-grey at base, tipped with light grey, giving grizzled appearance. In Western Australia, similar to *S. ooldea*, but

has distinctly grey (rather than brownish-grey) fur, especially on belly; shorter ears and hindfeet.

**Recent Synonyms** *Sminthopsis murina* (part).
**Other Common Names** None.
**Status** Common in suitable habitat.
**Subspecies** None.
**References**
Kitchener, D.J., J. Stoddart and J. Henry (1984). A taxonomic revision of the *Sminthopsis murina* complex (Marsupialia: Dasyuridae) in Australia, including descriptions of four new species. *Rec. West. Aust. Mus.* 11: 201–248.
Dickman, C.R. (1988). Body size, prey size and community structure in insectivorous mammals. *Ecology* 69: 569–80.

BABS & BERT WELLS

# Hairy-footed Dunnart

## *Sminthopsis hirtipes*

### Thomas, 1898

her'-tee-pez: 'hairy-footed mouse-like (animal)'

*In adaptation to its life in soft, sandy habitats, the Hairy-footed Dunnart has fringes of stiff hair extending from the soles of the feet, increasing their area.*

The fine bristles that cover the soles of the feet of this small marsupial are a very unusual feature, possibly assisting its locomotion on sandy surfaces. First described in 1898 from a specimen collected near Charlotte Waters,

Northern Territory, it has since been found in sandy, arid or semi-arid country westward to the coast and, recently, in western Queensland. Over most of its range the rainfall is irregular but in some semi-arid parts of Western Australia it is seasonal.

The Hairy-footed Dunnart has been recorded from a variety of plant communities associated with reddish sand plains and sand-dunes: open low woodlands of Marble Gum (*Eucalyptus gongylocarpa*), *Eucalyptus youngiana* or Desert Oak (*Allocasuarina decaisneana*) and/or shrublands of *Acacia*, *Thryptomene* or *Grevillea* over hummock

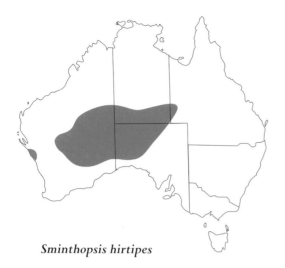

*Sminthopsis hirtipes*

grasslands of *Triodia* or *Plectrachne*. Towards the south-western limit of its range, it has been found in *Callitris* woodland, shrub mallee, heath and hummock grassland communities on plains or dunes of red to yellow sands. The single specimen collected from near the coast at Kalbarri, Western Australia, was in a shrubland of *Banksia* and *Grevillea* on grey-yellow sand plains.

The limited data on its biology suggest that populations increase after good rains. In the Great Victoria Desert and semi-arid Eastern Goldfields of Western Australia, the period of births is strongly seasonal, occurring in spring and summer. Females with pouch young have been recorded as early as October and lactating females as late as April. Individuals can live for at least three years in the wild. The diet includes insects and small lizards. Individuals have been found during the day in abandoned bull-ant nests (twice) and once in a deep burrow (possibly excavated by hopping-mice). The base of the tail fattens during good seasons.

The presumed rarity of the Hairy-footed Dunnart has reflected inappropriate collecting techniques in the past. Where present, it is readily captured in pitfall traps. Populations are known to exist in Kalbarri National Park and Wanjarri, Neale Junction and Queen Victoria Spring Nature Reserves in Western Australia; in

Uluru National Park in the Northern Territory; and in the 'Unnamed' Conservation Park in South Australia. The species does not appear to be threatened.

D.J. PEARSON AND N.L. MCKENZIE

## *Sminthopsis hirtipes*

### Size

*Head and Body Length*
72–83 (77) mm
*Tail Length*
72–101 (85) mm
*Weight*
13.0–19.5 (15.0) g

**Identification** Brownish-yellow to light yellowish-brown above, black-tipped hairs giving a peppered appearance; white below. Basal half of tail slightly swollen. Distinguished from other species of *Sminthopsis*, including *S. youngsoni*, by long (16–19 mm), broad feet, covered with fine silvery hairs, short on the pads, long elsewhere, forming a fringe around the sole.

**Recent Synonyms** None.

**Other Common Names**
Fringe-footed Sminthopsis, Hairy-footed Pouched Mouse.

**Status** Widespread and common in suitable habitat, habitat common.

**Subspecies** None.

**References**

Archer, M. (1981). Systematic revision of the marsupial dasyurid genus *Sminthopsis* Thomas. *Bull. Amer. Mus. Nat. Hist.* 168: 61–224.

Dickman, C.R., F.J. Downey and M. Predavec (1993). The Hairy-footed Dunnart *Sminthopsis hirtipes* (Marsupialia: Dasyuridae) in Queensland. *Aust. Mammal.* 16: 69–72.

# White-footed Dunnart

## *Sminthopsis leucopus*

### (Gray, 1842)

luke'-oh-poos: 'white-footed
mouse-like (animal)'

Studies of the White-footed Dunnart are cloud-
ed by many factors. It has not yet been assigned a
reliable suite of distinguishing field characteristics and may be confused with the Common
Dunnart, *S. murina*. The *S. murina–S. leucopus*
species pair may also prove to conceal other
species, as was found in Western Australia in the
1980s within *S. murina*. Studies of the species
suggest that its occurrence and density may not
be accurately assessed unless pitfall traps are
used extensively in trapping. Studies in East
Gippsland and North Queensland found that *S.
leucopus* was captured only in pitfall traps,
whereas a study near Melbourne caught *S. leucopus* in 'Sherman' aluminium traps and in

*The White-footed Dunnart may comprise more than
one species.*

Mumbulla State Forest in south-eastern New
South Wales, adults were readily captured and
recaptured in 'Elliott' brand traps, but the juveniles were caught only in pitfall traps. The
weight of present evidence provides reasonable
grounds for believing that this species is sparsely
and patchily distributed: analysis of Dog and Fox
scats in another study lends weight to this growing body of evidence.

A 1991 review in Tasmania showed that the
White-footed Dunnart occurred in 7 per cent of
the State and in most vegetation types, but was
infrequently recorded. In the south-eastern
mainland, it is an animal of open understorey
and low density vegetation. At Sandy Point, near
Melbourne, all but one individual were captured
on the grassy foredune complex, backed by
*Leptospermum laevigatum* scrub and woodland of
banksia and eucalypt.

A long-term study in Mumbulla State Forest
found that the White-footed Dunnart preferred
forest with sparse ground cover. It showed no
immediate response to logging or burning but
disappeared within three years of these events:
dense regeneration appears to be inimical to its

## Sminthopsis leucopus

**Size**

**NSW and Victoria**

*Head and Body Length*

82–98 mm (males)

67–74 mm (females)

*Tail Length*

82 mm (males)

67 mm (females)

*Weight*

26 g (males)

19 g (females)

**North Queensland**

*Head and Body Length*

unknown (males)

88–101 mm (females)

*Tail Length*

unknown (males)

87–96 mm (females)

*Weight*

unknown (males)

25+ g (females)

**Tasmania**

*Head and Body Length*

98–112 mm (males)

95–117 mm (females)

*Tail Length*

73–100 mm (males)

70–95 mm (females)

*Weight*

32 g (males)

24 g (females)

Note 1: Newly independent juveniles in early summer may weigh as little as 8 g.

Note 2: The samples upon which these figures are based are small and perhaps unrepresentative.

Note 3: A study at Anglesea in coastal Victoria found no weight difference between sexes, 16–32 g.

**Identification**

Fur on back and face light brown-yellow with black guard hairs, giving brown-grey pattern: belly off-white, gradually darkening along flanks. Eyes large, dark, protruding; large thin ears can be laid back against head; brown skin on top of muzzle. Superficially similar to House Mouse but has needle-shaped incisors and five toes on fore-foot. Feet pink with covering of fine, white hair. North Queensland specimens have grey fur on the cheeks. Similar to *S. murina* but usually with striations on interdigital footpads and slightly heavier. Distinguished from *Antechinus stuartii* by more slender body and much greater contrast between colour of upper and lower parts.

**Recent Synonyms** None.

**Other Common Names** White-footed Marsupial Mouse.

**Status** Uncommon in Victoria, infrequently recorded Tasmania, rare in North Queensland and officially listed as 'Vulnerable and Rare' in New South Wales.

**Subspecies** The Tasmanian population has been referred to *Sminthopsis leucopus leucopus* and the mainland population to *Sminthopsis leucopus ferruginifrons* but, since the differences between *Sminthopsis leucopus* and *Sminthopsis murina* are not clear, this is premature.

**References**

Lunney, D. and T. Leary (1989). Movement patterns of the White-footed Dunnart, *Sminthopsis leucopus* (Marsupialia: Dasyuridae) in a logged, burnt forest on the south coast of New South Wales. *Aust. Wildl. Res.* 16: 207–215.

Lunney, D., M. O'Connell, J. Sanders and S. Forbes (1989). Habitat of the White-footed Dunnart *Sminthopsis leucopus* (Gray) (Dasyuridae: Marsupialia) in a logged, burnt forest near Bega, New South Wales. *Aust. J. Ecol.* 14: 335–344.

Woolley, P.A. and S.L. Gilfillan (1991). Confirmation of polyoestry in captive White-footed Dunnarts, *Sminthopsis leucopus* (Marsupialia: Dasyuridae). *Aust. Mammal.* 14: 137–138.

survival. A more detailed investigation showed that it selects treeless ridges and mid-slopes with less than 50 per cent ground cover in the logged areas of burnt forest. It bred in this disturbed habitat but did not persist when regeneration led to dense, low vegetation. It was not found in gullies or where ferns were abundant in the ground cover. A management program for the species in this area would stagger logging and burning operations to always have some parts of the forest free of dense regrowth. In contrast, the limited records for North Queensland indicate that it is a rainforest resident in this region.

The study at Mumbulla showed that the White-footed Dunnart preys on a wide variety of terrestrial invertebrates ranging in length from less than 1 millimetre to more than 18 millimetres. Skinks of up to 1.5 grams were the only vertebrates known to be eaten. In a semi-captive study, a male and a female were placed in an enclosure and offered live cockroaches, moths and beetles, which they rapidly killed and ate. They attacked a large female huntsman spider by biting off a leg, withdrawing to eat this, then repeating the sequence until the spider was limbless; then they ate the body. When eating, the White-footed Dunnart holds its prey in the forepaws and breaks off morsels with its cheek-teeth. Prey items occur on ridges and in gullies and in both logged and unlogged forest, indicating that the Dunnart's habitat preference is not determined by the availability of food.

At Mumbulla, juveniles made their nests under strips of bark or small rotting logs on ridges in open areas near disturbed sites and roads, where they have been found by researchers. The nest consists of a shallow depression with a few leaves and blades of grass.

In New South Wales and Victoria, mating occurs in late July and August. Laboratory studies show that the female can breed more than once in a season but there is no evidence of this in the wild. The female usually has ten teats, which begin to enlarge in late July, when the pouch skin swells and reddens and the pouch hairs become white and shiny. From mid-August to mid-September, up to ten young are born:

***Sminthopsis leucopus***

they have a head-rump length of about 3 millimetres, which more than doubles in a week. At the age of about eight weeks, the young detach from the teats and are suckled in the nest for about a month.

The population at Mumbulla was discrete, occupying an area of about 500 hectares near the centre of the forest. Adult females occupied small home ranges and capture sites did not overlap with the home ranges of other females. Males did not have exclusive capture sites and their home ranges overlapped. Male movement patterns fell into two groups: those of 'explorer' males and 'resident' males. The largest movement of an explorer male was 1025 metres in 24 hours (average range-length for the group being 720 metres). Resident males had a range-length of 105 metres while adult females ranged 80 metres. The ability to travel long distances enables this species to find suitable habitat and to utilise suddenly abundant and transient resources, such as recently disturbed forest, which may occur naturally only as disjunct and temporary patches. The ecology of this species displays a pattern in vivid contrast to the well-known Brown Antechinus, which occurs in many of the same geographic areas. Its patchy distribution, low density and inability to persist in regrowth forest was sufficient to place it on the official list of endangered species in New South Wales.

D. LUNNEY

# Long-tailed Dunnart

## *Sminthopsis longicaudata*

### Spencer, 1909

lon'-jee-kaw'-dah-tah: 'long-tailed
mouse-like (animal)'

Prior to 1975 there were only three whole specimens of the Long-tailed Dunnart in museums
and nothing was known of its natural history.
The earliest known specimen was discovered by
C.W. Brazenor in or after 1916 in a collection
of mammals donated to the National Museum
of Victoria by Baldwin Spencer. An adult female
with five pouch young, it had been collected in
1895 in Central Australia, possibly near
Charlotte Waters.

Apparently unaware that he had this specimen, Spencer described the species in 1908
from a single adult male collected in Western
Australia. Once again, neither habitat nor
detailed locality data were recorded but the
specimen was probably collected soon after
1896, possibly near Marble Bar in the Pilbara.
The third specimen was collected at Pillendinnie, near Marble Bar and donated to the

Western Australian Museum in 1940, but again
no accessory data were recorded.

Although part of a cranium of unknown age
was subsequently found under a Ghost Bat roost,

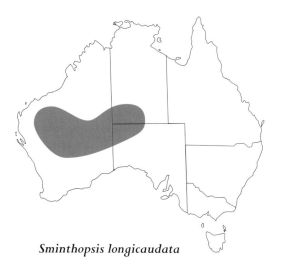

*Sminthopsis longicaudata*

## *Sminthopsis longicaudata*

**Size**
(based on limited data)
*Head and Body Length*
80–100 mm (males)
80–90 mm (females)
*Tail Length*
200–210 mm (males)
180–200 mm (females)
*Weight*
15–21 (18) g

## Identification

Grey above; pale cream to white below; legs
and feet white. Tail scaly with short hairs
except at the tip, where long hairs protrude to
make a fine brush. Head flattened; snout long.
Distinguished from all other *Sminthopsis* species
by tail being more than twice the length of the
head and body.

**Recent Synonyms** None.

**Other Common Names** Long-tailed
Marsupial Mouse.
Indigenous names: Tjarrtjalaranpa
(Ngaatjatjarra), Yarntala (Pintupi), Yarrutju
(Ngaanyatjarra).

**Status** Restricted to rocky outcrops in the
western arid zone; may be locally common at
times. Two populations known from national
parks, another from a nature reserve.

**Subspecies** None.

**References**

Burbidge, A.A. and N.L. McKenzie (1976). A
    further record of *Sminthopsis longicaudata*
    (Marsupialia, Dasyuridae). *West. Aust. Nat.*
    13: 144–145.

Woolley, P.A. and A. Valente (1986).
    Reproduction in *Sminthopsis longicaudata*
    (Marsupialia: Dasyuridae): Laboratory observations. *Aust. Wildl. Res.* 13: 7–12.

again near Marble Bar and bones were found in sub-fossil deposits from limestone caves in the Cape Range, Western Australia, it was not until 1975 that the fourth whole specimen was found. An adult female was flushed at late dusk from a spinifex hummock near Miss Gibson Hill in the Gibson Desert, Western Australia and ran with its tail held stiffly behind and horizontal to the ground. The site was a lateritic plateau with shallow, sandy gravel soil, supporting a hummock grassland of spinifex (*Triodia* sp. and *Plectrachne* sp.) with emergent Corkwood (*Hakea lorea*) and *Acacia* shrubs. It was about 10 metres from a large breakaway, below which were a scree and valley with low open *Acacia* woodland.

In 1981, an expedition mounted by the Western Australian Wildlife Research Centre to locate further populations of the Long-tailed Dunnart trapped nine animals in the Young Range, a series of flat-topped hills in the Gibson Desert Nature Reserve. Eight came from plateaus near breakaways and screes and one came from a particularly rugged area of scree. The plateaus, composed of boulders and stones, with a little fine red soil, were sparsely vegetated with Mulga (*Acacia aneura*) and Minni Ritchie (*A. grasbyi*) shrubs over spinifex (*Triodia pungens*). The scree supported a low, open woodland of Mulga.

Since then, additional specimens have been captured in the Clutterbuck Hills (Gibson Desert), near Mount Anderson (Murchison), near Glenayle Station (North-eastern Goldfields), near Paraburdoo (Pilbara), Mount Augustus (Ashburton) and the western MacDonnell Range. The Clutterbuck Hills specimens came from among boulders on the crest of a low sandstone range with scattered mulga. The Pilbara specimen was captured near the crest of a ridge, in an area with extensive rock pavement and an overstorey of open mulga over sparse spinifex. Sightings have been reported from 50 kilometres north-east of Sandstone and from Belele Station near Meekatharra.

Faecal analysis shows that, in winter, the Long-tailed Dunnart feeds on arthropods: mainly beetles and ants, but also spiders, cockroaches, centipedes, grasshoppers, flies and various larvae. The extremely long tail is strongly muscular

The Long-tailed Dunnart lives in rocky country. Its very long, brush-tipped tail probably acts as a balancer.

at its base and can be moved into a variety of positions. Under cold conditions a Long-tailed Dunnart may become torpid. Studies on captive animals suggest that breeding occurs from August to December.

The few records of the Long-tailed Dunnart come from widely scattered localities in the arid zone. Available evidence shows that it inhabits rugged, rocky areas. Its striated foot-pads, long tail and behaviour in captivity suggest that it is an active and capable climber.

A.A. BURBIDGE, N.L. MCKENZIE AND P.J. FULLER

# Stripe-faced Dunnart

## *Sminthopsis macroura*

### (Gould, 1845)

mak'-roh-yue'-rah: 'long-tailed mouse-like (animal)'

D. WHITFORD

**Sminthopsis macroura**

**Size**
*Head and Body Length*
70–100 (85) mm
*Tail Length*
80–110 (95) mm
*Weight*
15–25 (20) g

**Identification** Distinguished from other dunnarts by length of (usually fat) tail which is about 1.25 times the length of the head and body. Each interdigital pad on the hindfoot has a distinctly enlarged apical granule surrounded by a number of slightly enlarged granules.
**Recent Synonyms** *Sminthopsis froggatti*, *S. larapinta*.
**Other Common Names** Darling Downs Dunnart, Stripe-headed Sminthopsis, Froggatt's Sminthopsis, Larapinta.
**Status** Common.
**Subspecies** *Sminthopsis macroura macroura*, central eastern Australia.
*Sminthopsis macroura froggatti*, central Australia.
**Reference**
Woolley, P.A. (1990). Reproduction in *Sminthopsis macroura* (Marsupialia: Dasyuridae). Parts I and II. *Aust. J. Zool.* 38: 187–217.

This widespread inhabitant of much of inland central and northern Australia derives its common name from a prominent line of dark hair running from between the eyes to between the ears. The specific name refers to the tail, which is noticeably longer than the head and body and frequently very swollen with fat. Stripe-faced Dunnarts are found in many habitats in the arid and semi-arid parts of Australia: they occur in low shrublands of saltbush and bluebush, in tussock grasslands on clay, sandy or stony soils, in spinifex grasslands on sandy soils, among sparse *Acacia* shrublands and on low, shrubby, rocky

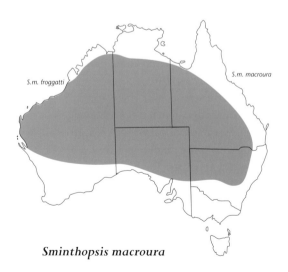

*S.m. froggatti*

*S.m. macroura*

**Sminthopsis macroura**

D. WHITFORD

*Like its fat-tailed relative, the Stripe-faced Dunnart stores fat in its tail. in most other respects it is a typical dunnart.*

D. WHITFORD

ridges. Despite the variety of environments in which the species has been found, it seems probable that the densest populations occur in tussock grasslands.

The behaviour of the Stripe-faced Dunnart in the wild is poorly known but it is probably strictly nocturnal. There is no evidence of specialisation upon any particular type of invertebrate prey. The rapidity with which previously deserted areas are colonised after rainfall suggests that it is quite mobile. It is almost certainly independent of drinking water and the extensive fat stores in the tail provide an energy reserve that may be utilised in times of food shortage. High temperature is avoided by sheltering during the day in cracks in the soil or under rocks and logs,

probably in nests, although these have not been seen. In captivity, the Stripe-faced Dunnart may become torpid during the day.

In captivity, females come into heat from June to February, during which time most individuals can raise at least two litters. The gestation period—the shortest known for any marsupial—is about 11 days. The usual litter size of eight is equal to the number of teats and the pouch fully encloses the newborn young. Females mature earlier than males, respectively at about four and nine months of age. Observations in the field suggest that breeding is confined to the same season as in captivity, but the number of litters raised by individual females is unknown.

Nothing is known of fluctuations in density in the wild but, because much of its range lies within areas of highly variable rainfall, populations almost certainly vary considerably. There is no evidence that the species has changed in status since European settlement, but grazing by domestic stock throughout the inland may have been detrimental in reducing the suitability of the tussock grassland which it prefers.

S.R. MORTON

D. WHITFORD

*The Common Dunnart extends, patchily, over much of eastern Australia.*

# Common Dunnart

## Sminthopsis murina

### (Waterhouse, 1838)

myue-ree'-nah: 'mouse-like mouse-like (animal)'

The Common Dunnart, first described in 1838, occurs in south-eastern Australia and north-eastern Queensland. A taxonomic revision including intensive studies of electrophoretic and morphometric characters has allocated animals from south-western Australia, previously recognised as forms or subspecies of the Common Dunnart, to separate species. *Sminthopsis murina* (in the strict sense) probably does not extend west of the Flinders Ranges, but has been confirmed to include the north-eastern Queensland population. It is most commonly found in woodland, open forest and heathland but has also been collected in transitional habitats close to rainforest. It is absent from such habitats in coastal regions of Victoria, where its place is taken by the White-footed Dunnart. A nocturnal species, it rests during the day in a cup-shaped nest of dried grass and leaves built in a fallen hollow log, a clump of grass or a grass-tree (*Xanthorrhoea*).

*Sminthopsis murina*

**Size**
*Head and Body Length*
76–104 (81) mm (males)
64–92 (76) mm (females)
*Tail Length*
70–99 (79) mm (males)
68–92 (76) mm (females)
*Weight*
16–28 (20) g (males)
10–22 (14) g (females)

**Identification** Mouse-grey above, predominantly white below. Slender, pointed muzzle; large ears and eyes. Most similar to *Sminthopsis dolichura*, *S. gilbert*, *S. griseoventer* and *S. aitkeni*, for which geographic location may be the best distinguishing feature as all four occur west of Flinders Ranges. Distinguished from *S. dolichura* by shorter tail and brownish dorsal pelage; from *S. gilberti* by shorter hindfeet and ears; from *S. griseoventer* by white rather than grey belly hair; from *S. aitkeni* by not having dark, sooty back hair; from *S. crassicaudata* by unfused, hairless interdigital pads on hindfeet; from *S. leucopus* by lack of striations on interdigital pads and small hallucal pad; from *S. macroura* by absence of head-stripe (although dark patch may be present). Tail length is variable but approximately equal to head and body (longer in *S. ooldea*): never incrassated (as sometimes in *S. macroura*, *S. crassicaudata* and *S. ooldea*).
**Recent Synonyms** None.
**Other Common Names** Common Marsupial Mouse, Mouse-sminthopsis, Slender Mouse-sminthopsis.
**Status** Common, limited, may be decreasing.
**Subspecies** Two subspecies are currently recognised:
*Sminthopsis murina murina*, south-eastern Australia.
*Sminthopsis murina tatei*, north-eastern Queensland.

**Sminthopsis murina**

D. WHITFORD

*A litter of 25-day-old Common Dunnarts, still attached to the mother's teats.*

It is insectivorous, feeding on beetles, roaches, cricket larvae and spiders. It readily eats the mixture of peanut butter and oatmeal which is commonly used as bait but is more readily caught in pitfall traps than in the metal (Elliott) traps that are usually used in mammal surveys. Local distribution of the Common Dunnart is very patchy but, where present, it may reach densities of six per hectare. Since such densities have been recorded only in areas which have been burnt in the previous two to four years, it seems that the Common Dunnart is adapted to a mid-successional complex of vegetation and that it therefore benefits from periodic burning of its habitat.

**References**

Fox, B.J. and D. Whitford (1982). Polyoestry in a predictable coastal environment: Reproduction, growth and development in *Sminthopsis murina* (Dasyuridae: Marsupialia) (in) M. Archer (ed.) *Carnivorous Marsupials*. Roy. Zool. Soc. NSW and Surrey Beatty and Sons, Sydney. pp. 39–48.

Kitchener, D.J., J. Stoddart and J. Henry (1984). A taxonomic revision of the *Sminthopsis murina* complex (Marsupialia: Dasyuridae) in Australia, including descriptions of four new species. *Rec. West. Aust. Mus.* 11: 201–248.

The species appeared to be rare along the central New South Wales coast from 1989 to early 1992 despite extensive surveys in areas where it had been captured previously in reasonable numbers. However, after additional surveys, individuals were again reported in some of these areas at the start of the 1992 breeding season and were abundant again in 1994.

With the onset of breeding, males become increasingly aggressive toward each other and may be wounded in combat. Vocalisation of females attracts males to them. The oestrous cycle is 24 days and two litters are normally born during the August to March breeding season. A high reproductive rate, which facilitates population increase following colonisation of a burnt area, is made possible by a short gestation (12.5 days), reduced parental care, rapid development of the young (which are weaned when about 65 days old) and litters of up to ten young on eight to ten teats. Adult size is reached at about 150 days.

Experimental encounters in outdoor enclosures have demonstrated strong competitive interactions with the Brown Antechinus, which is often found in similar habitats but usually exhibits clear spatial and ecological separation. The Brown Antechinus dominates the Common Dunnart to the point of killing an individual in one such encounter.

B.J. FOX

HANS & JUDY BESTE

# Ooldea Dunnart

## *Sminthopsis ooldea*

Troughton, 1965

ool-day'-ah: 'Ooldea (SA)
mouse-like (animal)'

The name of the species refers to a small settlement on the transcontinental railway line in western South Australia where the first specimens were collected, but it is much more widely distributed. It occurs in a variety of arid habitats

For many years the Ooldea Dunnart was considered to be a subspecies of the Common Dunnart but recent studies have established it as a distinct species. Like several other members of the genus, it may develop a fattened tail, but this is never as extreme as in the Fat-tailed or Stripe-faced Dunnarts. Most wild-caught examples show little sign of tail-fattening until they have been in captivity for some months.

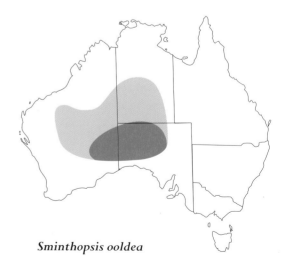

*Sminthopsis ooldea*

*Little is known of the biology of the Ooldea Dunnart.*

such as acacia and eucalypt woodland, mallee shrubland and hummock grassland. Most recent specimens have been caught in pitfall traps sunk in sandy soils and one individual was dug from a shallow burrow in a sand ridge. Common and Fat-tailed Dunnarts occur in the same habitats.

Little is known of its biology but it is reasonably assumed to be nocturnal and insectivorous, like other dunnarts. Captive individuals are nocturnal and fiercely attack and devour insects such as locusts and moths. Torpor is occasionally observed in captive individuals but it is not known whether this is a habitual feature of their daily cycle (as in the Stripe-faced Dunnart) or a response to temporary food shortage.

Females with small pouch-young have been caught in October. Lactating females and juveniles are encountered in December and January. Young thus appear to be born in spring and to become independent by mid-summer. Most females have eight teats and litters obtained from the wild or born in captivity comprise seven to eight young. The pouch develops prior to birth of the young and encloses them until they are about three weeks old, during which time they grow from a length of 3–15 millimetres. The young detach themselves from the teats at about 30 days of age, when they have sparse grey fur and a head–body length of 20 millimetres. They produce a scraping distress call when separated from their mother. Eyes open at around 45 days and at this age the young may cling tightly to their mother's back. By 70 days, having attained a weight of around 5 grams and a head–body length of 60 millimetres they are capable of independent existence.

It is not known whether young remain associated with their mother after weaning, or whether any other social aggregations occur in the wild. Adult females frequently produce a clicking call at or around the time they are in oestrus and this may serve to attract mates.

## *Sminthopsis ooldea*

**Size**

*Head and Body Length*
60–85 (72) mm (males)
55–85 (72) mm (females)

*Tail Length*
65–95 (79) mm (males)
60–90 (76) mm (females)

*Weight*
9–17 (11) g (males)
8–15 (11) g (females)

**Identification** Greyish-brown or greyish-yellow above; belly hair white with grey bases; slightly shaggy. Ears large and triangular. Similar to *Sminthopsis crassicaudata*, *S. macroura* and *S. murina* but, on average, adults are lighter in weight and shorter in head and body length than adults of these three species. Interdigital pads on sole are not fused, or are joined only at base.

**Recent Synonyms** *Sminthopsis murina ooldea.*

**Other Common Names** Troughton's Sminthopsis.

**Status** Common.

**Subspecies** None.

**Reference**

Archer, M. (1981). Systematic revision of the dasyurid marsupial genus *Sminthopsis* Thomas. *Bull. Amer. Mus. Nat. Hist.* 168: 61–224.

Although the Ooldea Dunnart was virtually unknown until the 1970s, it has since been captured regularly in pitfall traps set in arid regions and will probably prove to be a common species. It occurs in large areas of suitable habitat within Aboriginal reserves; in Uluru National Park, Northern Territory; in a number of nature reserves in south-eastern Western Australia; and in the huge unnamed conservation park in western South Australia.

H.J. ASLIN

# Sandhill Dunnart

## *Sminthopsis psammophila*

### Spencer 1895

sam-off'-il-ah: 'sand-loving
mouse-like (animal)'

BABS & BERT WELLS

BABS & BERT WELLS

*The little-known Sandhill Dunnart is an inhabitant of arid, sandy regions.*

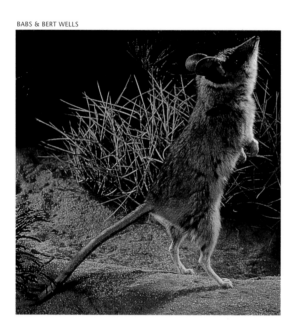

The Sandhill Dunnart has remained an enigma since it was first collected by the Horn Scientific Expedition in June 1894 near Lake Amadeus, Northern Territory. An adult male was flushed in broad daylight and avoided its captors with a spirited escape, until brought down by a well-directed throw of a boot. The Sandhill Dunnart then 'disappeared' for 75 years.

In April 1969, a farmer at Mamblyn on Eyre Peninsula, South Australia, dropped his hat on a second male as it ran from burning spinifex. He sent it to the South Australian Museum in a battered shoe-box loosely tied with string. The rediscovery of the Sandhill Dunnart generated great local publicity and two months later, four juveniles were captured as they fled from smouldering debris in front of a bulldozer, 80 kilometres south-east of Boonerdoo.

Since then, the Sandhill Dunnart has been captured in Western Australia; five specimens from the south-western edge of the Great Victoria Desert in 1985; and 12 between 1987 and 1991 from the nearby Queen Victoria Spring Nature Reserve. In South Australia, five were recorded in 1987 from Yarle Lakes, Ooldea and Mount Christie, during a wildlife survey of the Yellabinna sand-dunes. The Sandhill Dunnart has not been collected in the Northern Territory since 1894, although a skull was recovered from an owl pellet of unknown age found in a cave at Ayers Rock.

The species is known from a variety of climatic zones and habitat types. Low, unpredictable rainfall is characteristic of the Lake Amadeus and Great Victoria Desert localities, the former experiencing summer maxima. In contrast, Eyre Peninsula receives regular winter rains. All recorded habitats have sandy soils (sometimes with low dunes) and an understorey of hummock grass, (*Triodia* or *Plectrachne* spp.). Overstorey vegetation is variable. Groves of desert oaks occur near Lake Amadeus, mallee and tea-tree scrub on Eyre Peninsula and low, open eucalypt and *Callitris* woodlands with diverse shrub layers in the Great Victoria Desert and Yellabinna sand-dunes. Typically the habitat is long unburnt,

## Sminthopsis psammophila

**Size**

*Head and Body Length*
85–114 mm
*Tail Length*
107–128 (118) mm
*Weight*
30–44 (37) g (males)
25–35 (30) g (females)

**Identification** Dorsal fur drab grey to buff, brindled; head pale grey with black pencilling extending from shoulders to wedge between eyes. Black eye-rings. Cheeks and flanks buff; underside and feet white. Ears large. Tail pale grey above, dark grey below tapering towards tip, with a vertical crest of blackish-grey hairs on terminal quarter. Distinguished from all other *Sminthopsis* species (except *S. douglasi*) by large size and distinctive tail.

**Recent Synonyms** *Sminthopsis psammophilus.*

**Other Common Names** Large Desert Sminthopsis.

**Status** Little known. Rare in widely scattered localities.

**Subspecies** None.

**References**

Aitken, P.F. (1971). Rediscovery of the Large Desert Sminthopsis (*Sminthopsis psammophilus* Spencer) on Eyre Peninsula, South Australia. *Vic. Nat.* 88: 103–111.

Hart, R.P. and D.J. Kitchener (1986). First record of *Sminthopsis psammophila* (Marsupialia: Dasyuridae) from Western Australia. *Rec. West. Aus. Mus.* 13: 139–144.

Pearson, D.J. and A.C. Robinson (1990). New records of the Sandhill Dunnart, *Sminthopsis psammophila* (Marsupialia: Dasyuridae) in South and Western Australia. *Aust. Mammal.* 13: 57–59.

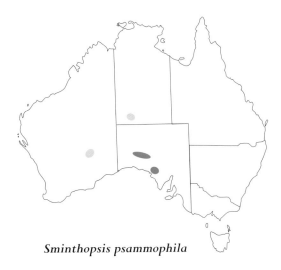

*Sminthopsis psammophila*

lactating female without pouch-young in December; and a female accompanied by four dependent juveniles in early January. The pouch has a wide perimeter flap of skin which covers the eight teats and the young during early pouch life.

Material in the faeces indicates a predominantly insectivorous diet. In captivity, the Mamblyn male ate adult and larval insects, spiders and chopped offal. It was very agile and active from the late afternoon until dawn. When threatened, the Sandhill Dunnart adopts a crouched defensive position, baring its teeth and issuing throaty hisses.

Queen Victoria Spring Nature Reserve is the sole conservation reserve in Western Australia where the Sandhill Dunnart is known to occur. The other Western Australian population is on vacant Crown land where the only existing land use is mineral exploration. In South Australia, populations are known from the Yellabinna Regional Reserve and the Maralinga Tjarutja Aboriginal Lands. There have been no records from Eyre Peninsula since 1969.

Southern desert regions have vast areas of habitat seemingly suitable for the Sandhill Dunnart, but low densities at all known localities suggest that it is rare. Research is necessary to establish its habitat requirements and its survival status.

although one male was captured in an area burnt three years earlier.

Limited data suggest that reproduction occurs in spring and early summer. A female with five small pouch-young was captured in October; a

D.J. PEARSON

# Red-cheeked Dunnart

## *Sminthopsis virginiae*

(Tarragon 1847)

ver-jin'-ee-ee: 'Virginia's mouse-like (animal)'

*Appropriately named, the Red-cheeked Dunnart has rufous fur on the sides of its head.* D. WHITFORD

Three forms of the Red-cheeked Dunnart, *Sminthopsis virginiae*, are currently recognised. *S. v. virginiae* occurs in Queensland, *S. v. nitela* in the Northern Territory and Western Australia and *S. v. rufigenis* in the southern lowlands of New Guinea and on the Aru Islands. They are inhabitants of savanna woodlands and probably nest on the ground under the cover of dense vegetation such as pandanus fronds or thick grass. Their food consists largely of insects but they are capable of killing and eating small lizards.

Growth appears to continue throughout life, albeit more slowly after sexual maturity: differences in age may explain the wide range in size recorded for wild-caught adults. The figures given are for the largest series, obtained from Western Province in Papua New Guinea, but Australian specimens fall within the range.

Red-cheeked Dunnarts from the Western Province and from Queensland have been bred at La Trobe University. The length of the oestrous cycle is around 30 days and the interval between mating and birth of the young is around 15 days. The number of young that can be reared is limited by the number of teats (eight in *S. v. nitela* and *S. v. virginiae*; six in *S. v. rufigenis*).

The young suckle for 65–70 days and are mature when four to six months old. In captivity, *S. virginiae* is capable of breeding throughout the year but whether or not it does so in the wild is uncertain. Females suckling young have been captured in March (Queensland); March and July to November (Northern Territory); and March, May, July to October and December (New Guinea).

There appears to be no barrier to breeding between Australian forms (a Queensland male sired young by a Northern Territory female) but a barrier may exist between the New Guinean and Australian forms. Three attempts, in successive oestrous cycles, to mate a Northern Territory male with a New Guinea female were unsuccessful.

P.A. WOOLLEY

Melville Is.

*S.v. nitela*

*S.v. virginiae*

*Sminthopsis virginiae*

## Sminthopsis virginiae

### Size

*Head and Body Length*
96–135 mm (males)
90–133 mm (females)
*Tail Length*
100–135 mm (males)
90–122 mm (females)
*Weight*
31–58 g (males)
18–34 g (females)

### Identification

Pale-tipped, coarse, dark grey fur above; white to buff below. The pale tips give the animal a speckled appearance. Distinct black face-stripe and rufous cheeks. Tail usually sparsely covered with short, dark hairs, but some specimens from the Northern Territory have pale tail.

**Recent Synonyms** *Sminthopsis lumholtzi*, *Sminthopsis rufigenis*, *Sminthopsis nitela*, *Phascogale rona*.

**Other Common Names** Lumholtz's Marsupial Mouse, Aru Islands Pouched Mouse, Forehead-striped Pouched Mouse, Daly River Sminthopsis, Queensland Stripe-faced Sminthopsis.

**Status** Uncommon in limited range.

**Subspecies** *Sminthopsis virginiae virginiae*, north-eastern Queensland.

*Sminthopsis virginiae rufigenis*, Papua New Guinea.

*Sminthopsis virginiae nitela*, Northern Territory and Western Australia. This may itself be a composite of several distinct forms.

**Extralimital distribution** New Guinea.

### References

Taplin, M.E. (1980). Some observations on the reproductive biology of *Sminthopsis virginiae* (Tarragon) (Marsupialia: Dasyuridae). *Aust. Zool.* 20: 407–418.

Braithwaite, R.W. and W.M. Lonsdale (1987). The rarity of *Sminthopsis virginiae* (Marsupialia: Dasyuridae) in relation to natural and unnatural habitats. *Cons. Biol.* 1: 341–343.

# Lesser Hairy-footed Dunnart

## *Sminthopsis youngsoni*

### McKenzie and Archer, 1982

yung'-sun-ee: 'Youngson's mouse-like (animal)'

*The Lesser Hairy-footed Dunnart, like its larger relative, has hairy soles and fringes on its hindfeet, assisting it to move on fine sand.* D. WHITFORD

First described in 1982 from the Great Sandy Desert, the Lesser Hairy-footed Dunnart ranges through the subtropical arid regions of Western Australia, the Northern Territory and Queensland. It is nocturnal, preying on insects and other small invertebrates in sandy environments that are the habitat equivalents of its generally more temperate counterpart, the Hairy-footed Dunnart, *Sminthopsis hirtipes*. Other similarities include a tail that fattens during good seasons and fine bristles covering the soles of its feet, probably enhancing traction on loose, sandy surfaces.

It has been recorded from a variety of plant communities associated with reddish desert sand plains, sand-dunes and inter-dune swales: open shrublands of *Acacia*, *Thryptomene*, *Grevillea* or *Melaleuca*, often with sparse *Owenia*, *Casuarina* or *Eucalyptus* trees, over hummock grasslands of

**Sminthopsis youngsoni**

*Triodia* or *Plectrachne*. In some places these are mixed with tussock or cane grasses such as *Eragrostis* and *Zygochloa*.

Young are born in spring. Pregnant females have been recorded in September and females with five to six pouch-young from September to January. Independent juveniles have been captured from November to February. Between April and June, populations include many subadults.

Where present, the Lesser Hairy-footed Dunnart is readily captured in pitfall traps. Its recent discovery reflects inappropriate collecting techniques used in the past. Populations are known to exist in the Rudall River (Western Australia) and Uluru (Northern Territory) National Parks. The species does not appear to be threatened.

N.L. MCKENZIE AND J.R. COLE

*Sminthopsis youngsoni*

### Size

*Head and Body Length*
66–71 (68.5) mm

*Tail Length*
62–68 (65.2) mm

*Weight*
8.5–12.0 (10.1) g

**Identification** Brownish-yellow above; white below. Basal half of tail slightly swollen. Smaller than the similar *S. hirtipes*, with proportionally shorter tail, hindfeet and ears in relation to body length. Both *S. hirtipes* and *S. youngsoni* are distinguished from other *Sminthopsis* by broad feet that, even on the soles and granular foot-pads, are entirely covered with fine, silvery hairs.

**Recent Synonyms** None.

**Other Common Names** Desert Dunnart. Indigenous names: Djudububa, Malabuba, Djudumuwa.

**Status** Widespread and common in suitable habitat; habitat common.

**Subspecies** None.

**References**

McKenzie, N.L. and M. Archer (1982). *Sminthopsis youngsoni* (Marsupialia: Dasyuridae) the Lesser Hairy-footed Dunnart, a new species from arid Australia. *Aust Mammal.* 5: 267–279.

Cole, J.R. and D.F. Gibson (1991). Distribution of Stripe-faced Dunnarts *Sminthopsis macroura* and Desert Dunnarts *Sminthopsis youngsoni* (Marsupialia: Dasyuridae) in the Northern Territory. *Aust. Mammal.* 14: 129–131.

McFarland, D. (1992). *Fauna of the Channel Country Biogeographic Region, South West Queensland.* Queensland National Parks and Wildlife Service, Department of Environment and Heritage, Brisbane.

*Spider burrow in the Simpson Desert, used as a home by a Lesser Hairy-footed Dunnart.*

# FAMILY MYRMECOBIIDAE
## *Numbat*

*Juvenile Numbats, still associating near a nest.*

Many terrestrial vertebrates—frogs, lizards, birds and mammals—catch and eat insects. Small numbers feed on ants and termites: social insects that usually live in vast numbers in elaborate nests. The anteaters of South America, the Aardvark of Africa, the pangolins of Asia and the Short-beaked Echidna of Australia and New Guinea have powerful forelimbs with strong claws that are used to dig into a nest. They feed by means of a long, mobile, sticky tongue that is inserted into the galleries of a nest when it has been breached. Insects that adhere to the tongue are retained in the mouth as the tongue is flicked in and out.

The Numbat feeds in this manner on termites but without breaching any but the weakest nests. Instead, it exposes the shallow, unfortified runways that radiate from a nest, just below the surface of the ground. In common with other mammals that are specialised for feeding on social insects, it has a degenerate dentition but this is polyprotodont. Together with the simple structure of the hindfoot, this indicates that the Numbat is a dasyuroid, a conclusion that is supported by serological evidence. It is the only living member of the Myrmecobiidae and it appears to be seriously endangered.

Given the prevalence of ants and termites in Australia, it is surprising that only one marsupial utilises this food resource and that it is a rather small animal with unspecialised forelimbs: there appears to be an opportunity—a 'situation vacant'—for an animal about the size and strength of the Giant Anteater of South America that could exploit the vast amount of food concentrated in the large and abundant termite mounds of tropical and subtropical Australia. It is possible that this niche was occupied to some extent by some of the large long-beaked echidnas which were once common in Australia but became extinct long before European settlement.

# Numbat

## *Myrmecobius fasciatus*

### Waterhouse, 1836

mer'-mek-oh-bee'-us  fas'-ee-ah'-tus:
'banded ant-eater'

The striking appearance and diurnal activity of the Numbat sets it apart from the less colourful and more secretive marsupials. It is so specialised for a diet of termites that it is placed in a family of its own.

At the time of European settlement, its distribution extended from western New South Wales, through South Australia and across much of the southern half of Western Australia. Now, only a few isolated populations remain in southwestern Western Australia, the largest of which are at Dryandra, near Narrogin and at Perup, near Manjimup.

Its present habitat is eucalypt forest and woodland, dominated by Wandoo, Powderbark Wandoo, or Jarrah. Areas with these vegetation types provide the Numbat with hollow logs and branches for its shelter and termites for food. It was previously found in a range of woodland types, including York Gum woodland in Western Australia, Mulga woodland in central Australia and scrub along the lower Murray River in South Australia. Termites are plentiful in these areas but the abundance of hollow logs was not of prime importance before the arrival of the Fox.

In its woodland habitat, the Numbat feeds in open areas near the cover of shrubs or within easy reach of hollow logs. Some logs are used merely as bolt-holes to avoid predators, but nests are built in other logs and hollow trees and the Numbat sleeps in these at night. Numbats also dig burrows in which they construct nests and sleep, particularly in winter. The burrow usually takes the form of a single narrow shaft, sloping downwards at a gentle angle, then opening out after 1–2 metres into a spherical

chamber about 250 millimetres in diameter that is packed with nest material consisting of grass, shredded bark, leaves and other soft material, such as flowers.

When feeding, a Numbat searches by scent for underground termite galleries in the woodland floor. Concentrating on areas clear of leaf litter, it makes small excavations in the soil and turns over sticks and small branches to expose the termites, before using its long, slender tongue to extract the insects from their galleries. This process is not selective, termite species being taken in proportion to their relative abundance in the area. Some ants are also taken, but there is no evidence that they are more than incidental to the diet.

C. ANDREW HENLEY/LARUS

*Numbat on a termite nest. Nests are seldom excavated, most termites being taken from shallow runways leading from the nest.*

Daily activity changes during the year, corresponding to the changing pattern of activity of termites in the upper soil layers. There are peaks of activity in the middle of the day in winter but in the morning and late afternoon in summer activity reduces, with a 'siesta' in the heat of the day. The night is spent in the nest.

D. WHITFORD

*Young Numbats in a nest.*

Females come into oestrus in January. Mating occurs up to 48 hours after the onset of oestrus and the young are born 14 days later. Most females carry young on all four teats until late July, when they are deposited in one of her burrows. The female continues to suckle them in the nursery burrow at night, but leaves the nest and forages alone during the day. In early September, the young begin to emerge each day with their mother but, after she departs for the day, remain basking and playing very close to the burrow mouth for an hour or two before returning inside. Over the next six weeks, their movements extend further from the nursery burrow (or another nest in a burrow or hollow log to which they were subsequently moved) and they are weaned by late October. By the end of November, most young spend the night apart from their mother and, by mid-December, they have dispersed from their maternal home ranges and established their own home ranges, in which usually they remain for life. Dispersal movements of more than 15 kilometres have been recorded. Females breed in their first year, but males are not sexually mature until their second year.

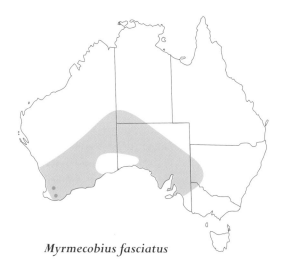

*Myrmecobius fasciatus*

D. WHITFORD

*Numbat basking in morning sun.*

Numbats are essentially solitary. Individuals of the same sex occupy exclusive home ranges but males and females overlap to some degree. Home range size varies from 25–50 hectares, but males roam in search of females over a much wider area prior to and during the breeding season. Males have been observed fighting during the breeding season.

Recent studies at Dryandra have shown that natural predators of the Numbat include the Carpet Python, Little Eagle, Brown Goshawk, Collared Sparrow-hawk and, occasionally, the Wedge-tailed Eagle. However, the forty-fold increase in the Numbat sighting rate at Dryandra that followed a program of poisoning of the Fox shows population numbers are regulated by this introduced predator. The contraction of range occurred largely during the first half of this century, but a population crash in the late 1970s brought numbers well below 1000. An intensive program of research and management by the Western Australian Government since 1980 has resulted in a much greater understanding of the Numbat's biology and a dramatic increase in its numbers, particularly in its stronghold at Dryandra. A population has been re-established at Boyagin Nature Reserve, where the Numbat became extinct in the 1970s. The recovery of the species is being promoted by a program of reintroduction to other areas where routine Fox control has been implemented.

J.A. FRIEND

## *Myrmecobius fasciatus*

**Size**

*Head and Body Length*
200–274 (239) mm (males)
200–272 (240) mm (females)

*Tail Length*
164–210 (170) mm (males)
161–195 (177) mm (females

*Weight*
300–715 (484) g (males)
320–678 (459) g (females)

**Identification** Red-brown above, paler below. Rump darker than upper back, often jet-black, with prominent, white, transverse bars. Head narrow, snout sharp, dark horizontal eye-stripe. Tail hairs long, often erected to give 'bottle-brush' appearance.

**Recent Synonyms** None.

**Other Common Names** Banded Anteater.

Indigenous names: Walpurti, Mutjurarranypa, Parrtjilaranypa, Waihoo (Weeoo).

**Status** Rare, occurring in isolated populations.

**Subspecies** *Myrmecobius fasciatus fasciatus*, south-western Western Australia. *Myrmecobius fasciatus rufus*, western New South Wales, South Australia, Gibson Desert, Rawlinson and Warburton Ranges, Western Australia.

**References**

Calaby, J.H. (1960). Observations on the Banded Anteater *Myrmecobius f. fasciatus* Waterhouse (Marsupialia), with particular reference to its food habits. *Proc. Zool. Soc. Lond.* 135: 183–207.

Friend J.A. (1990). Myrmecobiidae (in) D.W. Walton and B.J. Richardson (eds) *Fauna of Australia. Mammalia, Vol. 1B*. Aust. Gov. Publ. Serv., Canberra. pp. 583–590.

# FAMILY THYLACINIDAE
## *Thylacine*

The affinities of the Thylacine, largest of the predatory marsupials existing at the time of European settlement, have been subject to much debate. First described in 1808, at a time when little was known about the Australian marsupials, it was classified initially with the American opossums. Later, in recognition of the many characters shared with the dasyurids, it was included in the Dasyuridae and subsequently, because of the extent of its differences from this group, in an allied family of its own, the Thylacinidae.

Towards the middle of the twentieth century, its affinities again came into question. It had long been recognised as resembling a member of the wolf family but the similarities were clearly superficial and attributable to convergent evolution. However, it also bore a remarkable resemblance to the South American fossil marsupial *Borhyaena* and strong arguments were advanced for regarding the Thylacine as the sole Australian member of the Borhyaenidae. If this hypothesis had been correct, it would have had far-reaching implications regarding the origin of the Australian marsupials, but subsequent research failed to provide evidence in favour of a South American relationship and has reinforced the location of the Thylacinidae in the Dasyuroidea. One of the most interesting researches involved the extraction of albumin from a piece of untanned skin of a Thylacine and a serological comparison of this with albumins from diverse living marsupials. Results indicated only a very distant relationship of the Thylacine to Southern American marsupials and a very close affinity with dasyurids. A significant implication of these findings is that adaptation to a running, predatory way of life has led *three* animals groups—Thylacinidae, Borhyaenidae and Canidae—to a similar functional solution.

It should be noted, however, that although the head of the Thylacine is similar to that of a wolf, the proportions of its body and limbs resemble those of dasyurids more than those of canids. No revolutionary changes were involved in the course of the Thylacine's evolution from a dasyurid-like ancestor: it is merely—like *Borhyaena*—a product of the evolutionary steps necessary to become a running (pursuit) carnivore.

# Thylacine

## *Thylacinus cynocephalus*

### (Harris, 1808)

thile'-ah-seen'-us  sine'-oh-sef'-ah-lus:
'dog-headed pouched-dog'

The last known wild Thylacine was captured in 1933 and died in 1936. Since then, despite many supposed sightings, there has been no firm evidence of its survival. Many surveys have been made, the best being intensive, localised searches in areas of repeated alleged sightings. Procedures successfully used to record and study other animals occupying similar niches have been used extensively, including long-term surveillance with automatically triggered movie cameras. It is interesting that reports of apparently equal quality to the best from Tasmania are received from time to time from a number of places on the mainland— equally without substantiation. Some reports have been shown to be hoaxes but most are the outcome of honest error, often fuelled by expectation: reports tend to come in waves, one publicised report sparking a number of others.

Often referred to as the Tasmanian Tiger in reference to the stripes on its back and rump, the

*Thylacinus cynocephalus*

---

*Thylacinus cynocephalus*

**Size**
*Head and Body Length*
100–130 cm
*Tail Length*
50–65 cm
*Weight*
15–35 kg

**Identification** Sandy brown, coarse fur, parallel dark brown bands across back, increasing in width towards rump. Dog-like head and forequarters; rump tapering into semi-rigid tail. Eye-shine said to be bright pale yellow. Skull lacks enlarged lower premolar (present in Dog).
**Recent Synonyms** None.
**Other Common Names** Tasmanian Tiger, Tasmanian Wolf.
**Status** Extinct.
**Subspecies** None.
**Reference**
Rounsevell, D.E. and S.J. Smith (1982). Recent alleged sightings of the Thylacine (Marsupialia: Thylacinidae) in Tasmania (in) M. Archer (ed.) *Carnivorous Marsupials*. Roy. Zool. Soc. of NSW and Surrey Beatty and Sons, Sydney. pp. 233–236.
Smith, S.J. (1980). The Tasmanian Tiger— 1980. *Wildlife Division Technical Report* 81/1. N.P.W.S., Tasmania.
Guiler, E.R. (1986). *Thylacine—The Tragedy of the Tasmanian Tiger*. Oxford Univ. Press, Melbourne.

Thylacine has been more aptly described as a marsupial 'wolf', for its teeth, head and forequarters have a remarkably canine appearance, although the hindlegs and broad-based tail betray its marsupial nature. It had a rather stiff gait and could not run very fast. Locating its prey by scent and hunting at night, singly or in pairs, the Thylacine tired its prey by dogged pursuit until it was able to kill it with its powerful, wide-gaping jaws. Since kangaroos and wallabies were its main source of food, it was most abundant in open forest and woodland, retiring into lairs for shelter during the day but occasionally emerging during cold weather to bask in the sun.

Adult males are larger than females of similar age and in both sexes the head widens, becoming less wolf-like, with age. Old records indicate that breeding occurred mainly in winter or spring and that two to three young were reared, although the backwardly opening pouch had four teats.

After leaving the pouch, young were left in a nest until weaned, thereafter accompanying the mother until well-grown and able to hunt independently. Folklore and the behaviour of captive animals suggest that the sense of smell was well developed. Captive animals were normally mute, emitting a coughing bark only when anxious or disturbed. Wild animals are said to have communicated by terrier-like yapping while in pursuit of prey.

The Thylacine was once widespread on the Australian mainland (and in Papua New Guinea) but it appears to have declined in competition with the Dingo and became extinct in these regions not less than 2000 years ago (although an intriguing age of 0–80 years has been estimated for a Thylacine humerus from north-western Western Australia).

Tasmania became isolated from the mainland 10 000–8000 years ago, before the arrival of the Dingo in Australia and the Thylacine thrived there in the open forest and woodlands until Europeans introduced sheep. Thylacines killed sheep, bringing them into direct competition with Europeans. The Van Diemen's Land Company offered bounties for scalps in 1830, private landowners made private arrangements with 'doggers,' and the Tasmanian government became an official predator in 1888, making payments on more than 2000 scalps over the subsequent 20 years. The actual number of animals killed is uncertain, for cunning 'doggers' received multiple payment by submitting scalps for both private and government bounties and fur-trappers poisoned thylacines to reduce their raiding of snare-lines. Yet, despite

*Thylacines: exhibit in the South Australian Museum.*

this deliberate destruction, it is difficult to account for the rapid decline of the species between 1900 and 1920 only in terms of hunting pressure. It is possible that the same epidemic which, at the turn of century, appears to have decimated the Tasmanian Devil and, on the mainland, the Eastern Quoll, swept the Thylacine into oblivion. It is ironic that the species was not protected by Tasmanian law until 1936, by which time it was probably extinct.

Being at the apex of a food-pyramid, the Thylacine was probably always sparsely distributed: it may be significant that Europeans had been in Tasmania for two years before the first specimen was shot. Considering the size of the home ranges of comparable eutherian carnivores, it has been calculated that the Tasmanian population may never have exceeded 2000 individuals. The species was thus very vulnerable to alienation and fragmentation of its habitat and to persecution.

D.E. Rounsevell and N. Mooney

# ORDER PERAMELEMORPHIA

## *Bandicoots and Bilbies*

*Bandicoot embryos are attached to a placenta. When the young are born and move to teats, they remain connected for a while by tightly stretched umbilical cords. This illustration is of a litter of Northern Bandicoots.*

This group contains only one superfamily, the Perameloidea, which includes all the living bandicoots and the two known species of bilbies. Nomination to ordinal level serves to indicate the level of evolutionary separation of the bandicoot group from the Dasyuromorphia and Diprotodontia.

166

# Superfamily Perameloidea

The perameloids are not much more varied among themselves than the dasyuroids. They are strictly terrestrial and all have rather long, pointed heads and compact bodies. In typical bandicoots the forelimbs are short and the hindlimbs large, but in the Pig-footed Bandicoot, now extinct, all limbs were quite long and slender and the animal apparently ran on its toes. The legs of bilbies are longer than in typical bandicoots.

When foraging, a bandicoot digs conical holes with its forelimbs and explores the excavations with its long snout. A bilby employs its forelimbs also to dig substantial burrows, kicking out the soil with its hindlimbs. In adaptation to its digging function, the forefoot is elongate and bears strong curved claws on the second, third and fourth toes; the first and fifth toes, when present, are rudimentary and clawless. The process of reduction of the toes of the forefoot is taken further in the Pig-footed Bandicoot, in which only the second and third toes are well developed.

The resemblance of the hindlimbs to that of a macropod (more particularly, a potoroid) is marked. The thigh is powerful, the foot is elongate and its axis is continued into a very large, strongly clawed fourth toe. The first toe is reduced or absent. As in macropods and all other diprotodonts, the second and third toes are joined together in the conditions known as syndactyly. The hindlimbs can be used in leaping but the usual fast gait is a gallop. The slow gait is a 'bunny hop' in which the weight of the body is taken by the forelimbs while the hindlimbs are brought forwards together. Intermediate between the slow and fast gait is an ungainly quadrupedal run in which the hindlegs move alternately. This lack of grace is particularly noticeable in bilbies, which are more quadrupedal than typical bandicoots. The tail, which has little or no function in locomotion, is short in typical bandicoots but long and tufted in bilbies, where it may serve as a device for social communication.

The long jaws accommodate four to five pairs of upper incisors and three lower pairs: these are flattened at the tips, not pointed as in typical dasyuroids. A pair of well-developed canines is followed by three pairs of upper and lower premolars and four pairs of sharp-crowned upper and lower molars. This dentition is eminently suited to an insectivorous habit but is also appropriate to feeding on small vertebrates and succulent parts of plants. In fact, insects and other arthropods make up the greater part of the food of typical bandicoots but the diet may be supplemented, as opportunities arise, by infant rodents, fruit and soft tubers. Bilbies are reported to be the most carnivorous of the perameloids and to hunt small mammals and lizards. The Pig-footed Bandicoot appears to have included grass in its diet.

It is widely stated that one of the most significant features of marsupials is that their embryos do not develop with the help of a placenta and that the extent to which they can be nourished in the maternal uterus is therefore much less than in eutherian mammals. This is not strictly true, for the embryos of bandicoots (and the Koala) make a connection with the uterine wall that is structurally very similar to that of a placental mammal. However, the bandicoot placenta does not develop into a large structure for the exchange of materials between the blood of the mother and the embryo and the period of gestation is remarkably brief, even for a marsupial. One of the shortest gestations recorded in mammals—12.5 days from conception to birth—occurs in the Northern Brown Bandicoot and Long-nosed Bandicoot.

Most perameloids have eight teats but seldom carry more than four young in the pouch. This apparent under-occupancy is due to the increase in size of a teat when it is in use; one that has recently been vacated is too large for the attachment of a newborn animal. Thus, two successive litters cannot comprise a total of more than eight—an average of four per litter.

The perameloids occur in Australia and New Guinea. As mentioned above, they are polyprotodont (like the dasyuroids) but the hindfoot is syndactylous (as in diprotodonts). They thus appear to be, in some sense, intermediate between dasyuroids and diprotodonts but the question of their relationship to the other Australian marsupials has been bedevilled by opposing opinions on the relative significance of dentition and foot structure. One view is that the perameloids evolved from dasyuroids, retaining the primitive polyprotodont dentition and developing a syndactyl hindfoot quite independently of the diprotodonts. Another regards the independent evolution of such an unusual condition as syndactyly to be so unlikely that the perameloids must have evolved from the diprotodonts (polyprotodonty therefore having evolved twice). A more plausible compromise between these extremes is that an arboreal 'proto-perameloid' stock arose from the early dasyuroids and that the syndactylous condition arose in these animals. One line of evolution from these (syndactylous polyprotodont) proto-peramelids led to the modern insectivorous, terrestrial bandicoots; another to the herbivorous, arboreal possums, which modified the old polyprotodont dentition to the gnawing-grinding diprotodont pattern.

In the absence of linking fossils, all this remains conjectural and there is no satisfactory explanation for the origin of syndactyly. In modern species that have syndactylous hindfeet, the claws of the fused second and third toes are employed as a comb to groom the fur, a fact which led to the suggestion that syndactyly and polyprotodonty are *alternative* conditions. In this view, the many incisors of polyprotodont marsupials provided an adequate comb but, with reduction of their incisors, the diprotodont marsupials had to develop another grooming device. However, since bandicoots have *both* devices, this hypothesis is weak. Indeed, syndactyly is almost certainly an outcome of specialisation of the hindfoot for climbing.

In the first edition of this work (1983), the perameloids were divided into two families: the Thylacomyidae for the bilbies; the Peramelidae for the remainder. Recent studies have shown that it is more appropriate to erect the family Peroryctidae to include spiny bandicoots of the genus *Peroryctes* and the other New Guinean genera, placing all the Australian genera in the Peramelidae. In this classification, the bilbies fall into the subfamily Thylacomyinae, the other bandicoots into Peramelinae. Future studies may well show that the Pig-footed Bandicoot should be separated from the Peramelinae.

# FAMILY PERAMELIDAE
## *Bandicoots and bilbies*

Three genera of this family, *Isoodon*, *Perameles* and *Macrotis*, now occur in Australia. The monotypic *Chaeropus*, the Pig-footed Bandicoot, became extinct early in the twentieth century.

Species of *Perameles* and *Isoodon* are similar in appearance but the former are more slenderly built, have a longer

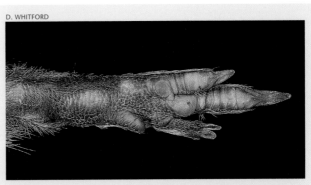

D. WHITFORD

*Hindfoot of the Northern Bandicoot (a typical bandicoot) showing the very small first toe, the conjoined second and third toes, the very large fourth toe and somewhat shorter fifth toe.*

head and, in some species, a pattern of transverse bars across the rump. One species of *Macrotis* is recently extinct, the other is endangered. Both are characterised by large, rabbit-like ears, silky fur and a long, well-furred tail with a terminal tuft. The day is spent in a deep burrow. Little is known of the Pig-footed Bandicoot but it appears to have been unusual in being quadrupedal and in including at least some grass in its diet.

In the first edition of this work, the perameloids were divided into the Peramelidae (all bandicoots) and the Thylacomyidae (bilbies). To retain a distinction between the two groups, they are treated here as subfamilies Peramelinae and Thylacomyinae.

*Southern Brown Bandicoot.*

## Subfamily Peramelinae

This group includes all the non-spiny bandicoots and (with strong reservations) the Pig-footed Bandicoot. It may well be that *Chaeropus* deserves subfamilial or even familial status.

# Pig-footed Bandicoot

## *Chaeropus ecaudatus*

### (Ogilby, 1838)

keer'-oh-poos ay'-kawd-ah'-tus:
'tail-less pig-foot'

The generic and common names of this singularly delicate bandicoot refer to the forefoot, which has two functional toes with hoof-like claws and thereby bears some resemblance to the cloven trotter of a pig or, more aptly, a diminutive deer. Only the fourth toe of the hindfoot was employed in locomotion: the tiny fused second and third toes may have been used, as in other bandicoots, for grooming. The fore- and hindlegs were long and slender, adapted to speedy locomotion.

The first specimen was collected by Major Thomas Mitchell in 1836 in Victoria near the junction of the Murray and Murrumbidgee rivers. It had lost its tail through some accident of life but the describer assumed that tailessness was normal and named it '*ecaudatus*'. In fact, the tail is one of the longest of any bandicoot.

Gerard Krefft, one of the few scientists to observe living Pig-footed Bandicoots, remarked in 1866 that the species moved '...like a broken-down hack in a canter, apparently dragging the hind quarters after it'. This description contrasts with those of Aborigines of central Australia who recall that it rested by day in grassy nests from which it exploded and departed with great speed when disturbed. When chased by dogs it sometimes ran into a hollow log.

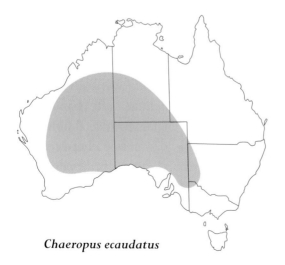

*Chaeropus ecaudatus*

---

### *Chaeropus ecaudatus*

**Size**
*Head and Body Length*
230–260 mm
*Tail Length*
100–150 mm
*Weight*
200 g (estimated)

**Identification** Long slender limbs; two functional toes on front feet. Orange-brown fur above, light fawn below. Long tail with terminal crest on last 70 mm. Long and conspicuous ears.
**Recent Synonyms** None.
**Other Common Names** Indigenous names: Bertie (Western Australia), Yirratji (Warlpiri), Kanytjilpa (Kartutjarra, Manytjilytjarra, Ngaanyatjarra, Ngaatjatjarra, Pintupi,

Krefft secured eight specimens during a six-month stay near Mitchell's site in 1857. Subsequently, only a few additional specimens were collected from South Australia, Western Australia and the Northern Territory until the last reliably dated museum specimen was taken in 1901. Aboriginal testimony indicates that it disap- peared from South Australia between 1910 and 1920 but the Pintupi people in the central deserts of Western Australia recall it surviving there until as recently as the 1950s.

The fur is coarse but not spiny and almost orange-brown above, merging to light fawn beneath. The tail is long and bears a terminal crest for the final 70 millimetres of its length. The pouch opens to the rear and although it con- tains eight teats, no more than two young have been recorded. Breeding in May and June has been assumed but this is based on very little information.

In the central deserts, the Pig-footed Bandicoot occupied sand-dunes and sand plains with hummock grassland and tussock grass,

*Pig-footed Bandicoot. Illustration in John Gould's* The Mammals of Australia.

sometimes with a mulga overstorey. In Victoria, it occurred on grassy plains. Elsewhere, it appears to have favoured open woodland with a shrub and grass understorey. It built a grass-lined nest and some Aborigines say that it dug a short, straight burrow with a nest at the end. It appears to have been primarily nocturnal.

Some very early explorers report seeing appreciable numbers of Pig-footed Bandicoots but collapse of the species appears to have been rapid after European settlement. It is now presumed to be extinct. Krefft remarked in 1866 that it was disappearing fast and likely to be dispersed by growing flocks of sheep and cattle. It had gone, or greatly declined, in much of south-western Western Australia by 1900, before the arrival of the Rabbit and Fox. The feral Cat had nevertheless become widely established by this time.

Aborigines report that it ate termites and ants and the explorer Charles Sturt commented that it was partial to flesh. In captivity it ate grass, lettuce, bulbous roots and grasshoppers: gut analyses of museum specimens have revealed vegetable matter. More than in any other bandi- coot, the tooth and gut structures indicate a herbivorous diet, even a degree of grazing.

Pitjantatjara, Wangkatjungka), Takanpa (Kukatja, Pintupi, Warlpiri), Kalatawirri (Walmatjari), Kalatawurru (Kukatja), Marakutju (Pintupi, Wangkatjungka), Parrtiriya (Kukatja), Watda (York area, Western Australia).

**Status** Extinct.

**Subspecies** None.

**References**

Jones, F.W. (1924). *The Mammals of South Australia, Part 2*. Government Printer, Adelaide, pp. 168–171.

Burbidge, A.A., K.A. Johnson, P.J. Fuller and R.I. Southgate (1988). Aboriginal knowledge of the mammals of the central deserts of Australia. *Aust. Wildl. Res.* 15: 9–39.

K.A. JOHNSON AND A.A. BURBIDGE

# Golden Bandicoot

## *Isoodon auratus*

## (Ramsay, 1887)

ie-soh'-oh-don or-ah'-tus: 'golden equal-tooth'

J. LOCHMAN

*Now rare on the mainland, one population of the Golden Bandicoot thrives on Barrow Island—where there are no Cats, Dogs or Foxes.*

First described in 1887 from a specimen collected near Derby, Western Australia, this species was named for the golden-brown colour of its back and sides. Stiff, almost quill-like, guard hairs lie flush over the head and body, completely hiding the softer, greyish underfur and giving the surface a sleek, crisp texture.

Golden Bandicoots have been recorded in a wide range of habitats: sand-dune and sand plain country with spinifex formations in the arid zone; sand plains with *Acacia* and *Eucalyptus* woodlands over tussock grasses in the tropical semi-arid zone; rugged sandstone–spinifex country and volcanic country (rainforest patches, low woodlands over tussock grasses) of the tropical, subhumid north-western Kimberley.

It is not a well-known animal; most knowledge comes from field observations of the abundant population on Barrow Island, Western

Australia. These suggest that it is nocturnal, with peaks of activity in the three to four hours after dusk and again before dawn, when it forages in open habitat between clumps of spinifex. It covers areas of up to 10 hectares each night. Food is taken in much the same way as other short-nosed bandicoots, by making small conical diggings. An 1895 report from near Broome, Western Australia, refers to the building of simple grass nests on the ground. On Barrow Island, it seeks shelter in limestone caves and large spinifex hummock grasses. It is most common in coastal areas of the island where it attains densities of ten adults per hectare. The diet of animals from north-western Kimberley includes termites, centipedes, insect larvae and plant material. Animals on Barrow Island are known to eat ants, termites, moths, turtle eggs, small reptiles and the Common Rock-rat, in addition to roots and tubers. It appears to be solitary and individuals are aggressive towards each other.

Females on Barrow Island give birth throughout the year. However, the proportion of females with pouch-young increases after substantial rains. Similarly, in the north-western Kimberley females with pouch-young have been found at the height of both the wet season (December–January) and of the dry season (August). Females have eight teats and it was

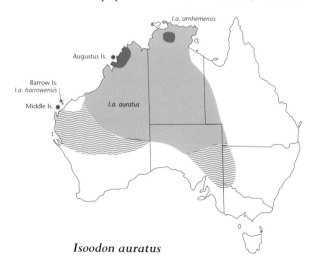

*I.a. arnhemensis*

Augustus Is.

Barrow Is.
*I.a. barrowensis*

Middle Is.

*I.a. auratus*

**Isoodon auratus**

reported from Broome in the nineteenth century that they gave birth to three young of which only one survived to weaning. On Barrow Island, two young are normally born and both may survive.

The species was widely distributed in arid central Australia as recently as the 1930s and extended further eastward prior to that time. Subfossil and historical remains suggest that its Late Pleistocene range included western New South Wales and the entire arid zone of Western Australia. It is allopatric with the Southern Brown Bandicoot, even if the Late Pleistocene subfossil records are included. It was extinct in New South Wales prior to 1870. Although numerous in the better watered coastal country near Broome in 1895, soon after the advent of a pastoral industry, it has not been recorded since 1898; skeletal remains from Aboriginal middens indicate a corresponding pattern of disappearance in the eastern Kimberley. By the 1950s it had virtually disappeared from central Australia, the only subsequent record being from the Granites, in the arid Northern Territory, in 1952. Similarly, the species seems not to have survived the period of European settlement in areas of medium to high rainfall in the Northern Territory where it was last reported in 1967 from the Goodparla Pastoral Lease that is now included in Kakadu Stage III.

The species still survives in Prince Regent Nature Reserve and on the nearby Augustus Island in the north-western Kimberley Region. Secure populations persist on Barrow and Middle Island off the Pilbara coast. It is the most ubiquitous mammal on Barrow Island where it is estimated that 60 000–80 000 occur. This island, although an important oilfield, is subject to strict environmental protection procedures; clearing and other vegetation damage are minimised and no exotic animals are permitted. The extinction of the Golden Bandicoot from Hermite Island (near Barrow Island) just prior to 1912 has been attributed to the introduced feral Cat.

## *Isoodon auratus*

**Size**

*Head and Body Length*
190–295 (245) mm
*Tail Length*
84–121 (105) mm
*Weight*
300–670 g Kimberley
250–600 g Barrow Island

**Identification** Distinguished from other bandicoots by golden-brown colour, pencilled with black. Barrow Island and north-western Kimberley forms are darker, approaching colour of *Isoodon obesulus*. White below.

**Recent Synonyms** *Isoodon barrowensis* is now included within *I. auratus* and it is possible that *I. auratus* is a form of *I. obesulus*.

**Other Common Names** Northern Golden Bandicoot, Northern Golden-backed Bandicoot. Indigenous names: Windaru, Wintaroo, Nyulu, Pakaru.

**Status** Barrow Island: common, limited, accessible for study.
North-western Kimberley: sparse, scattered, possibly underestimated.
South-western and eastern Kimberley, subhumid Northern Territory: uncertain.
Central Australia: probably extinct.

**Subspecies** *Isoodon auratus auratus*, mainland. *Isoodon auratus barrowensis*, Barrow Island, WA. *Isoodon auratus arnhemensis*, Arnhemland.

**References**

Dahl, K. (1897). Biological notes on North-Australian Mammalia. *Zoologist* Ser. 4, 1: 189–216.

Ellis, M., P. Wilson and S. Hamilton (1991). The Golden Bandicoot, *Isoodon auratus* Ramsay 1887, in western New South Wales during European times. *Aust. Zool.* 27: 36–37.

N.L. McKenzie, K.D. Morris
and C.R. Dickman

# Northern Brown Bandicoot

## *Isoodon macrourus*

(Gould, 1842)

mak'-roh-yue'-rus: 'long-tailed equal-tooth'

This is the common bandicoot of some suburban gardens on the east coast of Australia north of the Hawkesbury River. It extends, in higher rainfall zones, to the tip of Cape York in Queensland and across the Northern Territory to the north of Western Australia. It prefers

HANS & JUDY BESTE

*Northern Brown Bandicoot posed on a log.*

### *Isoodon macrourus*

**Size**

*Head and Body Length*
300–470 (400) mm (males)
300–410 (350) mm (females)
*Tail Length*
90–215 (170) mm (males)
80–185 (130) mm (females)
*Weight*
500–3100 (2100) g (males)
500–1700 (1100) g (females)

**Identification** Speckled brown-black above; whitish below. Similar in appearance to *Isoodon obesulus* but larger (largest member of genus). Skull distinguished from most *I. obesulus* by having no accessory palatal opening.
**Recent Synonyms** *Isoodon torosus, Thylacis macrourus, Thylacis torosus.*
**Other Common Names** Brindled Bandicoot, Giant Brindled Bandicoot, Long-tailed Short-nosed Bandicoot, Large Northern Bandicoot.
**Status** Common to abundant.

areas of low ground cover including tall grass and dense shrubbery, irrespective of the presence or absence of a tree canopy or tree spacing, but there seems to be a tolerance for sparser ground cover in the northern part of its range. A seasonal shift in habitat utilisation has been found in the Northern Territory. Habitats include grassland, woodland, open forest and, in a few districts, closed forest. It penetrates into the lower rainfall areas of inland Queensland along riverine fringing forests, to about the 625 millimetre isohyet, faring less well in grazed areas.

The day is spent in a well concealed nest consisting of a heap of ground litter over a shallow depression, providing an internal chamber with loose regions at each end for entry and exit. A layer of soil may be scraped over the top of the nest in rainy weather for waterproofing. Animals may also rest up in hollow logs, or under tussocks.

At night it moves over a home range of 1–6 hectares in search of food (supplemented, in males, by a more rapid and directed patrolling of the range). Although it prefers insects and eats other invertebrates such as spiders and earthworms, the Northern Brown Bandicoot is omnivorous, including berries, grass seeds and plant fibre such as sugar cane in its diet. Food is obtained mainly on the surface of the ground or by digging in the soil with the strong forepaws.

Males are larger than females, have larger canine teeth, are more aggressive and have larger home ranges. A gland behind the ear is used for marking the ground and vegetation during aggressive encounters. Males and females appear to come together only for mating, which is preceded by persistent following of the female until she accepts mounting.

The breeding season varies geographically. In central New South Wales breeding occurs mainly from August to March; in south-eastern Queensland it may occur throughout the year;

*Isoodon macrourus*

**Subspecies** The subspecific taxonomy is in need of revision but the following taxa are currently recognised.

*Isoodon macrourus macrourus*, Northern Territory and northern Western Australia.

*Isoodon macrourus torosus*, Cape York, Queensland, to Hawkesbury River, New South Wales.

*Isoodon macrourus moresbyensis*, New Guinea.

**Extralimital Distribution** Papua New Guinea.

**References**

Friend, G.R. (1990). Breeding and population dynamics of *Isoodon macrourus* (Marsupialia: Peramelidae): studies from the wet-dry tropics of Australia (in) J.H. Seebeck, P.R. Brown, R.L. Wallis and C.M. Kemper (eds) *Bandicoots and Bilbies*. Surrey Beatty and Sons, Sydney. pp. 357–365.

Gordon, G. (1974). Movements and activity of the Shortnosed Bandicoot *Isoodon macrourus* Gould (Marsupialia). *Mammalia* 38: 405–31.

and in the Northern Territory between August and April. There are eight teats and the litter size is one to seven, usually two to four. The gestation period is 12.5 days, as in the Long-nosed Bandicoot. The young leave the pouch permanently by about 55 days of age and are weaned by about 60 days. The rapid development of young is facilitated by the production of more concentrated milk (reaching 40–45 per cent total solids) and a greater supply of milk than in other marsupials. Some females may mate and become pregnant again while still carrying pouch-young, the new litter being born soon after weaning the previous one. Since females may become sexually mature at three to four months of age and can produce several litters during the breeding season, a high reproductive rate is possible.

European settlement has caused the withdrawal of the Northern Brown Bandicoot from much of its inland range in pastoral country and patchy extinction has occurred in coastal areas subject to intensive development, farming or grazing. Otherwise survival has been extremely good, especially in outer urban areas where it is common to abundant.

G. GORDON

# Southern Brown Bandicoot

## *Isoodon obesulus*

(Shaw, 1797)

oh-bes'-yue-lus: 'rather fat equal-tooth'

The Southern Brown Bandicoot prefers sandy soil with scrubby vegetation and/or areas with low ground cover that are burnt out from time to time. During the early stages of regeneration after fire, the diversity of growing vegetation supports abundant insect food and is a very favourable habitat. Later, as the vegetation approaches maturity, the food supply is reduced. For a particular area to support a more or less stable population, parts of it must be burned fairly regularly, creating a changing mosaic of suitable habitats.

### *Isoodon obesulus*

**Size**

*Head and Body Length*
300–360 (330) mm (males)
280–330 (300) mm (females)

*Tail Length*
90–140 (120) mm (males)
90–140 (110) mm (females)

*Weight*
500–1600 (850) g (males)
400–1100 (700) g (females)

**Identification**  Dark greyish or yellowish brown above (rather coarse to touch); creamy white below. Tail and upper surface of hindfeet usually dark brown. Adult scrotum usually darkly pigmented. Ears rounded. Large auditory bullae.

**Recent Synonyms**  None.

**Other Common Names**  Short-nosed Bandicoot, Southern Short-nosed Bandicoot, Brown Bandicoot.
Indigenous names: Quenda, Bung (Woiwurrung).

**Status**  Common in some parts of range.

**Subspecies**  *Isoodon obesulus obesulus*, New South Wales, Victoria, Southern Australia.
*Isoodon obesulus peninsulae*, Cape York.

**References**

Broughton, S.K. and C.R. Dickman (1991). The effect of supplementary food and home range of the Southern Brown Bandicoot, *Isoodon obesulus* (Marsupialia: Peramelidae). *Aust. J. Ecol.* 16: 71–78.

Lobert, B. and A.K. Lee (1990). Reproduction and life history of *Isoodon obesulus* in Victorian heathland (in) J.H. Seebeck, P.R. Brown, R.L. Wallis and C.M. Kemper (eds) *Bandicoots and Bilbies*. Surrey Beatty and Sons, Sydney. pp. 311–318.

It is nocturnal and prefers to stay close to cover when in search of food on the surface of the ground and in the shallow, conical holes that it digs with its powerful foreclaws. It feeds on earthworms and other invertebrates but mainly insects, both adult and larval. It also eats fungi and other subterranean plant material. During the day, it sleeps in a nest which it constructs on the ground by collecting grass and other plant material, sometimes mixed with earth. Some nests are extremely well concealed among litter and debris or among dense vegetation.

Reproduction is very opportunistic and, with rapid development of young, the species is capable of a high reproductive rate under favourable conditions. Breeding begins in winter and usually lasts six to eight months. There are eight teats in a rear-opening pouch, accommodating one to six (usually two to four) young in a litter. Two or three litters may be reared in a season, weaning of one litter being soon followed by the birth of another. The death rate of juveniles is usually high.

J. LOCHMAN

The prospects of young animals, which are weaned when about 60–70 days old, are determined by where they begin their independent lives. If a juvenile is fortunate to discover and claim a small area of newly regenerating vegetation, it will probably survive to an age of about three years. Another may have the luck to be weaned just at the time when a piece of habitat becomes vacant due to death, or movement

*Southern Brown Bandicoot.*

away, of an old animal. The few young females to establish themselves in suitable habitats mature quickly and produce a succession of litters.

Individual survival depends upon possession and defence of an adequate home range (a big adult may use an area as much as 7 hectares) but in times and places of good food supply, home ranges may overlap substantially. Individuals appear to be solitary in the wild. Captive animals are likely to attack each other if put in the same enclosure.

Before European settlement, the use of fire by Aborigines maintained a complex mosaic of habitats very suitable for the Brown Bandicoot. European settlement has led to clearance, the loss of dense vegetation due to the spread of sheep and cattle and a reduction in the frequency of small-scale fires, all detrimental to the species, which now has a patchy distribution over a reduced range.

R. W. BRAITHWAITE

**Isoodon obesulus**

BABS & BERT WELLS

# Western Barred Bandicoot

### *Perameles bougainville*

Quoy and Gaimard, 1834

pe'-rah-mel'-ayz  bue'-gan-veel:
'Bougainville's pouched-badger'

The Western Barred Bandicoot now exists only in the nature reserve constituted by Bernier and Dorre Islands in Shark Bay, Western Australia. It is probable that, prior to European settlement, it extended over much of the southern half of Australia but the situation is confused by the variety of names given by early taxonomists to local populations, based largely on pelage

*The Western Barred Bandicoot appears to be extinct on the mainland but is abundant on Bernier and Dorre Islands, in the absence of Cats, Dogs and Foxes.*

colour: *Perameles bougainville* from Shark Bay, *P. myosuros myosuros* from King George Sound, Western Australia, *P. myosuros notina* from St Vincent Gulf, South Australia and *P. fasciata* from the Liverpool Plains, New South Wales. Since these mainland forms are now extinct and few museum specimens exist, it is unlikely that the relationships between them will ever be resolved. For the purpose of this account, it is considered that all are forms of the Western Barred Bandicoot. The exact distributions and relationships of the Western Barred Bandicoot and the Desert Bandicoot remain unresolved.

An inhabitant of semi-arid areas, the Western Barred Bandicoot once lived in a variety of vegetation types. In south-western

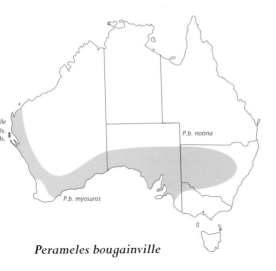

*Perameles bougainville*

P.b. notina

P.b. myosuros

Western Australia it was found in dense scrub and particularly favoured thickets of *Allo-casuarina* seedlings. In south-eastern Western Australia and the southern half of South Australia it lived in open saltbush and bluebush plains. It was found on stony ridges bordering scrub near the Murray River in South Australia and occurred along much of Murray and Darling River system.

On Bernier and Dorre Islands, it is especially common in the scrub associated with stabilised dunes behind the beaches, but it also occurs in open steppe associations. During the day it occupies a nest, made from local plant material, in a hemispherical hollow scrape beneath a low or prostrate shrub. Near the coast, where seagrass litter has accumulated, nests are filled with this material. Nests are very well concealed and the presence of an occupant is betrayed only by a slight disruption of the litter surface where the animal entered. The bandicoot enters its nest by pushing through the litter from one end, where the lip of the scrape is lower, forming a runway. Exit is through the same point, unless the animal is disturbed, when it emerges rapidly from the nest, disrupting it. Over a period of a week, most animals use the same nest repeatedly, but some use a different nest each night. The only nest-sharing observed on Dorre Island has been between females and their young.

## *Perameles bougainville*

**Size**
**Bernier and Dorre Islands**
*Head and Body Length*
171–236 (203) mm
*Tail Length*
60–102 (87) mm
*Weight*
172–286 (226) g
**Mainland (Nullabor)**
*Head and Body Length*
205–280 (236) mm
*Tail Length*
75–98 (84) mm

**Identification** Light grey to brownish-grey above, white below. Feet white. Two or three alternating paler and darker bars across the hindquarters, pronounced in some populations but muted in others (e.g. Bernier and Dorre Islands). Large, erect ears. Distinguished from *P. eremiana* by colour and by shorter tail.
**Recent Synonyms** *Perameles fasciata*, *Perameles myosura*.
**Other Common Names** Barred Bandicoot, Eastern Barred Bandicoot, Shark Bay Striped Bandicoot, West Australian Striped Bandicoot, Zebra Rat (Nullarbor), South Australian Striped Bandicoot, New South Wales Striped Bandicoot, Marl, Little Marl.
Indigenous name: Nyemmel (Albany, WA).
**Status** Abundant on Bernier and Dorre Islands, presumed extinct on mainland.
**Subspecies** *Perameles bougainville bougainville*, Shark Bay.
*Perameles bougainville myosuros*, south-western Western Australia.
*Perameles bougainville notina*, South Australia, south-eastern Western Australia.

**References**

Freedman, L. (1967). Skull and tooth variation in the genus *Perameles*. Part I: Anatomical features. *Rec. Aust. Mus.* 27: 147–66.

Ride, W.D.L. and C.H. Tyndale-Biscoe (1962). Mammals (in) A.J. Fraser (ed.) The Results of an Expedition to Bernier and Dorre Islands, Shark Bay, Western Australia in July 1959. *Fauna Bulletin No. 2*. Fisheries Dept., Perth.

At dusk, it emerges to forage for insects and other small animals and to feed on seeds, roots and herbs obtained by digging or hunting. Activity continues until dawn in winter, but ceases several hours before dawn in summer.

Home ranges overlap but core areas are generally exclusive. The size of a home range varies with population density. Near White Beach on Dorre Island, female home ranges average 6.2 hectares when population density is low and 1.4 hectares at high density. Male home ranges are about twice the size, averaging 14.2 hectares at low density and 2.5 hectares at high density.

On Bernier and Dorre Islands, young are produced continuously from April until October but there is a breeding hiatus, apparently extending from November until March, through the hottest months. Two young are usually carried in the pouch, but litters of one to three have been recorded. Up to four young have been recorded in *P. b. myosuros* from South Australia, where breeding is reported to occur from May to August.

The only predation recorded has been of a juvenile bandicoot by Gould's Monitor, *Varanus gouldii*, on Dorre Island, but Western Barred Bandicoot remains are common in owl roost deposits in the mainland part of their former range and have been found in a concrete tank on Dorre Island that had been occupied at times by a Barn Owl.

J.A. FRIEND AND A.A. BURBIDGE

# Desert Bandicoot

## *Perameles eremiana*

### Spencer 1897

e'-rem-ee-ah'-nah: 'desert pouched-badger'

Among its adaptations to life in an arid, sandy environment, the Desert Bandicoot has long ears and noticeably hairy feet, the soles of the hindfeet being covered with hairs up to the base of the two largest toes. It was first described in 1897 by Baldwin Spencer from specimens from the Burt Plain, about 25 kilometres north of Alice Springs, Northern Territory and from the type locality, sandhills about 64 kilometres north-east of Charlotte Waters, Northern Territory. Some specimens referred to this species have a strong rufescent tone in their dorsal and lateral coloration, hence the name 'Orange-backed Bandicoot', whereas others have brown and fawn colouration. The taxonomy of the *P. bougainville*, *P. eremiana* group is confused and the Desert Bandicoot may not be a valid species.

Many of the arid zone bandicoots disappeared before very much was known about them. Our knowledge of this species rests mainly on the accounts of the South Australian

*An imaginative reconstruction of the extinct Desert Bandicoot.* K. WYNN-MOYLAN

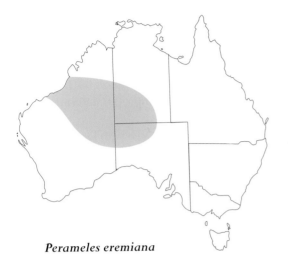

*Perameles eremiana*

zoologist Hedley Finlayson and information gathered recently from Aborigines. The extent of its former range in central Australia is uncertain. Hedley Finlayson mentions unverified reports of it from near the Granites in the Northern Territory and the northernmost specimens of *Perameles* in Western Australia, from Well 35 on the Canning Stock Route, appear to belong to this species. According to Finlayson, it was common in north-western South Australia, the south-west of the Northern Territory and adjacent parts of Western Australia in the 1930s. Its range extended as far north as the Tanami Desert in the Northern Territory. The most recent specimen was collected in 1943 from Well 35 and it is uncertain if any more remain alive.

Its habitat was sand plain and sand ridge desert, vegetated with spinifex grassland and tussock grass flats. It occurred sympatrically with the Golden Bandicoot, *Isoodon auratus*. Like other bandicoots, it was nocturnal and rested during the day inside a nest in a shallow, oval depression scooped out in the ground and lined with grass and twigs. The nest had an entry and exit hole. Aborigines were said to catch it by placing one foot on the nest to pin the animal down and pulling it out by hand. The pouch has eight teats but Baldwin Spencer stated that 'apparently two young [are] produced at a time'. Other reports include three to 'many' young.

## *Perameles eremiana*

**Size**
(2 adults)
*Head and Body Length*
180–285 mm
*Tail Length*
77–135 mm

**Identification**  In some regions a dull orange or rufescent body colour dorsally and laterally, otherwise brown to dark brown and white ventrally. Dark mid-dorsal area extends in one or two dark bands on either side of rump. Tail very dark brown dorsally, white laterally and ventrally.
**Recent Synonyms**  None.
**Other Common Names**  Orange-backed Bandicoot.
Indigenous names: Mulgaruquirra (Alice Springs), Iwurra (Charlotte Waters), Walilya (Warburton region).
**Status**  Presumed extinct.
**Subspecies**  None.
**Reference**
Spencer, W.B. (1897). Description of two new species of marsupials from Central Australia. *Proc. Roy. Soc. Vic.* (new series) 9: 5–11.

Food is said to have included termites and ants, including honey-pot ants and beetle larvae.

The cause of its decline is uncertain. Its disappearance from the south-west of the Northern Territory coincides approximately with the arrival of the Fox in this area. However, it vanished from the Tanami Desert, farther north, well before the appearance of this predator and, at least here, extinction has been attributed to changes in the fire regime following the movement of Aborigines out of the desert areas and the cessation of Aboriginal firing practices. This factor seems to correlate best with its decline, which is thought to have occurred from the 1940s through to the 1960s, taking place earlier in the more southern areas.

G. GORDON

D. WATTS

# Eastern Barred Bandicoot

## *Perameles gunnii*

### Gray, 1838

gun'-ee-ee: 'Gunn's pouched-badger'

With three to four pale bars on its hindquarters, the Eastern Barred Bandicoot is one of the most conspicuously marked of the peramelids. Of the native mammals that prefer a grassland habitat, it is one of the few to have survived European settlement of south-eastern Australia. It is widely distributed in Tasmania, particularly in the northern and eastern part of the island, but its mainland range has been drastically reduced to one relict colony at Hamilton, Victoria. Fossils have been reported from Hunter and Flinders Islands, Bass Strait, and south-eastern South Australia.

In Tasmania, it prefers open grasslands and flourishes near areas of pastoral development. In Victoria, particularly around Hamilton, it survives in suburban gardens and nearby grasslands and grassy woodlands—now mostly improved

*The Eastern Barred Bandicoot is now rare on the mainland but reasonably abundant in Tasmania.*

pasture—developed on the basalt soils. The availability of shelter is an important determinant of its local distribution.

The forefeet are armed with long claws, used to dig shallow, conical pits from which it extracts earthworms, insects and their larvae, bulbs and tubers. It is an opportunistic feeder, even including berries in its diet, but it is reported to avoid some abundant invertebrate species. Sufficient water may be obtained from its food to make drinking unnecessary.

Emerging from its grass nest at about dusk, it forages with several peaks of activity throughout the night. It usually moves at a rapid gallop or

---

### *Perameles gunnii*

**Size**

*Head and Body Length*
270–350 (310) mm (Victoria)
270–350 (300) mm (Tasmania)
*Tail Length*
70–110 (94) mm (Victoria)
70–95 (80) mm (Tasmania)
*Weight*
500–1100 (750) g (Victoria)
500–1450 (990) g (Tasmania)

**Identification** Grizzled, yellowish-brown above, slaty grey below. Tail white above except at base. Three or four pale bars on hindquarters.
**Recent Synonyms** None.
**Other Common Names** Tasmanian Barred Bandicoot, Gunn's Bandicoot, Striped Bandicoot.
**Status** Victoria: very rare, critically endangered.
Tasmania: locally common within fragmented distribution.
**Subspecies** None.

**References**

Heinsohn, G.E. (1966). Ecology and reproduction of the Tasmanian bandicoots (*Perameles gunni* and *Isoodon obesulus*). *Univ. Calif. Publ. Zool.* 80: 1–107.

Clark, T.W. and J.H. Seebeck (eds) (1990). *Management and Conservation of Small Populations.* Chicago Zoological Society, Brookfield, Illinois. [Includes 10 papers on *Perameles gunnii.*]

Robinson, N.A., W.B. Sherwin and P.R. Brown (1991). A note on the status of the eastern barred bandicoot, *Perameles gunnii* in Tasmania. *Wildl. Res.* 18: 451–457.

Backhouse, G. (1994). *Recovery plan for the Eastern Barred Bandicoot, Perameles gunnii 1994–1997.* Dept. of Conservation and Natural Resources, Flora and Fauna Branch, Heidelberg, Victoria.

bound, using all four feet and can clear more than a metre in a single leap. Vocalisation is limited to a series of snuffles, squeaks and hisses.

Males are larger than females, have a longer tooth row and larger canine teeth. In Tasmania, breeding occurs throughout most of the winter, spring and early summer. Wild and captive Victorian specimens breed throughout the year, but mostly from July to November. The onset of breeding may be related to availability of food (more abundant with increasing rainfall), falling temperature, or both.

Females have eight teats in the rear-opening pouch and litters comprise one to five young, usually two to three: spring litters are usually larger than those born at other times. Gestation probably lasts about 12.5 days. A female may become pregnant again while lactating and give birth to another litter immediately after the preceding one has been weaned. Females may breed when three to three-and-one-half months old and males when four to five months old. Young animals disperse at independence, three to five months after birth.

Males can be very aggressive towards each other but interactions tend to be avoided. In the wild, the life span is usually one-and-one-half to two years but some live and continue to breed, to an age of more than three years.

Population densities of about 8.5 per hectare have been reported in Tasmania but, in Victoria, the figure is usually around 2 per hectare. Since European settlement, populations appear to have increased in parts of Tasmania but to have decreased in others; in Victoria, the remnant wild population is on the verge of extinction. The absence of the Fox from Tasmania may, perhaps, be significant, although an array of causes have been identified for the decline on the mainland, including disease.

Three protected colonies have been established in conservation reserves in Victoria, as part of a statewide recovery program which involves large-scale captive breeding, habitat restoration and the reintroduction of the species to parts of its former range. In Tasmania, there are no plans for specific reserves. Its distribution is primarily in agricultural land and is fragmented. Although it may be locally common, its presence in conservation areas is peripheral only and dependent on adjacent farmland habitat.

J.H. SEEBECK

Note: two easternmost populations are reintroduced

*Perameles gunnii*

K. ATKINSON

# Long-nosed Bandicoot

## *Perameles nasuta*

## Geoffroy, 1804

nah-zute'-ah: 'notably-nosed pouched-badger'

*The Long-nosed Bandicoot is one of the few members of its family to be common and widespread.*

Although the natural home of the Long-nosed Bandicoot ranges from rainforest through wet and dry woodland to areas with little ground cover, it is probably best known from the conical holes that it makes at night when foraging in sub-urban lawns. These holes, which are dug with the forefeet, are usually vertical and large enough to accommodate the animal's snout when it is searching for insects and succulent plant material which it appears to locate by smell. Between bursts of digging it moves about sniffing the ground: a shrill, grunt-like squeak uttered while foraging often indicates its presence.

A Long-nosed Bandicoot spends the day in a nest, usually a shallow hole on the surface of the ground, lined with grass and leaves which it scrapes together with its forelegs. The upper surface of the nest, which is sometimes flattened and partly covered with soil, may be well concealed under debris. When the nest is occupied the entrance is closed, making it difficult to perceive. The two small joined toes on the hindfeet are used to groom the fur, which nevertheless carries many ectoparasites, particularly ticks.

The Long-nosed Bandicoot is a solitary animal, contact between males and females being restricted to the minimum necessary for successful reproduction. Males are attracted to a female for several nights prior to the peak of oestrus and the attraction builds up to persistent following. Mating takes place at night, the male placing his forelegs on the female's hindquarters and mounting her from behind. Breeding occurs throughout the year, at least in the Sydney region, with a trough in breeding activity from late autumn to mid-winter. Some females start to breed at about five months, before they are fully grown.

There are eight teats but the litter size is only one to five, usually two to three. Birth takes place during the daylight hours after a gestation of only 12.5 days. The newborn young are about 13 millimetres long and weigh about 0.25 grams. The pouch opens backwards, making the journey to the teats quite short, but the young remain connected to the placenta by long umbilical cords until some time after they have secured themselves to teats. They grow rapidly. The first hairs emerge on the body at about 40 days, eyes open at 45–50 days and weaning occurs at about 60 days after birth. When the young are about 50 days old the mother may

**Perameles nasuta**

### Perameles nasuta

**Size**

(males somewhat larger than females)
*Head and Body Length*
310–425 mm
*Tail Length*
120- 155 mm
*Weight*
850–1100 g

**Identification**  Drab greyish-brown above, creamy white below. Forefeet and upper surface of hindfeet creamy white. Distinguished from other *Perameles* by absence of distinct dark and light bars on rump, except in some juveniles and fewer adults which show a faint barred pattern. Muzzle long and pointed and ears distinctly longer and more pointed than in short-nosed bandicoots of the genus *Isoodon*.
**Recent Synonyms**  None.
**Other Common Names**  None.
**Status**  Common, widespread.
**Subspecies**  *Perameles nasuta nasuta*, east coast, south of Townsville, Queensland.
*Perameles nasuta pallescens*, Townsville to Ravenshoe, Queensland.
**Reference**
Stodart, E. (1977). Breeding and behaviour of Australian bandicoots (in) B. Stonehouse and D. Gilmore (eds) *The Biology of Marsupials*. Macmillan, London.

mate again and produce another litter several days after the previous one has been weaned. The older litter then has to fend for itself. Since litters can be produced in quick succession and adults begin breeding at an early age, the Long-nosed Bandicoot has a potentially high reproductive rate: it follows that there must be a high juvenile mortality.

E. STODART

## Subfamily Thylacomyinae

The characteristics of this group are essentially those of the single surviving species, notably long ears; long, silky fur; and a well-furred, long tail.

# Bilby

## *Macrotis lagotis*

### (Reid, 1837)

mak-roh'-tis lah-goh'-tis:
'hare-eared long-ear'

The long, soft, blue-grey fur and delicate features of the Bilby seem out of character with the harsh deserts it now occupies. These features, together with its large ears, long pointed snout and black and white tail that is crested throughout its length, carried like a stiff banner during the cantering gait, led Finlayson to comment that the species had carried 'a number of structural peculiarities to grotesque lengths yet

manages to reconcile them all in a surprisingly harmonious and even beautiful, whole'. The tail tips were used extensively by Aborigines for decoration.

Once inhabiting the arid and semi-arid regions throughout most of the Australian mainland south of about latitude 18° south, it is now confined to the deserts of central Australia, from the Tanami Desert in the Northern Territory west to Broome and south to Warburton in Western Australia. Satellite populations occur north of Birdsville in south-western Queensland. It occupies a variety of habitats ranging from the clayey and stony downs soils of

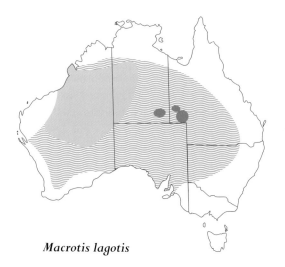

*Macrotis lagotis*

---

### *Macrotis lagotis*

**Size**
*Head and Body Length*
300–550 mm (males)
290–390 mm (females)
*Tail Length*
200–290 mm (males)
200–278 mm (females)
*Weight*
1000–2500 g (males)
800–1100 g (females)

**Identification** Light and delicate in build. Hair soft and silky. Ears long and rabbit-like. Tail black on proximal half, changes abruptly to white; prominent dorsal crest; extreme tip of tail naked. Muzzle long and pointed. Hindfoot lacks first toe.

Queensland with sparse ground cover to the sands, sometimes containing laterite, with spinifex (*Triodia* and *Plectrachne* spp.) and massive red earths with *Acacia* shrubland. Free surface water is rarely available and the Bilby must derive most of its water from food.

The strong forelimbs with stout claws are used to dig for food and to burrow. Feeding areas are characterised by numerous scattered excavations, up to 10 centimetres deep, from which the soil has been scattered on all sides. The diet includes insects and their larvae, seeds, bulbs, fruit and fungi. Seeds are licked

K. ATKINSON

*The Bilby digs a deep burrow in which it spends the day.*

from the ground with the long, slender tongue but a great deal of soil is usually consumed with the food and faeces may comprise 20–90 per cent sand. Vision is poor but the senses of hearing and smell are acute and are of primary importance in locating food.

It is a powerful burrower, constructing a system that may be 3 metres long and up to 1.8 metres deep but which contains no nest material. Attempts to dig an animal out often result in its frantic and more rapid extension of the burrow from the closed end. The entrance, often against a termite mound, spinifex tussock or small shrub, is always left open. A Bilby remains in its burrow throughout the day and does not venture out until well after dark. Bilbies live singly or in pairs, sometimes in the company of a recently independent young. Home ranges in the sandy deserts are usually temporary in location and may suddenly shift in response to changing availability of food.

In captivity, the Bilby breeds throughout the year. The backwardly opening pouch contains eight teats but is rarely occupied by more than two young. Pouch life lasts about 80 days. Bilbies housed in large outdoor yards cached their newly independent young in burrows for

**Recent Synonyms** *Thylacomys lagotis.*
**Other Common Names** Greater Bilby, Rabbit-eared Bandicoot, Rabbit Bandicoot, Pinkie.
Indigenous names: Dalgyte, Ninu (Western Aranta, Kukatja, Luritja, Manytjilytjarra, Ngaanyatjarra, Ngaatjatjarra, Pintupi, Pitjantjatjara, Warlpiri); Marrura (Kartutjarra, Kukatja, Manytjilytjarra, Ngaanyatjarra, Pintupi, Putitjarra, Wangkatjungka, Warlpiri); Walpatjirri (Warlpiri).
**Status** Rare, scattered.
**Subspecies**
*Macrotis lagotis lagotis*, Warburton Range, Western Australia.
*Macrotis lagotis sagitta*, Central Australia (major population).
*Macrotis lagotis cambrica*, New South Wales (extinct).
*Macrotis lagotis nigripes*, South Australia (extinct).
*Macrotis lagotis grandis*, South Australia (extinct).
*Macrotis lagotis interjecta*, southern Western Australia (extinct).
**Reference**
Johnson, K.A. (1989). Thylacomyidae (in) D.W. Walton and B.J. Richardson (eds) *Fauna of Australia Volume 1B Mammalia.* Aust. Gov. Publ. Ser., Canberra. pp. 625–635.

about two weeks, returning at regular intervals throughout the night to suckle them.

The Bilby was common throughout most of its range until the early 1900s when there was sudden and widespread contraction. The distribution appears to be still contracting and fragmenting. Direct and indirect effects on food by a changing fire regime and the grazing of Rabbits and livestock, together with predation by introduced Foxes and feral Cats were probably responsible for the decline. Manipulation of habitat diversity by the use of fire and the control of Rabbits and predators is fundamental to the conservation of this singularly attractive marsupial.

K.A. JOHNSON

*The Bilby has a shuffling quadrupedal walk but can be quite fast if surprised or pursued.*

G. STEER/TERRA AUSTRALIS

CAT. MARS. B. M.

Pl. II.

Peragale leucura.

J. Smit del. et lith.

Mintern Bros. imp.

# Lesser Bilby

## *Macrotis leucura*

### (Thomas, 1887)

luke-yue'-rah: 'white-tailed long-ear'

The Lesser Bilby, a diminutive and less colourful relative of the Bilby, has been recorded, very rarely, from the deserts of north-eastern South Australia and south-eastern Northern Territory where free surface water is scarce. It was last reported alive in 1931 near Cooncherie in north-eastern South Australia where it was considered to be reasonably common. Its burrows were restricted to sandhills, whereas the Bilby, which occurred in the same area, burrowed only in loamy flats between the dunes.

*Lesser Bilby. Illustration in Oldfield Thomas'* Catalogue of the Monotremes and Marsupials in the British Museum (Natural History).

It is well known to Aborigines of the central deserts who note its former distribution well into the Great Sandy and Gibson Deserts of Western Australia. The species appears to have survived in some of these areas until the 1960s. It usually occurred in sand plain or sand-dune country with spinifex, but also occupied mulga and tussock grass country.

The Lesser Bilby resembles the Bilby in resting in a burrow but differs from it in closing the entrance while in residence. The only evidence of its presence may be a slight dimple in the sand. Like the Bilby, it apparently did not construct pop-holes or vent shafts and had no nest or dwelling within its burrow.

## *Macrotis leucura*

**Size**

*Head and Body Length*
240–270 mm (males)
200–240 mm (females)
*Tail Length*
125–170 mm (males)
120–150 mm (females)
*Weight*
360–435 g (males)
311 g (females)

**Identification**  Long ears and pointed snout. Smaller than *Macrotis lagotis* and differs in having white fur along the entire upper surface of the tail (black on the underside of the proximal two-fifths).

**Recent Synonyms**  *Macrotis minor, Thylacomys minor, Thylacomys leucurus.*

**Other Common Names**  Lesser Rabbit-eared Bandicoot, Yallara, Lesser Rabbit Bandicoot. Indigenous names: Atnunka (Aranda), Nantakarra (Pintupi, Wangkatjungka, Warlpiri), Tjunpi (Kartutjarra, Kukatja, Manytjilytjarra, Ngaanyatjarra, Ngaatjatjarra, Pintupi), Ngatukutiri (Kartutjarra, Manytjilytjarra, Kukatja, Putitjarra).

**Status**  Probably extinct.

**Subspecies**  None.

**Reference**

Finlayson, H.H. (1935). On mammals from the Lake Eyre Basin Part II. The Peramelidae. *Trans. Roy. Soc. S. Aust.* 59: 227–36.

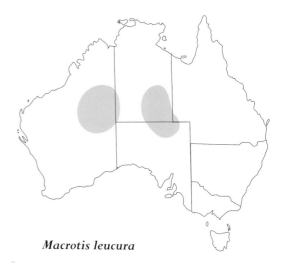

*Macrotis leucura*

It was strictly nocturnal and remained in its burrow by day, taking refuge from predators and from intense solar radiation.

Finlayson reported that Lesser Bilbies '...completely belied their delicate appearance by proving themselves fierce and intractable and repulsed the most tactful attempts to handle them by repeated savage snapping bites and harsh hissing sounds and one member of the party, who was persistent in his attentions, received a gash in the hand three-quarters of an inch long from the canines of a male'. The same male also attacked and killed a native 'mouse' which was placed in its cage.

Stomach contents from a limited sample were found to include large quantities of skin and fur of rodents, seeds (probably from *Solanum* sp.) and sand. No insect fragments could be recognised.

Females possess eight teats in two rows of four. The pouch opens to the rear and it appears that two young were usually reared at a time. Little is known of the social organisation of the species but females excavated from burrows were either alone or with dependent young. It seems likely that the Lesser Bilby had a social structure based on single individuals or pairs. To humans, the Bilby has a strong (although not offensive) smell, whereas the Lesser Bilby had little odour.

The most recent record is a skull of unknown age found in 1967 in a Wedge-tailed Eagle's nest south-east of Alice Springs at the edge of the Simpson Desert.

K.A. JOHNSON

# FAMILY PERORYCTIDAE
## *Rufous Spiny Bandicoot*

Three genera of New Guinean bandicoots—*Peroryctes*, *Microperoryctes* and *Echymipera*—constitute this family. In contrast to species of the Peramelidae, which inhabit relatively dry country, peroryctids live in rainforest. The skull is more or less cylindrical (flattened in peramelids). Other skull features also distinguish the two families.

## Rufous Spiny Bandicoot

### *Echymipera rufescens*

(Peters and Doria, 1875)

ek'-ee-mip'-er-ah  rue-fes'-enz:
'reddish pouched-hedgehog'

G. STEER/TERRA AUSTRALIS

*The Rufous Spiny Bandicoot, which also occurs in New Guinea, is the only Australian member of the Peroryctidae.*

Far more common in Papua New Guinea than in Australia, the Rufous Spiny Bandicoot, like the Common Spotted Cuscus and Striped Possum, is an outlying member of a Melanesian species which became isolated in Australia in the late Pleistocene when the land bridge across Torres Strait was submerged. The first specimen to be described was found in the nineteenth century on the Kei Islands but it was not discovered in Australia until 1932 when it was collected by the biogeographer, P.J. Darlington on the upper Nesbit River, Queensland. (The original description by Tate in 1948 gave the locality, in error, as 'Rocky Scrub, Rocky River'.) It is now known from eight localities between the tip of Cape York and the McIlwraith Range, Queensland and

appears to be the common rainforest bandicoot of Cape York Peninsula.

In New Guinea, it occurs mainly in low altitude rainforest and at rainforest margins, but may also be found in adjacent habitats including grassland. In Queensland, its distribution coincides with closed forest areas including mesophyll vine forest, notophyll vine forest and gallery forest and gallery forest. It is not confined strictly to this habitat, being also recorded in eucalypt grassy woodland and coastal closed heath near closed forest and in low layered open forest.

Little is known of its biology but it is nocturnal and, in captivity, has a preference for insect food while able to accommodate to an omnivorous diet. Males are considerably larger than females and breeding seems to be seasonal, having been recorded in February in two captive females. Females taken in the wild in October showed no sign of breeding.

This spiny bandicoot and other species of its genus possess a number of distinctive characteristics. With the New Guinean *Rhynchomeles prattorum*, they are the only bandicoots to lack the fifth upper incisor. Also with the latter species, they show the most pronounced elongation of the muzzle of any bandicoots, possibly a specialisation for extracting food from small crevices. Observations suggest that there is even a slight development of a mobile proboscis. The colouration is unusual for a bandicoot, with a blackish crown and rufous body. The spiny guard hairs from which it takes its common name are much stiffer and broader (greater than

### *Echymipera rufescens*

**Size**
*Head and Body Length*
300–400 mm
*Tail Length*
75–100 mm
*Weight*
500–2000 g (males)
500–1400 g (females)

**Identification**  Blackish above on muzzle and crown; short, almost naked, black tail. Four pairs of upper incisor teeth. (All other Australian bandicoots have five pairs of upper incisor teeth and lack this distinctive colouration).
**Recent Synonyms**  None.
**Other Common Names**  Spiny Bandicoot, Rufescent Bandicoot.
**Status**  Common.
**Subspecies**  The Australian subspecies is *Echymipera rufescens australis*.
**Reference**
Gordon, G and B.C. Lawrie (1977). The rufescent bandicoot, *Echymipera rufescens* (Peters and Doria) on Cape York Peninsula. *Aust. Wildl. Res.* 5: 41–45.

0.5 millimetres) than those of other Australian bandicoots and the short, almost naked tail is also distinctive.

Failure to find the species after the initial Australian record led to the belief that it was rare on the continent but collecting since 1970 suggests that it is relatively widespread and common.

G. GORDON

*Echymipera rufescens*

# ORDER DIPROTODONTIA

*Koala, Wombats, Possums and Macropods*

Marsupials in this group are characterised by possession of only one pair of incisors in the lower jaw; a second, small, non-functional pair may be present. There are usually three pairs of upper incisors but wombats have only one pair (a condition parallel to that of rodents). Lower canine teeth are never present; upper canines, vary from a low, smooth-cusped condition (bunodont) to having many sharp, curved ridges (selenodont). A pair of sectorial premolars is present in the Mountain Pygmy-possum and rat-kangaroos. Most of the features of the dentition are clear adaptations to a herbivorous diet, culminating in the efficient grazing habits of the larger kangaroos. Nevertheless, the extensive evolutionary radiation within the order has given rise to some secondarily insectivorous species and even to some feeders on fluids such as sap and nectar.

All members of the Diprotodontia resemble peramelemorphs in having syndactyl second and third toes on the hindfoot: the two small digits are fused together except at the tip, where a pair of slender claws protrude. These are employed in grooming but it is most unlikely that the structure arose for this purpose. Considering the function of the foot, it seems clear that, as in some primates, the second and third digits became reduced as a means of improving a pincer-like grip between the very mobile first digit and those on the outer side of the foot. Another arboreal adaptation present in most members is an ability to oppose the first two digits of the forefoot against the other two. This is absent in terrestrial species such as wombats and kangaroos, in the wrist-winged gliders (in which the structure of the forefoot is strongly influenced by the attachment of the gliding membrane) and, for no apparent reason, in the brushtail possums. It is very apparent in the Koala and ringtail possums.

The Diprotodontia is a successful group with many diverse subgroups. Living species can be divided into two suborders, the Vombatiformes (Koala and wombats) and the Phalangerida (possums, rat-kangaroos and kangaroos). It seems that the vombatiforms represent an early branch from the diprotodontian stock.

# SUBORDER VOMBATIFORMES

To emphasise the degree of difference between wombats and the Koala, this group is divided into two infraorders: the Vombatomorphia and the Phascolarctomorphia. In respect of living forms, these can be regarded as indistinguishable from the families Vombatidae and Phascolarctidae.

The two groups share the following features. The tail is so short as to be almost invisible. The pouch of females opens to the rear and encloses two teats. Cheek-pouches are present. The stomach has a unique gastric gland. Sperms have a very different shape from those of other marsupials.

Previous page: *The Western Grey Kangaroo may move about during the day but does not normally become active until around dusk.* G.I. LITTLE

# FAMILY PHASCOLARCTIDAE
## *Koala*

*The Koala does not use a den or a nest. It sleeps during the day in the fork of a tree.* C. ANDREW HENLEY/LARUS

The single living species, the Koala, is an arboreal folivore. It has three pairs of upper incisors. Its teeth do not grow continuously like those of wombats.

# Koala
## *Phascolarctos cinereus*

(Goldfuss, 1817)

fas'-koh-lark'-tos sin'-er-ay'-us:
'ash-coloured pouched-bear'

C. ANDREW HENLEY/LARUS

*Where the forest canopy is discontinuous, a Koala must descend to the ground to move from tree to tree, becoming vulnerable to Dogs and to traffic.*

Extending from the temperate south to the tropical north, the Koala occupies a vast but fragmented range in eastern Australia. While the harsh winter excludes it from the montane forests along the backbone of the Great Dividing Range, it is widely distributed on either side. To the south and east it is most abundant in the forest remnants of the foothills and coastal plains. To the west it extends well inland, following the River Red Gum (*Eucalyptus camaldulensis*) forests that skirt the mosaic of rivers and watercourses draining this region.

This association with eucalypt forests is characteristic throughout the range for the Koala

### *Phascolarctos cinereus*
**Size**
Animals from the southern part of the range are significantly larger than those from the northern part.
**Victoria**
*Head and Body Length*
750–820 (782) mm (males)
680–730 (716) mm (females)
*Weight*
9.5–14.9 (12.0) kg (males)
7.0–11.0 (8.5) kg (females)
**Queensland**
*Head and Body Length*
674–736 (705) mm (males)
648–723 (687) mm (females)
*Weight*
4.2–9.1 (6.5) kg (males)
4.1–7.3 (5.1) kg (females)

feeds almost entirely on the foliage of species of this genus. There are, however, marked local and regional preferences. In the south, the preferred species are Manna Gum (*E. viminalis*), Swamp Gum (*E. ovata*) and Blue Gum (*E. globulus*), while the Red Gums (*E. camaldulensis* and *E. tereticornis*), Tallowwood (*E. microcorys*) and Grey Gum (*E. punctata*) are important in the north. The more abundant populations tend to be linked to species growing on higher-nutrient soils, but Koalas also occur in forests on the poorer coastal soils.

Dependence on eucalypts is one of the enigmas of Koala biology. Regardless of the level of soil nutrients, eucalypt foliage is poor-quality food. It contains high concentrations of indigestible compounds, such as fibre and lignin and potentially toxic phenolics and terpenes. It is also low in protein. Koalas rely on a suite of behavioural, anatomical and physiological adaptations to exist on this diet. Foremost among these is a very low energy requirements compared with most other mammals. A Koala is usually inactive for 20 hours a day and this sedentary habit significantly lessens its metabolic requirements. Food (about 500 grams of eucalypt leaf each day) is very efficiently processed. The powerfully muscled jaws

C. ANDREW HENLEY/LARUS

*In the absence of a den or nest, the young koala is carried by its mother until quite large.*

**Identification** Arboreal; large; vestigial tail; woolly fur. Animals from the north have short, pale grey fur; animals from the south have longer grey-brown fur.

**Recent Synonyms** None.

**Other Common Names** Koala Bear, Native Bear, Monkey Bear.

**Status** Common, limited.

**Subspecies** Although three subspecies have been described, these may represent arbitrary selections from a cline.

*Phascolarctos cinereus cinereus*, New South Wales.

*Phascolarctos cinereus adustus*, Queensland.

*Phascolarctos cinereus victor*, Victoria.

**References**

Lee, A.K. and R.W. Martin (1988). *The Koala: a Natural History*. University of New South Wales Press, Sydney.

Lee, A.K., K.A. Handasyde and G.D. Sanson (eds) (1991). *Biology of the Koala*. Surrey Beatty and Sons, Sydney.

and sharply ridged teeth masticate the leaves to a fine paste, releasing the plant cell contents (the principal source of energy) for immediate absorption in the stomach. Toxic compounds are inactivated in the liver and excreted. The fine digesta which passes beyond the stomach is retained in the hind gut for a long period of microbial fermentation. The major organ involved in this, the caecum, is one of the most capacious of any mammal. Products of fermentation make an additional contribution to the Koala's overall energy requirements.

A Koala spends most of the day resting in a low fork, usually climbing into the canopy around dusk to commence feeding. It is an accomplished climber and can rapidly ascend the bole of a tree by grasping it with the sharp claws of its forepaws, then bringing the hindfeet up together in a bounding movement. It moves comfortably along smaller branches by gripping these with the opposable digits of the forepaw and the first toe of the foot. In the course of the night, an individual often changes trees and, because of the open canopy of eucalypt woodlands, this usually involves descending to the ground. An individual

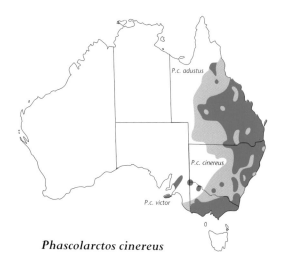

*P.c. adustus*

*P.c. cinereus*

*P.c. victor*

**Phascolarctos cinereus**

may walk several hundred metres to a favoured feeding tree. On the ground, it moves in an unhurried quadrupedal walk but, if the situation demands it, a Koala can move quickly, bounding along at a gallop: a photograph of a swiftly moving Koala showed it to be supported on the tip of the third digit of one hindfoot.

The Koala is solitary and individuals spend most of their time in distinct home ranges, the size of which varies according to the density of the population and the abundance of mature food trees in an area. In the denser populations these ranges overlap but they appear to be discrete at lower densities. Males are not territorial but there is a dominance hierarchy and dominant males chase and attack subordinates if they are encountered. Most adult males carry scars on the forearms, face and ears as a result of such encounters. Much of this activity occurs during the summer breeding season, the onset of which is heralded by bellowing of the males. These low-pitched, snoring inhalations and exhalations can be heard for up to 800 metres and serve to advertise the presence of one male to others and to receptive females.

Females commence breeding when two years old and healthy animals are able to produce one young a year until more than 14 years of age. Longevity is around 18 years for females, probably several years less for males. Females give birth to a single young after a gestation of 35

days. Twins are rare. After birth, the neonate climbs from the urogenital opening to the pouch where it attaches to one of the two teats and remains for about six months. It is first seen when its head appears at the pouch opening and the weaning process begins around this time. The infant's change of diet, from milk to eucalyptus leaves, is initiated by its feeding on soft faeces which is produced by the mother at this time. This 'pap', which is thought to be of caecal origin, serves to inoculate the hindgut of the young with the micro-organisms that it needs in order to digest foliage. After weaning commences, the young develops rapidly and leaves the pouch permanently at about seven months. It continues to associate with the mother, usually travelling around on her back, feeding on both milk and eucalyptus leaves until it is fully weaned at 12 months of age. At this time, it becomes independent and the mother usually gives birth again. Young males do not disperse from their natal ranges until two to three years old, while young females often remain and breed nearby. There is undoubtedly some mortality associated with the dispersal of the young males, but most appear to establish themselves in new areas, distant from their natal ranges.

The Koala is still widespread but there are management problems with many populations. Remnant populations living at high densities in isolates of habitat are at greatest risk. The impact of diseases of the eye and reproductive tract, resulting from infection with the endemic bacterium *Chlamydia psittaci*, are exacerbated in many of these populations. In others, despite *Chlamydia* infection, fecundity remains high, leading to overpopulation. As there are now no natural predators of koalas, the density of these populations has to be artificially reduced to avoid the defoliation and death of scarce food trees. Habitat conservation, particularly the preservation of large tracts of habitat and the linking of isolates with habitat corridors, is now the key issue in Koala conservation.

R.W. MARTIN AND K.A. HANDASYDE

# FAMILY VOMBATIDAE
## *Wombats*

*Although usually nocturnal, the Common Wombat forages during the day in very cold weather.*

Wombats, like kangaroos, are primarily grazers. The two upper incisors work against the lower pair as efficient cutters of grass and forbs. The cheek-teeth are broad and have grinding ridges. All teeth have open roots and grow continuously, a feature not found in any other marsupial. The other major adaptation of wombats is to a burrowing habit. The powerful limbs, with short broad feet and flattened claws, can burrow into very firm soil: the flattened head and hindquarters are employed in compacting the walls of the burrow.

The three living species fall into two genera. The single species of *Vombatus* lives in forests and grazes in clearings. The hairy-nosed wombats of the genus *Lasiorhinus* live in grassland or savanna.

# Northern Hairy-nosed Wombat

## *Lasiorhinus krefftii*

### (Owen, 1872)

lah'-zee-oh-rine'-us  kref'-tee-ee: 'Krefft's hairy-nose'

This species was first described from a fossil skull found in the Wellington Caves, New South Wales, in 1869. Fossil remains have since been found in south-western New South Wales, Victoria and Queensland. Living wombats were found near Jerilderie, New South Wales, in 1884 and at two sites near St George in southern

Queensland in 1890s, but these populations appear to have become extinct very early in this century. In 1937, the Queensland Museum located a population at Epping Forest station near Clermont in central Queensland. Several persistent and independent reports suggest it also occurred near Injune and Tambo in central Queensland, but no specimens are available to confirm these reports and it is not now known outside a small area reserved as Epping Forest National Park.

Local landholders recall that Hairy-nosed Wombats were much more widespread in the Epping Forest region earlier in this century and it seems that there were at least two discrete populations. One of these has disappeared entirely and the other has contracted quite severely. The major declines occurred during severe droughts, probably because heavy cattle grazing at these times left little pasture for wombats. The total area now occupied by the Northern Hairy-nosed Wombat is only 300 hectares and the total population is little more than 70. However, the situation has improved since cattle were excluded from its habitat in 1982. A large proportion of the population now consists of young animals born since 1982 and there is evidence of a steady increase in population size. There is sufficient unoccupied habitat in the Park to allow this increase to continue.

---

### *Lasiorhinus krefftii*

**Size**

*Head and Body Length*
1020 ± 49 mm (males)
1070 ± 36 mm (females)
*Weight*
30.1 ± 3.2 kg (males)
32.5 ± 2.6 kg (females)

**Identification**  Predominantly brown-grey, or grey mottled with fawn, brown or black. Distinguished from *Vombatus* by silky hair and longer ears; from *Lasiorhinus latifrons* by large, very square muzzle, nasal bones (measured in mid-line) being longer than frontal bones.
**Recent Synonyms**  *Wombatula gillespiei, Lasiorhinus latifrons gillespiei, Lasiorhinus gillespiei, Lasiorhinus latifrons barnardi, Lasiorhinus barnardi.*
**Other Common Names**  Queensland Wombat, Queensland Hairy-nosed Wombat, Moonie River Wombat.
Indigenous name: Yaminon (St George area).
**Status**  Rare, limited, endangered.
**Subspecies**  None.

*Lasiorhinus krefftii*

C. ANDREW HENLEY/LARUS

*The Northern Hairy-nosed Wombat is extremely rare and endangered but its population appears to be stable.*

*Hairy-nosed wombats differ from the Common Wombat in having a short-haired rhinarium (bare in the Common Wombat)*

C. ANDREW HENLEY/LARUS

choosing the most comfortable times of night for excursions. It feeds almost exclusively on perennial native grasses and, as far as possible, grazes only in close proximity to a burrow. Movements between burrow clusters are very rare but most adult females eventually leave the burrows in which they were born and may possibly move several times in their lives. Females breed, on average, twice every three years. Most young are born in spring and summer and spend 10 or 11 months in the pouch.

C.N. Johnson and G. Gordon

Epping Forest has a vegetation of flat grassland and eucalypt woodland, with some patches of closed scrub on deep, sandy soil. Burrows have from one to seven entrances and are distributed in loose clusters which may contain up to 20 burrows within a few hectares and are occupied by up to ten wombats. Each individual makes use of most of the burrows in a cluster (which are connected by well-used trails) but avoids encounters with other wombats and spends most of its time alone. Burrows in use are carefully maintained and marked around the entrance with piles of dung and splashes of urine; dung-piles are also placed at intervals along trails.

The Northern Hairy-nosed Wombat lives in a harsh climate with very hot summers and long periods of dry weather. It copes by minimising the amount of time spent above ground and

**References**

Johnson, C.N. (1991). Utilization of habitat by the northern hairy-nosed wombat, *Lasiorhinus krefftii*. *J. Zool. Lond.* 225: 495–507.

Johnson, C.N. and D.G. Crossman (1991). Dispersal and social organization in the Northern Hairy-nosed Wombat, *Lasiorhinus krefftii*. *J. Zool. Lond.* 225: 605–615.

Crossman, D.G., C.N. Johnson and A.B. Horsup (1994). Trends in the population of the Northern Hairy-nosed Wombat, *Lasiorhinus krefftii* in Epping Forest National Park, central Queensland. *Pac. Conserv. Biol.* 1: 141–149.

# Southern Hairy-nosed Wombat

## *Lasiorhinus latifrons*

(Owen, 1845)

lah'-tee-fronz: 'broad-headed hairy-nose'

First described in 1845 from a skull sent to the British Museum, the Southern Hairy-nosed Wombat was later discovered living in colonies west of the Murray River in South Australia. Its range has diminished since European colonisation and it is now confined to semi-arid regions with an annual rainfall of 200–500 millimetres.

Being sedentary, it has to cope with extreme local conditions. It rarely has access to water and in summer will select green shoots, when available, in a pasture with a generally high fibre and low water content. In this way it maximises intake of water and protein, its rodent-like incisor teeth and split lip enabling it to pick emerging shoots from around the base of the perennial grasses. During the day it rests in a humid burrow, allowing its body temperature to fall, thereby conserving both water and energy. In summer, animals generally emerge in the evening when temperature and humidity are at levels that minimise water loss by evaporation. Physiologically it is frugal. The rate of metabolism when resting is only two-thirds that of most marsupials and, although the food is very finely milled by the continuously growing molars, it may take up to eight days to pass through the gut. The rate of water turnover and the dietary nitrogen requirement are both low.

*Lasiorhinus latifrons*

**Size**
*Head and Body Length*
772–934 mm
*Tail Length*
25–60 mm
*Weight*
19–32 kg

**Identification** Stout head and body and short, powerful limbs. Differs from *Vombatus* in having softer, silky grey fur and white hair on rhinarium (bare in *Vombatus*). Differs from *Lasiorhinus krefftii* in having nasal bones longer than frontal bones (both measured along midline of skull).
**Recent Synonyms** None.
**Other Common Names** Hairy-nosed Wombat.
**Status** Common; limited.
**Subspecies** The relationships of the recent hairy-nosed wombats are not well understood. While some authorities recognise three species, others include all extant forms within *Lasiorhinus latifrons*.
**Reference**
Wells, R.T. (1989). The Vombatidae (in) D.W. Walton and B.J. Richardson (eds) *Fauna of Australia. Vol 1B. Mammalia.* Aust. Gov. Publ. Serv., Canberra. pp. 755–568.

*Lasiorhinus latifrons*

E. BEATON/TERRA AUSTRALIS

*The Southern Hairy-nosed Wombat has an extensive range in the southern and semi-arid mainland.*

Although it appears to be a slow, bumbling animal, it is exceedingly alert to the slightest sound or unusual scent and, when disturbed, can bound as fast as 40 kilometres an hour over short distances.

It constructs extensive burrow systems, the entrances to which are usually clustered to form a large warren, often with smaller warrens or single burrows at a radius of 100–150 metres. Warrens are connected by a network of trails and these are marked at intervals with a combination of urine, faeces and scratchings. Trails also lead to rubbing posts and wallows that are similarly marked, suggesting that olfactory communication is important to this species.

Each warren system may be occupied by five to ten wombats, but all individuals may not be present at the same time. Some burrows may be preferred by individuals but there is no evidence to support ownership among the occupants of a warren. Females show greater burrow-preference than males. There are well-defined dominance relationships among males, while females appear to be subordinate to all adult males. Inter-warren disputes seem to be confined to males and involve the establishment of territories. The highest levels of social activity occur in the breeding season, aggressive behaviour being most evident between males and often resulting in bites to the ears, rump and flanks. There is also an increase in vocalisations at this time, the most common being a harsh cough which may become quite strident, particularly when an animal is alarmed. These calls are emitted by oestrous females when pursued by males; less frequently by subordinate males in male–male encounters. Females may emit a softer call to summon young.

Copulation occurs in the burrow, the male rolling the female onto her side and mounting from behind for a prolonged intromission. Most births occur from late September to December and the single young remains on one of the two teats in the pouch for six to nine months. Weaning occurs at approximately one year and both sexes are mature at three years. In times of drought, ovulation and sperm formation may cease. In captivity, the Southern Hairy-nosed Wombat may mate again following weaning or loss of young.

An increase in the adult population requires a minimum of three consecutive years of effective rainfall, critical factors being the ability of the female to provide milk, the availability of green feed at the time of weaning and good pasture at the time of maximum growth of the juvenile. Along the Murray River valley, effective rainfall has occurred in only 25 of the past 100 years, so longevity (in excess of 20 years in captivity) is highly adaptive for a species in which recruitment is potentially low.

Pasture composition is critically important to the recruitment of young. Overgrazing by the Rabbit and domestic stock can shift pasture composition to favour annual and exotic plants whose period of maximum productivity is out of phase with weaning of young, leading to high infant mortality and dwindling of wombat populations.

R.T. WELLS

K. GRIFFITHS

HANS & JUDY BESTE

*Common Wombats can be affectionate pets when juvenile but become dangerously aggressive when mature.*

*Over much of its range, the Common Wombat inhabits forests, moving at night into clearings and forest edges to graze.*

# Common Wombat

## *Vombatus ursinus*

### (Shaw, 1800)

vom-bah'-tus  er-seen'-us: 'bear-like wombat'

The Common Wombat is a large, herbivorous, burrowing marsupial. Early settlers often called it a 'badger' because of its burrowing behaviour but it shows more convergence with marmots. Its main habitat is the forest-covered, often mountainous areas of south-eastern Australia; its requirements include a temperate, humid microclimate, suitable burrowing conditions and native grasses for food. In southern Queensland and northern New South Wales, it occurs only in sclerophyll forest above 600 m but, further south, particularly in South Australia and Tasmania, it also occurs at lower altitudes and in more open vegetation such as woodland, coastal scrub and heathland.

Burrows vary considerably in location and size, patterns of use and functions. The smallest, most numerous, 'minor' burrows usually represent an abortive excavation at an unfavourable site. 'Medium' burrows (2–5 metres long) are mainly used for temporary refuge but some of these are eventually developed into 'major' bur-rows (up to 20 metres long) for diurnal shelter. Many major burrows contain more than one bedding chamber, divide or connect underground and have several entrances. Slopes above creeks and gullies are favoured sites.

Each Common Wombat visits from one to four burrows within its home range each night and up to 13 over several weeks. Others also use these burrows if their home ranges overlap, either at separate times or simultaneously. Home range (5–23 hectares) varies according to the distribution of burrows in relation to feeding areas but, despite overlap, individuals maintain separate feeding areas through scent-marking, vocalisation and aggressive behaviour. The characteristically rectangular dung may frequently be seen on logs, rocks and other prominent objects.

During the summer, the Common Wombat is mainly nocturnal, emerging above ground at night when the air is cool. In winter and other cool periods, some emerge during daylight to bask or graze. The main food is native grasses but also includes sedges, rushes and the roots of shrubs and trees. During three to eight hours of grazing each night, an individual may travel 3 kilometres.

Populations are limited by availability of food. Densities of 0.3–0.6 individuals per hectare occur in areas of native forests adjoining open, grassy areas but are much lower in less favourable habitats. Stable populations are maintained by emigration of immature animals or their relegation to transient status pending the death of resident adults.

Breeding may occur at any time of the year. The rear-opening pouch of the female contains two teats but usually only one young is born. It remains in the pouch for approximately six months and at heel for up to another 11 months. Common Wombats become sexually mature after two years and may live at least five years in the wild and up to 20 in captivity.

Fossil evidence indicates that the range of the Common Wombat has contracted southwards and eastwards since the Pleistocene. European settlement has accelerated the process and its distribution is now far more discontinuous, the colonies in south-western Victoria and south-eastern South Australia being only remnants. The smaller subspecies, *Vombatus ursinus*

### *Vombatus ursinus*

**Size**
*Head and Body Length*
900–1150 (985) mm
*Tail Length*
circa 25 mm
*Weight*
22–39 (26) kg

**Identification**  Large, naked nose; nasal bones narrower than *Lasiorhinus*. Coarse, thick fur. Ears short, slightly rounded.
**Recent Synonyms**  *Phascolomis* or *Phascolomys mitchelli*, *Phascolomis* or *Phascolomys ursinus*, *Vombatus hirsutus*.
**Other Common Names**  Naked-nosed Wombat, Coarse-haired Wombat, Island Wombat, Tasmanian Wombat, Forest Wombat.
**Status**  Common.
**Subspecies**  *Vombatus ursinus ursinus*, Flinders Island.
*Vombatus ursinus tasmaniensis*, Tasmania.
*Vombatus ursinus hirsutus*, south-eastern mainland.
**References**
McIlroy, J.C. (1976). Aspects of the ecology of the Common Wombat, I. Capture, handling, marking and radio-tracking techniques. *Aust. Wildl. Res.* 3: 105–116.
Triggs, B. (1988). *The Wombat: Common Wombats in Australia*. University of New South Wales Press, Sydney.

*ursinus*, the first wombat to be studied by Europeans, is now extinct on all Bass Strait islands except Flinders Island. The species is protected in most States except eastern Victoria where it is classed as vermin, mainly because of its damage to rabbit-proof fences.

**Vombatus ursinus**

J.C. McILROY

# SUBORDER PHALANGERIDA

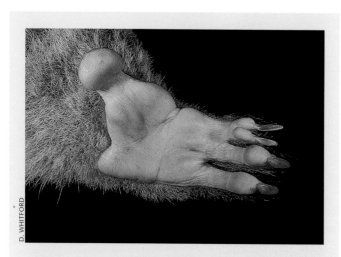

*Hindfoot of the Green Ringtail Possum, showing the typical possum arrangement of a powerful, opposable first digit, conjoined second and third digits and fourth and fifth digits of comparable size.*

It has long been recognised that possums and kangaroos are more similar to each other than to wombats and the Koala but the classification of the marsupials did not reflect this. The recently erected suborder Phalangerida is well supported on biochemical grounds but evades rigid definition on anatomical characters. In respect of living species it could be said that its members are tailed diprotodonts.

The group falls into five superfamilies: the Phalangeroidea (cuscuses, brushtail possums and the Scaly-tailed Possum); the Burramyoidea (pygmy-possums); the Petauroidea (ringtail possums, wrist-winged gliders and the Greater Glider); Tarsipedoidea (Honey Possum and, possibly, the Feathertail Glider and Possum) and the Macropodoidea (rat-kangaroos and kangaroos). This classification implies that the differences between various possum groups may be no less than between any possum and the kangaroo group. Not all authorities agree with this postulate.

# Superfamily Burramyoidea

In the first edition of this work, the pygmy-possums and Feathertail Glider were included in the family Burramyidae. Since then, the Feathertail Glider and the New Guinean Feathertail Possum have been removed to the family Acrobatidae. To indicate the evolutionary separation between the pygmy-possums and other members of the suborder Phalangerida, it was necessary to erect the superfamily Burramyoidea, the characteristics of which are those of the family Burramyidae. Some authorities would retain Acrobatidae within the Burramyoidea, but there are no strong grounds for this (see the entry for the superfamily Tarsipedoidea, page 257).

# FAMILY BURRAMYIDAE
## Pygmy-possums

*The Eastern Pygmy-possum has a brush-tipped*
*tongue and feeds mostly on nectar.*

This family which includes the smallest of the possums (10–50 grams) comprises the terrestrial Mountain Pygmy-possum, *Burramys,* and four arboreal species of the genus *Cercartetus,* one of which also occurs in New Guinea. The tail is long, slender, lightly haired and strongly prehensile. Pygmy-possums occupy a wide range of environments, including cool heath and savanna, alpine regions, temperate wet sclerophyll forest and tropical rainforest.

Pygmy-possums are primarily insectivorous but the Eastern Pygmy-possum includes much nectar in its diet and the Little Pygmy-possum can eat small lizards. The Mountain Pygmy-possum includes hard seeds in its diet. The molar teeth of all species have low, smooth cusps.

All pygmy-possums construct nests, usually in a tree hollow, but the nest of the Mountain Pygmy-possum is made on the ground and, in winter, lies under snow. All are able to become torpid for varying periods: the Mountain Pygmy-possum hibernates for as long as seven months in the year. Females have four teats (six in the Western Pygmy-possum). The forwardly opening pouch is least well developed in the Little Pygmy-possum, which is the smallest species.

# Mountain Pygmy-possum

## *Burramys parvus*

bu'-rah-mis par'-vus: 'small burra-burra (stony place)-mouse'

The Mountain Pygmy-possum is the only Australian mammal limited in its distribution to alpine and subalpine regions, where there is a continuous period of snow cover for up to six months. Although other burramyids are known to enter periods of torpidity in winter, this is the

*Burramys parvus*

only marsupial known to undergo long periods of hibernation, similar to those of classical placental hibernators of the northern hemisphere. It was first described in 1895 from a fossil found at Wombeyan caves, New South Wales, with further fossils discovered at Buchan Caves, Victoria, in the early 1960s. There was great excitement when in August, 1966, a live animal was collected in a ski hut at Mount Hotham, Victoria. The skull and mandible fragments had aroused great interest because of the distinctive, high sectorial upper premolar with a serrated cutting edge and grooved sides, a type of tooth which was found among living mammals only in

## *Burramys parvus*

### Size
*Head and Body Length*
110 mm (males)
111 mm (females)
*Tail Length*
138 mm (males)
136 mm (females)
*Weight*
30–54 (41) g (males)
30–82 (42) g (females)

**Identification** Grey-brown above, sometimes darker in mid-dorsal area continuing to top of head; dark ring around eye; pale grey-brown to cream below, developing to bright fawn-orange in ventral area and flanks of adults (especially males) during the breeding season. Fur fine, dense. Tail long, thin, scaly with sparse, short hairs, prehensile.
**Recent Synonyms** None.
**Other Common Names** Burramys.
**Status** Rare, vulnerable, but locally moderately common to abundant in very restricted areas.
**Subspecies** None.

### References
Geiser, F. and L.S. Broome (1993). The effect of temperature on the pattern of torpor in a marsupial hibernator. *J. Comp. Physiol. B.* 163: 133–137.

Mansergh, I.M and L.S. Broome (1994). *The Mountain Pygmy-possum of the Australian Alps.* Univ. NSW Press, Sydney.

Mansergh, I.M. and D. Scotts (1990). Aspects of the life history and breeding biology of the Mountain Pygmy-possum *Burramys parvus* (Marsupialia: Burramyidae) in alpine Victoria. *Aust. Mammal.* 13: 179–191.

L. BROOME

*The Mountain Pygmy-possum inhabits alpine areas and spends the winter hibernating in boulder deposits, under snow.*

some potoroids. The fossil record and further fossils found at Jenolan Caves, New South Wales, indicate that its range has been decreasing with the receding snowline. It is now known to occur in four isolated populations, separated by river valleys, on the highest peaks of south-eastern Australia at altitudes above 1400 metres and up to 2230 metres (Mount Kosciusko). Three populations occur between Mount Bogong and Mount Hotham, Victoria and a fourth in Kosciusko National Park.

The habitat requires accumulations of boulders, which have formed below mountain peaks (blockfields) and in gullies (blockstreams) from periglacial weathering processes and associated shrubby heathland. The boulders ameliorate temperature extremes and provide deep hibernacular and sheltered nesting sites. Temperatures at ground level during winter are 0–2°C. Boulders are also used as summer aestivation sites by the migratory Bogong Moth, *Agrotis infusa*, which, along with other arthropods such as caterpillars, millipedes, beetles and spiders,

form a major part of the diet. The seeds and fruits of heathland shrubs (including Mountain Plum-pine, *Podocarpus lawrenceii*, which is mostly restricted to the boulderfields), and Snow-beard Heath, *Leucopogon montanus*, become increasingly important in late summer and autumn. Many of the boulder patches are small (less than half a hectare) and often more than a kilometre apart, but there appears to be substantial movement between them, especially by males and dispersing juveniles. Radio-tracked males frequently travelled 1.5 kilometres in a night and even females with large pouch-young commuted up to 1 kilometre from nest sites to the highest peaks where Bogong moths are most abundant.

The size of home ranges, population density, sex ratios and longevity vary considerably and are related to the distribution and quality of habitat. In the highly productive, basalt boulder-heaths at Mount Higgenbotham, Victoria, adult females are sedentary and have small (0.06 hectare) overlapping home ranges. Densities can be as high as 116 individuals per hectare. The oldest female recorded was over 12 years old. Home ranges are larger in the granite boulder-heaths of New South Wales, varying from 0.12 and 7.7 hectares, with densities averaging about 6.5 individuals per hectare. The oldest female trapped was five years old.

The single litter is born between late October and early November in Victoria and up to a month later in NSW, following snowmelt. Gestation is 13–16 days. Supernumary young are born and most females carry a full complement of four young in the pouch. Growth is rapid, in adaptation to the short, alpine summer. Pouch life lasts approximately 30 days, followed by a nestling period of 30–35 days, after which young become

*In cold weather or when food is scarce, the Mountain Pygmy-possum curls into a ball and hibernates.*

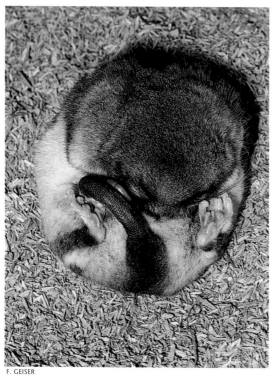

F. GEISER

independent. They breed in the following spring and reach adult weight by the end of their second summer. Around the time that young leave the pouch, most adult males disperse from the habitats occupied by the breeding females to surrounding areas of lower habitat quality.

Because of the distribution of resources (for example, shallower boulderfields and fewer Bogong Moths), these 'male habitats' are often at lower altitudes. Older males are followed, a month or two later, by juvenile males and some dispersing juvenile females. In the highest quality habitats (for example, Mount Higginbotham) sex ratios during the breeding season are heavily biased towards females because of the lower survival of males. Males usually live for less than two years; the oldest recorded individual being four years old.

Following breeding, animals fatten extensively. Body weight is approximately doubled before the period of winter snowfalls. Adult females enter hibernation as early as February at Mount Higgenbotham but as late as April in New South Wales. Juveniles and males are active for an additional one to two months, those at the lowest altitudes entering hibernation as late as May or early June. Hibernation lasts for up to seven months in adults and five to six months in juveniles. Body temperatures are regulated at around 2°C for periods lasting up to 20 days, interspersed by short periods of normal temperature of less than one day.

Like some potoroids, the Mountain Pygmy-possum caches food, burying seeds that it may or may not excavate to eat during periods of food shortage. This behaviour is well illustrated in captivity but less so in the wild. captive animals have hibernated for up to 185 days but caches might be used by individuals which hibernate for lesser periods.

The total extent of the Mountain Pygmy-possum habitat is less than 10 square kilometres and the total population may be no more than 2600 adults.

L.S. BROOME

*The Long-tailed Pygmy-possum inhabits tropical rain-forest, climbing with the aid of a long, prehensile tail.*

D. WHITFORD

# Long-tailed Pygmy-possum

## *Cercartetus caudatus*

(Milne-Edwards, 1877)

ser'-kar-tay'-tus  kaw-dah'-tus: 'notably tailed cercartetus' (meaning unknown)

Nocturnal and mainly arboreal, this tiny marsupial is seldom seen. In Australia, it has been found only in the Queensland rainforests between Townsville and Cooktown—south of the Daintree River in rainforests and fringing casuarina forests at an altitude of 300 metres or more and, to the north of the river, on the coastal plain and in fringing eucalyptus—melaleuca forests. Since it also occurs in New Guinea, its apparent absence from the rainforests of Cape York Peninsula is puzzling but this may reflect inadequate collecting.

Little is known of its normal diet although it has been seen to feed on nectar from the flowers of the Bumpy Satin Ash. Captive animals have been maintained on a semi-liquid 'Phalanger Mix' developed by the Melbourne Zoo, supplemented by a wide selection of chopped fruits. Tinned dog food and dog biscuits are eaten, as are grasshoppers and moths.

A more or less spherical nest is constructed from leaves or fern fronds. Such nests have been found in Australia in a clump of ferns and in a hollow stump; in New Guinea in a tree hollow, beneath the dead fronds of pandanus palms; and attached to the stems of tall pit-pit grass. On cool winter days an animal may become torpid: stiff and cold to the touch and with the lips pulled back over the teeth, it appears to be dead. Torpor has not been observed to extend overnight.

In Australia, births have been recorded in January and February and from late August to early November. A coincident cycle of change in the size of the testes of males is further evidence that breeding takes place twice a year. A somewhat similar pattern has been observed in New Guinea, with a period of reproductive inactivity from late May to early August. The female has four teats and may have from one to four young

M. TRENERRY

*The Long-tailed Pygmy-possum is usually solitary.*

at a time. The gestation period is unknown but young leave the pouch when about 45 days old and weighting 5–7 grams. They make quiet 'sik-sik-sik' calls when distressed. A mother defends her young by standing over them on her hindlegs with the forelegs extended, uttering a quiet hiss.

The Long-tailed Pygmy-possum usually forages alone, but feeding aggregations of up to four individuals have been seen. Studies in New Guinea showed that lactating females share a nest with their offspring but not with other adults. However, males and to a lesser extent, non-lactating females, sometimes share a nest, the largest known group found comprising four adult males and a sub-adult female. In Australia, three males and a female have been collected from one nest, but some nests contain only a single individual.

### Cercartetus caudatus

**Size**
*Head and Body Length*
103–108 (106) mm
*Tail Length*
128–151 (135) mm
*Weight*
25–40 (30) g

**Identification** Brownish-grey above, pale grey below; black patch around eye. Large crinkly ears. Distinguished from superficially similar *Pogonomys* by typical possum hindfoot and by tail which is slightly thickened and furred at base.

**Recent Synonyms** *Eudromicia macrura.*
**Other Common Names** Queensland Pygmy-possum.
**Status** Common, limited.
**Subspecies** *Cercartetus caudatus caudatus*, New Guinea.
*Cercartetus caudatus macrurus*, Australia.
**Extralimital Distribution** Papua New Guinea.
**References**
Dwyer, P.D. (1977). Notes on *Antechinus* and *Cercartetus* (Marsupialia) in the New Guinea Highlands. *Proc. Roy. Soc. Qld*. 88: 69–73.
Atherton, R.G. and A.T. Haffenden (1982). Observations on the reproduction and growth of the Long-tailed Pygmy-possum, *Cercartetus caudatus* (Marsupialia: Burramyidae), in captivity. *Aust. Mammal*. 5: 253–260.

*Cercartetus caudatus*

Longevity in the wild is unknown but an animal born in captivity lived for 38 months.

There are no direct observations of predation in the wild but skeletal remains have been found in owl pellets in New Guinea and a captive individual was eaten by a free-living Children's Python. The Spotted-tail Quoll is very probably a predator in Australia. Clear-felling of rainforest is the greatest threat to the survival of the Long-tailed Pygmy-possum.

R.G. ATHERTON AND A.T. HAFFENDEN

A.C. ROBINSON

# Western Pygmy-possum

## *Cercartetus concinnus*

### (Gould, 1845)

kon-sin'-us: 'neat cercartetus'

Asleep in its gum-leaf nest during the day, the Western Pygmy-possum is a minute ball of soft, red-brown fur. The thin ears are folded forwards, covering the eyes, the head rests on the flat-coiled tail, the nose touches the chest and limbs are tucked in under the belly. Taken from the nest, it feels cold and almost weightless. It does not arouse easily even in the warmth of one's hand and it can be returned to its nest unaware of having been disturbed. Towards evening it wakes and, soon after dark, scurries from the nest, its quick movements and wide eyes in sharp contrast to its previous unresponsive immobility. The many very long vibrissae are never still and the large ears turn to catch the faintest rustling.

*Because it lives close to the ground, the Western Pygmy-possum is frequently taken by the feral Cat.*

The Western Pygmy-possum is found in mallee heath and in dry sclerophyll forest, especially where there is an undergrowth of shrubs, such as banksias, grevilleas, callistemons and melaleucas: in mallee and woodland it is restricted to shrubby areas. It may sometimes range beyond these preferred habitats, one individual having been found living in a mulga–saltbush association.

Being small and nocturnal, it is seldom seen. Although it frequently moves about on the ground, it is mainly arboreal and is very agile among small branches which it grips with its well-developed toe-pads and its strongly prehensile tail. The forefoot, with tiny claws on the upper surface of the toes, has some resemblance to a human hand and is used to grasp food while it is being eaten, inedible portions such as the wings and hard parts of insects being discarded. The daytime shelter is usually a leaf-lined nest in a tree hollow or in the leaves of a grass-tree (*Xanthorrhoea*) but individuals have been found sleeping in the disused nests of Babblers and on the ground, amongst leaves or under a stump. It

W.G. BREED

## Cercartetus concinnus

### Size
*Head and Body Length*
71–106 (81) mm
*Tail Length*
71–96 (86) mm
*Weight*
8–20 (13) g

**Identification** Fawn or reddish-brown above, white below; tail finely scaled and without fur for all but the proximal one-sixth. Claws on second to fifth fingers do not extend beyond distal pads. Smaller than *Cercartetus nanus*. Distinguishable from *C. lepidus* on dentition: three molar teeth each side above and below, upper third premolar with a single point, lower third premolar tiny.
**Recent Synonyms** *Dromicia concinna*.
**Other Common Names** South-western Pygmy-possum, Lesser Dormouse-phalanger, Elegant Dormouse Opossum.
Indigenous name: Mundarda.
**Status** Common, limited.
**Subspecies** None.
**Reference**
Wakefield, N.A. (1963). The Australian Pygmy Possums. *Vict. Nat.* 80: 99–116.

*The Western Pygmy-possum lives mostly in heathland, feeding on nectar and insects.*

Feral and domestic cats find the Western Pygmy-possum an easy prey but the reproductive rate of the species is adapted to heavy predation. A greater threat comes from rural or urban development within its range, for this almost inevitably leads to clearing of the scrub patches that are its preferred habitat and hence to reduction or extinction of a local population.

M.J. Smith

is frequently caught in pitfall traps during faunal surveys and occasionally seen during agricultural activities such as clearing and fencing.

Breeding occurs throughout most of the year and females may rear two or even three litters in close succession. The pouch contains six teats and as many as six young may be reared in a litter. Within a day or two of giving birth, the female may mate again, the resultant embryos growing slowly while she suckles the litter in the pouch and being born more than 50 days later, following the weaning of the older litter. The litter remains in the pouch for about 25 days, after which the blind, semi-naked young are left in the nest.

Kangaroo Is.

*Cercartetus concinnus*

D. WATTS

# Little
# Pygmy-possum
## *Cercartetus lepidus*

### (Thomas, 1888)

lep'-id-us: 'scaly (-tailed) cercartetus'

Smallest of all the possums, this species was once thought to be peculiar to Tasmania but has recently been found to occur in two areas of south-eastern mainland Australia. It occupies a wide range of tree-dominated habitats from the mallee scrub of Victoria, with an annual rainfall of 300 millimetres, to dry sclerophyll forests on Kangaroo Island and eastern Tasmania and the wet sclerophyll forests of western Tasmania, where annual precipitation may reach 1200 millimetres. It is not known from rainforests.

Because of the Little Pygmy-possum's minute size and secretive, mostly nocturnal behaviour, little is known of its life and habits.

Individuals that come to hand are usually discovered accidentally, as when a split log or stump discloses a rough nest made of strips of fibrous bark in a cavity in the dead wood. Nests have also been found beneath overturned turf on ploughed ground, in the decaying centre of a green tree, in wall cavities and in various man-made objects which provide shelter and

*The Little Pygmy-possum eats some nectar but is primarily a predator on arthropods and very small lizards.*

## *Cercartetus lepidus*

**Size**
*Head and Body Length*
50–65 (64) mm
*Tail Length*
60–75 (71) mm
*Weight*
6–9 (7) g

**Identification** Fur soft, pale fawn above, grey below. Distinguishable from other burramyids by small size and grey belly. The presence of a small fourth molar distinguishes it from *Cercartetus concinnus* and *C. nanus*.
**Recent Synonyms** *Dromicia lepida, Eudromicia lepida.*
**Other Common Names** Tasmanian Pygmy-possum.
**Status** Sparse.
**Subspecies** None.
**References**
Green, R.H. (1980). The Little Pygmy-possum, *Cercartetus lepidus* in Tasmania. *Rec. Queen Vict. Mus.* No. 68: 1–12.
Green, R.H. (1993). *The Fauna of Tasmania: Mammals.* Potoroo Publishing, Launceston.
Turner, V. and G.M. Mackay (1989). Burramyidae (in) D.W. Walton and B.J. Richardson (eds) *Fauna of Australia, Vol. 1B, Mammalia.* Aust. Gov. Publ. Serv., Canberra. pp. 652–664.

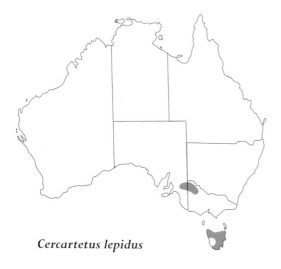

*Cercartetus lepidus*

mother forages but, if for any reason they have to be moved, they cling to the fur of the mother's back. At the age of three months, having attained almost adult size, they become independent.

The principal predators on the Little Pygmy-possum are carnivorous mammals, owls and snakes. Fossil remains in eastern New South Wales demonstrate that the range of the species was contracting southwards long before the arrival of Europeans and, since then, extensive forest clearing has removed much of the sclerophyll

D. WATTS

*After quitting the pouch, young Little Pygmy-possums are suckled in a well-made spherical nest.*

seclusion. The abandoned nests of small birds may be used as temporary sleeping places.

The observation that, in captivity, the Little Pygmy-possum eats honey and other sweet food led to the belief that it is a nectar-feeder. In fact, it is primarily insectivorous and feeds on a wide range of insects, spiders and even small lizards, securing the prey with its forepaws and tearing away edible portions with its teeth.

Although an excellent climber and possessing a prehensile tail that can support its weight, it normally lives close to the ground or in dense scrub, shunning the higher branches of trees where it would be vulnerable to attack by owls.

Like other burramyids, the Little Pygmy-possum undergoes periods of torpor as a means of reducing energy expenditure, particularly during the winter when food is scarce. The torpid animal curls into a tight ball with its face buried in its lower abdomen and its temperature drops almost to that of its surroundings. It may remain immobile for up to six days. Held in the warmth of cupped hands, it may take as long as 20 minutes to return to full activity.

The pouch is shallow when no young are present and encloses four teats. The few available records from Tasmania indicate that litters of about four are born between September and January. When half-grown and too large for the pouch, the young are left in a nest while the

forest that may have provided habitat for it on the mainland. Sufficient suitable habitat remains in Tasmania to ensure the short-term future of the species but the long-term effect of clear-felling forestry practice is difficult to predict.

R.H. GREEN

# Eastern Pygmy-possum

## *Cercartetus nanus*

## (Desmarest, 1818)

nah'-nus: 'dwarf cercartetus'

Once thought to be principally insectivorous, the Eastern Pygmy-possum is now known to feed largely on nectar and pollen which it gathers from banksias, eucalypts and bottlebrushes with a brush-tipped tongue. When pollen is abundant in the habitat it may supply the possum with most, if not all, of its protein requirement. It is not particularly destructive to the flowers and may play an important role in pollination. When flowers are unavailable, soft fruits may be eaten. Insects are consumed throughout the year and may be important in the diet of possums inhabiting wet forest where fruit and blossoms are less abundant. Captive animals readily eat spiders, mantises, termites, grasshoppers, beetles and beetle larvae. Flying moths are caught with the forepaws, the wings are bitten off and the body eaten. Feeding occurs in short bursts followed by elaborate grooming.

The Eastern Pygmy-possum is found from rainforest through sclerophyll forest to tree heath. Banksias and myrtaceous shrubs and trees are favoured as food sources and nesting sites in the drier habitats. Its minute size allows it to nest in very small spaces during the day. Hollows in trees are favoured, but spherical nests, of about 6 centimetres in diameter, constructed of short, shredded bark, have been found between the wood and bark of eucalypts; abandoned bird-nests and shredded bark in the forks of tea-trees are other nest sites. Nest-building seems to be

*Like other members of its genus, the Eastern Pygmy-possum is an agile climber. Its claws are reduced to rather nail-like structures.*

P. GERMAN

## Cercartetus nanus

**Size**

*Head and Body Length*
70–110 (91) mm

*Tail Length*
75–105 (89) mm

*Weight*
15–43 (24) g

**Identification** Fawn above, white below. Larger than *Cercartetus lepidus*; juveniles with longer hair. Smaller and paler than *C. caudatus*. Distinguished from *C. concinnus* by having ventral hairs tipped with white (completely white in *C. concinnus*).

**Recent Synonyms** None.

**Other Common Names** Pygmy-possum, Common Dormouse-phalanger, Dormouse Opossum, Possum Mouse.

**Status** Common.

**Subspecies** Two subspecies have been proposed: *Cercartetus nanus nanus* from Tasmania and *Cercartetus nanus unicolor* from the mainland but the difference between the two populations may not be greater than within the mainland populations.

**Reference**

Ward, S.J. (1990). Life history of the eastern pygmy-possum, *Cercartetus nanus* (Burramyidae: Marsupialia) in south-eastern Australia. *Aust. J. Zool.* 38: 287–304.

*Cercartetus nanus*

restricted to mothers with young. The Eastern Pygmy-possum appears to be mainly solitary, each individual using several nests. The home ranges of males (about 0.68 hectares) are larger than those of females (0.35 hectares) and not exclusive. One individual was observed to move at least 125 metres in one night.

It is generally nocturnal, becoming active soon after dusk. Although not speedy, it is a very agile climber, making use of its grasping tail and feet. It is occasionally caught in pitfall traps or postholes, indicating that it comes to the ground at times. Activity is reduced in winter when much time is spent in torpor; the body tightly curled, ears folded and the internal temperature close to that of the surroundings. Towards the end of summer it increases in weight and the base of the tail may be noticeably thickened with stored fat. Even under seemingly ideal conditions, a captive animal may regularly become torpid for periods from several days or up to nearly two weeks.

On the mainland, births may occur at any time of year if abundant food supplies are available, but most occur in late spring to early autumn. In Tasmania, breeding appears to be confined to late winter and spring. Usually there are six teats, of which four are functional: four young comprise the normal litter. Young remain in the pouch for about 30 days, after which they are left in a nest and are weaned when 65 days old. At weaning, the young are about half the adult weight and are independent of their mother, but they may continue to associate with their siblings and other juveniles. Sexual maturity may be attained as early as four-and-one-half to five months. Females do not have embryonic diapause but may return to oestrus late in lactation and often give birth to the next litter immediately after the previous litter is weaned. Most females produce two litters within a season and three litters within a 12 month period have been recorded. Maximum longevity in the wild is at least four years and one captive animal lived for eight years.

The Eastern Pygmy-possum has little vocalisation but may hiss loudly when provoked.

V. TURNER AND S.J. WARD

# Superfamily Petauroidea

The erection of this superfamily serves to indicate that the member families, Petauridae and Pseudocheiridae are more closely related to each other than to any other possums.

*Restricted to Mountain Ash forest, Leadbeater's Possum is rare and endangered.*

## FAMILY PETAURIDAE

*Striped Possum, Leadbeater's Possum and wrist-winged gliders*

In the first edition of this work, the Petauridae was regarded as comprising the ringtail possums and Greater Glider in addition to the wrist-winged gliders and their non-gliding relatives. It is now restricted to the latter groups. The lower incisors of petaurids are long, sharp and protuberant; the molars have low, smooth cusps. All species have a prominent dark dorsal stripe that extends onto the forehead. The tail is long, well-furred and prehensile.

Two subfamilies are recognised. The Petaurinae comprises the seven species of *Petaurus* (four of which occur in Australia) and Leadbeater's Possum, *Gymnobelideus*, which has only vestiges of a gliding membrane. The term, 'wrist-winged glider' serves to distinguish species of *Petaurus* from the Greater Glider, drawing attention to the extension of the gliding membrane to the wrist (actually to the tip of the fifth finger). The Dactylopsilinae comprises the four species of striped possums from New Guinea, one of which occurs in North Queensland rainforest.

The dactylopsilines are unique among marsupials in having an extremely long fourth finger, terminating in a powerful curved claw. They are also characterised by longitudinal stripes. The dark dorsal stripe is broad and bordered on each side by a white stripe extending from the head to the base of the tail; the upper flanks and outer surfaces of the limbs are dark; the belly and lower flanks are white. The general appearance is skunk-like.

Petaurids vary considerably in their diet. Petaurines feed mainly on insects and other arthropods when these are abundant but, particularly in winter, also consume the sap and gum of eucalypts and acacias, promoting the flow of fluids by maintaining incisions in the bark of these trees. Dactylopsilines are exclusively insectivorous, feeding mainly in adult and larval insects that burrow in trees. These are exposed by biting into the wood and are extracted with the aid of the tongue and the long fourth finger.

The forwardly opening pouch of petaurids enclosed two or four teats. In some species, the pouch has a median septum, dividing it, incompletely, into right and left compartments.

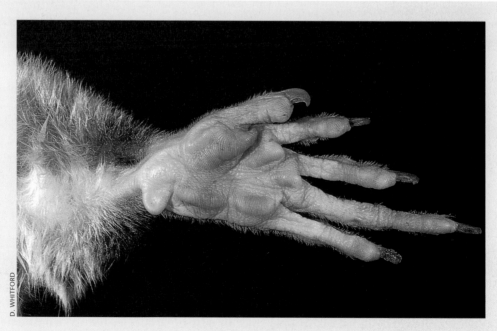

*Forefoot of the Common Striped Possum, showing the enormous development of the fourth finger.*

*The Yellow-bellied Glider.*
D. WHITFORD

## Subfamily Dactylopsilinae

Members of this group have longitudinally striped fur and a very elongate fourth finger.

# Striped Possum
## *Dactylopsila trivirgata*

### Gray 1858

dak'-til-op'-sil-ah trie'-ver-gah'-tah: 'three-striped naked-finger'

M. TRENERRY

While the initial impact of this spectacular animal stems from its impressively bold, skunk-like appearance, more lasting memories usually remain of its clinging, pungent odour. Although uncommon, this slightly built possum is widely distributed throughout rainforests and adjacent woodlands in Queensland from Mount Spec, near Townsville to the Iron Range. It is widespread in lowland New Guinea.

It is amazingly agile, its erratic movements being sometimes so fast that, were it not so noisy, it would be impossible to observe for long. The strange method of walking along horizontal branches is characterised by a peculiar 'rowing' motion resulting from the simultaneous swinging movement of diagonally opposite limbs. In this manner it moves with a lithe, flowing gait even during high-speed chases between rivals. If, in such a pursuit through the forest canopy, an animal reaches the outer end of the limb along which it is rushing, it may either leap toward a neighbouring tree, hurling itself into the leafy branches or, after a lateral swaggering of the body, catapult itself from the branch to an over-hanging vine on the next tree. When the rustling, snorting and scratching sounds that accompany the possum's movements cease, its position can usually be traced by the noises of its slurping and chewing or from the shower of debris that falls from its perch.

Most food-finding activities involve the investigation of tree trunks, limbs and fallen rotting logs in search of wood-boring grubs and other insects. To expose a potential meal the possum uses its two long, bayonet-like lower incisors to stab and gouge away the bark. If the food morsel is lying protected in a deep hole or crack, it is

*Dactylopsila trivirgata*

M. TRENERRY

*The Striped Possum feeds mainly on tree-burrowing larvae of beetles, biting into the wood with its powerful incisors and extracting the insect with its fourth finger.*

shrieks of 'gar-gair, gar-gair'. Copulation in this case lasted approximately ten minutes and was characterised by the intertwining and thrashing of tails, similar to that seen in disgruntled domestic cats, both animals screaming throughout the performance. Mating may occur from February to August and up to two young are born. There are two teats in the pouch.

Although the Striped Possum is a conspicuous animal, its low densities and shyness have made it one of the least known of Australian possums.

S.M. VAN DYCK

extracted either with the long tongue or by using the sharp claw on its elongated fourth finger as a probe and skewer. Leaves, fruits, small vertebrates and the honey of native bees are reported to be eaten, but insects provide the greater part of the diet.

During the day the Striped Possum sleeps in a leafy nest made inside a tree hollow or amongst clumped epiphytes. Almost nothing is known of its social life and the reason for its strong, sweet, musty odour, which resembles newly hardened fibreglass, is not understood. One observation recording the mating of wild animals noted intense raucous rivalry between two males over an oestrous female, with all three gurgling their continuous, night-shattering, rolling, guttural

### *Dactylopsila trivirgata*

**Size**
*Head and Body Length*
256–270 (263) mm
*Tail Length*
310–340 (325) mm
*Weight*
246–528 (423) g

**Identification** Slightly built; with variable pattern of black and white stripes along head, body and tail. Toes very lightly furred and elongated, fourth finger particularly elongated.
**Recent Synonyms** *Dactylopsila picata*.
**Other Common Names** Striped Phalanger, White-footed Phalanger, Common Striped Possum.
**Status** Sparse, limited (in Australia). More widespread in New Guinea.
**Subspecies** Three subspecies in New Guinea. Australian form is *Dactylopsila trivirgata picata*.
**Extralimital Distribution** Papua New Guinea.
**References**
Smith, A. (1982). Is the Striped Possum (*Dactylopsila trivirgata*, Marsupialia, Petauridae) an arboreal anteater? *Aust. Mammal.* 5: 229–234.
Van Dyck, S. (1990). Picture adjustments to the old black and white. *Wild. Aust.* 27: 8–9.

## Subfamily Petaurinae

Members of this group are most simply designated as non-dactylopsiline petaurids.

# Leadbeater's Possum

## *Gymnobelideus leadbeateri*

### McCoy 1867

jim'-noh-bel-id'-ay-us led'-beet-er-ee:
'Leadbeater's naked-*Belideus*'
(*Belideus* is a junior synonym of *Petaurus*.)

The cool, misty mountain forests of the Central Highlands of Victoria are home to the rare and endangered Leadbeater's Possum. This shy, fast-moving inhabitant of our tallest forests is difficult to observe, except for a brief period when it emerges from its nest at dusk and moves rapidly through the forest canopy, often making spectacular leaps from tree to tree. Colonies of up to eight animals build a communal nest of shredded bark in the hollow centre of a large dead or living tree, 6–30 metres above ground. The nest tree is close to the centre of a territory of 1–2 hectares that is actively defended from members of adjoining colonies. Each colony usually comprises a monogamous breeding pair and one or more generations of offspring, but may also include one or more apparently superfluous males. There is much visitation between colonies and membership turnover is high. Cohesion within a group is maintained by spread of salivary odours during extensive grooming in the nest and by mutual licking of the tail-base by the dominant breeding pair. Females are more socially aggressive than males and readily attack and pursue animals of either sex from another colony. Females also attack and forcibly disperse their daughters when they reach sexual maturity, but leave their sons unmolested. In consequence, female young disperse earlier (ten months) than males (15 months) and suffer a higher mortality due to exclusion from established colonies. Mean reproductive life span in wild adult females is 1.6 years (seven years in captivity). Dispersing males may be absorbed into neighbouring colonies as non-breeding residents or form bachelor groups while awaiting mating opportunities. Social encounters are often accompanied by a repertoire of hissing and chattering vocalisations, including an 'alarm hiss' which elicits a mobbing attack response from other sexually mature members of a colony, when emitted by an individual grappling with a potential competitor or predator.

The diet includes crickets, beetles, spiders and other arthropods, harvested beneath the shedding bark of eucalypts and a range of exudates including the gums of Silver and Hickory

### *Gymnobelideus leadbeateri*

**Size**
*Head and Body Length*
150–170 (160) mm
*Tail Length*
145–180 (172) mm
*Weight*
100–135 (122) g (spring)
110–166 (133) g (autumn)

**Identification** Grey to greyish-brown above, with prominent dark mid-dorsal stripe; pale below. Distinguishable from the related Sugar Glider by absence of gliding membrane and club-shaped tail, broader near tip than at base.
**Recent Synonyms** None.
**Other Common Names** Fairy Possum.
**Status** Rare, endangered.
**Subspecies** None.

*Leadbeater's Possum has the appearance of a Sugar Glider without its gliding membrane.*

E. BEATON/TERRA AUSTRALIS

Wattles, manna produced at sites of insect-caused and other physical damage on eucalypt leaves and branches and honeydew secreted by sap-sucking lerp insects. Arthropods are an important source of protein, needed for reproduction and are thought to comprise approximately 20 per cent of daily energy intake, while exudates licked from plant surfaces comprise the remaining 80 per cent. Food is most abundant during spring and summer, but the presence in the diet throughout the year of an undescribed species of tree cricket which shelters under the shedding bark of Mountain-Ash eucalypts, may be an important factor permitting breeding in winter as well as spring.

Most births occur in May and June or October and November. Females have four teats and produce one to two young in a litter. Average litter size is 1.5. Young remain in the pouch for 80–93 days and first emerge from the nest after 111 days. They forage alone after leaving the nest and at this time are especially vulnerable to predation by owls.

Leadbeater's Possum was known from only five specimens between 1867 and 1909 and had long been presumed to be extinct until its rediscovery in 1961, near Marysville, Victoria. Since then, extensive surveys by numerous volunteers using a specially designed technique ('stag-watching' in which observers stand beneath large trees—stags—and count animals emerging from tree hollows at dusk) have detected the species at more than 180 scattered localities throughout a range of 4000 square kilometres in the Central Highlands. Less than 3 per cent of this range occurs in nature reserves; 75 per cent is in timber-production forests and 20 per cent in water catchment forest where future timber harvesting is a possibility. The future abundance of Leadbeater's Possum will depend on the extent to which timber-harvesting practices take its special needs into consideration.

Information collected during stag-watching surveys has been used to develop models for predicting future availability of habitat.

### References

Smith A.P. (1984). Demographic consequences of reproduction dispersal and social interaction in a population of Leadbeater's Possum (in) A.P. Smith and I.D. Hume (eds) *Possums and Gliders*. Surrey Beatty and Sons, Sydney. pp. 359–373.

Smith A.P. and D.B. Lindenmayer (1992). Forest succession and habitat management for Leadbeater's Possum in the State of Victoria, Australia. *Forest Ecol. and Manag.* 49: 311–332.

Optimum habitat (1.5–3 animals per hectare) occurs in regrowth (15–50 year old) and uneven-aged Mountain Ash forests with a dense understorey of Silver or Hickory Wattles and an overstorey of more than four large trees of more than 1 metre diameter per hectare. This habitat is created by infrequent wildfire and/or selective logging in old growth Ash forests. Extensive areas of suitable habitat developed in the Central Highlands as a result of wildfire in 1939, giving rise to an estimated peak population size of around 7500 in the early 1980s. Future population size is expected to undergo a massive decline (approximately 90 per cent) over the next 30 years, followed by a bottleneck lasting until approximately 2075, caused by a decline in the availability of tree hollows in regrowth forests. Hollows suitable for Leadbeater's Possum do not develop until trees are in excess of 120–200 years in age and 1 metre in diameter. The supply of hollows is threatened by natural decay and collapse of large, dead trees (killed in the 1939 fires); clear felling and salvage logging of remnant mature and old growth forest; and clear felling of timber production forests on rotations too short (40–80 years) to permit recruitment of tree hollows.

A.P. SMITH

*Gymnobelideus leadbeateri*

# Yellow-bellied Glider

## *Petaurus australis*

### Shaw 1791

pet-or'-us ost-rah'-lis: 'southern rope-dancer'

Although a yellow belly is a notable feature of the adults of this species, the underparts are white to cream in juveniles, subadults and in adults in north Queensland. It is the largest of the Australian petaurids and the most vocal. Its fluffy tail is relatively much larger than those of other petaurids. It occurs patchily in tall, mature wet eucalypt forest, at a density 0.05–0.14 individuals per hectare in its preferred habitat (in contrast to the leaf-eating Greater Glider which has a density of about 0.8 per hectare). The lower density of the Yellow-bellied Glider is probably a reflection of less available food. Plant and insect exudates (sap, nectar, honeydew and manna) make up the bulk of its diet and, in Queensland, eucalypt sap is consumed throughout the year. Eucalypt blossom provides valuable food when available; insects, spiders and pollen provide most of the protein in the diet.

Sap is obtained by biting out small patches of bark of the trunk or main branches of a eucalypt. When the flow from a site dries up, a new excision is made, so a well-used tree becomes heavily scarred after a few years. In southern Australia about 24 species of eucalypt are used as food trees but in North Queensland only one, the Red Mahogany, *Eucalyptus resinifera*, is exploited.

The Yellow-bellied Glider has several distinctive calls, most characteristic of which is a short, high-pitched shriek that subsides into a throaty rattle. This territorial call can be heard at a distance of 400 metres. It is an active and very mobile climber, often running along the

northern part, a male may associate with two or three adult females and up to three young. The adult male has a scent gland on the head, chest and underside of the tail (a few centimetres from the cloaca) and scent exchange involves application of the secretion from the male's head gland to the underside of the recipient's tail. Other members of the group obtain scent for their heads by rubbing against the tail gland of the adult male. Mating can occur while the pair is clinging to the underside of a stout branch.

D. WHITFORD

*The Yellow-bellied Glider is the largest petaurid.*

underside of a branch. Grooming of the long, non-prehensile tail and lower body is undertaken while hanging head-down from a slender support. During the day, it rests in a den in a hollow branch, usually in a living, smooth-barked eucalypt. It emerges at night to forage, sometimes travelling more than 2 kilometres from the den. The home range is of the order of 35 hectares.

Males and females are similar in appearance but males are significantly heavier. There is a high degree of sociality. In the southern part of the range it is usual for an adult male to share a den with an adult female and one young. In the

*Petaurus australis*

*P.a. reginae*

*P.a. australis*

ern parts (although some births have been recorded throughout the year). Pouch life is 90–100 days, after which the young is suckled in the den for 40–60 days. Under suboptimal conditions, breeding may occur only in alternate years.

In general, the Yellow-bellied Glider gives the impression of a species that is hard-pressed over much of its distribution. The home range of an individual is remarkably large and it may spend 90 per cent of its waking hours in foraging for food, much of which (sap) is not very nutritious. Its numbers appear to be diminishing and its long-term survival depends upon maintaining the integrity of large areas of forest. Conservation of the Yellow-bellied Glider requires the preservation of food resources, of den trees and of cohorts of younger trees to provide future dens and adequate feeding ranges.

The female's pouch is divided by a septum into two compartments, each with a teat, but twins are rare. Young are born mostly from August to September in the southern part of the range and from May to September in the north-

R. RUSSELL

## *Petaurus australis*

**Size**
*Petaurus australis reginae*
*Head and Body Length*
270–300 (280) mm
*Tail Length*
420–480 (433) mm
*Weight*
450–700 g

**Identification** Grey above, whitish to orange below; with gliding membrane extending from wrists to ankles. Oblique black stripe on thigh. Head and body much longer than in Sugar or Squirrel Gliders, shorter than in Greater Glider. Tail relatively longer (about 1.5 times length of body) than in other marsupial gliders. Large, bare ears.
**Recent Synonyms** None.
**Other Common Names** Fluffy Glider.
**Status** Rare, limited.

**Subspecies** *Petaurus australis australis*, eastern Australia from Portland, Victoria, to central coastal Queensland.
*Petaurus australis reginae*, northern Queensland to western slopes of rainforest between Mount Windsor and Yamanie, on the north bank of the Herbert River Gorge.
**References**
Russell, R. (1980). *Spotlight on Possums*. Univ. Queensland Press, Brisbane.
Russell, R. (1984). Social behaviour of the Yellow-bellied Glider, *Petaurus australis reginae* in North Queensland (in) A.P. Smith and I.D. Hume (eds) *Possums and Gliders*. Surrey Beatty and Sons and Aust. Mamm. Soc., Sydney. pp. 343–353.
R.L. Goldingay and R.P. Kavangah (1991). The Yellow-bellied Glider: a review of its ecology and management considerations (in) D. Lunney (ed.) *Conservation of Australia's Forest Fauna*. Roy. Zool. Soc. NSW., Sydney. pp. 365–375.

YVONNE DYMOCK

# Sugar Glider

## *Petaurus breviceps*

Waterhouse, 1839

brev'-ee-seps: 'short-headed rope-dancer'

Intermediate in size between the tiny Feathertail and the Greater Glider, the Sugar Glider can volplane for at least 50 metres. Thrust by its hindlegs, it leaps from a tree, spreading the membranes which extend on each side of the body from the fifth finger to the first toe of the foot, steering and maintaining stability by varying the curvature of the right or left membrane. When about 3 metres from a target tree it brings its hindlegs in towards the body and, with an upward swoop, lands with four feet in contact with the bark. Gliding is an efficient way of exploiting patchy food resources and may also help the animal to avoid predators.

The Sugar Glider is locally common (up to at least 10 per hectare) where tree hollows are available for shelter and there is abundant food such as the gum produced by acacias; nectar and pollen; the sap of certain eucalypts; invertebrates; and invertebrate exudates. The quantity of each item in the diet varies according to seasonal availability. Nectar and pollen are used extensively when available and green seeds of Golden Wattle may also be taken. The Sugar Glider can tolerate a wide range of temperatures and, in extreme cold, it conserves energy by huddling with others in its spherical or bowl-shaped nest of leaves, or by becoming torpid.

*The Sugar Glider is the commonest petaurid: its success may be due to its very varied diet, including, as shown here, sap from excavations that it makes in trees.*

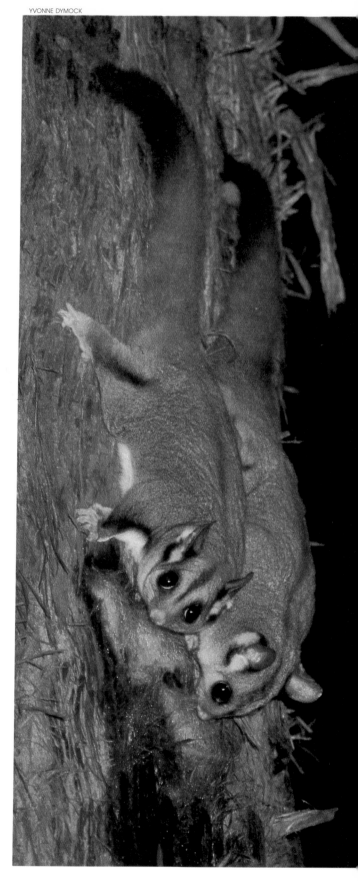

## *Petaurus breviceps*

**Size**

***Petaurus breviceps breviceps***

*Head and Body Length*
160–210 (170) mm

*Tail Length*
165–210 (190) mm

*Weight*
115–160 (140) g (males)
95–135 (115) g (females)

**Identification** Blue-grey to brown-grey above, dark mid-dorsal stripe from between eyes to mid-back; cream to pale grey below. Tail grey to black, sometimes tipped with white. Very similar to *Petaurus norfolcensis* but usually much smaller, with blunter face and less bushy base of tail.

**Recent Synonyms** None.

**Other Common Names** Sugar Squirrel, Lesser Flying Squirrel, Short-headed or Lesser Flying Phalanger, Lesser Glider.

**Status** Common.

**Subspecies** Seven subspecies are recognised, four from New Guinea. Australian species are:
*Petaurus breviceps breviceps*, New South Wales, Victoria, Tasmania.
*Petaurus breviceps longicaudatus*, Queensland.
*Petaurus breviceps ariel*, Northern Territory.

**Reference**
Smith, M.J. (1973). *Petaurus breviceps. Mammal. Species* 30: 1–5.

*The patagium of the sugar glider extends from the wrist to the ankles, for which reason it and its relatives are known as 'wrist-winged gliders'.*

Most live in social groups containing up to seven adults and their young of the season, sharing a common nest. Some individuals may be members of more than one group. Scent-marking glands are well developed, particularly in the male and individuals are recognised by odour. In south-eastern Australia, breeding usually begins in June or July, each female normally producing two young which remain in the pouch for about 70 days and are then deposited in the group nest for a further 40–50 days. When 110–120 days old, the young begin to leave the nest to forage, usually in company with their mother.

When seven to ten months old, most young animals are displaced from their maternal groups and may travel across open pasture to reach isolated forest habitats. Groups may recruit female offspring following the deaths of older females in the group, but males that die are usually replaced by solitary adults from adjacent territories. New groups may be formed by displaced young animals but many are unable to establish themselves and mortality is high during the first 12 months of life due to predation by owls, kookaburras, goannas the Fox and the Cat. Longevity in the wild is usually four to five years but ages of up to nine years have been recorded.

The Sugar Glider uses a range of calls, including a shrill yapping that warns other individuals of danger and is often heard when a owl or Cat is nearby. There is also a sharp, threatening scream which is occasionally heard when animals fight and a defiant gurgling chatter from animals disturbed in their nest. The Sugar Glider is by no means timid and may audaciously defend a food source against an intruding phascogale or even a Common Brushtail.

Populations appear to be reasonably stable. In south-eastern Australia the density is often highest in open forest habitats where animals have access to dense patches of *Acacia*, a habitat that was probably much more widespread

*Petaurus breviceps*

Meville Is.
Bathurst Is.
*P.b. ariel*
Augustus Is.
Groote
Eylandt
*P.b. longicaudatis*
*P.b. breviceps*

before European settlement. However, the Sugar Glider often thrives in strips and patches of forest remaining on cleared agricultural land and has therefore not suffered as greatly as some other possums. Sugar Gliders have been successfully introduced into re-established habitat in southern Victoria and can inhabit young forest and woodland areas if nest boxes are provided. Retention of interconnected systems of suitable forest and woodland habitat is essential for its conservation in agricultural areas.

G.C. SUCKLING

231

# Mahogany Glider

## *Petaurus gracilis*

(de Vis, 1883)

grah'-sil-is: 'slender rope-dancer'

Although relatively large, the Mahogany Glider was lost to science for more than a century. It was described as a new species, *Belideus gracilis*, in 1883 by Charles de Vis, then Director of the Queensland Museum but, as early as 1888, it had been regarded as identical with the much better known Squirrel Glider, *Petaurus norfolcensis*. In the apparent absence of any specimens to which to refer, this view was perpetuated in subsequent systematic literature, including the *Zoological Catalogue of Australia, Mammalia* (1988). However, the distinctness of the species was demonstrated in 1986 by the discovery in the Queensland Museum of three poorly preserved skins of specimens collected near Cardwell, north-eastern Queensland, in 1886. A number of searches by the Museum in this general area were fruitless but interest was rekindled when a 1989 census of the Museum's

*Petaurus gracilis*

---

**Petaurus gracilis**

**Size**
*Head and Body Length*
247–265 (254) mm (males)
215–261 (244) mm (females)
*Tail Length*
335–380 (351) mm (males)
300–390 (352) mm (females)
*Weight*
330–410 (363) g (males)
255–407 (343) g (females)

**Identification** Similar to *Petaurus norfolcensis* but distinguished externally by much larger size and longer tail which is narrow at its base. Differs cranially in much larger zygomatic width, snout height, basicranial length and narrower interorbital width. Colour varies from overall mahogany brown (dorsal and ventral) to grey-brown dorsally with buff to apricot belly.
**Recent Synonyms** *Petaurus norfolcensis gracilis*.
**Other Common Names** None.
**Status** Rare, endangered.
**Subspecies** None.
**References**
Van Dyck, S. (1991). Raising an old gliders' ghost … a devil of an exorcise. *Wildlife Australia* 28: 10–13.
Van Dyck, S. (1992). Lasting impressions of Mahogany Gliders. *Aust. Nat. Hist.* 24: 32–40.
Van Dyck, S. (1993). The taxonomy and distribution of *Petaurus gracilis* (Marsupialia: Petauridae), with notes on its ecology and conservation status. *Mem. Qld. Mus.* 33: 77–122.

mammal holdings revealed an unregistered specimen, apparently of this species, that had been collected in 1973 at Barratt's Lagoon, near Tully, 65 kilometres north of Cardwell. An expedition was immediately mounted to the locality and

*Apparently lost since 1883, the Mahogany Glider was rediscovered in 1989. It is rare and endangered.*

B. COWELL/QLD MUSEUM

living representatives were seen, feeding in flowering Moreton Bay Ash near the lagoon. Sadly, this habit was cleared one month later for banana and pineapple plantations. Not until two years later was another population located nearby. A survey of the area recorded the glider from a mosaic of habitats dominated by medium to low woodland on swampy coastal plains, beach ridges and swales. It is also found in *Melaleuca* swamps and *Xanthorrhoea* woodlands, where it nests either singly or in pairs inside the hollows of both dead and living trees.

The diet includes nectar from eucalypt blossoms and grass-trees and exudates licked from gashes incised in bloodwoods, the Large-fruited Red Mahogany and three species of wattles. It also eats invertebrates, lichens and gum. The gum is tapped from the spears of the grass-tree, *Xanthorrhoea johnsonii* by biting a chunk from just below the flower head or at the spear's base. This wound slowly exudes an amber mucilaginous jelly that is eaten the following evening. If a particular spear is growing at some distance from trees, a glider may bound through long grass in order to reach it.

In summer, when ripe pods of the wattle, *Acacia crassicarpa*, burst open to hang their seeds on coiled strings of gum (arils), female gliders, which may have up to two young ready to leave the nest, consume these lipid-rich, gummy 'rubber-bands' almost to the exclusion of other foods.

Although elusive, virtually silent and difficult to trap, the presence of the Mahogany Glider in an area can often be detected by the large footprints left by an animal on the heavy natural coating of white powdery 'bloom' (similar to that on black grapes) on the spear of a flowering grass-tree.

At present, the Mahogany Glider is known to occur coastally between Bambaru near Ingham, 100 kilometres north to the Hull River near Tully. In this narrow strip of woodland it occurs between sea-level and an elevation of 120 metres. Approximately 80 per cent of this habitat type has been cleared and destruction and degradation continues apace. Clear-felling and drainage of woodland for planting of sugar cane, bananas, pineapples, improved pasture, Carribean pine or for aquaculture, leaves the species endangered at most known sites. Few Australian mammals are more in need of immediate attention.

S.M. Van Dyck

C. ANDREW HENLEY/LARUS

*Like other petaurids, a Squirrel Glider marks its territory with scent. The scent gland on the forehead is visible in this photograph*

# Squirrel Glider

## *Petaurus norfolcensis*

(Kerr, 1792)

nor'-fol-ken'-sis: 'Norfolk (Island) rope-dancer' (wrong locality)

K. ATKINSON

*The facial markings of the Squirrel Glider are more distinct than those of the Sugar Glider.*

Although up to twice the size of the Sugar Glider, the Squirrel Glider is otherwise similar to it in appearance and gliding ability. The Squirrel Glider is generally rarer and more restricted in range than the Sugar Glider, but in some habitats the larger species is far more common (up to 3 per hectare) than the smaller. The Squirrel Glider inhabits dry sclerophyll forest and woodland in south-eastern Australia where it is absent from the dense coastal ranges. However, in northern New South Wales and Queensland it occurs in coastal forest and in some wet forest areas bordering on rainforest.

The Squirrel Glider has a diet similar to that of the Sugar Glider. In Central Victoria the Squirrel Glider feeds on insects (principally beetles and caterpillars); the gum produced by acacias; the sap of certain eucalypts; nectar; and pollen. It also eats the green seeds of the Golden Wattle. Nectar and pollen are important dietary items; when nectar is unavailable, greater use is made of eucalypt sap and wattle gum.

Squirrel Gliders nest in a bowl-shaped, leaf-lined nest in a tree hollow. Typically, a family group comprises one mature male (more than two years old), one or more adult females and their associated offspring of the season. Occasionally one or more young males (less than two years old) may also be associated with a group of up to ten animals, including as many as five adults. Scent-marking glands are well developed in males. Squirrel and Sugar Gliders breed at similar times of the year and growth and development of their

D. WHITFORD

young are strikingly similar. The two species have interbred in captivity, producing fertile offspring.

The Squirrel Glider lacks the shrill, yapping call characteristic of the Sugar Glider but a defiant gurgling chatter is common to both, although considerably deeper and more throaty in the larger species. Squirrel Gliders also utter a soft, nasal grunt and a repetitive, short gurgle, similar to the last part of the call of the Yellow-bellied Glider.

While the Squirrel Glider is usually more common than the Sugar Glider in places where the species coexist, the Squirrel Glider may be endangered in the southern part of its range. In northern Victoria, an important habitat for the larger species is remnant vegetation along roadsides and creek and river frontages in cleared pasture: preservation of such disjunct habitats may be vital for its continued survival. It is likely

*The Squirrel Glider is the larger relative of the Sugar Glider.*

that clearing of open woodland areas for agriculture, as well as forest operations which have reduced the abundance of tree hollows, have had a dramatic effect on the Squirrel Glider, which has been less able to adapt to the changed habitats than its smaller relative.

G.C. SUCKLING

## *Petaurus norfolcensis*

**Size**

*Head and Body Length*
180–230 (210) mm
*Tail Length*
220–300 (270) mm
*Weight*
190–300 (230) g

**Identification** Similar to *Petaurus breviceps* but with a longer and more pointed face, lengthier and narrower ears and a much bushier, softly furred tail; facial markings are often more distinct. Belly usually a rich white or creamy white. Molar teeth much larger than in *P. breviceps*.
**Recent Synonyms** *Petaurus sciurus*.
**Other Common Names** Flying Squirrel, Sugar Squirrel, Squirrel Flying Opossum, Squirrel Flying Phalanger.
**Status** Rare.
**Subspecies** None.
**Reference**
Fleay, D. (1947). *Gliders of the Gum Trees*. Bread and Cheese Club, Melbourne.

*Petaurus norfolcensis*

# FAMILY PSEUDOCHEIRIDAE
## *Ringtail possums and the Greater Glider*

In the first edition of this work, ringtail possums were grouped with the wrist-winged gliders and their kin in the family Petauridae. Intensive anatomical and biochemical studies in the 1980s and early 1990s have shown that, although these two groups are more closely related to each other than to other possums, they warrant distinction as separate families. Within what we now recognise as the family Pseudocheiridae, there had also been a strong tendency to override earlier generic distinctions and to 'lump' most species in the genus *Pseudocheirus*, but we now recognise the ringtail possums themselves to be a quite divergent group.

Of the six currently recognised genera, all occur in Australia. *Pseudocheirus*, the most familiar genus, includes two species. Its closest relatives appear to be the mono-typic *Petropseudes* or Rock Ringtail and *Pseudochirops*, a primarily Melanesian assemblage with one Australian species, the Green Ringtail Possum. Distantly related to all these, but surprisingly closely to each other, are the monotypic *Hemibelideus*, the Lemuroid

Ringtail Possum and *Petauroides*, the Greater Glider. The primarily Mel-anesian genus *Pseudochirulus* appears to be most distantly related to all of the foregoing: it has two species in Australia, the Daintree and Herbert River Ringtail Possums. This situation could well justify two subfamilies but these have not yet been erected.

Most pseudocheirids have a strong-ly prehensile, short-furred tail that is about the same length as the head and body and so strongly prehensile as to be able to support the body without aid from the limbs. One notable exception is seen in the semi-terrestrial Rock Ringtail, the tail of which is less than two-thirds the head–body length and, although prehensile, cannot support the body. The other is the Greater Glider, which has a tail that is some 15 per cent longer than the head and body but is only weakly prehensile: its primary function appears to be related to gliding. All pseudocheirids are capable of opposing the first two digits of the forefoot against the other three to make a firm grip on a branch.

The molar teeth of pseudocheirids bear a series of sharp, crescentic or triangular ridges (a condition known as selenodonty) that enable leaves to be finely ground before

I. MORRIS

Left: *Common Ringtail in its spherical nest.*          *A family of Rock Ringtails.*

being swallowed. A large caecum provides a compartment in which micro-organisms can break down fibre into assimilable nutrients. In the Common Ringtail Possum—and possibly in other species—the efficiency of nutrition is enhanced by coprophagy. Once a day the contents of the caecum are voided as 'soft' faecal pellets that are then eaten. After passing again through the alimentary canal and providing further assimilable products, the remainder of this material is disposed of as 'hard' faecal pellets.

Both species of *Pseudocheirus* inhabit sclerophyll forest, as does the monotypic *Petauroides*. The single species of *Petropseudes* has a disjunct tropical distribution in rocky environments. Both species of *Pseudochirulus* inhabit tropical rainforest.

All pseudocheirids have a forwardly opening pouch. Species of *Pseudocheirus* have four teats and frequently rear two young. Other Australian petaurids have two teats and usually rear a single young.

# Lemuroid Ringtail Possum

## *Hemibelideus lemuroides*

### (Collett, 1884)

hem'-ee-bel-id'-ay-us  leem-yue-roy'-dayz:
'lemur-like half-*Belideus*' (*Belideus* is a junior
synonym of *Petaurus*)

*Hemibelideus lemuroides*

Seen by spotlight high in a tree, this ringtail
appears as a charcoal-grey ball of fur without
markings to indicate whether it is male, female
or even juvenile. Invariably arboreal, it lives in

### *Hemibelideus lemuroides*

**Size**
*Head and Body Length*
315–360 (344) mm (males)
313–400 (342) mm (females)
*Tail Length*
320–365 (345) mm (males)
300–373 (346) mm (females)
*Weight*
810–1060 (925) g (males)
750–1140 (980) g (females)

**Identification** Body usually charcoal-brown,
darker above with brownish tinge on shoulders,
yellowish tinge bellow. A second, less common
colour phase is creamy-white in colour, with a
touch of orange on the shoulders. Tail slightly
bushy and only slightly tapered, with short, fin-
ger-like bare tip; short pug-like face, brilliant
white-yellow eyeshine. Leaps onto foliage.
**Recent Synonyms** *Pseudocheirus lemuroides*.
**Other Common Names** Brush-tipped
Ringtail Possum, Lemur-like Ringtail Possum.
**Status** Limited distribution, vulnerable.
**Subspecies** None.

rainforest above an elevation of 450 metres
between Ingham and Cairns, with a small isolated
population above 1100 metres elevation on the
Mount Carbine Tableland west of Mossman.
The latter population contains particularly high
proportion of white individuals, about 1 : 6 compared with 1 : 1500 in the larger population.

The Lemuroid Ringtail is almost exclusively
a folivore, specialising on leaves with relatively
low fibre content. Young leaves are preferred
but, when these are not available, it will eat the
mature leaves of the favoured trees, rather than
switch to other species. Fermentation of masticated leaves probably takes place in the caecum
and large intestine. Particularly favoured species
are the Queensland Maple, *Flindersia brayleyana*
and White Carabeen, *Sloanea langii*. Other
species eaten include the Brown Quandong,
*Elaeocarpus ruminatus*; Brown Tamarind, *Castan-
ospora alphandii*; and Buff Walnut, *Endiandra* sp.
The flowers of Bollywood, *Litsea leefeana* and the
fleshy outer covering of the Yellow Walnut,
*Beilschmedia bancroftii*, are also eaten.

Strictly nocturnal, it spends the day in a tree
hollow, emerging just after dark to forage. It may
not return until dawn is actually breaking, much
later than the Coppery Brushtail and Herbert
River Ringtail which share its habitat. Leaps of
2–3 metres are frequently made from the tip of
one branch to another: with legs outstretched
like those of a glider, it lands heavily on a cushion
of foliage. The sound of such a crash is usually the

*The Lemuroid Ringtail feeds on the leaves of rainforest trees.* M. TRENERRY

first sign that the Lemuroid Ringtail has begun its nocturnal wanderings.

There are two teats in the pouch but usually only one young is reared. The breeding season appears to be more restricted than in other rainforest ringtails, most pouch-young being recorded from August to November; young riding on their mother's backs from October to April. A young animal separated from its mother signals its distress with a thin, high-pitched, hissing squeak; adults appear to be silent. Adults of both sexes have a strong, musky odour—readily detected in the forest at night—produced by a sticky cream-coloured fluid, often excreted from the vent when an animal is handled.

More gregarious than other rainforest ringtails, it is frequently seen in groups of two (mother and young, or male and female) or in family groups of three. When disturbed, individuals of a group come together to sit in a tight group, in contact with each other. Up to three individuals may share a den and feeding aggregations of as many as eight have been recorded from one tree. Densities of two to five per hour of spotlighting are common but up to ten per hour have been recorded at elevations above 900 metres—greater than in any other rainforest ringtail.

The Carpet Python, *Morelia spilota*, is a known predator. However, clearing and disturbance to rainforest is the greatest threat to Lemuroid Ringtail. It is the least able of the rainforest leaf-eating possums to survive in remnant patches, disappearing from fragments of 40–80 hectares within 35–60 years and from smaller fragments within nine years. It appears also to have been the most susceptible to rainforest logging, possibly because it is the most canopy-loving of the ringtails, favours mature phase trees in its diet and is reliant on tree hollows for shelter. The species appears secure, because, although more than 23 per cent of the original range of the main population has been cleared,

substantial tracts of its rainforest habitat remain, most of which are now within the Wet Tropics of Queensland World Heritage Area. Negligible clearing of the Mount Carbine Tableland population habitat has occurred and all is now within the World Heritage Area.

J.W. WINTER AND N.J. GOUDBERG

### References

Winter, J.W. and R.G. Atherton (1984). Social group size in north Queensland ringtail possums of the genera *Pseudocheirus* and *Hemibelideus* (in) A.P. Smith and I.D. Hume (eds) *Possums and Gliders*. Surrey Beatty and Sons and Aust. Mamm. Soc., Sydney. pp. 311–319.

Goudberg, N.J. (1990). *The feeding ecology of three species of upland rainforest ringtail possums,* Hemibelideus lemuroides, Pseudocheirus herbetensis *and* Pseudocheirus archeri. PhD thesis, James Cook University, Townsville.

Laurance, W.F. (1990). Comparative responses of five arboreal marsupials to tropical forest fragmentation. *J. Mammal.* 71: 641–653.

The Greater Glider feeds on eucalypt leaves.

# Greater Glider

## *Petauroides volans*

### (Kerr, 1792)

pet'-or-oy'-dayz voh'-lahnz: 'flying
Petaurus-like (animal)'

*Petauroides volans*

Largest of the gliding possums, the Greater Glider is adapted to an almost exclusive diet of eucalypt leaves. Like the Koala, it has a greatly enlarged caecum in which much of the cellulose of the leaves is broken down to assimilable substances by bacterial fermentation. It also has a low field metabolic rate which is not only a reflection of the efficiency of gliding but also the need to cope with the low nutritive value of *Eucalyptus* foliage. In common with most arboreal leaf-eaters, it does not normally need to drink.

The Greater Glider lives in a variety of eucalypt-dominated habitats, ranging from low, open forests on the coast to tall forests in the ranges and low woodland westwards of the Dividing Range; it does not penetrate into rainforests. In any particular area it feeds on only one or two species of eucalypt but, over its entire range, the number of species eaten is considerably greater. Strictly nocturnal and essentially solitary, it rests during the day in a tree hollow, usually high in an old tree. After emerging, it moves by a series of glides, often along established routes, to a feeding area. Bouts of feeding in the terminal clusters of leaves are interspersed with periods of rest while the animal sits in a fork or on a large horizontal branch.

It is an agile climber and like cuscuses, ringtails and the Koala, can effect a pincer-like grip on a branch by opposing the first two toes of the forefoot to the other three. Since the leading edge of the patagium extends only to the elbow (not to the wrist, as in other marsupial gliders) it does not extend the whole forelimb when gliding, but flexes it at the elbow, bringing the paws under the chin.

A glide may cover a horizontal distance of up to 100 metres and involve changes of direction of as much as 90 degrees. Just before reaching a target tree, the flight is directed upwards so that the animal loses speed and lands with all four feet against the trunk: habitually used trees may be recognised by the scratches made when landing. The Greater Glider moves clumsily in a loping gait when on the ground, where it is known to fall prey to the Dingo and Fox. The Powerful

D. WHITFORD

240

E. BEATON/TERRA AUSTRALIS

*The Greater Glider has thick fur that increases its apparent size.*

Owl is a known predator on adults.

Unlike species of *Petaurus* and particularly the Yellow-bellied Glider which often occurs with it, the Greater Glider is silent. Greater Gliders occupy individual home ranges with no overlap between males although female home ranges may overlap with each other and with those of males. Males and females normally share a den from the onset of the breeding season until the young emerge from the pouch. It seems likely that territories are defined by the scent marks deposited from the large anal glands, which are particularly well developed in adult males.

The breeding season is from March to June. Females have two teats but only one young is born. This emerges from the pouch when three to four months old and, for the next three months, may be carried on the mother's back or left in the nest while she is feeding. Juveniles become independent at the age of about nine months but sexual maturity and breeding do not occur until the second year of life.

The abundance of the Greater Glider in undisturbed forests is in strong contrast to its absence from pine plantations and its paucity is regenerated forest which lacks old trees with hollows suitable for nesting.

Its conservation is utterly dependent upon sympathetic forest management which retains buffer strips of old forest between coupes and preserves old 'habitat' trees and their potential successors in small unlogged areas.

G.M. McKay

---

## *Petauroides volans*

**Size**

*Head and Body Length*
350–450 mm
*Tail Length*
450–600 mm
*Weight*
900–1700 g

**Identification** Dark grey, cream, mottled cream and grey or dusky brown above; whitish below. Long, furry tail, pale below on basal half. Short snout; very large ears. Tail not prehensile.
**Recent Synonyms** *Schoinobates volans*.
**Other Common Names** Greater Flying Phalanger, Dusky Glider, Greater Glider-possum, Squirrel.
**Status** Abundant.
**Subspecies** *Petauroides volans volans*, Victoria to about Tropic of Capricorn.
*Petauroides volans minor*, north of about Tropic of Capricorn.
It is possible that these forms represent distinct species.

**References**
Henry, S.R. (1984). Social organisation of the Greater Glider (*Petauroides volans*) in Victoria (in) A.P. Smith and I.D. Hume (eds) *Possums and Gliders*. Surrey Beatty and Sons and Aust. Mamm. Soc., Sydney. pp. 221–228.
Hume, I.D., W.J. Foley and M.J. Chilcott (1984). Physiological mechanisms of foliage digestion in the Greater Glider and Ringtail Possum (Marsupialia: Pseudocheiridae) (in) A.P. Smith and I.D. Hume (eds) *Possums and Gliders*. Surrey Beatty and Sons and Aust. Mamm. Soc., Sydney. pp. 247–251.
Kehl, J. and A. Borsboom (1984). Home range, den tree use and activity in the Greater Glider, *Petauroides volans* (in) A.P. Smith and I.D. Hume (eds) *Possums and Gliders*. Surrey Beatty and Sons and Aust. Mamm. Soc., Sydney. pp. 229–236.

# Rock Ringtail Possum

## *Petropseudes dahli*

### (Collett, 1895)

pet'-roh-sude'-ayz  dah'-lee:
'Dahl's rock-*Pseudocheirus*'

When Aboriginal people first described the Rock Ringtail to Knut Dahl in 1895, he thought that it must be an unknown form of tree-kangaroo. With further information, he soon realised that the animal was a possum new to science. Dahl found his first specimens near the headwaters of the Mary River, Northern Territory and the species is now known most commonly from the Alligator Rivers region of the Northern Territory. It is also found in the Kimberley, on Groote Eylandt, in the Katherine and Roper Rivers region and across the Gulf Fall country into north-western Queensland, the most easterly location being Lawn Hill National Park.

It lives exclusively in rocky outcrops and appears to prefer areas of deeply fissured rock and large boulders to those with broken, rocky slopes. Although common within this preferred

*Petropseudes dahli*

---

**Petropseudes dahli**

**Size**
*Head and Body Length*
334–375 mm (males)
349–383 mm (females)
*Tail Length*
200–220 mm (males)
207–266 mm (females)
*Weight* (unsexed)
1280–2000 g

**Identification**  Grey to reddish-grey above, paler below. Mid-dorsal stripe from crown of head to about middle of back. Pupil vertical. Fur long and woolly. Tail with thickly furred basal half and almost naked terminal half, often held at a sharp angle to basal half; undersurface naked for terminal two-thirds. Distinguished from the other rock-dwelling possum, *Wyulda squamicaudata*, by absence of scales on the terminal third of tail.
**Recent Synonyms**  *Pseudocheirus dahli*.
**Other Common Names**  Rock-haunting Ringtail, Rock Possum.
Indigenous name:  Wogoit.
**Status**  Common, limited.
**Subspecies**  None.
**Reference**
Dahl, K. (1897). Biological notes on north-Australian Mammalia. *Zoologist* Ser. 4, 1: 189–216.

habitat, it is very secretive in its behaviour and difficult to trap, so little is known of its biology. Although not frequently observed, its presence can be readily detected by its distinctive scats, which resemble a slightly bent cigar, 15–25 millimetres long, about 5 millimetres wide and reddish-brown or black in colour.

It is strictly nocturnal, climbing trees to feed at night. In a spotlight, its eyeshine is very bright but, unlike most possums, it does not 'freeze' when caught in the beam. Instead, it retreats from a food tree or rock ledge into the protec-

*The Rock Ringtail spends the day among rocks, climbing into trees to feed at night. It has a short tail.*

S. SWANSON

tion of rock clefts and crevices where it sometimes may 'hide' with its head in a crevice but its body exposed. It is not known to make a nest and can occasionally be observed sleeping on well-protected rock ledges during the day. This behaviour, unique among ringtails, indicates the extent to which it has become adapted to a terrestrial existence. The longer snout, shorter tail, shorter legs and shorter claws are also indicative of a reduction in arboreal adaptation but, as in other ringtails, its tail is prehensile.

The Rock Ringtail feeds on flowers, fruits and leaves of a variety of trees and shrubs found in and around its rocky habitat. In the Northern Territory it has been observed to eat blossoms of Darwin Woollybutt, *Eucalyptus miniata* and Darwin Stringybark, *E. tetrodonta*; the fruits of *Zyziphus oenoplia*, *Vitex glabrata*, Billy Goat Plum, *Terminalia fernandiana*, *Owenia vernicosa* and probably *T. carpentariae;* and the leaves of Vine Reed-cane, *Flagelleria indica* and *Pouteria sericea*. At Lawn Hill in Queensland, it has been observed, presumably feeding, in *Terminalia canescens*. It usually remains among rocks when feeding but has been observed as far as 100 metres from the nearest outcrop. On such occasions the animals *do* freeze when spotlit but retreat to the rocks at the first opportunity.

Breeding does not appear to be restricted to a particular season. Females with large pouch-young and pickaback young have been observed in March, July, August and September. The female has two teats but normally rears only one young. It is frequently seen in pairs, the two adults often accompanied by a juvenile at heel and sometimes with one also on the back. Of all the Australian possums, the Rock Ringtail has the most tightly-knit family group, in that adults and their young will stay within 2–3 metres of each other throughout the night and may spend an hour or more sitting in contact, on the rocks. It is probably monogamous, establishing long-term pair-bonds (the only ringtail to do so). Aggregations of up to nine individuals have been observed.

Communication is probably primarily by scent—adults have a distinct sternal gland on the chest and males (at least) have very large para-cloacal glands (approximately 2 centimetres in diameter). Areas of a dark, glossy coating on the rocks, smelling of urine and faecal accumulation at commonly visited localities act as marking posts. Possible cloacal marking of a tree branch has been observed. Tapping of the near-naked terminal half of the tail may be a form of communication within the rocky habitat. A quiet screech and grunt is the only known vocalisation.

Potential predators are the Dingo, Olive Python and humans. The Rock-Ringtail species does not appear to have suffered significant diminution of range since its discovery by Europeans but it may be affected by frequent human disturbance: it seems often to temporarily desert an area after being disturbed by spotlighting.

J.A. KERLE AND J.W. WINTER

HANS & JUDY BESTE

# Green Ringtail Possum

## *Pseudochirops archeri*

### (Collett, 1884)

sude'-oh-kire'-ops arch'-er-ee: 'Archer's
*Pseudocheirus*-like (animal)'

This remarkably beautiful ringtail is aptly
named: a fine banding of black, yellow and
white on the hairs confers a most unusual lime-
green colour to its thick, soft fur. It lives in
dense upland rainforest between Paluma (near
Townsville) and the Mount Windsor Tableland,
west of Mossman. It is not found at elevations
below about 300 metres. Markedly arboreal, it

*The Green Ringtail does not use a den or a roost,
simply hunching its body into a ball to sleep on
a branch.*

descends to the ground only to traverse a gap
between trees and it appears to favour areas of
rainforest with many tangled, thornless vines. It
may occasionally be seen high in the tree canopy,
usually in a Strangler Fig.

The diet consists almost exclusively of
leaves. The Green Ringtail has the most specialised diet of all rainforest ringtails, selecting
leaves with much fibre but little protein, mainly
from several species of fig—the only possum
known to eat these. Even when younger foliage
of figs (with a lower fibre content) is available,
it eats mature foliage in quantity. The Green
Ringtail has a relatively small caecum and a particularly large colon, where digestion of the
leaves probably occurs. Like the Common

## Pseudochirops archeri

### Size

*Head and Body Length*
344–371 (353) mm (males)
285–377 (335) mm (females)

*Tail Length*
309–372 (330) mm (males)
315–333 ((325) mm (females)

*Weight*
880–1190 (1064) g (males)
670–1350 (1119) g (females)

**Identification**  Greenish fur, two silvery vertebral stripes; white patches under eye and ear. Thick base to tail, tail shorter than head and body. Characteristic resting posture.
**Recent Synonyms**  *Pseudocheirus archeri.*
**Other Common Names**  Striped Ringtail Possum.
Indigenous name: Toolah.
**Status**  Sparse, limited.
**Subspecies**  None.

time roost which, unlike that of other possums, is on a branch amongst foliage where its green fur blends perfectly with the surrounding vegetation. It sleeps upright, curled into a tight ball, gripping the branch with one or both hindfeet and sitting on the base of its coiled tail, with the forefeet, face and curled tail tucked tightly into its belly. When seen in a spotlight at night, it usually assumes this posture, staring down at the intruder in apparent composure. However, this resting posture may be an attempt at concealment. Should it seek evasion, often at the precise moment when an observer has turned away, it moves rapidly through the canopy, running along branches and up narrow, swaying vines but avoiding leaps—except across the forks of major branches. It is perhaps the fastest of the ringtails at this type of travel. Sightings of Green Ringtails, unlike those of Lemuroid and Herbert River Ringtails, are not reduced by cold weather—either because they are physiologically adapted to cooler temperatures from sleeping on exposed branches, or because they do not use a

Ringtail, it is known to pass food through its gut twice by eating special faecal pellets produced during the day. Other favoured food species are Sankey's Walnut, *Endiandra sankeyana*; Rose Maple, *Cryptocarya rigida*; Bollywood, *Litsea leefeana*; Red Oak, *Carnarvonia* sp. 231; Lamington's Silky Oak, *Helicia lamingtoniana*, Northern Red Ash, *Alphitonia whiteii*; Candlenut, *Aleurites moluccana*; Tulip Oaks, *Argyrodendron* sp.; Flame Kurrajong, *Brachychiton acerifolius*; Water Vine, *Cissus hypoglauca*; and a vine, *Elaeagnus triflora*. Included in its diet are noxious or irritating plants such as the highly alkaloidal *Solanum viride* and the Shining-leafed Stinging Tree, *Dendrocnide photoinophylla*. A leaf is usually bitten off at the petiole then held in the paws while being eaten. Ripe figs are the only non-leaf item known to be eaten.

Although essentially nocturnal, the Green Ringtail may feed and move during the day, often after having been disturbed from its day-

### References

Russell, R. (1980). *Spotlight on Possums.* Univ. Queensland Press, Brisbane.

Goudberg, N.J. (1990). *The feeding ecology of three species of upland forest ringtail possums,* Hemibelideus lemuroides, Pseudocheirus herbertensis *and* Pseudocheirus archeri. Ph.D. thesis, James Cook University, Townsville.

Proctor-Grey, E. (1984). Dietary ecology of the Coppery Brushtail Possum, Green Ringtail Possum and Lumholtz's Tree-kangaroo in North Queensland (in) A.P. Smith and I.D. Hume (eds) *Possums and Gliders.* Australian Mammal Society and Surrey Beatty and Sons, Sydney. pp. 129–135.

Laurance, W.F. (1990). Comparative responses of five arboreal marsupials to tropical forest regeneration. *J. Mammal.* 71: 641–653.

Y. DYMOCK

*A combination of black, white and yellow hairs gives an unusual colour to the Green Ringtail.*

of varying intensity when handled.

The female has two teats in the pouch but normally gives birth to a single young. Pouch-young have been recorded from August to November and back-riding young from October to April, suggesting a peak of births in the latter half of the year. We do not know how long the young stays with its mother once it begins to emerge from the pouch, but it may have the longest back-riding stage of any ringtail, probably reflecting the lack of a den.

Known predators are the Rufous Owl and Spotted-tailed Quoll. In the past, the Green Ringtail was hunted by Aboriginal people but direct human predation has not posed a serious threat. It is the most secure of the rainforest ringtails because it can continue to exist in fragments of rainforest and appears to have been relatively unaffected by past rainforest logging. Although about 18 per cent of its past rainforest habitat has been cleared, substantial tracts remain, most of these now in the Wet Tropics World Heritage area.

J.W. WINTER AND N.J. GOUDBERG

den and therefore remain visible even when inactive. Nevertheless, activity is curtailed during clear nights, possibly because of their greater visibility to potential predators.

Evenly but sparsely distributed throughout its range, the Green Ringtail is the most solitary of the arboreal ringtails. When two are seen together, these are usually a mother with her young at heel—only occasionally an adult pair (presumably at the time of mating). It is the most silent of the ringtails, adults having never been reported to make a sound, even when handled. The only known vocalisation is that of pouch-young which give a quiet 'tssk-tssk-tssk'

*Pseudochirops archeri*

# Daintree River Ringtail Possum

## *Pseudochirulus cinereus*

### (Tate, 1945)

sude'-oh-kire'-ul-us  sin'-er-ay'-us:
'ash-coloured little-*Pseudocheirus*'

M. TRENERRY

*The Daintree River Ringtail is a cautious climber, seldom leaping between branches.*

The Daintree River Ringtail Possum lives in the rainforests at the northern end of the wet tropics of Queensland above 420 metres in elevation. It consists of three separate populations; respectively on the Thornton Peak massif, the Mount

---

### *Pseudochirulus cinereus*

**Size**
*Head and Body Length*
346–360 (353) mm (males)
335–368 (352) mm (females)
*Tail Length*
320–395 (360) mm (males)
325–362 (344) mm (females)
*Weight*
830–1450 (1092) g (males)
700–1200 ( 908) g (females)

**Identification** Body pale caramel-fawn to dark brown, darker on top, grading to creamy white below; dark longitudinal stripe from between eyes to lower back. Tail tapering and (usually) terminal one-third white. Naked skin pinkish colour. Eyes light brown to honey coloured; bright red eye-shine in spotlight. Juveniles same colour but paler.
**Recent Synonyms** *Pseudocheirus herbertensis cinereus.*
**Other Common Names** Cuscus.
**Status** Sparse, restricted distribution.
**Subspecies** None.

---

Windsor Tableland and the Mount Carbine Tableland. Reports of 'cuscuses'—a local name for the ringtail—from the Finlayson Range north of the Bloomfield River and below 100 metres elevation at the base of Thornton Peak need confirmation. The Daintree River Ringtail was first described in 1945 as a subspecies of the Herbert River Ringtail, but was made a separate species in 1989 on the basis of its possession of 16 chromosome pairs (in contrast to 12 in the Herbert River Ringtail).

It is mainly arboreal and climbs cautiously, making only small leaps between large, stable branches. Occasional observations have been made of adult males crossing narrow rainforest

### References

Winter, J.W. (1984). Conservation studies of tropical rainforest possums (in) A.P. Smith and I.D. Hume (eds) *Possums and Gliders*. Surrey Beatty and Sons, Aust. Mamm. Soc., Sydney. pp. 469–481.

Murray, J.D., G.M. McKay, J.W. Winter and S. Ingleby (1989). Cytogenetics of the Herbert River Ringtail Possum, *Pseudocheirus herbertensis* (Diprotodonta: Pseudocheiridae): evidence for two species. *Genome* 32: 1119–1123.

Nix, H.A. and M.A. Switzer (1991). Rainforest Animals: Atlas of Vertebrates Endemic to Australia's Wet Tropics. *Kowari 1*. Aust. Nat. Parks and Wildl. Serv., Canberra.

roads on the ground, even when the canopy was continuous overhead.

Leaves of rainforest trees form the bulk of the diet. Two favoured species are the Bleeding Heart, *Omalanthus novo-guineensis* and Pink Almond, *Alphitonia petriei*, which grow along roadsides. Other species eaten are Buttonwood, *Glochidion* sp.; a tamarind, *Arytera* sp.; Northern Quandong, *Elaeocarpus ferruginiflorus*; Big-leaf Planchonella, *Planchonella macrocarpa*; Timonius, *Timonius singularis*; a Blush Walnut, *Endiandra* sp., Hyland code 15; Needlebark, *Macaranga subdentata*; Brown Cudgerie, *Canarium australasicum*; White Basswood, *Polyscias murrayi*; a boxwood, *Chrysophyllum* sp.; and Paperback Satinash, *Syzygium papyraceum*. Fruits of several species of fig are eaten, Green-leaved Moreton Bay Fig, *Ficus watkinsiana* being a favoured fruit in spring. One individual has been seen to carry a large, reddish-pink fruit in a forepaw while climbing.

The Daintree River Ringtail is primarily nocturnal and is presumed to use tree hollows and epiphytic clumps as daytime shelter (like the Herbert River Ringtail). One has been seen carrying a bunch of leaves, curled in its tail, into a tree hollow about 10 metres above the ground. However, there are observations of one animal

*The Daintree River Ringtail feeds on soft leaves of rainforest trees.*

sleeping during the day on an exposed branch and another moving about during the mid-afternoon.

Little is known about the reproduction and behaviour of the Daintree River Ringtail, but it is presumed to resemble the closely related Herbert River Ringtail in these respects. The female has two teats in the pouch and usually two young are reared. There appears to be an extended breeding season, since pouch-young have been recorded from July, November and December; back-riding young in July, September, October, November, December and January and young at heel in July. Mating or attempted matings have been recorded in July, October, December and January. Young weighing 280–400 grams have been recorded as 'parked' in low foliage in October, November, December and February.

It is usually solitary, pairs probably being restricted to a consort period prior to mating. Males are more aggressive and more likely to bite than females when handled. Vocalisations include a short, hissing grunt given when handled or when disturbed by proximity to observers and a 'sik-sik-sik' call given by fully-furred pouch-young when handled.

***Pseudochirulus cinereus***

Greater densities occur at higher elevations within the range, up to 15 animals being seen per hour of spotlighting above 1000 metres on the Mount Carbine Tableland. Predation on the species is probably greatest during the 'parked' period of development. The Lesser Sooty Owl takes only individuals within this size range, as does the Spotted-tailed Quoll, taking advantage of their smallness and closeness to the ground at this stage of development. Other known or probable predators include the Amethystine Python, Carpet Python, Dingo and Grey Goshawk, taking animals that are exposed during the day.

J.W. WINTER AND M. TRENERRY

# Herbert River Ringtail Possum

## *Pseudochirulus herbertensis*

(Collett, 1884)

herb'-ert-en'-sis: 'Herbert (River) little-*Pseudocheirus*'

*Pseudochirulus herbertensis*

The Herbert River Ringtail lives mainly above an elevation of about 350 metres in the dense rain-forests of north-eastern Queensland between the Mount Lee area west of Ingham and the Lamb

---

### *Pseudochirulus herbertensis*

**Size**

*Head and Body Length*
350–400 (370) mm (males)
301–376 (344) mm (females)
*Tail*
290–470 (373) mm (males)
335–410 (363) mm (females)
*Weight*
810–1530 (1177) g (males)
800–1230 (1053) g (females)

**Identification** Dark, almost black, body with varying amounts of white on the chest, belly and around the upper forearm. Adults may lack white markings (males usually white around scrotum). Distinguished from Lemuroid Ringtail by tapered tail and pointed face with 'roman nose' in profile. Juveniles pale fawn with longi-tudinal median stripe on head and upper back, shadow of adult pattern of black and white marking visible. Eyeshine bright pinkish-orange.
**Recent Synonyms** *Pseudocheirus herbertensis*.
**Other Common Names** Indigenous name: Mongan (Atherton Tableland).
**Status** Limited distribution.
**Subspecies** None.

---

Range west of Cairns. It is occasionally found in the tall open forests of Flooded Gum that fringe the western edge of the rainforests. It is evenly and sparsely distributed throughout its range, 0.5–3.0 individuals being seen per hour of spot-lighting. The estimated density at Longlands Gap is about 9 per hectare.

It is arboreal, rarely descending to the ground and climbs cautiously and methodically. It makes small leaps, but only from substantial support and usually across the angle between a major branch and the trunk of a tree.

Leaves of rainforest trees form the bulk of the diet. In comparison with other rainforest ring-tails it selects leaves containing significantly more protein. The masticated leaf meal is probably digested in the relatively large caecum. Both old and new leaves of the Pink Ash, *Alphitonia petriei* (which produces new leaves continuously), are its most important food source. Other favoured food species include the Brown Quandong, *Elaeocarpus ruminatus*; White Basswood, *Polyscias murrayi*; Bumpy Satinash, *Syzygium cormiflorum*; Paperbark Satinash, *Syzygium papyraceum*; *Acronychia cras-sipetala*; *Planchonella brownlessiana*; Red Eungella Satinash, *Acmena resa;* and a vine, *Eleagnus triflora*. It is also known to eat the fruits or flowers of the Silver Quandong, *Elaeocarpus angustifolius*; Bumpy Satinash; Paperbark Satinash; and a vine *Melodinus bacellianus*. In fringing open forest the leaves of Cadaghi, *Eucalyptus torelliana;* and Pink Blood-wood, *Eucalyptus intermedia,* are eaten.

*Apart from a variable amount of white on the belly, adult Herbert River Ringtails are black.*

Strictly nocturnal, the Herbert River Ringtail usually emerges from its den approximately ten minutes after astronomical 'last light' but emergence may be earlier during overcast conditions. Animals return 50–100 minutes before 'first light'. On emergence, much of the first hour is spent in grooming, before moving off to feed. Most of the middle hours of the night are spent feeding and stationary in a selected tree. Movement back to the den usually begins after 3.30 a.m. and the den is entered about half an hour later. Activity is reduced on clear moonlit nights (presumably because it is more readily seen by predatory owls) and at ambient temperatures below 14–16°C. Tree hollows and large epiphytic ferns are the favoured daytime resting sites in rainforest but the Herbert River Ringtail is also capable of constructing a rudimentary nest like that of the Common Ringtail. Nesting material is carried in the curled tail. The home range of males and females overlap and range between 0.5 and 1 hectare.

M. TRENERRY

### References

Haffenden, A. (1984). Breeding, growth and development in the Herbert River Ringtail Possum, *Pseudocheirus herbertensis herbertensis* (Marsupialia: Petauridae) (in) A.P. Smith & I.D. Hume (eds) *Possums and Gliders*. Aust. Mamm. Soc., and Surrey Beatty and Sons, Sydney. pp. 277–281.

Laurance, W.F. (1990). Effects of weather on marsupial folivore activity in a North Queensland upland tropical rainforest. *Aust. Mammal.* 13: 41–47.

Spear, R., A.T. Haffenden, P.W. Daniels, A.D. Thomas and C.D. Seawright (1984). Disease of the Herbert River Ringtail, *Pseudocheirus herbertensis* and other North Queensland rainforest possums (in) A.P. Smith and I.D. Hume (eds) *Possums and Gliders*. Aust. Mammal. Soc. and Surrey Beatty and Sons, Sydney. pp. 283–302.

Although males are usually solitary, a consort individual remains close to a female prior to mating. Immediately on emerging from its den, a consort male travels to his female and remains within a few metres of her throughout the night, occasionally approaching to sniff at the base of her tail, but not coming into contact; the female appears to ignore the male. Two consort periods were observed to last for 44 and 48 days respectively, but consort-following may be as short as a single night. The occasional groups of three adults seen together probably result from two males being attracted to a female on heat. Extended consort periods, probably culminating in mating, have been observed only in March and April, but matings or attempted matings have also been seen in June, October and January.

Two functional teats are present in the forward-opening pouch and two young are usually reared. Three-quarters of births occur from May to July, none having been recorded between January and March. Young begin to leave the pouch soon after 105 days and depart permanently by 120 days of age. Young ride on the mother's back for about two weeks, a shorter period than in any other ringtail. Thereafter, they are left in

HANS & JUDY BESTE

*Female Herbert River Ringtail with young.*

# Western Ringtail Possum

### *Pseudocheirus occidentalis*

(Thomas, 1888)

ox'-id-ent-ah'-lis: 'western false-hand'

the nest and make increasingly longer forays alone as they increase in age. Young weighing only 300 grams may be found motionless in low foliage, 'parked' by their mothers. A very quiet, high-pitched, rapid chatter is given by young when separated from the mother, who responds by moving towards them. Adults give a buccal click when under stress and a variety of relatively quiet explosive screeches and grunts in antagonistic situations.

Aboriginal people used to hunt this species by ascending trees with the aid of a climbing cane to search in fern clumps. Known predators are the Rufous Owl and the Carpet Snake. However, the greatest threat to the Herbert River Ringtail is the clearing of rainforest: numbers decline sharply in forest fragments less than 20 hectares in size. The species appears to be secure because, although 21 per cent of its habitat was cleared for pasture and agriculture, most of what remains is now within the Wet Tropics of Queensland World Heritage Area.

J.W. WINTER AND N.J. GOUDBERG

In the nineteenth century the Western Ringtail Possum was widely distributed throughout the south-western forests of Western Australia from Perth to Albany. It was first reported to be in decline in 1909 and was given 'rare and endangered' status by the Western Australian government in 1983. The pattern of decline has been local and patchy. Local extinctions have been most extensive in inland areas, occurring as recently as 1950–1970.

Most populations now occur in near-coastal areas from Bunbury to Albany where forests include Peppermint, *Agonis flexuosa*. It is relatively common and abundant in a small part of the southern Swan Coastal Plain near Busselton and viable groups live in some parts of urban Busselton. Within this area, the highest densities occur in habitats with dense, relatively lush vegetation, usually associated with drainage lines. An inland population (90 kilometres from the coast) occurs in eucalypt forest in the Warren River catchment area, along with several other rare mammal species. This is the only viable population known at present from outside the Peppermint range.

The Western Ringtail is highly arboreal: feeding, resting and socialising in the canopy as much as possible, it is rarely seen on the ground. Peppermint leaves are the major dietary component but those of several other trees are eaten in small amounts or seasonally. In the inland forest, Jarrah, *Eucalyptus marginata* and Marri, *E. callophylla*, are the main food. In coastal areas, dreys are the usual daytime rest site, but populations occurring more than 4 kilometres from the

*Pseudocheirus occidentalis*

*The Western Ringtail lives mainly in the canopy of peppermint or eucalypt forests.*

coast rely mainly on eucalypt hollows that provide cool rest-sites on hot summer days. Adults usually rest alone but females may be accompanied by young (usually only one) although twins occur in about 10 per cent of births. Births are mostly in winter, but some populations breed throughout the year. Young emerge from the pouch at about 3 months, (weighing about 125 grams) and suckle until six to seven months (550 grams). Captive females have reached adult body weight at about eight months and produced their first young at about ten months. The normal life span is unknown, but some females have been known to breed in three successive years.

The Western Ringtail Possum has a relatively small and stable home range. In dense, coastal Peppermint forest, home ranges are about 0.5–1.5 hectares and in eucalypt forest about 2.5 hectares.

Regardless of forest type, individuals use three to eight different rest sites (dreys or hollow trees) in the course of a year. Adjacent home ranges may overlap as much as 70 per cent, but there is usually temporal separation in the use of shared areas. Social activity occurs at night, primarily by investigation of scent trails on tree limbs (marked with urine) and males may visit adjacent female home ranges. In some populations, most young disperse to home ranges adjacent to the natal range but, in high density groups, young travel across at least several home ranges.

In most coastal populations the Fox is the main predator of the Western Ringtail. In coastal

BABS & BERT WELLS

*Once common in the forests of south-western Australia, the Western Ringtail is now rare and endangered.*

## Pseudocheirus occidentalis

**Size**

*Head and Body Length*
300–400 mm
*Tail Length*
300–400 mm
*Weight*
900–1100 g

**Identification** Usually very dark brown dorsally, occasionally dark grey, with cream or grey ventral pelage. Ears short and rounded. Tail slender with terminal white tip of variable length.

**Recent Synonyms** *Pseudocheirus peregrinus occidentalis*.

**Other Common Names** Indigenous name: Nguara (Nyoongar).

**Status** Rare and endangered.

**Subspecies** None.

**References**

Ellis, M. and B. Jones (1992). Observations of captive and wild Western Ringtail Possums *Pseudocheirus occidentalis*. *West. Aust. Nat.* 19: 1–9.

Jones, B.A., R.A. How and D.J. Kitchener (1994). A field study of *Pseudocheirus occidentalis* (Marsupialia: Petauridae). I. Distribution and habitat. *Wildl. Res.* 21: 175–187.

Jones, B.A., R.A. How and D.J. Kitchener (1994). A field study of *Pseudocheirus occidentalis* (Marsupialia: Petauridae). II. Population studies. *Wildl. Res.* 21: 189–201.

# Common Ringtail Possum

## *Pseudocheirus peregrinus*

### (Boddaert, 1795)

pe'-re'-green'-us: 'foreign false-hand'

This colourful and variable ringtail is familiar to many residents of the eastern mainland as a persistent thief of rosebuds. Although here considered as a single species, it probably comprises a group of closely related species that are still in the process of diverging from each other. The prehensile tail, with its long friction pad, is used as a fifth limb while the animal is climbing among small branches in search of food and also serves to carry nesting material.

It occupies a variety of vegetation types, including rainforest, where shrubs form dense, tangled foliage. Spherical nests lined with shredded bark or grass are made in a hollow limb, in a bunch of mistletoe, or dense undergrowth. In southern Australia, as many as eight nests may be included in the home range of an individual but, in the north, a nest is rarely constructed and the animal sleeps in a tree hollow which may or may not be lined with dry leaves. Availability of nest

habitats which are characterised by an extensive or dense canopy or mid-stratum, the possum is difficult prey. In habitats where the canopy is discontinuous, individuals descend to the ground more frequently and are more exposed to terrestrial predators. The pattern of extinction in the more open inland forests appears to implicate Fox predation in the decline of the species.

B. JONES

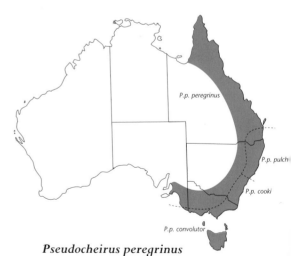

*Pseudocheirus peregrinus*

K. GRIFFITHS

*Although primarily a leaf-eater, the Common Ringtail also feeds on flowers and fruits*

In major Australian cities, it has adapted its behaviour to live in close association with humans and their gardens and makes use of a wide variety of introduced flowers and fruits. It is strictly nocturnal and most active, foraging and grooming, during the first half of the night. Later, it rests, clinging to a branch or in a secondary nest, returning to feed again before dawn. Family groups tend to nest and forage together until the young disperse. The male–female bond usually carries over from one breeding season to the next, although the male may mate and associate with a second female in a different part of its home range after the first

***Pseudocheirus peregrinus***

**Size**
*Head and Body Length*
300–350 mm
*Tail Length*
300–350 mm
*Weight*
700–1100 g

**Identification** Ears short with white patch behind. Tapering prehensile tail with white tip.
**Recent Synonyms** *Pseudocheirus rubidus* (part), *Pseudocheirus laniginosus* (part), *Pseudocheirus convolutor* (part), *Pseudocheirus occidentalis* (part), *Pseudocheirus victoriae* (part).
**Other Common Names** Grey Queensland Ringtail, Banga, Rufous Ringtail, South-eastern Ringtail, Tasmanian Ringtail.
**Status** Common.
**Subspecies** *Pseudocheirus peregrinus peregrinus*, Cape York to Kangaroo Island.
*Pseudocheirus peregrinus cooki*, coastal scrubs of south-eastern mainland.
*Pseudocheirus peregrinus convolutor*, Tasmania and Bass Strait islands.
*Pseudocheirus peregrinus pulcher*, south-eastern Queensland and north-eastern New South Wales rainforest.

sites is probably a major factor limiting distribution and abundance.

Although anatomically specialised as a leaf-eater, the Common Ringtail also feeds upon flowers (particularly those of eucalypts) and fruits. Its dietary preferences vary geographically but nowhere extend to insects. The Common Ringtail is one of a few marsupials able to feed on eucalypt leaves, which are a poor source of nutrition and difficult to digest. The gut has a large colon and a very large caecum in which masticated food is fermented by symbiotic bacteria; fluid and fine particles are retained in the caecum longer than coarser particles. During the day, while the Ringtail is in its nest, the contents of the caecum are evacuated as soft faeces and immediately eaten. Hard faeces, partly derived from material that has been twice digested, are produced at night.

P. GERMAN

*The Common Ringtail is one of the few marsupials to frequent suburban gardens, often eating fruits and buds.*

been observed to assist females in caring for the young. The putative fathers carry the young on backs while travelling to and from the nest or while foraging; groom the young; stay with the young in the nest while the mother forages alone; nest with the young during the day; and vocalise when potential predators approach the young or call the young back to the nest.

Abundance varies geographically, the Cape York populations being particularly rare. Although the availability of nest sites seems to affect local abundance (as in Victoria), the rarity of this form at the extremes of the range presumably has some other cause—habitat destruction and predation by the Cat and the Fox being two possible factors.

G.M. McKay and P. Ong

female gives birth. The home range of a male usually includes the ranges of two females but there is little overlap between the ranges of adjacent males. It is a vocal animal, but much less so than the Common Brushtail. Its presence is often detected by its soft, high-pitched, twittering call.

The Common Ringtail breeds from April to November throughout its range. In Victoria and northern Queensland, most females give birth at the beginning of the season but in the vicinity of Sydney there is a high incidence of breeding in spring.

In the south-eastern suburbs of Melbourne, there is an initial consort period during late April when the male follows, grooms, forages with and later nests with, the female. Mating occurs in May and the first litter is born in May or early June; the first litter is born earlier to established pairs. The female has four teats but two young are usually produced. These leave the pouch when about four months old but are not weaned until after six months of age. The pair may then mate again, in late October, to produce a second litter in mid- to late November. Usually, juveniles remain in the nest or are carried on the mother's back while she forages. However, males have

**References**

Hughes, R.L., J.A. Thomson and W.H. Owen (1965). Reproduction in natural populations of the Australian Ringtail Possum, *Pseudocheirus peregrinus* (Marsupialia: Phalangeridae) in Victoria. *Aust. J. Zool.* 13: 383–406.

Thomson, J.A. and W.H. Owen (1964). A field study of the Australian Ringtail Possum *Pseudocheirus peregrinus* (Marsupialia: Phalangeridae). *Ecol. Monogr.* 34: 27–52.

Cork, S.J. and L. Pahl (1984). The possible influence of nutritional factors on diet and habitat selection by the ringtail possum (*Pseudocheirus peregrinus*) (in) Smith, A.P. and Hume, I.D. (eds) *Possums and Gliders*. Surrey Beatty and Sons and Aust. Mamm. Soc., Sydney. pp. 269–276.

How, R.A., J.L. Barnett, A.J. Bradley, W.F. Humphreys and R. Martin (1984). The population biology of *Pseudocheirus peregrinus* in a *Leptospermum laevigatum* thicket (in) Smith, A.P. and Hume, I.D. (eds) *Possums and Gliders*. Surrey Beatty and Sons and Aust. Mamm. Soc., Sydney. pp. 261–268.

# Superfamily Tarsipedoidea

The Honey Possum differs in so many respects from most other members of the Phalangerida, that it has long been placed in its own superfamily, with one family: the Tarsipedidae. Some authorities regard its evolutionary separation as worthy of even greater taxonomic distinction. However, it has recently been proposed that the Acrobatidae (Feathertail Glider and Feathertail Possum) share sufficient characters with the Honey Possum, to warrant their inclusion within the Tarsipedoidea. This is the position taken in the latest (1987) taxonomic review of the marsupials, but, although it has not won general acceptance from researchers in the field of possum biology, it has not been formally refuted. In the opinion of the Editor, the appropriate response to this situation would be to raise the Acrobatidae to superfamilial level (thus, for the time being, to leave the relationships of this group indeterminate within the Phalangerida). However, any such taxonomic decision must be argued at length in an appropriate journal, not here.

Therefore, the Acrobatidae is here treated as a family within the Tarsipedoidea—but with strong misgivings.

*The tiny Honey Possum is restricted to richly flowering heathland in south-western Australia. It feeds exclusively on nectar and pollen.* R.L. SMITH

## FAMILY TARSIPEDIDAE
### *Honey Possum*

The family Tarsipedidae contains only one member, the tiny Honey Possum, which is specialised for a diet of nectar. The characteristics of the family are those of the species.

# Honey Possum

## *Tarsipes rostratus*

### Gervais and Verraux, 1842

tar'-sip-ez ros-trah'-tus: 'notably snouted Tarsier-foot'

*Tarsipes rostratus*

The Honey Possum was described in 1842 by French and British naturalists within five days of each other, but little was known about it until very recently. One of the few truly nectarivorous mammals, it has a long, pointed snout and an elongate, protrusible, brush-tipped tongue with which it probes flowers for its exclusive diet of nectar and pollen. It is notable for having fewer teeth than other marsupials (no more than 11 in a row) and for less obvious micro-anatomical and biochemical characteristics that indicate that it may be no more closely related to possums than to bandicoots or quolls. It shares some characters with the Burramyidae and Acrobatidae but it is an anomalous animal, apparently the sole survivor of a long-extinct marsupial group which owes it survival to a highly specialised way of life.

The size of a mouse, it has coarse fur, grizzled greyish-brown above and cream below. Dorsal markings consist of a distinct, dark brown central stripe, flanked by two lighter brown stripes. Females are larger than males. The eyes are close together, directed upwards and forwards. The tail is long, tapering and prehensile.

Most of the day is spent in sleep, except in cooler weather. It does not make a nest but may shelter in abandoned bird nests or the hollow stems of grass-trees. In cold weather, or when food is scarce, it frequently becomes torpid: captive animals often huddle together to conserve heat. Unlike most mammals, it does not climb with the aid of claws, for (except on the conjoined grooming toes of the hindfoot) these are reduced to nail-like structures on the upper surface of the expanded tips of the fingers and toes with which it grips branches, like a primate: its generic name refers to a supposed resemblance of its feet to those of the Tarsier, a primitive primate. The first clawless digit of the hindfoot is opposable to the other four and the

### *Tarsipes rostratus*

**Size**
*Head and Body Length*
40–94 (68) mm
*Tail Length*
45–110 (83) mm
*Weight*
7–12 (9) g

**Identification** Light brown or grey above with central dark stripe flanked by paler bands. Very pointed snout; eyes towards top of head.
**Recent Synonyms** *Tarsipes spenserae*, *Tarsipes spencerae*.
**Other Common Names** Honey Mouse. Indigenous name: Noolbenger.
**Status** Common, limited.
**Subspecies** None.

R.H. GREEN

*When resting, a Honey Possum curls up to reduce loss of body heat. It does not make a nest but sometimes uses an abandoned birds' nest.*

fourth digit is much longer than either the fifth or the syndactylous second and third digits. The long, slender, prehensile tail is able to support the weight of the body but is seldom employed to do so. When foraging, it moves in excited darts from blossom to blossom.

Laboratory studies suggest that scent plays a part in guiding the Honey Possum to its food,

most of which is found in low heath, although animals may ascend several metres to reach banksia blossoms. The snout is inserted into a flower and the long (18 millimetres) mobile, brush-tipped tongue is protruded through a funnel formed by flanges on the upper and lower lips. The brush-tip of the tongue is a series of filiform papillae and its anterior half is triangular in cross-section with a stiffened ventral keel. Pollen is licked rapidly from the protruding anthers and, as the tongue is retracted into the mouth, pollen is scraped off against the canine teeth and a series of ridges on the palate. The stomach has a diverticulum (about half the size of the main chamber) which may act as a storage organ. The intestine is short and there is no caecum. It is not noticeably specialised for the digestion of pollen, which forms an important part of its diet and is passed through the gut in about six hours. There is no evidence that insects form a part of the normal diet although mealworms are eaten by captive animals. The Honey Possum is common only on the coastal sand plain heaths of south-western Western Australia where a rich diversity of nectar-producing plants (especially members of the Proteaceae and Myrtaceae) provide it with food throughout the year. It is probably a significant pollinator of these plants.

## References

Wooller, R.D., M.B. Renfree, E.R. Russell, A. Dunning, S.W. Green and P. Duncan (1981). Seasonal changes in population of the nectar-feeding marsupial *Tarsipes spencerae* (Marsupialia: Tarsipedidae). *J. Zool. Lond.* 195: 267–279.

Renfree, M.B., E.M. Russell and R.D. Wooller (1984). Reproduction and life history of the honey possum, *Tarsipes rostratus* (in) A.P. Smith and I.D. Hume (eds) *The Biology of Possums and Gliders.* Aust. Mamm. Soc., and Surrey Beatty and Sons, Sydney, Australia. pp. 427–437.

Russell, E.M. and M.B. Renfree (1989). Family Tarsipedidae (in) D.W. Walton and B. Richardson (eds) *Fauna of Australia. Vol. 1B, Mammalia.* Government Printer, Canberra. pp. 769–782.

There is no obvious breeding season. Females with pouch-young are found throughout the year, but predominantly in early autumn, winter and spring when pollen and nectar are most abundant. The Honey Possum gives birth to the smallest of all mammalian young (weighing less than 5 milligrams) but the species is also notable for having the longest mammalian spermatozoa (about one-third of a millimetre). The testes and epididymides are suspended in a large, furry scrotum: testes weigh about 5 per cent of adult male body weight, perhaps reflecting a promiscuous mating system. Four to six eggs are ovulated but no more than four develop into neonates. Females have a well-developed pouch with four teats, but usually carry only two to three young, which grow rapidly and remain in the pouch until about half the size of the mother. For the last two weeks of the ten-week suckling period, the young are left in a shelter or nest while the mother is foraging. Juveniles reach adult size at the age of about eight months and females may breed in their first year. Some females breed at least twice a year. Females are dominant to males and several males compete for access to a receptive female.

Like most kangaroos and wallabies, the female Honey Possum has a post-partum oestrus and carries uterine blastocysts that remain dormant for at least three months while a female is lactating. The uterine blastocysts grow from 1.2–2.0 millimetres diameter before entering diapause. What stimulates these quiescent embryos to resume development after the young has left the pouch is not yet understood, but is presumed to be under environmental control.

At present, the Honey Possum is common in limited winter rainfall areas of sand plain heathland, rich in plant species of the family Proteaceae. Provided that large areas of this habitat are maintained in reasonable succession, the species appears to be secure, despite predation by the Fox and Cat.

M.B. RENFREE

# FAMILY ACROBATIDAE
## *Feathertail Glider*

*The Feathertail Glider is an agile climber, aided by claws and by adhesive pads on the tips of the toes and fingers.* P. GERMAN

In the first edition of this work, the Feathertail Glider was placed with the pygmy-possums in the family Burramyidae. Since then, it has been set apart, with the Feathertail Possum, *Distoechurus*, of New Guinea, as a distinct family, the Acrobatidae. The most obvious characteristic of the family is the arrangement of long, stiff hairs on either side of the tail to form a feather-like structure. The hindfoot has a sixth pad on the sole, not present in burramyids.

Since the structure of the tail is obviously an adaptation to flight, it must be assumed that the Feathertail Possum, which lacks a gliding membrane, is descended from a gliding ancestor. The affinities of the Acrobatidae with other families of the Phalangerida are unclear, but the most recent classification of the marsupials places them in the superfamily Burramyoidea. This opinion is respected here, but with reservations. It may well be that Acrobatidae deserves separate superfamilial status.

# Feathertail Glider

## *Acrobates pygmaeus*

(Shaw, 1794)

ak'-roh-bah'-tays pig-mee'-us:
'pygmy acrobat'

The common name of this species draws attention to the unusual tail, which has very short fur on the upper and lower surfaces and a conspicuous fringe of long, stiff hairs on either side, resembling the barbs of a feather. Its closest relative is a ground-dwelling possum, *Distoechuris pennatus*, from New Guinea, which has a similar tail.

The Feathertail Glider is the world's smallest gliding mammal. It is widely distributed in

*The Feathertail Glider is an agile climber, aided by claws and by adhesive pads on the tips of the toes and the fingers.*

tall forests and woodlands of eastern Australia and exhibits a remarkable suite of adaptations to this environment. For example, its molar teeth have the shape and cusps typical of an insect-eater, while its brush-tipped tongue is typical of a nectar-feeder. The large, forwardly directed eyes probably provide nocturnal binocular vision (advantageous to a leaping animal) and there are large, serrated pads on each toe (reminiscent of those of tree-frogs and geckos) which provide adhesion to smooth surfaces, such as the bark of some eucalypts. Sharp claws assist its climbing on softer surfaces. The feather-like tail is somewhat prehensile and provides a grip on twigs and small branches.

There is a gliding membrane between the elbows and the knees, relatively thicker than in other gliding marsupials. Once it leaps, the membrane is extended and the feathery tail assists in steering and braking before landing.

Among its other specialisations is the existence of a small, bony disc in front of the tympanic membrane and an unusual alignment of the bones that surround the ear: this may provide a filtering mechanism that leads to selective sensitivity to very low and very high frequencies, but its significance has yet to be determined. Its agility and small size make it difficult to find in its habitat. It is presumed to be abundant but very few surveys have attempted to quantify this. It is more frequently located in tall, mature and moist forests than elsewhere and it has been suggested that it may require access to complex or mature forests to meet all its food and nesting requirements and sufficient security from predators. Unfortunately, the availability of suitably mature and unlogged forests is in decline and only a few pockets will remain by the end of the next decade. There is, however, little knowledge of related changes in its distribution or abundance.

---

### *Acrobates pygmaeus*

**Size**
*Head and Body Length*
65–80 mm
*Tail Length*
70–80 mm
*Weight*
10–14 g

**Identification** No other Australian mammal has a feather-like tail (although this is found in related species from New Guinea).
**Recent Synonyms** None.
**Other Common Names** Pygmy Glider, Pygmy Phalanger, Flying Mouse.
**Status** Common.
**Subspecies** A supposed subspecies *Acrobates pygmaeus frontalis* from northern Queensland was based on poorly described juvenile specimens.

Captive animals demonstrate the need for complex furnishings in enclosures. It is important that the vegetation used offers dense cover, many vertical and horizontal climbing surfaces, many feeding points and a surplus of nest cavities. The combination essential to maintaining a breeding colony includes an overabundance of climbing surfaces, an overabundance of basic resources (food and shelter), a diet of nectar substitute and live insects and a large social group. This combination has supported successful breeding programs in zoos spanning several generations. Breeding has not occurred in enclosures that were more sparsely furnished or where the animals were maintained in pairs. It is important to determine which of the conditions is most critical to this breeding success and to follow this through with ecological research in the forests.

The Feathertail Glider is normally active at night, except when rearing young. In captivity, lactating females often emerge to feed or drink during the late afternoon. In the wild, this may expose them to a wide range of avian predators such as currawongs and kookaburras. Evidence in Fox and Cat scats also indicates the relatively high frequency with which Feathertail Gliders fall prey to introduced terrestrial predators. Other predators probably include reptiles and large dasyurids. In some parts of their range, Ghost Bats may feed on Feathertails, locating them by their high-pitched social calls.

Feathertails are often found nesting or torpid in groups of up to 20 individuals but these are not stable associations. Groups have been located in virtually any available enclosed space including tree hollows, telephone interchange boxes, bird boxes, old bird nests or deserted possum dreys. They make spherical nests (dreys) of vegetation, particularly casuarina and eucalypt leaves, bark and tree-fern fibre. Nests are often lined with fresh leaves, feathers or other soft and flexible material. In captivity, nests are maintained by many members of the group, which tend to use more than one nest at a time. Group-nesting behaviour may be part of group-bonding behaviour and is more likely to occur among relatives in the wild. So far, this behaviour has been noted in captive groups ranging in size from 7–22 individuals.

*Acrobates pygmaeus*

In the wild, it feeds on nectar, pollen and insects in a variety of locations and canopy levels, including bushes, large trees and on the ground. It feeds on the tall inflorescences of grass-trees (*Xanthorrhoea* sp.) when in bloom and has been found nesting among the dead fronds of this plant. It is known to feed in groups on *Banksia integrefolia* at the same time as other species such as the common Blossom Bat, *Syconycteris australis*.

Not all females in a group breed at the same time. In captivity, breeding can occur at any time of the year, although the most pouch-young appear in spring. In the wild, this species appears to breed throughout the year in the northern part of the range and in late winter, spring and summer at higher latitudes. In southern Victoria, females regularly have two litters of three to four young, most of which survive the 65-day pouch life and are weaned at approximately 100 days. There is embryonic diapause, whereby females can become pregnant again immediately after giving birth. The resultant embryos become quiescent until late in the lactation stage of the previous litter and are born soon after that litter is weaned.

Growth is relatively slow for a small marsupial and maternal investment is high. Communication between mother and young involves a number of high-frequency calls and marking

**References**

Bennett, A.F., L.F. Lumsden, J.S.A. Alexander, P.E. Duncan, P.G. Johnson, P. Robertson and G.E. Silveira (1991). Habitat use by arboreal mammals along an environmental gradient in north-eastern Victoria (Australia). *Wildl. Res.* 10: 125–146.

Harman, A.M., L.A. Coleman and L.D. Beazley (1990). Retinofugal projections in a marsupial, *Tarsipes rostratus* (honey possum). *Brain Behav. Evol.* 36: 30.

Ward, S.J. (1990). Life history of the Feathertail Glider, *Acrobates pygmaeus* (Acrobatidae: Marsupialia), in south-eastern Australia. *Aust. J. Zool.* 30: 503–510.

Flemming, M.R. and M. Frey (1984). Aspects of the natural history of Feathertail Gliders (*Acrobates pygmaeus*) (in) A. Smith and I.D. Hume (eds) *Possums and Gliders*. Aust. Mamm. Soc. and Surrey Beatty and Sons, Sydney. pp. 403–408.

with urine. There is reason to believe that young may receive care from more than one female, but this has not been proven.

Males and females show few differences, but females can be slightly larger. Males are tolerated by females even during lactation in the captive colonies and very little aggression has been observed among males. Males appear to become sexually mature in their second year whereas females can breed in their first year. Both sexes live about three years in the wild and up to five years in captivity.

Feeding in groups is common in captivity and has been observed once in the wild. Behavioural trials indicate a tendency for one animal to respond by joining another one feeding. The sounds of an individual scratching over the surface of plants such as *Banksia* appears to excite other members of a group.

D.P. WOODSIDE

# Superfamily Phalangeroidea

This group contains only one living family, the Phalangeridae, including the brushtail possums and their relatives. Superfamily status denotes the level of distinction between these and the other possums.

## FAMILY PHALANGERIDAE

### Brushtail possums, cuscuses and the Scaly-tailed Possum

The Phalangeridae comprises four species of brushtail possums in the genus *Trichosurus*; the monotypic *Wyulda* or Scaly-tailed Possum; 16 species of cuscuses in the genera *Phalanger*, *Spilocuscus* and *Strigocuscus*; and the monotypic *Ailurops* or Bear Cuscus. *Trichosurus* and *Wyulda* are endemic to Australia. *Phalanger* and *Spilocuscus* are widespread through Melanesia, with one species each in Australia. *Ailurops*, which is restricted to Sulawesi, appears to be the least specialised member of the family and is placed in a subfamily of its own, the Ailuropinae. All other phalangerids are allotted to the subfamily Phalangerinae.

Phalangerids have a rather short face (notably so in many cuscuses) with eyes directed forward. The prominent rhinarium is moist. Ears vary from relatively large in brushtails to minute in many cuscuses. There is considerable variation in the relatively long tail. It is well-furred in *Trichosurus*, except for a granulated strip along most of the undersurface and is not strongly prehensile. In *Wyulda*, only the extreme base of the tail is furred, the remainder being covered with small, conical, horny tubercles: it is strongly prehensile. The tails of cuscuses range between these extremes and are always strongly prehensile.

All phalangerids are excellent climbers, usually moving slowly and deliberately. In common with most arboreal diprotodonts, the first two digits of the forefeet are opposable to the other three in a powerful grip—except in the case of the brushtail possums which lack this useful adaptation. The Scaly-tailed Possum is semi-terrestrial, sleeping by day among rocks and seeking shelter there when disturbed. In this respect it resembles the Rock Ringtail Possum.

Brushtail possums feed predominantly on leaves but take flowers and fruits opportunistically: they have a large caecum. The Scaly-tailed Possum and the Australian cuscuses are generalised herbivores.

The forwardly opening pouch encloses two or four teats. Most phalangerids rear only one young at a time but twins are said to be not uncommon in the Australian cuscuses.

# Common Spotted Cuscus

## *Spilocuscus maculatus*

(Desmarest, 1818)

spile'-oh-kus'-kus mak'-yue-lah'-tus: 'spotted spotted-cuscus'

In the forests of Cape York Peninsula north of Coen, a casual observer could be startled to see a round, bare-skinned face with large, forward-looking eyes, staring through the foliage. Tales of monkeys in this region arise from misidentification of the Common Spotted Cuscus which spends the day on a branch or in a clump of foliage. It occurs mainly in rainforest, from sea-level to the summit of the McIlwraith Range which reaches an altitude of 820 metres, but it may also be seen in nipa palms of the mangrove fringe, in freshwater and saline mangroves, in large paperbarks in thin riparian forest strips and in open forest up to half a kilometre from the nearest rainforest. Although predominantly arboreal, it is capable of travelling long distances on the ground: a radio-tracked subadult male traversed about 150 metres across a dirt road and recently burnt grassland. This dispersing ability and its lack of need for tree-hollow shelters, enables the Spotted Cuscus to invade rainforest regrowth, rainforest fragments and many other habitats. It may also account for its wider distribution than that of the Grey Cuscus.

It is mainly nocturnal but may continue its activity for short periods after sunrise. It usually sleeps in the thick canopy of a rainforest tree, reputedly building a rudimentary sleeping platform of leaves by drawing twigs under itself (making it very difficult to see, even when its location is pinpointed by a fitted radio collar). It seldom returns to the tree that it slept in the previous day. A slow and cautious climber, it forages in the canopy, maintaining a strong grip with its

---

### *Spilocuscus maculatus*

**Size**

*Head and Body Length*
348–580 mm
*Tail Length*
310–435 mm
*Weight*
1.5–4.9 kg

**Identification** Fur dense and woolly; snout short; ears almost invisible; rim of yellow to reddish skin around eye; body skin yellowish-pink. Distinguished from *Phalanger intercastellanus* by absence of mid-dorsal stripe, small ears and skin colour. Males have grey and white blotches above, white below. Juveniles may be uniformly grey. Females are uniformly grey above but sometimes with white, tinged yellow, on rump.
**Recent Synonyms** *Spilocuscus nudicaudatus, Phalanger maculatus*.
**Other Common Names** Spotted Phalanger. Indigenous name: Ampoiyu (Lockhart region).
**Status** Australia: limited and sparse; New Guinea: widespread.
**Subspecies** Three subspecies in New Guinea and the Moluccas. Australian form is *Spilocuscus maculatus nudicaudatus*.
**Extralimital Distribution** New Guinea, the Moluccas.

---

feet and tail, the distal two-thirds of which is naked, strongly prehensile and has rough papillae on the undersurface. It is, however, capable of leaping across a small gap between adjacent trees.

Cuscuses have been likened to sloths—other tropical canopy dwellers not using dens—in that they have a low metabolic rate and retain their body heat with thick, insulating fur, but the comparison should not be taken too far, since cuscuses are much more active. The relatively thick fur of the Spotted Cuscus is probably a thermoregulatory adaptation to sleeping in the windy rainforest canopy. During winter a radio-collared individual was seen sunning itself on several occasions. In

## References

George, G.G. (1987). Characterisation of the living species of cuscus (Marsupialia: Phalangeridae) (in) M. Archer (ed.) *Possums and Opossums*. Surrey Beatty and Sons and Roy. Zool. Soc. New South Wales, Sydney. pp. 507–526.

Flannery, T.F. (1995). *Mammals of New Guinea*, 2nd edn. Reed Books, Sydney.

Dawson, T.J. and R. Degabrielle (1973). The cuscus (*Phalanger maculatus*)—a marsupial sloth? *J. Comp. Physiol.* 83: 41–50.

hot weather, it loses heat by panting and licking saliva onto the bare-skinned feet and face for evaporative cooling.

Under natural conditions it is herbivorous, feeding on a variety of fruits of rainforest trees such as the Leichhardt Tree, *Nauclea orientalis*, native Star-apple, *Planchonella ripicola* and a Fig, *Ficus hispida*; on various flowers; and selectively on rainforest leaves such as those of Buttonwood, *Glochidion* sp.. In captivity, the young leaves of several *Acacia* species and wild Mango, *Mangifera indica*, are preferred. Nevertheless, the large canine teeth suggest a diet that is at least partly carnivorous and, in captivity, it readily eats beef, chicken, eggs and tinned dog food.

Little is known of the social behaviour of the Spotted Cuscus other than that it is usually solitary. A female with a bulging pouch was seen in August, a pair (presumably consorting prior to mating) in September and another carrying an infant on its back in December. These observations suggest that, as in New Guinea, there is an extended breeding season. There are four teats and, although three young have been recorded in a pouch, it seems that it is usual for only one to be reared. Males are aggressive and cannot be housed together in captivity. A blind male found

*The Common Spotted Cuscus is a mainly New Guinean species. Only the males are spotted.*

P. GERMAN

**Spilocuscus maculatus**

# Southern Common Cuscus

## *Phalanger intercastellanus*

### Thomas, 1895

fal'-an-jer in'-ter-kas'-tel-ah-nus: 'webbed-
toe from D'Entrecasteaux Islands'

dead on the forest floor had head wounds,
apparently inflicted by another cuscus. Females
make a peculiar high-frequency hiss which is
thought to be a mating call. Other calls given by
both sexes are hisses, screeches and buccal
clicks. When stressed, the Spotted Cuscus
secretes a red-brown substance on the bare skin
of its face, particularly around the eyes: when
dry, this stains the skin a reddish colour.

The Spotted Cuscus is widespread on Cape
York Peninsula, on the east coast south to the
Stewart River, in the Centre and on the west
coast south to the Coen-Archer River system. Its
density is low, as estimated by spotlighting—
only one or two animals, at most, being seen per
hour. However, spotlighting probably leads to an
underestimate because animals tend to hide in
thick foliage and look away from the spotlight,
providing little or no eye-shine.

The Amethystine Python is a potential
predator and one was seen following the path of
a Spotted Cuscus, apparently stalking it. Aboriginal people once hunted this species by looking
for the animals during the day, usually along thin
strips of gallery forest where it is much easier to
see than in tall, thick primary forest. Aboriginal
people rarely hunt it at present.

J.W. WINTER AND L.K-P. LEUNG

At the time of its discovery in Australia in 1938,
this mammal was assigned to the Grey Cuscus,
*Phalanger orientalis*, which was long recognised to
be extremely variable over its extensive range in
Melanesia. Recent studies demonstrate that it
comprises at least two (possibly three) species
and that, as redefined, *P. orientalis* is not represented in Australia. The species on Cape York—
also occurring in southern New Guinea and a

C. ANDREW HENLEY/LARUS

P. GERMAN

*The Southern Common Cuscus (long known as the Grey Cuscus) is more brown than grey.*

but it penetrates the acacia fringes of rainforest. Reports of cuscuses seen well to the south in the rainforests between Mossman and Cooktown can be attributed to the superficial resemblance of the Grey Cuscus to the Daintree River Ringtail.

The Southern Common Cuscus is more strictly nocturnal than the Spotted Cuscus and rests in a den during the day rather than on an exposed branch. Its late discovery in Australia, 90 years after the Spotted Cuscus, reflects its more secretive daytime behaviour, greater confinement to rainforest and more restricted geographical range within Australia. Population densities are low; on average, one individual is seen in about every two-and-one-half hours of

number of offshore islands—is *P. intercastellanus*. For consistency in nomenclature between New Guinea and Australia, it is appropriate to refer to *P. orientalis* as the Northern Common Cuscus and to *P. intercastellanus* as the Southern Common Cuscus. Apart from these considerations, the name 'Grey Cuscus' is not descriptive of the colour of the latter species and has led to much confusion, since females of the Common Spotted Cuscus, *Spilocuscus maculatus* are frequently greyish.

The Southern Common Cuscus is quite different from the Spotted Cuscus. In many features, such as its lighter build, longer snout, more prominent ears and general body colouration (including a dark dorsal stripe) it superficially resembles the Daintree River Ringtail Possum. Nevertheless, the characteristic naked tail, long canine teeth and many less obvious anatomical features mark it as a cuscus.

In Australia, it is restricted to the rainforests of eastern Cape York Peninsula, from the McIlwraith Range in the south to Iron Range in the north, at all elevations. It is more confined to primary rainforest than the Spotted Cuscus,

*The Southern Common Cuscus is a slow and careful climber in the rainforest canopy.*

## Phalanger intercastellanus

**Size**
*Head and Body Length*
350–400 mm
*Tail Length*
280–350 mm
*Weight*
1.5–2.2 kg

**Identification** Greyish-brown above, off-white below, brownish mid-dorsal stripe from between ears to rump. Males have distinct yellowish chest gland and yellowish tinge to side of neck. Skin greyish-brown. Distinguished from *Spilocuscus maculatus* by mid-dorsal stripe, larger ears, longer snout and skin colour. Distinguished from *Pseudocheirus cinereus* by largely naked tail.
**Recent Synonyms** *Phalanger orientalis.*
**Other Common Names** Grey Cuscus, Grey Phalanger.
Indigenous name: To-ili (Lockhart region).
**Status** Australia: limited distribution, sparse. New Guinea: widespread.
**Subspecies** None.
**Extralimital Distribution** New Guinea.

*Phalanger intercastellanus*

dog food is also taken. In comparison with the Spotted Cuscus, it has a greater appetite for leaves and less interest in animal food.

Its arboreal habits are facilitated by a long, prehensile tail, furnished with horny papillae on the undersurface. It is a slow, deliberate climber but, compared with the Spotted Cuscus, it has a less tenacious grip, moves faster through the canopy and is more likely to jump across gaps. On the ground, it moves in a slow, bounding gait.

The Southern Common Cuscus is solitary. Females have four teats in the pouch and, in New Guinea where the species is better known, twins appear to be usual. In Australia, one female with a single young was seen in June and another with two naked pouch-young in September.

Newly caught individuals may react aggressively, giving harsh screeches and lashing out with the forelimbs, which are at all times held low in front of the body (not raised in the outstretched threat-posture typical of brushtails). Captive animals, which can become very tame, have been heard to make repetitive buccal clicks, probably indicative—as in brushtails—of anxiety.

Amethystine and Carpet Pythons are likely to be the major predators. Predation by humans has probably always been light on Cape York Peninsula because the Aboriginal people appear to be far less familiar with the Southern Common Cuscus than the Spotted Cuscus. The latter is hunted by searching in its leafy retreats during the day, whereas the Southern Common Cuscus's use of a den and its strictly nocturnal behaviour save it from detection. This relative lack of familiarity with the species is reflected in its omission from the comprehensive list of mammals made for the Lockhart district by Donald Thomson, an anthropologist and zoologist who spent many months with the Aboriginal people of the area between 1928 and 1932. Because of its secretive behaviour and since the greater part of its original habitat remains intact, it seems to be in no immediate danger.

spotlighting, both in upland and lowland rainforest. Unlike the Spotted Cuscus, it makes no attempt to hide from an observer at night, but stares continuously into the spotlight, giving a bright red eye-shine that makes detection easy.

Mainly herbivorous, it has been seen feeding on the green fruit of the Red Cedar, *Toona australis*; buds and flowers of Corky Bark, *Carallia brachiata*; and on the leaves of a variety of trees. The stomach of one animal was filled with a paste apparently derived from the seeds of the Black Bean tree, *Castanospermum australe*. In captivity, it eats a variety of leaves and fruits: tinned

J.W. WINTER AND L.K-P. LEUNG

### References

Menzies, J.I. and Pernetta, J.C. (1986). A taxonomic revision of cuscuses allied to *Phalanger orientalis* (Marsupialia: Phalangeridae). *J. Zool. Lond.* (B), 1: 551–618.

Flannery, T.F. (1995). *Mammals of New Guinea*, 2nd edn. Reed Books, Sydney.

Colgan, D., T.F. Flannery, J. Trimble and K. Aplin (1993). Electrophoretic and morphological analysis of the systematics of the *Phalanger orientalis* (Marsupialia) species complex in Papua New Guinea and the Solomon Islands. *Aust. J. Zool.* 41: 355–378.

J. M<sup>c</sup>CANN

*Female Mountain Brushtail Possum with young.*

# Mountain Brushtail Possum

## *Trichosurus caninus*

trik'-oh-sue'-rus  kah-neen'-us:
'dog-like hairy-tail'

The Mountain Brushtail Possum, or Bobuck, is abundant in the tall, open and closed forests which occur patchily throughout its range from south-eastern Queensland to southern Victoria, east of the Great Dividing Range. It is a robust possum which, apart from a melanic form in the closed forests of north-eastern New South Wales, has a remarkably uniform steely grey colour throughout its range. Although arboreal, it spends considerable time on the ground and on fallen logs where it feeds extensively on leaves of mesophyllic shrubs, fruits, buds, fungi, lichens and occasionally bark. Dens are located in hollow spouts, branches, trunks, or logs and occasionally in epiphytes.

## Trichosurus caninus

**Size**
*Head and Body Length*
400–500 mm
*Tail Length*
340–420 mm
*Weight*
2.5–4.5 kg

**Identification** Generally steely grey above, whitish below. Short rounded ears (40–45 mm); bushy tail tapering to tip; long terminal naked area underneath.
**Recent Synonyms** None.
**Other Common Names** Bobuck, Short-eared Brushtail Possum.
**Status** Common.
**Subspecies** None.
**Reference**
How, R.A. (1981). Population parameters in congeneric possums, *Trichosurus* spp. in north-eastern NSW. *Aust. J. Zool.* 29: 205–215.

The sexes are similar in appearance but females live longer (up to 17 years) than males (up to 12 years). Vocalisation resembles that of the Common Brushtail and scent marking from the chin, sternal and anal glands is important in delineating occupied areas. Since the secretion from the sternal gland is a clear fluid, it does not colour the fur of the chest as in the Common Brushtail. Known predators include the Carpet Python and Dingo; the Spotted-tailed Quoll may take younger animals. When pressed, the Mountain Brushtail is a competent swimmer.

The breeding season extends from March to May with only occasional births occurring outside this period. Two-year-old females can reproduce but seldom rear their young beyond the pouch stage, being more successful in the third and subsequent years. Some adult females do not reproduce every year; others may have a second young if the first dies. After a gestation period of 15–17 days a single young is born. It spends five to six months in the pouch and continues to suckle from one of the two teats for a further two to five months. After weaning, juveniles remain in the area in which they were born until they are 18–36 months old. Dispersal occurs during this period, females leaving earlier than males. The presence of a dependent infant appears to adversely affect the development of young in the pouch. As a result, many pouch-young die before weaning. Adult males and females occur in equal numbers in an area and the ranges of a male and female may overlap considerably. This and the fact that a male and female are sometimes caught together in a trap, suggests that a degree of pairing occurs. The density in north-eastern New South Wales is about one individual per 10 hectares. Greatest numbers are found in forest gullies and where there are abundant hollow-bearing trees.

In parts of its range, the Mountain Brushtail causes considerable damage to pine plantations by stripping off the bark or even ringbarking trees. During past open seasons on possums in eastern Australia, considerable numbers of this species were taken, its pelt being preferred to that of the Common Brushtail by many trappers and furriers.

R.A. How

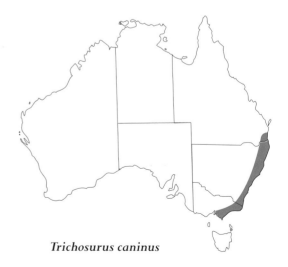

*Trichosurus caninus*

D. WHITFORD

*Common Brushtail Possum eating a Hairy Walnut fruit.*

# Common Brushtail Possum

## *Trichosurus vulpecula*

vool-pek'-yue-lah: 'little-fox-like hairy-tail'

This is the most familiar of the Australian possums, frequently cohabiting with humans. Over its extensive range it manifests considerable variation in size and colour: in the tropical closed forest of central eastern Queensland there is a short-haired, copper-coloured form; the larger more woolly Tasmanian form has a blackish as well as a grey phase. Males are generally larger than females and often have rufous fur on the shoulders. The northern form was briefly treated as a separate species (as in the first edition of this work).

Its abundance varies significantly throughout its range. It is most common in cities, in Tasmania and in New Zealand (where it was introduced in 1840) but it is rare in arid central Australia. It generally occurs where there are trees, especially in open forests and woodlands. A nocturnal animal, it spends the day in a den in a hollow dead branch, tree trunk, fallen log, rock cavity, or even a hollowed termite mound. In urban areas, almost any dark recess may be utilised, the space between a ceiling and roof being commonly favoured. Although it travels extensively on the ground, it is an arboreal animal, climbing by means of sharp claws, the

---

### *Trichosurus vulpecula*

**Size**
*Head and Body Length*
350–550 mm
*Tail Length*
250–400 mm
*Weight*
1200–3500 g (females)
1300–4500 g (males)

**Identification** Generally silver-grey above, white to pale grey below. Long oval ears (50–60 mm); tail varies from being bushy to sparsely furred with short, terminal, naked area underneath.

**Recent Synonyms** *Trichosurus fuliginosus, Trichosurus arnhemensis.*

**Other Common Names** Silver-grey Possum, Brushtail Possum, Northern Brushtail Possum.

**Status** Abundant in south-eastern Australia and Tasmania, declining elsewhere.

**Subspecies** *Trichosurus vulpecula vulpecula,* south-eastern and south-western Australia. *Trichosurus vulpecula arnhemensis,* tropical northern Northern Territory and Western Australia. *Trichosurus vulpecula eburacensis,* Cape York. *Trichosurus vulpecula johnsoni,* central eastern Queensland. *Trichosurus vulpecula fuliginosus,* Tasmania.

**Extralimital Distribution** New Zealand (introduced).

D. WATTS

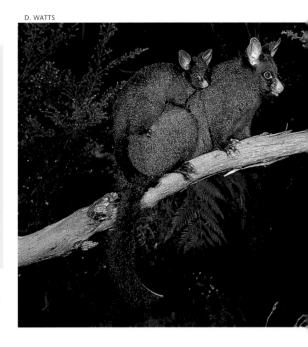

**References**

Kerle, J.A. (1984). Variation in the Ecology of *Trichosurus vulpecula*: its adaptive significant (in) A.P. Smith and I.D. Hume (eds) *Possums and Gliders*. Aust. Mamm. Soc. and Surrey Beatty and Sons, Sydney. pp. 115–128.

Kerle, J.A., G.M. McKay and G.B. Sharman (1991). A systematic analysis of the Brushtail Possum, *Trichosurus vulpecula* (Kerr 1792) (Marsupialia: Phalangeridae). *Aust. J. Zool.* 39: 313.

opposable first toe on the hindfoot and a moderately prehensile tail.

Because the Common Brushtail Possum occupies a vast range of habitats throughout Australia, its preferred foods are also very varied. Leaves comprise the bulk of its diet and some of the favoured species are very toxic. These include the Cooktown Ironwood in the tropics, *Gastrolobium* species in Western Australia, as well as eucalypts. The choice of leaves is determined mainly by their nutrient and fibre content and demonstrates a remarkable tolerance to plant toxins. Flowers and fruits form an important component of the diet, particularly in the arid and tropical areas where these foods are essential to successful breeding. Meat may be

consumed by captive animals if offered but it is eaten only very occasionally in the wild.

Most populations have a major autumn and a minor spring breeding season but some—including those in the tropics and arid regions—breed continuously if the required food supplies are available. Females usually begin to reproduce when about one year old. Where breeding is seasonal, over 90 per cent of females breed annually and in some populations 50 per cent may breed in both seasons. A single young is born 16–18 days after copulation and spends four to five months in the well-developed pouch, attached to one of two teats and developing rapidly. A further one to two months are spent in suckling and riding on the mother's back before weaning is completed. Few young die in the pouch but considerable mortality can occur at 6–18 months of age when juveniles are dispersing from the area of their birth in an endeavour to establish home ranges. The home ranges of individuals vary between sexes and between habitats and are related to the population density which varies from 0.2–4.0 individuals per hectare. Densities twice as high as this have been recorded in New Zealand, where there is a large introduced population.

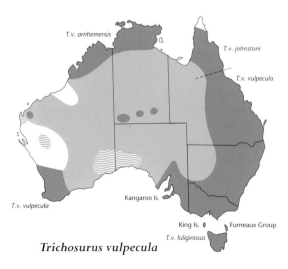

*T.v. arnhemensis*

*T.v. johnstoni*

*T.v. vulpecula*

*T.v. vulpecula*

Kangaroo Is.

King Is.

Furneaux Group

*T.v. fuliginosus*

**Trichosurus vulpecula**

P. GERMAN

Left: *Female Common Brushtail Possum with young.*

Right: *The Common Brushtail Possum is the most familiar marsupial in eastern Australia.*

Below: *A reddish-brown phase of the Common Brushtail Possum is not uncommon.*

G. WEBER

Communication is by sound and scent. Deep guttural coughs and sharp hisses are frequent, particularly in the breeding season and extensive use is made of glands under the chin, on the chest and near the anus, to mark areas and define occupancy. The reddish secretion from the chest gland of an adult male stains the surrounding fur. In some populations, males may establish territories, but vocal and olfactory signs are usually sufficient to establish den ownership and to maintain sufficient inter-individual distances to minimise direct aggression. The Dingo, Cat and Fox, large pythons and large monitors are known to prey on the Common Brushtail and can significantly affect numbers when population size is low. Longevity is usually less than 11 years but one individual is known to have lived for 13 years.

These population characteristics and their substantial variability allow the Common Brushtail to use a variety of habitats and to rapidly colonise suitable new areas and build up its numbers. In Australia, it can cause damage to pine plantations and in New Zealand, it has become a major pest and a host to bovine tuberculosis, but its skin supports a lucrative fur industry. In most Australian States, large numbers were once killed for their pelts but the species is now protected on the mainland and partly protected in Tasmania. In some parts of the mainland, the range is contracting, many populations having disappeared and others become vulnerable.

R.A. HOW AND J.A. KERLE

# Scaly-tailed Possum

## *Wyulda squamicaudata*

### Alexander, 1919

wie-ool'-dah, skwah'-mee-kaw-dah'-tah:
'scaly-tailed wyulda (an Aboriginal name for
brushtail possums)'

The Scaly-tailed Possum was first described from
an animal sent to the Perth Zoo in 1917 from
what is now the Violet Valley Aboriginal Reserve
in the Kimberley of Western Australia. A specimen was collected from Kunmunya Mission in
1942 and another from the Wotjulum Mission in
1954 but it was not until intensive systematic
zoological exploration of the north-western
Kimberley began in the 1960s that the status of
this unusual possum began to be established. It
resembles brushtail possums in some respects; in
others, such as the nakedness and prehensility of
its tail, it is similar to cuscuses. Electrophoretic
studies of its blood proteins confirm its affinities
with both *Trichosurus* and *Phalanger*.

It inhabits very rugged, rocky country, often
with rainforest vegetation elements and apparently lives deep in rock piles during the day,
emerging at night to feed in trees. In the wild, it

*Wyulda squamicaudata*

### *Wyulda squamicaudata*

**Size**
*Head and Body Length*
310–395 mm
*Tail Length*
300 mm
*Weight*
1350–2000 g

**Identification** Pale grey, tipped with black
above; dark mid-dorsal stripe from shoulders to
rump; rufous tinge to base of tail; creamy white
below. Proximal fifth of tail furred, remainder
covered with scales.

**Recent Synonyms** None.

**Other Common Names** Indigenous names:
Watmuna (Ungaringyin); Ilangalya (Worora);
Lilangnai, Yilgal, Ilanglal (Wunambai). *Wyulda*
comes from the Aboriginal 'wayurta' a name
for brushtail possums in many Western Desert
dialects.

**Status** Rare to common in limited area;
habitat secure.

**Subspecies** None.

**References**

Humphreys, W.F., R.A. How, A.J. Bradley,
C.M. Kemper and D.J. Kitchener (1984). The
biology of *Wyulda squamicaudata* Alexander
1919 (in) A. Smith and I. Hume (eds) *Possums
and Gliders*. Surrey Beatty and Sons and Aust.
Mamm. Soc., Sydney. pp. 162–169.

McKenzie, N.L., A. Chapman and W.K.
Youngson (1975). Mammals of the Prince
Regent River Reserve, North-west
Kimberley, Western Australia. *Wildl. Res.
Bull. West. Aust.* No. 3: 69–74.

has been observed to feed on blossoms but, in
captivity, it also eats fruits, nuts, leaves and
insects.

Apart from the type specimen, all records are
from the coastal portion of the north-western
Kimberley region which receives in excess of 900
millimetres annual rainfall. Violet Valley has an

*The Scaly-tailed Possum is a little-known species from the Kimberley. It shares resemblances with brushtail possums and cuscuses.* BABS & BERT WELLS

annual rainfall of only about 600 millimetres and this, together with the failure to find any more animals there or in similar country in the Kimberley or Northern Territory, casts some doubt on the accuracy of the type-locality. A sight record from Broome has not been substantiated and seems unlikely on a habitat basis.

Even within its known range, distribution of the Scaly-tailed Possum is patchy. The Wotjulum specimen was the only one seen in two months of field work in 1954 and the species was not recorded during an intensive mammal survey of the Mitchell Plateau area in 1976, although a skull was picked up there two years later. In 1981 a small population was located at Mitchell Plateau and studied for over a year.

During a biological survey of the Prince Regent Nature Reserve in 1974, it was recorded at only one site (later named Wyulda Creek)

despite searches in similar habitat elsewhere on the Reserve. The possums at Wyulda Creek were seen by spotlight in an area of extremely rugged sandstone and, when disturbed, usually descended rapidly from a tree and sought shelter in the rocks below. The Short-eared Rock-Wallaby and Monjon occupy the same habitat.

In 1970, the Scaly-tailed Possum was relatively abundant at Kalumburu where more than 20 animals have been seen in the course of two nights of spotlighting and two have been caught in four trap-nights. It has also been seen at the nearby Theda Station. Its present status is unknown.

Females give birth mainly in the dry season between March and August, although one from Kalumburu had a small pouch-young in January, suggesting that breeding may sometimes extend beyond August. Only one juvenile is carried. It remains in the pouch for 150–200 days and is weaned after eight months. Females are sexually mature at about two years of age.

A.A. BURBIDGE

# Superfamily Macropodoidea

Kangaroos and their kin are characterised by powerful hindlimbs and long hindfeet. The superfamily Macropodoidea is divided into two families: the Potoroidae, including the rat-kangaroos, potoroos and bettongs; and the Macropodidae, comprising the more specialised kangaroos, wallabies, hare-wallabies, nailtail wallabies, rock-wallabies, pademelons, Swamp Wallaby, Quokka, tree-kangaroos and the forest wallabies of New Guinea. All of these are collectively known as macropods (although 'macropodoids' would be a better term).

With the exception of the Musky Rat-Kangaroo (which bounds), macropods move at speed by hopping on their hindlimbs. At slow speed, potoroids progress like slow-moving rabbits, taking the weight on the forelimbs and bringing the hindlimbs forward in a 'shuffling' gait but the slow gait of most macropodids is uniquely 'pentapedal'. The weight of the body is taken on the forelimbs and the down-turned tail while the hindlimbs are rather awkwardly swung forward. A kangaroo cannot walk backwards or move its hindlegs independently except when lying on its side or swimming. This arrangement may seem ineffective but it has been demonstrated that, when hopping, a kangaroo expends less energy than a galloping quadruped of equivalent size.

There can be little doubt that the earliest marsupials were arboreal. The macropods are a group that became secondarily terrestrial and, with the notable exception of the Musky Rat-kangaroo, lost the opposable first digit of the hindfoot (as in many terrestrial dasyuroids and bandicoots).

In all macropods, the fourth toe is the longest and strongest: aligned with the axis of the foot, it is the structure through which the thrust of hopping is transmitted from the leg to the ground. It is much more strongly developed in typical macropodids than in potoroids but is secondarily shortened in the rock-wallabies and tree-kangaroos, which have a broad and relatively shorter foot. Also associated with arboreal life, the tail of most marsupials (Australian and American) is prehensile, a capacity that has been lost in the essentially terrestrial dasyuroids, bandicoots and macropodids. Interestingly, however, the tail of potoroids is sufficiently prehensile to be employed in carrying nesting material. In macropodids it has a balancing function in the fast hopping gait and, as mentioned earlier, it acts as a fifth limb in the slow 'pentapedal' gait. Tree-kangaroos would be much more efficient as arboreal animals if their tails were prehensile but it seems that, having evolved from terrestrial macropodids, they were unable to redevelop this ability.

Macropods are basically herbivorous, eating a wide range of plants. Potoroids tend to feed mainly on underground fungi and tubers but some also take seeds and insects; macropodids can be browsers or grazers.

Most macropods are nocturnal but some are active in the early morning and late afternoon. Potoroids make nests and one species constructs a short burrow. Macropodids do not make nests but many retire to dense vegetation during the day. Some from arid environments shelter during the day in caves or rock-clefts, thus reducing heat stress and the loss of water by evaporation, but the Red Kangaroo requires no more shelter than the shade of a low tree or bush.

In common with the other phalangeroids, female macropods have a well-developed forwardly opening pouch. This contains four teats but, with the exception of the Musky Rat-kangaroo, which usually bears twins, it is usual for only one young to be born at a time. In most macropods the female becomes sexually receptive immediately after giving birth (post-partum oestrus) and mating usually results in the fertilisation of an ovum. Since gestation usually occupies no more than three to five weeks, this would seem to lead to the production of young every month or so. However, the presence of one suckling young exercises an effect on the female endocrine system such that the newly fertilised ovum ceases development after it has become a simple spherical sac about 0.25 millimetres in diameter, a stage of development known as the blastocyst. The blastocyst normally remains in this quiescent state until a few weeks before its older sibling permanently vacates the pouch (at an age ranging from about 90 days in the Brushtailed Bettong to about 200 days in the Red Kangaroo). The phenomenon, which is known as embryonic diapause, is not restricted to macropods, being also found in the Honey Possum and some eutherian mammals such as bats and seals.

If a pouch-young should die or be lost from the pouch, the quiescent blastocyst resumes development and proceeds to birth, quickly replacing the lost individual.

The situation is complicated by the fact that a young macropod does not cease to suckle immediately after leaving the pouch: weaning extends over several weeks, during which time it suckles by inserting its head into the pouch. From the time of birth to the time of weaning, it feeds from the same teat, so its younger sibling must feed from one of the other three. It is very unusual for a mammal to cope with suckling two young of very different ages and it is even more remarkable that the mammary glands of a macropod supplying two young of different ages produce milk of quite different composition.

With the exception of the Tammar Wallaby, *Macropus eugenii*, which has a fixed annual reproductive cycle, all macropods that have been studied have the potentiality for continuous reproduction. It is normal for a female to be pregnant from the time it reaches sexual maturity until its death. After the birth of its first young, it is also continuously lactating. However, malnutrition arising from drought may cause a female to terminate its oestrous cycle temporarily. In this condition, it does not mate after giving birth but may nevertheless produce one more young from its quiescent blastocyst.

# FAMILY POTOROIDAE

## *Potoroos, bettongs and the Musky Rat-kangaroo*

D. WATTS

*The Rufous Bettong, a fairly typical potoroid, has smaller hindquarters than a kangaroo or wallaby.*

Potoroids appear to be a rather conservative branch from the evolutionary tree that led to the Macropodidae. For example, they retain a somewhat prehensile tail and the stomach is less elaborately sacculated than in macropodids. The second and third upper incisors are smaller and more laterally placed; there are well-developed upper canines; and there is no forward movement of the molars or loss of the most anterior ones. The disparity between the fore- and hindlimbs is less than in typical macropodids and the hindfoot is proportionately shorter.

There are two subfamilies. The Hypsiprymnodontinae, contains only the monotypical *Hypsiprymnodon* or Musky Rat-kangaroo. The Potoroinae comprises the genera *Potorous* (potoroos); and *Bettongia, Caloprymnus* and *Aepyprymnus* (bettongs).

*Hypsiprymnodon* is small (average weight about 530 grams). Potoroines vary in average weight from about 900 grams in *Caloprymnus* to a little over 3 kilograms in *Aepyprymnus*. The head is long and slender in *Hypsiprymnodon* and *Potorous*; short and broad in bettongs, particularly *Caloprymnus*.

*Hypsiprymnodon* is omnivorous, feeding on insects, fruits and large seeds of tropical rainforest trees. Tough seeds and hard-bodied insects are held in the paws at the side of the mouth and cut up by the large, blade-like premolars. Its stomach is an unspecialised chamber. Potoroines feed largely on underground fungi, succulent bulbs and tubers, insects and their larvae. *Bettongia lesueur* includes native figs in its diet and has been seen to scavenge dead fish from a beach. *Aepyprymnus* eat roots, tubers, grasses and herbs. The stomach of potoroines varies in structure but is characterised by a large diverticulum, a sacciform forestomach and a small hindstomach. Micro-organisms in the forestomach break down cellulose to assimilable food. There is no firm evidence that any potoroid needs to drink.

*Hypsiprymnodon* is diurnal and *Potorous tridactylus* is sometimes active in the late afternoon; other potoroids are nocturnal. Potoroids are solitary and frequently aggressive, although small feeding aggregations have been reported for *Hypsiprymnodon*, *Aepyprymnus* and *Potorous tridactylus*.

In historical times, potoroids have proved to be far less successful than macropodids: they appear to be remnants of a group that had its peak prior to the dominance of the macropodids and there is fossil and subfossil evidence that some species were declining in range prior to European settlement. Over the past two centuries, two of the nine known species have become extinct; two have become extinct on the mainland and survive only on offshore islands; one has declined from common to rare; and one, which was not described until 1980, is both rare and endangered.

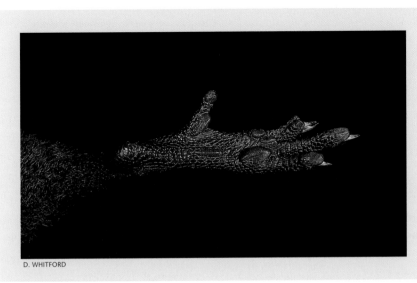

D. WHITFORD

*The hindfoot of the Musky Rat-kangaroo differs from those of all other macropods in having a first digit ('big toe'). This is a possum-like feature.*

## Subfamily Hypsiprymnodontinae

The characteristics of this subfamily are those of the single species, most notably the presence of five toes on the hindfoot and a bounding (rather than hopping) gait.

# Musky Rat-kangaroo

## *Hypsiprymnodon moschatus*

### Ramsay, 1876

hip'-see-prim'-noh-don mos-kah'-tus:
'musky *Hypsiprymnus*-tooth'
(*Hypsiprymnus* is a junior synonym of *Potorous*)

The tropical rainforests of northern Queensland are home to the smallest and, in some respects, most unusual macropod. Although the Musky Rat-kangaroo is predominantly terrestrial, it occasionally climbs through fallen trees and branches using the extra grip afforded by its unique, opposable 'big toe'. In this, it resembles possums, as it does in its bounding quadrupedal gait; the usual birth of twins (sometimes triplets); limb proportions; simple stomach; and possession of two pairs of lower incisors (one pair vestigial).

By day, the Musky Rat-kangaroo forages in leaf litter on the forest floor and along creek banks, utilising networks of roots, logs and fallen trees as pathways. At night, it sleeps in a well-constructed nest of leaves, usually hidden

*Hypsiprymnodon moschatus*

### *Hypsiprymnodon moschatus*

**Size**
*Head and Body Length*
153–273 (230) mm (males)
212–252 (233) mm (females)
*Tail Length*
132–159 (145) mm (males)
123–153 (140) mm (females)
*Weight*
360–680 (529) g (males)
453–635 (511) g (females)

**Identification** General body colour rich brown, intermingled with dark brown hairs. Head, grey brown. Tail, dark brown and covered with small scales giving naked appearance. Distinguished from other macropods by having five toes on hindfoot.
**Recent Synonyms** None.
**Other Common Names** None.
**Status** Common, limited.
**Subspecies** None.

D. WHITFORD

*Much of the diet of the Musky Rat-kangaroo is made up of seeds and nuts of rainforest trees and palms.*

in tree buttresses, lawyer vine thickets or suspended in fallen branches. Material for the nest is collected in the mouth and forepaws before being placed on the ground and transferred to the coiled tail with a kick of the hind legs. It is then carried to the nest site where successive bundles are added to the interior of the developing nest. Each animal uses several nests and may rest in these at various times during the day. Maternal nests are much larger than others and, as the young mature, they may continue to use this nest for several months after the mother has abandoned it.

The diet comprises fruits, insects and fungi in varying proportions throughout the year. The mainstay is fruit, particularly the flesh of large and often colourful fruit from trees such as quandongs, figs, palms, *Diploglotis*, *Endiandra* and *Fontania*. When sufficient fruit is not available, Musky Rat-kangaroos may lose a debilitating amount of weight. The seeds of some species are eaten but most remain viable after the flesh has been stripped away. Some seeds, particularly those falling late in the season, are scatter-hoarded: buried haphazardly in the leaf litter to be dug up and eaten later. Not all of these are relocated, so scatter-hoarding probably facilitates their dispersal and contributes to maintenance of the forest ecosystem. Insects are eaten throughout the year and become more important as the

**References**

Johnson, P.M. and R. Strahan (1982). A further description of the Musky Rat-kangaroo, *Hypsiprymnodon moschatus* Ramsay, 1876. (Marsupialia, Potoroidae), with notes on its biology. *Aust. Zool.* 21: 27–46.

Seebeck, J.H., A.F. Bennet and D.J. Scott (1989). Ecology of the Potoroidae—a review (in) G. Grigg, P. Jarman and I. Hume (eds) *Kangaroos, Wallabies and Rat-Kangaroos*. Aust. Mammal Soc. and Surrey Beatty and Sons, Sydney.

D. WATTS

*The Musky Rat-kangaroo lives on the floor of the rainforest, but it can climb in low vegetation.*

availability of fruit declines. Epigeal fruiting bodies of fungi also proliferate during the late wet season, providing another important alternative food source.

The Musky Rat-kangaroo is generally solitary but up to eight individuals may aggregate under fruiting trees. Occasionally animals may fight over favoured fruit, engaging in dramatic chases and emitting quiet guttural grunts. However, animals tend to avoid contact when feeding together.

Males are sexually active from October to April in the wild and as late as July in captivity. Mating is preceded by several days of courtship in which the male approaches the female from in front and both animals stand erect, each touching the other's head and neck with its forepaws. Usually two (sometimes three) young are born and remain in the pouch for 21 weeks. After this, they are left in the maternal nest and visited periodically to be suckled by the mother. Gradually the young spend longer and longer periods away from the nest. By November, juveniles weighing about 140 grams may be seen wandering independently or as sibling pairs on the forest floor. In captivity, young remain close to their mothers at this stage. Sexual maturity is reached at just over two years of age in the wild, while captive animals can mature at just over one year.

Within the north Queensland wet tropics, the Musky Rat-kangaroo is distributed from sea-level to an altitude of over 1000 metres. Latitudinally it is found along the Great Dividing range from Mount Lee (near Ingham) in the south to Mount Amos (south of Cooktown) in the north: its distribution includes the Atherton, Carbine and Windsor Tablelands and the coastal forests around Mission Beach, Yarrabah and Cape Tribulation.

The Musky Rat-kangaroo has undergone a severe range reduction because of its dependence on tropical rainforest, much of which has been cleared for agriculture and grazing. Nevertheless, unlike most other potoroids, it remains relatively common in areas of suitable habitat and seems able to tolerate selective logging practices. The Musky Rat-kangaroo does not cope well with rainforest fragmentation: it is absent from all but the largest fragments or those very close to continuous forest on the Atherton Tableland.

A. J. DENNIS AND P.M. JOHNSON

## Subfamily Potoroinae

The living potoroines are smaller and have relatively shorter hindfeet than members of the Macropodinae. The tail is sufficiently prehensile to be used in carrying nesting material. They have powerfully clawed forefeet, employed in excavating underground fungi, which form much of the diet.

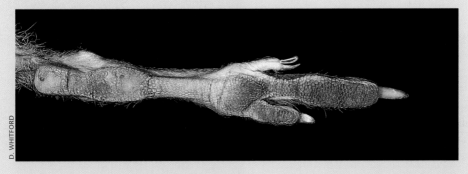

D. WHITFORD

*The hindfoot of the Rufous Bettong is much larger than that of the Musky Rat-kangaroo and lacks a first toe: the third digit is proportionately much longer.*

# Rufous Bettong

## *Aepyprymnus rufescens*

(Gray, 1837)

eep'-ee-prim'-nus  rue-fes'-enz:
'reddish high-rump'

The Rufous Bettong is now perhaps the most widely distributed potoroid, although this was certainly not the case prior to European settlement. As with all other potoroids it has suffered a dramatic reduction in distribution and numbers but, possibly due to low intensity land-use practices throughout much of its range, it has remained quite common. It occupies a variety of habitats from coastal eucalypt forests, through tall, wet sclerophyll, to low, dry open woodland westward of the Great Dividing Range. It seems only to occupy areas with a sparse or grassy understorey, adjacent areas of dense undergrowth being frequented by the Long-nosed Potoroo in the south and tropical Bettong in the north.

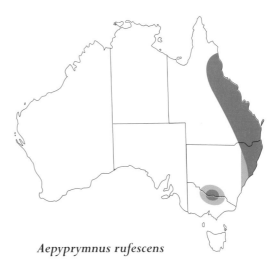

*Aepyprymnus rufescens*

*When moving fast, the Rufous Bettong bounds on its hindlegs: the forelegs are sometimes employed in turning.*

G.A. HOYE

Nests are built of fibrous vegetation in a shallow excavation and consist of a conical dome with a single entrance. These are generally placed against a log, tree or grass clump but may be in open ground. Up to five nests may be used at any one time and new nests are added as old ones are abandoned on a regular (monthly) basis. While nests are rarely shared by neighbouring adults, one that has been abandoned may be rebuilt and used by a neighbouring individual.

The Rufous Bettong emerges from its nest about 40 minutes after sunset and spends the night browsing on herbs and grasses and digging for roots and tubers. Often, entire plants including flowers, seeds, roots and leaves are devoured. Occasionally, it eats exudates from trees such as *Melaleuca nervosa*. Relatively more time is spent feeding on roots and tubers during dry times when browse is sparse. In areas where underground fungi are abundant, these also contribute to the diet. Bones of dead animals are visited and chewed on a regular basis. Although not usually needing to drink free water, the Rufous Bettong has been known to do so during drought.

Home ranges are large: those of males covering 75–110 hectares; females, 45–60 hectares. Individuals may travel 2–4.5 kilometres in a normal night's feeding, progressing with a fast bipedal hop between feeding areas and at a slow bound, with only slight tail support, while feeding.

Although it was previously thought to be solitary, increasing evidence suggests that the Rufous Bettong may form loose, polygynous associations. Males invest much of their time associating with one or two females that occupy overlapping ranges. These females associate exclusively with (and may be defended by) that male, even though other males may frequent the same area. One male and two associated females may spend much time as a trio.

## *Aepyprymnus rufescens*

**Size**

*Head and Body Length*
375–390 (385) mm

*Tail Length*
338–387 (360) mm

*Weight*
up to 3 kg (males)
up to 3.5 kg (females)

**Identification** Distinguished from other macropods of comparable size by reddish-brown fur and hairy muzzle.

**Recent Synonyms** None.

**Other Common Names** Rufous Rat-kangaroo.

**Status** Common.

**Subspecies** None.

## References

Johnson, P.M. (1978). Reproduction in the Rufous Rat-kangaroo (*Aepyprymnus rufescens*) (Gray) in captivity with age estimation of pouch young. *Qld J. Agric. Anim. Sci.* 35: 69–72.

Dennis, A.J. (1988). *Behaviour and movements of Rufous Bettongs,* Aepyprymnus rufescens, *at Black Rock, North Queensland*. Hons. Thesis, Zoology Department, James Cook University of North Queensland.

Breeding may occur at any time during the year. Females reach sexual maturity at about 11 months and thereafter enter oestrus at about three-weekly intervals. Males are capable of successful breeding at 12–13 months of age. They attempt to mount females at any time, but a female that is not in oestrus growls loudly, throws herself on her side and kicks. The male, meanwhile, stamps one outstretched foot near the female and threshes its tail vigorously. As the time of ovulation approaches, a female permits the male to smell her cloacal and pouch areas and eventually permits copulation. After a gestation of 22–24 days, a single young is born and attaches itself to one of four teats. The female mates again within a few hours, producing a quiescent blastocyst. The pouch young leaves the teat at the age of seven to eight weeks and vacates the pouch permanently when about 16 weeks old. For another seven weeks, the young animal remains with its mother, accompanying it when feeding, being groomed by it and sharing its nest.

At present, the Rufous Bettong seems to have a secure survival status. This should remain so if land-use practices do not intensify and the Rabbit and Fox do not become more abundant in the northern part of its range.

A. J. DENNIS AND P. M. JOHNSON

# Tasmanian Bettong

## *Bettongia gaimardi*

### (Desmarest, 1822)

bet-ong'-ee-ah gie-mard'-ee: 'Gaimard's bettong' (an Aboriginal name for a small wallaby)

The Tasmanian Bettong has the most secure status of any member of the seriously depleted genus *Bettongia*. Probably extinct on the Australian mainland by the early twentieth century, it remains in reasonable numbers in Tasmania.

It inhabits open, dry, fire-prone forests with a grassy or heath understorey on poor soil. These forests are often of Peppermint Gums and Silver Wattle. Where the habitat contains moist pockets of denser vegetation, the Tasmanian Bettong may occur in the same general areas as the Long-nosed Potoroo but the two do not appear to compete.

Along with some seeds, roots and bulbs, the diet consists predominantly of the fruiting bodies of underground fungi which exist in symbiotic relationship with the forest trees. It seems highly likely that the bettong is a vector in the dispersal of these useful fungi. A gummy substance, probably an acacia exudate, is occasionally found in stomach contents.

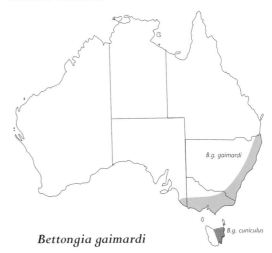

*Bettongia gaimardi*

D. WATTS

*Female Tasmanian Bettong and young. This species appears to be the most secure of the bettongs.*

A densely woven nest of dry grass and bark is commonly sited under a fallen limb or among short bushes or tussocks. Ovoid in shape, about 300 millimetres long and 200 millimetres wide, it has walls 20–30 millimetres thick and a small opening at one end. It is constructed in a depression about 150 millimetres in diameter and 50 millimetres deep, which is dug with the strongly clawed forelimbs. Material for the nest is transported by the tail as in other bettongs.

A strictly nocturnal animal, it rests in its nest throughout the period of daylight. When leaving the nest at nightfall, it travels some distance before pausing. A nest may be used regularly for at least a month and may be reoccupied after being vacant for a similar period. The home range is 65–135 hectares and an individual sometimes travels as much as 1.5 kilometres between its nest and a feeding area. In contrast with the larger macropods of Tasmania, it is rarely seen on pastures.

Gestation lasts 21 days and the young remains in the pouch for about 105 days. It is weaned 40–60 days after leaving the pouch and becomes sexually mature at about 12 months. Breeding is continuous and females may produce two to three young each year. Although the Tasmanian Bettong is normally solitary, two individuals may occasionally occupy the same nest. In captivity, adults are relatively aggressive and both males and females fight by biting and kicking the opponent.

It has often been suggested that it survived in Tasmania because the Fox has not become established there. An alternative or contributory factor may be the relatively low density of Rabbits on the island and the consequent preservation of the grassy habitat which the Tasmanian Bettong requires.

R. ROSE AND K.A. JOHNSON

### *Bettongia gaimardi*

**Size**

*Head and Body Length*
315–332 (323) mm
*Tail Length*
288–345 (326) mm
*Weight*
1.2–2.25 (1.66) kg

**Identification** Brownish-grey above pencilled with white, merging to a greyish-white beneath, fur rather harsh. Tail well furred, usually with white tip.
**Recent Synonyms** None.
**Other Common Names** Eastern Rat-kangaroo, Eastern Bettong, Gaimard's Rat-kangaroo, Gaimard's Bettong.
**Status** Common.
**Subspecies** *Bettongia gaimardi gaimardi*, south-eastern Australia (extinct).
*Bettongia gairmardi cuniculus*, Tasmania.

**References**

Rose, R.W. (1987). The reproductive biology of the Tasmanian Bettong (*Bettongia gairmardi*: Macropodidae). *J. Zool. Lond.* 212: 59–67.

Johnson, C.N. (1992). Distribution of feeding activity in the Tasmanian Bettong (*Bettongia gairmardi*) in relation to vegetation patterns. *Wildl. Res.* 21: 249–256.

# Burrowing Bettong

### *Bettongia lesueur*

## (Quoy and Gaimard, 1824)

le-swer': 'Lesueur's bettong'

Although often referred to as the only macropod to inhabit burrows, the Burrowing Bettong is more correctly distinguished as the only one to inhabit burrows on a regular basis (the Rufous Hare-wallaby lives in burrows during the summer). It once occurred in suitable country through much of the continent west of the Great Dividing Range and south of the tropical savanna. Although not recorded from Queensland, it was widespread in other mainland States and Territories, except in areas of dense vegetation and higher rainfall. Having had one of the largest geographic ranges of any Australian mammal, it is now extinct throughout mainland Australia and on Dirk Hartog Island (the type-locality), but remains on Bernier and Dorre Islands in

*The Burrowing Bettong, once widespread on the mainland is now extinct there, surviving only on some of Western Australia's offshore islands.*

J. LOCHMAN

## Bettongia lesueur

### Size*

**South-western Western Australia**
*Mean Head and Body Length*
400 mm
*Mean Tail Length*
300 mm
*Weight*
circa 1.5 kg

**South Australia**
*Mean Head and Body Length*
370 mm
*Mean Tail Length*
300 mm

**Bernier And Dorre Islands**
*Mean Head and Body Length*
360 mm
*Mean Tail Length*
300 mm

**Central Australia**
*Mean Head and Body Length*
345 mm
*Mean Tail Length*
295 mm

**Barrow Island**
*Mean Head and Body Length*
280 mm
*Mean Tail Length*
215 mm

*Size and relative proportions of tails, ears, etc., vary considerably over the range.

**Identification** Yellow-grey above (grey on islands), light grey below. Small, thickset; short, rounded ears; lightly haired, fat tail (with white tip in south-western Western Australia).

**Recent Synonyms** None, but sometimes improperly referred to as *Bettongia lesueuri*.

**Other Common Names** Lesueur's Rat-kangaroo, Lesueur's Bettong, Burrowing Rat-kangaroo, Boodie, Tungoo.

Indigenous names: Boordee (south-western Western Australia); Yalva (Broome, Western Australia); Yerki (Adelaide, South Australia);

Pudkurra/Pudkurru (Adnyamathanha); Aluta (Alywarra); Urtaya (Anmatjara); Tnunka (Arunta); Aluta (eastern Aranta); Atjurta, Kapwapwa (Kaititja); Tjilpuku (Karatjari); Mitika, Nurrtu, Tjungku, Yalanmunku, Yunkupalyi (Kartutjarra); Kunayuna, Mitika, Nuttru, Purtaya/Putaya, Tjunku, Walkatju, Yiilkirra (Kutatja); Purtaya (Luritja); Pitikariti, Purtaya, Walkatju, Yiikita, Yilyikarra, Yunkupalyi (Mangala); Mitika, Nurrtu, Tjiliku, Wilyinpa, Yunkupalyi (Manytjilytjarra); Yelki (Narranga); Mitika, Tjungku (Gnaanyatjarra); Mitika, Murluyuna (Ngaatjatjarra); Bukurra (Ngadjeri); Yarlki (Gnarunga); Kunayuna, Mitika, Murluyuna, Purtaya, Tjungku, Yungkupalyi (Pintupi); Mitika, Tjungku (Pitjantjatjarra); Nurrtu (Putitjarra); Purtaya, Walkaru, (Tjaru); Purtaya, Tjalpalyi, Walkatju, Yiilkita, Yunkupalyi (Walmatjari); Mitika, Nurtu/Nurrtu, Purtaya, Tjunku, Walkatju, Yiilkirra (Wangkatjungka); Purlana, Purtaya, Walkatju, Wirlana (Warlpiri); Minilka/Minirlka (Warnman); Kananka (Wonkanura, Dieri).

**Status** Abundant on Barrow, Bernier and Dorre Islands. Extinct on Dirk Hartog and Boodie Islands; reintroduced to Boodie Island, 1993. Presumed extinct on mainland.

**Subspecies** The following have been described as subspecies but their validity is not generally accepted.

*Bettongia lesueur lesueur*, Shark Bay islands.
*Bettongia lesueur graii*, south-western Western Australia (extinct).
*Bettongia lesueur harveyi*, Eyre Peninsula, South Australia (extinct).

### References

Burbidge, A.A., K.A. Johnson, P.J. Fuller and R.I. Southgate (1988). Aboriginal knowledge of the mammals of the central deserts of Australia. *Aust. Wild. Res.* 15: 9–39.

Stodart, E. (1966). Observations on the behaviour of the marsupial *Bettongia lesueuri* (Quoy and Gaimard) in an enclosure. *CSIRO Wildl. Res.* 2: 91–99.

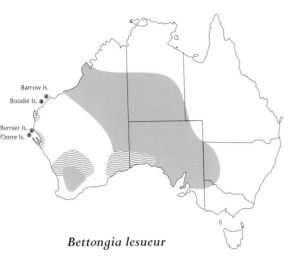

*Bettongia lesueur*

Barrow Is.
Boodie Is.
Bernier Is.
Dorre Is.

The Burrowing Bettong is nocturnal and gregarious. It does not emerge from its burrow until after sunset and disappears again before sunrise. Locomotion is always bipedal, the forelimbs and tail not being used for support except when the animal is stationary. Food is procured largely by digging and appears to be located by smell: feeding animals move slowly with their nostrils very close to the ground. On the mainland, the Burrowing Bettong ate tubers and bulbs as well as seeds, nuts and the green parts of some plants. On Barrow Island, food includes the fruit of native figs, seeds, roots, termites and fungi. From what we now know of other bettongs it is likely that the mainland population also ate fungi. The Burrowing Bettong is very vocal and makes a variety of grunts, hisses and squeals.

On Bernier and Dorre Islands, females have a short oestrous cycle (23 days) and a short gestation period (21 days). The single young remains in the pouch for 115 days and reaches sexual maturity at five months. Breeding occurs throughout the year and up to three young may be raised in 12 months.

Early naturalists in various parts of Australia noted that the Burrowing Bettong was common and, in many areas, the most abundant mammal. Its decline began in the nineteenth century and it had disappeared from Victoria by 1863 but, unlike some other mammals of similar size, it persisted in some parts of Australia until the 1930s and 1940s. Rabbits took over its disused warrens but in some places the two species cohabited for several decades and often could be found in the same warren. The disappearance of the Burrowing Bettong from central and Western Australia seems to have coincided with the establishment of the Fox. On Dirk Hartog Island, where there are no Rabbits or Foxes, the Cat has been implicated in its extinction.

A.A. BURBIDGE

Shark Bay and on Barrow and Boodie Islands off the Pilbara coast of Western Australia.

The burrow in which it spends the day may be a simple structure consisting of one or two entrances and a short, shallow, curving tunnel. More often there are many entrances to complex warrens with interconnecting deep passageways. One warren on Barrow Island had over 120 entrances and counts have revealed that there are at least half as many individuals in a warren as there are entrances. Nests made from vegetation are constructed in burrows. On the mainland, warrens were constructed in most types of country where the soil was deep enough. Loams were favoured and in the sand ridge deserts, burrows were in the damper low-lying areas. Burrows were often dug into slight outcrops of limestone or gypseous rock and rises in salt lake systems were a favoured habitat. Another favoured site was under boulders or capping rock. On Barrow Island, warrens are almost always associated with limestone cap-rock on slopes and the top of ridges; some are in the floors of caves. Other mammals, such as the Brush-tailed Possum and the Western Quoll, also sheltered in the warrens.

# Brush-tailed Bettong

## *Bettongia penicillata*

### Gray, 1837

pen'-is-il-ah'-tah: 'brushed (-tailed) bettong'

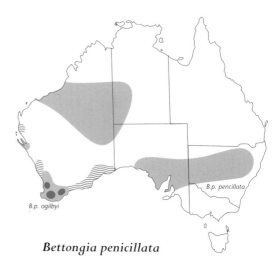

*Bettongia penicillata*

Widespread at the time of European colonisation, the Brush-tailed Bettong, or Woylie, is now an endangered species. It appears to have been primarily an animal of open forests and woodlands and a common factor in all habitats occupied by the surviving populations is a clumped, low understorey of tussock grasses or clumped' low woody scrub. Studies on Western Australian populations indicate that it does not drink and that no green

---

### *Bettongia penicillata*

**Size**
**South-western Australia**
*Head and Body Length*
300–380 (330) mm
*Tail Length*
290–360 (310) mm
*Weight*
1.1–1.6 (1.3) kg

**Identification** Yellowish-grey above, paler below; black crest on tail. When flushed from nest, bounds with head held low, back arched and tail almost straight.
**Recent Synonyms** None.
**Other Common Names** Brush-tailed Rat-kangaroo.
Indigenous name: Woylie.
**Status** Rare.
**Subspecies** Two subspecies are tentatively recognised.
*Bettongia penicillata penicillata*, south-eastern Australia (extinct).
*Bettongia penicillata ogilbyi* southern and south-western Australia.

---

material is eaten, its food consisting largely of the fruiting bodies of underground fungi, supplemented by bulbs, tubers, seeds, insects and resin, probably from *Hakea* shrubs. The proportion of fungal material in the diet is greatest in summer and autumn. Being deficient and imbalanced as a source of amino-acids, fungus is not a good food for mammals but the Brush-tailed Bettong does not utilise it directly. Bacteria in the large sacculated forestomach consume the fungus and these bacteria and their by-products (which constitute a more balanced diet) are digested in the posterior part of the stomach and small intestine.

The daylight hours are spent in an elaborate domed nest made of grass or shredded bark, built over a shallow depression scraped in the ground under a bush or other cover. As in other potoroines, the tail is used to convey nesting material. Between dusk and an hour or two before dawn, it forages for food. It is relatively slow moving, except when disturbed and hops with its head held low, back arched and tail extended.

Males and females are similar in appearance and occupy distinct individual home ranges, each including a nesting and a feeding area. Nesting areas appear to be territorial but the feeding areas of adjacent animals may overlap considerably. Breeding is continuous, a female giving birth to its first young at the age of 170–180 days and approximately every 100 days thereafter for the rest of its life of four to six years. The single young (rarely two, of which only one survives the

C. ANDREW HENLEY/LARUS

*The Brush-tailed Bettong is now rare and endangered. Small populations exist in Western Australia.*

first few weeks of pouch life) remains in the pouch for about 90 days and subsequently accompanies its mother at heel, sharing her nest until the next infant vacates the pouch and displaces it. Like many other macropods, the Brush-tailed Bettong exhibits embryonic diapause.

Where food resources are adequate, density of cover is the limiting factor on population size and, since the advent of Europeans, this has assumed prime importance. Where the Brush-tailed Bettong has not been excluded from its normal habitat by agricultural clearing or the impact of grazing animals, its distribution is now restricted by the Fox to a few isolated areas. Reduction in Fox populations by baiting with '1080' (sodium fluoroacetate) leads to a dramatic increase in bettong numbers. Sodium fluoroacetate occurs naturally in certain legumes native to south-western Australia and native vertebrates have evolved considerable tolerance of the poison, making selective baiting practicable. The last remaining

populations in Western Australia are in the Perup and Dryandra forests and the Tuttanning Reserve, in which areas the habitat is under active management directed to their survival.

P. CHRISTENSEN

### References

Christensen, P. and T. Leftwich (1980). Observations on the nest-building habits of the Brush-tailed Rat-kangaroo or Woylie (*Bettongia penicillata*). *J. Roy. Soc. West. Aust.* 63: 33–38.

Christensen, P.E.S. (1980). The biology of *Bettongia penicillata* Gray, 1837 and *Macropus eugenii* (Desmarest 1817) in relation to fire. *Forestry Dept. of W.A. Bull.* No. 91.

Burbidge, A.A., K.A. Johnson, P.J. Fuller and R.J. Southgate (1988). Aboriginal Knowledge of the Mammals of the Central Deserts of Australia. *Aust. Wild. Res.* 15: 9–39.

# Northern Bettong

## *Bettongia tropica*

### Wakefield, 1967

trop'-ik-ah: 'tropical bettong'

Until 1976 the Northern Bettong was known from only nine specimens, one collected from the Dawson Valley inland from Rockhampton in 1884, the others from the Cairns hinterland at Ravenshoe and Mount Carbine Tableland between 1922 and 1932. In 1976 it was found at Davies Creek east of Mareeba by P.M. Johnson and C.M. Weaver and subsequent records have confirmed the existence of two populations, isolated from each other: on the Lamb Range, east of Mareeba; and Mount Windsor Tableland, west of Mossman. Occasional unconfirmed sightings suggest that sparse populations may still exist on the Mount Carbine Tableland and in the Atherton to Ravenshoe area. No further records have been obtained from central coastal Queensland since 1884. Its status as a species is the subject of debate and some authorities regard it is a subspecies of the Brush-tailed Bettong, *Bettongia penicillata tropica*.

*Bettongia tropica*

| *Bettongia tropica* | |
| --- | --- |

**Size**
*Head and Body Length*
267–345 (313) mm (males)
277–404 (313) mm (females)
*Tail*
318–355 (342) mm (males)
317–364 (341) mm (females)
*Weight*
1.0–1.3 (1.2) kg (males)
0.9–1.4 (1.2) kg (females)

**Identification** Body colour grey, darker above and lighter below; tail grey, end third with short black crest on upper side. Light build; low springy hop.
**Recent Synonyms** *Bettongia penicillata tropica*.
**Other Common Names** Brush-tailed Bettong.
**Status** Restricted, endangered.
**Subspecies** None.

The Lamb Range population inhabits grassy, open forest and woodland on the western slopes of the range—from tall, open forest adjacent to rainforest, to eucalypt woodland characterised by Scented Gum and or Poplar Gum, between elevations of 400–1100 metres. On the Mount Windsor Tableland, records are restricted to tall open eucalypt forest characterised by Flooded Gums and/or Red Stringybarks and in adjacent rainforest, on the western side of the tableland between elevations of 900–1200 metres. Its presence several kilometres into rainforest may result from its use of roads. It is solitary and strictly nocturnal, active at night and retiring during the day in a grass nest built under a tussock or a grass-tree. Following fire, hollow logs are used as daytime shelters. Up to three nests are known to be used by the same individual. It ranges widely at night: one male travelled 590 metres from point of capture to its nest.

P.M. JOHNSON

*When moving fast the Northern Bettong bounds on its hindlegs. The forelegs can be employed in turning.*

Females have two teats in the pouch and raise one young at a time. Observations on wild and captive animals, suggest that, like the very closely related Brush-tailed Bettong, it is a continuous breeder. Other aspects of its reproductive biology appear identical to that of the Brush-tailed Bettong. A major item of diet is underground fruiting bodies of mycorrhizal fungi which are obtained by digging shallow holes in the ground. The Bettong eats the firm but soft spore-mass from within the fruiting body, leaving the tough outer husk beside the digging. It is probable that the Northern Bettong like the Brush-tailed Bettong, has a specialised digestive system for obtaining a balanced diet from fungi. The underground portion of Cockatoo Grass is also eaten, the fibrous portion of the grass being left beside the digging as a thoroughly chewed pellet.

The geographical ranges of the Northern and Rufous Bettongs appear to abut, rather than overlap, along the western edge of the Northern Bettong's range, the Rufous Bettong occurring in drier habitat. The two species appear to be mutually exclusive: anecdotal evidence from the Ravenshoe area suggests that the Rufous Bettong's range has expanded at the expense of the

Northern Bettong. The Lamb Range population of the Northern Bettong is substantial (up to 28 captures from 100 trap-nights) but that on the Mount Windsor Tableland is extremely sparse. Numbers may fluctuate in response to availability of fungal fruiting bodies, which may be determined by the frequency of fire, an association that has been documented for the Tasmanian and Brushtailed Bettongs.

It must be regarded as an endangered species because of its extremely restricted geographical range and discrete populations within its range, enhancing the possibility of accumulative local extinctions. The Windsor population occurs predominantly within the Wet Tropics World Heritage area, while the Lamb Range population is in State Forest, timber reserve and the small Davies Creek National Park, mostly outside the World Heritage area. Management of the discrete populations depend on obtaining a better understanding of its ecology, particularly in response to fire and its interactions with the common Rufous Bettong.

J.W. WINTER AND P.M. JOHNSON

**References**

Sharman, G.B., C.E. Murtagh, P.M. Johnson and C.M. Weaver (1980). The chromosomes of a rat-kangaroo attributable to *Bettongia tropica* (Marsupialis: Macropodidae). *Aust. J. Zool.* 28: 59–63.

Wakefield, N.A. (1967). Some taxonomic revision in the Australian marsupial genus *Bettongia* (Macropodidae) with description of a new species. *Vict. Nat.* 84: 8–22.

# Desert Rat-kangaroo

## *Caloprymnus campestris*

(Gould, 1843)

kal'-oh-prim'-nus  kam-pes'-tris:
'open-country beautiful-rump'

With only a flimsy nest as protection from heat and intense sunlight, this small potoroid once inhabited one of the hottest, driest and most exposed environments in Australia. Claypans,

gibber plains and sand ridges were the distinctive features of the habitat of the Desert Rat-kangaroo in the stony transition zone between true gibber plain and loamy flats where the sparse vegetation includes Saltbush, other chenopods, Emu Bush and, rarely, a clump of stunted Corkwood. It spent little time in the sandhills.

Unlike some potoroids it did not burrow but formed a nest in a shallow excavation under a bush or in the open. The cavity was lined with leaves and grasses and covered with twigs and grass stems laid across the top. The entrance was at the side and, from a hole in the roof of the nest, the animal was able to protrude its head and survey the surroundings. How long a nest

*The Desert Rat-kangaroo. Illustration in John Gould's* The Mammals of Australia.

was occupied is not known but it was used long enough for the occupant to beat a set of well-defined pads radiating in several directions. No nest was ever found to be occupied by more than one animal, except a female with its suckling young. Juveniles did not associate with their mothers after weaning.

The Desert Rat-kangaroo fed at night but its preferred food is not known. It seems to have been quite independent of surface water and, moreover, to have shunned the succulent plants which grow in the sandhills.

Its hindlimb is about the same length as the head-and-body and the foot is much longer than the thigh or the lower leg. Although its thigh muscles were not well developed, a Desert Rat-kangaroo, when pressed, could move at a speed fast enough to tire a galloping horse and keep this up for several kilometres. It hopped with an unfaltering, easy stride, with the trunk leaning well forward and the tail almost straight. Tracks show that, when hopping at speed, the right hindfoot touched the ground in front of the left, which was rotated outwards at about 30 degrees. The forelimb is small and feeble. Between the

forelimbs of both males and females there is a patch of hairless skin, of unknown function.

The deep pouch contains four teats. Females with one large, furred pouch-young have been found in December and August.

First described in the 1840s, the Desert Rat-kangaroo was not recorded again until 1931 and has not been seen since 1935. It has not ever occurred abundantly. The Yalliyanda Aborigines were able to catch it in its nest by creeping up from the side opposite its entry hole, silently slipping their hands over the top and grabbing the occupant.

M.J. SMITH

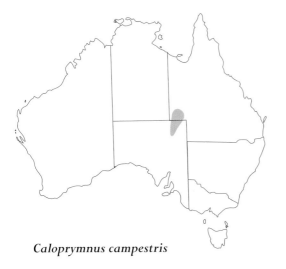

*Caloprymnus campestris*

### *Caloprymnus campestris*

**Size**
*Head and Body Length*
255–282 (268) mm (males)
254–277 (267) mm (females)
*Tail Length*
297–333 (314) mm (males)
310–377 (335) mm (females)
*Weight*
637–850 (797) g (males)
743–1060 (981) g (females)

**Identification**  Short muzzle, very large upper lips. Foot longer than lower leg; tail longer than head plus body.
**Recent Synonyms**  None.
**Other Common Names**  Plains Rat-kangaroo, Bluff-nosed Rat-kangaroo.
Indigenous name: Oolacunta.
**Status**  Presumed extinct.
**Subspecies**  None.
**Reference**
Finlayson, H.H. (1932). *Caloprymnus campestris.* Its recurrence and characters. *Trans. Roy. Soc. S. Aust.* 56: 146–67.

# Long-footed Potoroo

## *Potorous longipes*

Seebeck and Johnston, 1980

pot'-oh-roh'-us lon'-jee-pez: 'long-footed potoroo'

*Potorous longipes*

The Long-footed Potoroo occurs in an area of about 1600 square kilometres in east Gippsland and a small area of south-eastern New South Wales. As this book was in press, it was discovered in a small area south of Mount Buffalo in north-eastern Victoria. Larger than the Long-nosed Potoroo, it has proportionately longer hindfeet. It inhabits temperate rainforest, riparian forest and damp or wet sclerophyll forest, associated with a sparse to dense understorey of sclerophyllous shrubs and a dense field layer of wiregrass, ferns and sedges growing on friable clayey soils. The annual rainfall throughout its range is 1100–1200 millimetres. It has been found at altitudes of 150–800 metres.

The Long-footed Potoroo is terrestrial and obtains its food by excavating characteristic conical pits with its forefeet. Its diet is principally the fruiting bodies of a wide variety of underground and partially underground fungi. Some invertebrates and a small amount of vascular plant material are also eaten. Fungi provide more than 80 per cent of the diet and over 30 species are known to be eaten. Most of these fungi form symbiotic mycorrhizal associations with the roots of forest trees and shrubs and assist in the uptake of nutrients from the soil. Potoroos are believed to be important in the dispersal of spores from the fungi, which pass through the gut and are spread about the forest in its droppings.

Despite the long hindfeet, its slow and fast gaits are similar to those of the Long-nosed Potoroo with a bipedal hop. Vocalisation is restricted to low 'kiss kiss' sounds, uttered when under stress and between mother and young.

Breeding probably takes place throughout the year in the wild, but most young are produced in the winter, spring and early summer. The gestation period is not known but seems to

---

### *Potorous longipes*

**Size**

*Head and Body Length*
380–415 (400) mm

*Tail Length*
315–325 (320) mm

*Weight*
2.0–2.2 (2.1) kg (males)
1.6–1.8 (1.7) kg (females)

**Identification** Grey-brown above, grey below. Distinguished from *Potorous tridactylus* by larger size, hindfoot larger than head, hallucal pad on sole, 24 chromosomes. (*P. tridactylus*: 12 chromosomes in females, 13 in males).

**Recent Synonyms** None.

**Other Common Names** None.

**Status** Endangered.

**Subspecies** None.

*Described in 1980, the Long-footed Potoroo remains a rare and little-known species.*

J.B. COOPER

be similar to that of the Long-nosed Potoroo—about 38 days. Although four teats are present in the well-defined pouch, only a single young is produced. Pouch life is about 140–150 days. Reproductive maturity may not be reached until about two years. Reproductive potential is probably 2.5–3 young per year. In its growth and development, the Long-footed Potoroo resembles the Long-nosed Potoroo, the characteristic length of the hindfeet becoming apparent at about 100 days, at which time also the robust tail also begins to lengthen.

Populations of the Long-footed Potoroo seem to be of low density. The home range is probably less than 10 hectares and it is possible that the species is territorial and monogamous. It is nocturnal and shelters during the day in a temporary nest in ferns or thickets of wire-grass. The Dingo, feral Dog and Fox are

important predators and may contribute significantly to the observed low densities.

The geographic range in Victoria is mainly between the Snowy and Cann Rivers, but may extend further east to Mount Drummer. Its distribution in north-eastern Victoria is not yet known. It has been found sympatrically with the Long-nosed Potoroo, but the two species do not usually occur together. In New South Wales, the only localities reported are in the Rockton section of the Bondi State Forest.

In Victoria, about 13 per cent of the known range of the Long-footed Potoroo is in conservation reserves. The remaining habitat is within state forests, in which timber harvesting may occur. Management guidelines devised to protect habitat in key areas throughout the Victorian range have been established. A managed captive colony is maintained at Healesville Sanctuary, Victoria.

In New South Wales, forest management protocols directed at conservation of the Long-footed Potoroo are being developed.

J.H. Seebeck

**References**

Scotts, D.J. and J.H. Seebeck (1989). Ecology of *Potorous longipes* (Marsupialia: Potoroidae); and preliminary recommendations for management of its habitat in Victoria. *Arthur Rylah Institute for Environmental Research Technical Report* Series No. 62. Department of Conservation, Forests and Lands, Victoria.

Seebeck, J.H. (1992). Breeding, growth and development of captive *Potorous longipes* (Marsupialia: Potoroidae); and a comparison with *P. tridactylus*. *Aust. Mammal.* 15: 37–45.

Saxon, M.J., S.R. Henry and J.H. Seebeck (1994). Management strategy for the conservation of the Long-footed Potoroo *Potorous longipes*, in Victoria. *Arthur Rylah Institute for Environmental Research Technical Report* Series No. 127. Dept. of Conservation and Natural Resources, Victoria.

# Broad-faced Potoroo

## *Potorous platyops*

### (Gould, 1844)

plat'-ee-ops: 'flat-faced potoroo'

in the London Zoo in 1908, supposed to have been a Broad-faced Potoroo from Margaret River, Western Australia, was probably a juvenile Quokka.

Subfossil remains indicate that the Broad-faced Potoroo once had an extensive coastal distribution from South Australia westwards to Albany, Western Australia and northwards through the wheatbelt of Western Australia,

Regret for the extinction of the Broad-faced Potoroo is heightened by the fact that almost no observations on its natural history were recorded in the 36 years during which it was known to Europeans. The earliest specimen was collected from an unknown locality in 1839 and the type-specimen was obtained by John Gilbert near Goomalling, Western Australia, three years later. Five more had been collected from south-western Australia by 1866 and another five 'from Western Australia', sold by a dealer to the National Museum of Victoria in 1875, were the last known to be taken from the wild. An animal

*Potorous platyops*

*The Broad-faced Potoroo. Illustration in John Gould's*
The Mammals of Australia.

# Long-nosed Potoroo

## *Potorous tridactylus*

### (Kerr, 1792)

trie-dak'-til-us: 'three-toed potoroo'

possibly to North West Cape. It seems never to
have occurred in forested areas and the localities
of the six specimens of known origin suggest that
its distribution in the mid-nineteenth century
was to the north and east of the forests of south-
western Australia. It is very likely that its range
had contracted considerably prior to European
settlement of Western Australia and it may well
have been rare by then.

The only information about its habitat is a
statement attributed to John Gilbert in an unpublished manuscript: '...all I could glean of its
habits was that it was killed in a thicket surrounding one of the salt lagoons in the interior'.

D. J. KITCHENER

One of the first mammals recorded from Australia
was the Long-nosed Potoroo, a description and
illustration of which appeared in Governor
Phillips' account of the settlement at Botany Bay
in 1789. A further description of the 'Poto Roo',
given in Surgeon-General White's *Journal of a
Voyage to New South Wales*, referred to the foot as
having only three toes (the conjoined second and
third toes being counted as one) and this error
was perpetuated in the specific name.

Until recently thought to be rare in south-
eastern Australia, it is now known to be quite
widely distributed there and common in Tasmania. It is rare in south-western Western
Australia where no living specimens had been collected for more than a century, until 1994, when
three individuals were captured in Two Peoples
Bay Nature Reserve. Generally restricted to areas
with an annual rainfall greater than 760 millimetres, it inhabits coastal heath and dry and wet
sclerophyll forests. A major habitat requirement is
relatively thick ground cover and it seems to be

---

### *Potorous platyops*

**Size**
*Head and Body Length*
243 mm
*Tail Length*
183 mm

**Identification** Hairs of the back with grey
base passing into yellow-brown, then white and
tipped by black: upper parts of body pencilled
with white. Dusky white below, feet dirty
white, grizzled with brown.
**Recent Synonyms** None.
**Other Common Names** None.
**Status** Extinct.
**Subspecies** None.
**Reference**
Butler, W.H. and D. Merrilees (1971). Remains
of *Potorous platyops*, (Marsupialia,
Macropodidae) and other mammals from
Bremer Bay, Western Australia. *J. Roy. Soc.
West. Aust.* 54: 53–58.

*P.t. tridactylus*

*P.t. tridactylus*

King Is.
*P.t. apicalis*

Furneaux Group

**Potorous tridactylus**

C. ANDREW HENLEY/LARUS

*The Long-nosed Potoroo remains common in limited areas of the mainland and is reasonably common in Tasmania.*

concentrated in areas where the soil is light and sandy. It often digs small holes in the ground in much the same way as bandicoots and it is known to eat roots, tubers, fungi, insects and their larvae and other soft-bodied animals in the soil. It begins to feed at dusk and rarely ventures far from cover.

The Long-nosed Potoroo becomes sexually mature when about one year old. Females carrying pouch-young are found throughout the year but most frequently after the two breeding seasons, one towards the end of winter or early spring, the other in late summer. There are four teats but only a single young is reared at a time, the duration of pouch life being about four months. Although it is solitary and sedentary, trapping records indicate that individuals tend to aggregate in small groups. It is known to reach an age of seven years in the wild and at least 12 years in captivity.

Populations show considerable variation. Over a distance of only 200 kilometres from the western to the eastern coast of northern Tasmania, the average size nearly doubles and coat colour changes from rufous brown to grey-brown. The snout is relatively short in animals from Queensland and increases proportionally in size in more southern populations, being most elongate and narrow in Tasmania. The proportion of individuals with a white tip to the tail increases from nil at the northern extremity of the range to 80 per cent in Tasmania.

## *Potorous tridactylus*

**Size**

*Head and Body Length*
(380) mm (males)
(340) mm (females)
*Tail Length*
204–262 (235) mm (males)
198–254 (228) mm (females)
*Weight*
740–1640 (1180) g (males)
660–1350 (1020) g (females)

**Identification** Brown to grey above, paler below. Nose long and tapering; naked patch of skin extending onto snout from nose. Foot shorter than head. Smaller than *Potorous longipes*.
**Recent Synonyms** *Potorous apicalis*, *Potorous gilbertii*.
**Other Common Names** Long-nosed Rat-kangaroo, Wallaby Rat.
**Status** Common, limited.
**Subspecies** *Potorous tridactylus tridactylus*, south-eastern Queensland, coastal New South Wales, Victoria, south-eastern South Australia and south-western Western Australia. *Potorous tridactylus apicalis*, Bass Strait islands and Tasmania.
**References**
Heinsohn, G.E. (1968). Habitat requirements of the macropod marsupial *Potorous tridactylus* in Tasmania. *Mammalia* 32: 30–43.
Seebeck, J.H., A.F. Bennett and D.J. Scotts (1989). Ecology of the Potoroidae—A Review (in) G. Grigg, P. Jarman and I. Hume (eds) *Kangaroos, Wallabies and Rat Kangaroos*. Aust. Mammal Soc. and Surrey Beatty and Sons, Sydney. pp. 67–88.

The frequent occurrence of subfossil remains in cave deposits indicate that the Long-nosed Potoroo was much more common in the past. It is not clear to what extent its decline is the result of human activities but it is obvious that very large areas of suitable habitat along the eastern coast of Australia have been removed by land clearing.

P.G. JOHNSTON

# FAMILY MACROPODIDAE
## *Wallabies, kangaroos and tree-kangaroos*

*Eastern Grey Kangaroo.* B.J. NOLAN

Largest of the diprododont families, the Macropodidae comprises two subfamilies. The Sthenurinae was a successful group which had at least 20 species in the Pleistocene but is now represented only by the Banded Hare-wallaby, *Lagostrophus fasciatus*, restricted to Bernier and Dorre Islands, off Shark Bay. One diagnostic feature of sthenurines is that the lower incisors bite against the row of upper incisors (not behind these, as in other macropodids).

All other living macropodids are members of the subfamily Macropodinae. These appear to fall into five (unnamed) sub-groups: (a) 'typical' kangaroos, wallaroos and wallabies of the genus *Macropus* (divided by some authors into four subgenera) plus the

Swamp Wallaby, *Wallabia*; (b) rock-wallabies of the genus *Petrogale*, pademelons of the genus *Thylogale* and the Quokka, *Setonix*; (c) nailtail wallabies of the genus *Onychogalea* and the 'typical' hare-wallabies of the genus *Lagorchestes*; (d) tree-kangaroos of the genus *Dendrolagus*; (e) New Guinean forest wallabies of the genera *Dorcopsis* and *Dorcopsulus*.

Powerful hindlimbs with a long, narrow hindfoot and long, powerful fourth toe are characteristic of most macropodids, but there are exceptions. In rock-wallabies and tree-kangaroos, the hindfoot is secondarily shortened and broader. In most macropodids, the tail is broad at the base, long and tapering, serving as a balancer in the hopping gait. In the larger species it supports the rear part of the body in slow 'pentapedal' locomotion; and, together with the extended hindlimbs, forms a tripod that enables an animal to stand and move about in a vertical posture when fighting. However, the tail of *Setonix* is too short for it to contribute to locomotion or support. In *Onychogalea*, the tail has a horny tip which is sometimes pressed vertically into the substrate: similar behaviour is shown by New Guinean forest wallabies but no function has yet been demonstrated. The long, non-prehensile and almost untapered tail of *Dendrolagus* serves only as a balancer.

The forelimbs of most macropods are small and rather weakly developed but, in kangaroos and the larger wallabies, where males are significantly larger than females, the forelimbs of males are disproportionately larger and employed in fighting. The forepaws are capable of only a simple grip (all digits towards the palm) but this suffices to bring vegetation to the mouth. In *Dendrolagus*, the forepaws are also employed in climbing and are able to grip on vines.

Although the macropodines include some basically browsing species, the most successful are those that have a capacity to crop, chew and digest grasses that have a high content of fibre (mainly cellulose) and silica. The second and third upper incisors, which tend to lie behind the central incisor in other diprotodonts, lie alongside these in macropodines, forming a continuous cutting edge (dental arcade) at the front of the mouth. The lower incisors do not bite directly against these (as in sthenurines) but press against a tough pad on the roof of the mouth, just behind the arcade. Canine teeth are absent or vestigial and there is a long gap (diastema) between the incisors and the grinding teeth. This permits the tongue to arrange cut grass stems into packages that can be passed backwards for mastication.

In a young kangaroo there are two blade-like premolars on each side of the upper jaw but both of these are deciduous ('milk') teeth that are subsequently shed and replaced by a third premolar. The fate of the molars is unusual, for they not only erupt in slow succession (the anterior-most appearing first), but move forward along the jaw as the animal grows, eventually falling out. Of the total complement of four molars on each side of the upper and lower jaws, a young animal may have only the first two in use, these lying behind the premolar. An animal in mid-life may have all four in use, the anterior-most now lying in the position originally occupied by the premolar. As it nears the end of its life, a kangaroo may have no more than the last one or two molars in the series, these having progressed to the original front of the tooth row. A notable exception is seen in the Nabarlek, in which molars continue to erupt, move forward and be shed throughout life.

The molars of kangaroos are high-crowned. With use, they become flattened with serpentine ridges of hard enamel projecting only slightly above a matrix of dentine. Serial replacement of molars (which also occurs in elephants) allows a kangaroo to cope with very abrasive food by bringing new teeth into use successively rather than wearing down a complete set uniformly. Moreover, as in elephants, the total grinding area brought into action in the course of an individual's life is considerably greater than could be accommodated at one time in the jaws.

Grinding of food into small particles resolves only one of the problems facing a grass-eater. Another is the liberation of the food energy bound up in cellulose and other plant fibre. Very few animals possess the enzymes required to digest these polymerised carbohydrates but, among those that do, termites use a vast population of protozoans in their gut; mammals rely primarily upon bacteria. These micro-organisms may be accommodated in a large caecum, as in leaf-eating possums and the Koala, or in a large, sacculated stomach in such ruminants as cattle and sheep. The anatomy of the stomach of a kangaroo is quite different from that of a eutherian ruminant but each has a very large chamber in which fibrous plant material is fermented by micro-organisms. The end-products of fermentation (largely fatty acids) are absorbed into the bloodstream and transformed by the liver into glucose.

The success of kangaroos, as a group, seems to be due largely to their ability to utilise fibrous plant material, particularly grasses. They may, indeed, have evolved in response to the evolution of grasses as a major component of the Australian vegetation.

The impact of European settlement has little adverse effect on the larger kangaroos. These may have increased in range and numbers in response to clearing of forest and woodland and the provision of watering points for cattle and sheep in arid areas. Smaller species from woodlands have fared less well, due to loss of habitat, competition from introduced herbivores such as the Rabbit, sheep and Goat and predation by introduced carnivores such as the Fox and Cat. Hare-wallabies, nailtail wallabies, the Parma Wallaby and the Quokka have suffered severely. Of the medium-sized macropodids, only the Toolache Wallaby has become extinct in historic times, probably due more to loss of habitat than to hunting pressure.

# Subfamily Macropodinae

All but one of the extant species of kangaroos and wallabies are included in this large group. On the whole, their success appears to be related to adaptations to a diet of grass, but some species are also browsers. The macropodines are now the dominant macropods, having largely replaced the sthenurines and potoroids.

## TREE-KANGAROOS

### *Dendrolagus*

Why kangaroos—descended from arboreal ancestors and eminently adapted to terrestrial life—returned to the trees is not clear but, as suggested a century ago by Alfred Russell Wallace, it may have enabled them 'to feed on the vast forests of New Guinea, as these form the great natural feature which distinguishes that country from Australia'. Certainly the existence of an unutilised source of food often provides a spur to evolution but it is surprising that the ringtails and phalangers did not pre-empt the niche that tree-kangaroos came to occupy. If, as seems likely, tree-kangaroos came from a primitive browsing stock, it may be that colonisation of trees constituted no more than a (vertical) progression in search of their preferred food.

Whatever the pressures of natural selection that directed the evolution of this group towards arboreal life, these have involved the reversal of some major trends in macropod evolution, for the relative lengths of the fore- and hindlegs of tree-kangaroos are closer even than in the Musky Rat-kangaroo to those of ringtail and brushtail possums. The forelegs are stouter and more muscular than those of typical kangaroos and the hindfoot, instead of being markedly elongate, is almost rectangular in shape, with a uniformly granular sole. Despite these adaptations, tree-kangaroos are ungainly in trees and their success can only be explained by an absence of predators or of competitors of equal size.

Fossil hindlimb bones from the Wellington Caves, New South Wales, (at least 50 000 years old), appear to be from a tree-kangaroo as large as a mature Red Kangaroo. It is difficult to imagine a macropod of this size being truly arboreal but it could perhaps have been a browser with a limited ability to ascend sloping tree trunks (like modern rock-wallabies). In 1984, a tree-kangaroo from West Irian was discovered, showing many indications of a secondary adaptation to terrestrial life.

# Bennett's Tree-kangaroo

## *Dendrolagus bennettianus*

### De Vis, 1887

dend'-roh-lah'-gus ben'-et-ee-ah'-nus:
'Bennett's tree-hare'

L.J. ROBERTS

*Bennett's Tree-kangaroo is the largest of Australia's arboreal mammals. Compared with terrestrial kangaroos, it has longer forelegs and shorter hindfeet.*

Part of the remnant New Guinean fauna left on Cape York Peninsula, Bennett's Tree-kangaroo is patchily distributed over a small range which extends north from the Daintree River to the vicinity of Mount Amos (a distance of 75 kilometres) and west from the coast to the Mount Windsor Tablelands (about 50 kilometres). Although comparative abundances are not known, it appears to be as much at home in lowland vine forests as in montane rainforest.

It has the distinction of being the largest arboreal mammal in Australian tropical rainforest. Bennett's Tree-kangaroo is often mistakenly thought of as a primarily terrestrial macropod which occasionally and ineptly climbs trees. In reality, it spends most of its time in the canopy where, despite the apparent limitations of its macropodid anatomy, it moves around with surprising facility. However, it has lost none of its terrestrial abilities and, when necessary, can hop rapidly along the ground. Overall, it is superbly equipped to utilise the spatially and temporally scattered resources of the vine and gallery forests at the northern edge of Australia's wet tropics. In some of these areas it achieves abundances of 0.3 animals per hectare. It is predominantly nocturnal, spending most of the day sitting high in the canopy. In winter it often sits in a sunny position on top of a carpet of vines which also serve to conceal it from view. Generally, in all of its behaviour, it is a wary and very cryptic animal.

It feeds mainly on leaves but takes some fruit when seasonally available. In lowland rainforest, foliage from a limited number of species is preferred, particularly trees of the *Ganophyllum*, *Aidia* and *Schefflera*, the vine *Pisonia* and the fern *Platycerium*. Fruit is taken from *Chionanthus*, *Olea* and several species of figs.

### *Dendrolagus bennettianus*

**Size**

*Head And Body Length*
720–750 mm (males)
690–705 mm (females)

*Tail Length*
820–840 mm (males)
730–800 mm (females)

*Weight*
11.5–13.7 kg (males)
8–10.6 kg (females)

**Identification**  Overall dark brown, greyish tinge to forehead and snout, shoulders, neck and back of head rusty brown; light fawn below. Fore- and hindfeet black. Tail with black patch at base and light patch on dorsal surface. Distinguished from *Dendrolagus lumholtzi* by tail markings and absence of pale band across forehead and sides of face.

**Recent Synonyms**  None.

**Other Common Names**  Tree-climber, Grey Tree-kangaroo, Tree Wallaby.
Indigenous names: Jarabeena, Tcharibeena.

**Status**  Sparse.

**Subspecies**  None.

**References**

Rothschild, Lord and G. Dollman (1936). The genus *Dendrolagus*. *Trans. Zool. Soc. Lond.* 21: 477–548.

Waite, E.R. (1894). Observations on *Dendrolagus bennettianus* (De Vis). *Proc. Linn. Soc. NSW*, Ser. 2, 9: 571–582.

Bennett's Tree-kangaroo is one of the few macropods to defend a discrete territory. Adult males are extremely intolerant of each other and all carry numerous scars from fighting, some to the extent of missing an ear. Each male territory overlaps with that of several adult females and the mating system appears polygynous. Adult males are predominantly solitary and occupy territories of up to 25 hectares. At night, a male moves actively through his territory and is often seen in close proximity to females, particularly around food trees. Females occupy discrete ranges and are usually accompanied by one or two young. There is little detailed information on reproduction. Females breed annually and may exhibit embryonic diapause. Pouch life is around nine months and young accompany the mother for up to two years.

The major predators are now the Dingo, *Canis vulpes dingo* and Amethystine Python, *Morelia amethistina*, the latter taking a heavy toll of dependent juveniles in lowland forests. In the past, hunting pressure by Aborigines appears to have been heavy in lowland forests and tree-kangaroos were only abundant on the upper slopes of the mountains. Many of these areas were not visited by Aboriginal hunters because of taboos. Over the past forty years, traditional hunting has declined and in some areas, particularly in some vine forests in the northern part of their range, Bennett's Tree-kangaroo is now relatively common. The coastal lowland forests have been greatly disturbed by roads and other developments associated with tourism and the species is seldom seen in these areas.

R.W. MARTIN AND P.M. JOHNSON

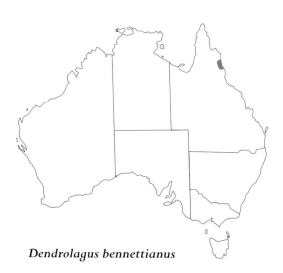

*Dendrolagus bennettianus*

# Lumholtz's Tree-kangaroo

## *Dendrolagus lumholtzi*

### Collett, 1884

lum'-holt-zee: 'Lumholtz's tree-hare'

Tree-kangaroos are predominantly Melanesian in distribution and the ancestors of the Australian species probably migrated over Torres Strait to exploit the rainforests of Cape York. Reported to have once been common in coastal lowland rainforests of northern Queensland (very little of which remains), Lumholtz's Tree-kangaroo is still found at higher altitudes between Kirrama and Mount Spurgeon. It is primarily a leaf-eater and, although leaves have a low nutrient value, its large, sacculated stomach permits large quantities to be ingested. The full range of its diet has not been determined but it is known to feed on leaves of the Ribbonwood and (introduced) Wild Tobacco, many fruits and even maize from farms on the rainforest edge. When eating foliage, a tree-kangaroo grasps a twig or small branch, bending it towards its head so that the leaves can be bitten off. Fruit is held in both forepaws while being eaten.

*Dendrolagus lumholtzi*

D. WATTS

*Lumholtz's Tree-kangaroo is an accomplished climber in rainforest.*

Lumholtz's Tree-kangaroo is nocturnal, spending the day asleep in a crouched sitting posture in the crown of a tree or on a branch: it has been suggested that the very obvious parting of the fur in a whorl just behind the shoulders is a device to shed water both forwards and backwards when the sleeping animal is deluged by rain. The fur is groomed, as in other kangaroos, with the syndactylous toes of the hindfeet, the stout claws of the forefeet and the tongue.

Despite its clumsy build it is an efficient climber, gripping a branch with the stout, recurved claws of its long, powerful forelimbs and walking forwards or backwards (or running slowly) with alternate movements of the short, broad hindfeet. Tree-kangaroos are the only kangaroos able to move their hindlegs independently of each other in normal locomotion. On broad, more or

H.& J. BESTE

*Lumholtz's Tree-kangaroo sleeps in a crouched posture on a branch but may rest in a tree fork. Juvenile.*

less horizontal, branches it may progress in the typical kangaroo hopping gait with the forelegs against the body and the tail held out rather stiffly behind. A prehensile tail would be a boon to such an animal but, as in terrestrial kangaroos, from which it presumably evolved, this organ serves only as a counterbalance. Descent from a tree is far from graceful: it moves tail-first, clasping the trunk with its forelimbs which are moved alternately down the tree while the soles of the hindfeet slide against the bark until the animal is about 2 metres from the ground. It then releases its grip and, kicking off from the trunk, twists in mid-air to land upright, ready to bound off. An individual disturbed in a tree may jump to another one or leap to the ground from a height of as much as 15 metres. Locomotion on the ground is by a quadrupedal walk or a fast, bipedal hop.

The sexes are similar in colour but males are larger than females. Lumholtz's Tree-kangaroo is a solitary animal but feeding aggregations of as many as four have been observed. Groups of females with a single male can be maintained amicably in captivity but two males in the presence of females will fight savagely. There appears to be no definite breeding season. A male investigating a receptive female stands in front of her, uttering a soft clucking sound and softly pawing her head and shoulders. When the female moves away, the male follows, pawing at the base of her

tail. Copulation is as in other kangaroos. There are four teats in the pouch and the single young attaches to the one that has become enlarged prior to birth. Studies on captive animals indicate a pouch life of approximately 230 days.

The range of Lumholtz's Tree-kangaroo was early reduced by extensive clearing of lowland rainforest. The logging of highland rainforest continues to reduce its distribution but it is present in reasonable numbers in a number of national parks and reserves.

P.M. JOHNSON

## *Dendrolagus lumholtzi*

**Size**

*Head And Body Length*
520–650 (591) mm (males)
533–603 (554) mm (females)
*Tail Length*
655–736 (698) mm (males)
670–732 (699) mm (females)
*Weight*
5.4–8.6 (7.5) kg (males)
5.1–7.0 (6.0) kg (females)

**Identification** Blackish-brown, sprinkled with lighter coloured fur on the lower part of back. Lighter coloured band across forehead and down each side of face distinguishes this species from *Dendrolagus bennettianus*. Forearms long and heavily muscled; hindfeet short and broad. Tail long with terminal half blackish-brown.
**Recent Synonyms** None.
**Other Common Names** Tree-climber.
Indigenous name: Boongarry.
**Status** Common, limited.
**Subspecies** None.
**References**
Lumholtz, C. (1890). *Among Cannibals. An account of four years' travels in Australia and of camp life with the Aborigines of Queensland.* Murray, London.
Rothschild, Lord and G. Dollman (1936). The genus *Dendrolagus. Trans. Zool. Soc. Lond.* 21: 477–502.

HARE-WALLABIES

*Lagorchestes*

It is unfortunate that similarities in behaviour led to the same common name being given to hare-wallabies of the genus *Lagorchestes* (subfamily Macropodinae) and to the monotypic *Lagostrophus* (subfamily Sthenurinae) dealt with on page 406. It being too late to create a new 'common' name for the sthenurine species, it is perhaps best to refer to species of *Lagorchestes* as 'true' hare-wallabies.

These small macropods received their scientific and common names from a supposed resemblance to the Hare. Most live in tropical woodland or grassland but the Rufous Hare-wallaby was once common in south-western Australia. All are primarily browsers but may eat some grass. Having an efficient water economy, they can also live without access to free water. During the day they shelter in dense vegetation, a cavity dug into or under a tussock, or a short burrow.

The distribution of hare-wallabies has declined since European settlement and only one species is relatively common.

# Central Hare-wallaby

## *Lagorchestes asomatus*

### Finlayson, 1943

lag'-or-kes'-tayz ah'-soh-mah'-tus: 'bodiless dancing-hare'

Probably the smallest and certainly the most enigmatic of all the hare-wallabies, this species is known to science from only one unsexed, adult skull extracted from a fresh carcass. It was collected by Michael Terry at an undetermined locality along a 130-kilometres strip of country between Mount Farewell and Lake Mackay, Northern Territory, during a prospecting expedition in 1932. Terry's pencil-written trip log and daily diary reveal that his party took an almost direct route from the western end of Mount Farewell, via Surprise Rock-hole. McEwin then retraced their steps to Mount Farewell, between 13 and 26 August, but no more exact date or locality can be determined. His entries for the relevant period contain many notes on 'spinifex rats', referring to their local abundance, nesting habits, vocal sounds, appeal as camp pets, methods of hunting by Aborigines and value as food for his tame Dingo. Nowhere, however, does he describe their appearance in sufficient detail to identify either the various species involved or the individual from which he took the skull that was later presented, undated, to the South Australian Museum. His 'spinifex rats' probably comprised at least three species: the Central Hare-wallaby, the Rufous Hare-wallaby and a bettong. Which, if any, of his notes referred specifically to the Central Hare-wallaby remains a mystery.

Western desert Aborigines remember an animal that was almost certainly the Central Hare-wallaby. It was small and similar in size to the Burrowing Bettong and hopped 'like a kangaroo'. It had long, soft, grey fur (similar in colour to the Burrowing Bettong and the Bilby), hairy feet and a relatively short and thickened tail. It inhabited sand plains and dunes with spinifex and sheltered in a scrape under a spinifex hummock, sometimes digging a short burrow similar to that of the Rufous Hare-wallaby. Many Aborigines referred to it as the 'quiet one,' 'deaf one', or sometimes 'stupid one' because it did not flush from its shelter. It was hunted by tracking to its hide and killed by spearing, although sometimes it could be caught by hand. As was the case with other long-haired mammals such as the Rufous Hare-wallaby and brushtail possums, its fur was used to make belts. It ate grass leaves and seeds (including those of spinifex) and the fruits of the Quandong, *Santalum acuminatum*.

Finlayson noted that the skull morphology of the type-specimen was not entirely consistent with the general form of *Lagorchestes* but lack of material caused him to be conservative and place it in that genus. The extremely quiet behaviour described by Aborigines contrasts with the very

---

### Lagorchestes asomatus

**Size**
About the size of the Burrowing Bettong (see page 289). Greatest length of skull 66 mm; greatest width of skull 42.8 mm.

**Identification**  Small wallaby with long, soft, grey fur; a relatively short, thickened tail; and hairy feet.

**Recent Synonyms**  None.

**Other Common Names**
Indigenous names: Kalanpa/Kananpa, Pungkurrpa, Raputji (Kartutjarra); Kalanpa, Kuluwarri, Tjuntatarrka, Yamari (Kutatja); Kalanpa, Pukurl-pukurl (Mangala); Kalanpa/Kananpa, Pungkurrpa, Raputji (Manytjilytjarra); Kuluwarri (Ngaatjatjarra); Kulkuma, Kuluwarri, Pilakarratja, Tjinapawulpa, Tjuntatarrka, Warrkuntjilpa, Yamarri (Pintupi); Kuluwarri (Pitjantjatjarra); Pukurl-pukurl (Walmatjari); Kalanpa, Kuluwarri, Tjuntatjarrka, Yamari (Wangkatjunka); Kulkuma, Nantjwayi, Yamari/Yamarri (Warlpiri).

**Status**  Presumed extinct.

**Subspecies**  None.

**Reference**
Finlayson, H.H. (1943). A new species of *Lagorchestes* (Marsupialia). *Trans. Roy. Soc. S. Aust.* 67: 319–321.

---

excitable nature of its supposed congeners. It is possible that, with its extinction, Australia lost an entire genus of mammals.

Desert Aborigines report that the Central Hare-wallaby disappeared from their country between about 1940 and 1960. Near Kiwirrkurra (south-eastern Great Sandy Desert) it was still present in 1960. It appears to have survived in the Tanami Desert until the late 1940s.

P.F. AITKEN, A.A. BURBIDGE, K.A. JOHNSON
AND P.J. FULLER

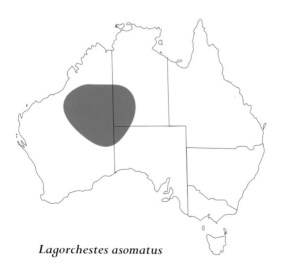

*Lagorchestes asomatus*

# Spectacled Hare-wallaby

## *Lagorchestes conspicillatus*

### Gould, 1842

kon-spis'-il-ah'-tus: 'spectacled dancing-hare'

The Spectacled Hare-wallaby is an inhabitant of tropical grasslands. Although once more common than today, it has not declined in range and abundance to the same extent as other hare-wallabies.

In Queensland, it is a widespread inhabitant of open forests, open woodlands, tall shrublands over tussock grass and hummock grassland; in the Mount Isa area it lives in hummock grass. Most of its range in Queensland is in grazing areas. Numbers appear to have declined in many areas during the 1980s. In the Northern Territory it was formerly common in shrub thickets and hummock grasslands among the central ranges but it is no longer found there. It still occurs sparsely in the Tanami Desert in acacia shrubland and spinifex, while in most of the remainder it is moderately common, especially between 16° and 18° south. There are only three records from the Kimberley region of Western Australia, but Aboriginal informants say it was once widespread and has declined. In the Pilbara

*The Spectacled Hare-wallaby is so-called because of the rufous rings around its eyes.* P.M. JOHNSON

## Lagorchestes conspicillatus

**Size**

*Head and Body Length*
400–470 mm
*Tail Length*
370–490 mm
*Weight*
1.6–4.5 kg

**Identification** Brown above, hairs white tipped; white below. Bright orange ring around eyes. White hip-stripe. Feet pale grey-brown. Tail with sparse, short, grey-brown hairs, darker near tip, no tuft or crest. Stocky, thickset, short-necked.

**Recent Synonyms** None.

**Other Common Names** Indigenous names: Kwalpa (Anmatjara); Kwalpa/Kwarlpa (eastern and western Aranta); Irlyaku, Kwalpa (Katitja); Marrkapurr, Pitang (Karatjari); Tjantjipuka, Tjunngaru (Kartutjarra); Mirlpatiri, Tingarri, Wampana (Kutjata); Wampana (Kurintji), Ukalpi (Luritja); Mirlpatiri, Pitan, Warngaru (Mangala); Milparti, Tjantjipuka (Manytjilytjarra); Wampana (Mutpura); Ukalpi/Mukalpi/Pukalpi (Ngaatjatjarra); Tjantjipuka (Nyangamarta); Yukalpi/Mukalpi, Wampana (Pintupi); Ukalpi (Pitjantjatjarra); Mangkapan, Mirlpatiri, Wampana (Tjaru); Kalama, Wampana (Tjinkili); Mangkapan, Wampana (Wangkatjunka); Mirlpatiri, Wampana, Yulkaminyi (Warlpiri).

**Status** Queensland, moderately common; Northern Territory, moderately common in the northern half, extinct in the south and Western Australia; rare, scattered in the Pilbara; extinct in the Great Sandy and Gibson Deserts; Barrow Island, abundant; Hermite Island, extinct.

**Subspecies** *Lagorchestes conspicillatus conspicillatus*, Barrow Island.
*Lagorchestes conspicillatus leichhardti*, mainland Australia.
*Lagorchestes conspicillatus pallidior*, coastal north Queensland (possibly invalid).

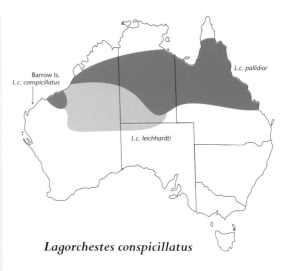

*Lagorchestes conspicillatus*

region of Western Australia, although still reported from some areas, it appears to have declined drastically, possibly because frequent burning of spinifex grassland, particularly on cattle stations, has prevented the development of the large hummocks required for shelter. It is abundant on Barrow Island.

On Barrow Island, Spectacled Hare-wallaby numbers were estimated at about 10 000 (42 per square kilometre) in 1988. There it spends the day in hides tunnelled into large spinifex hummocks where temperatures seldom rise above

**References**

Ingleby, S. (1991). Distribution and status of the Spectacled Hare-wallaby, *Lagorchestes conspicillatus*. *Wildl. Res.* 18: 501–519.

Short, J. and B. Turner (1991). Distribution and abundance of Spectacled Hare-wallabies and Euros on Barrow Island, West. Aust. *Wildl. Res.* 18: 421–429.

Johnson, P.M. (1993). Reproduction of the Spectacled Hare-wallaby *Lagorchestes conspicillatus* Gould (Marsupialia: Macropodidae) in Captivity, with Age Estimation of the Pouch Young. *Wildl. Res.* 20: 97–101.

30°C. Laboratory experiments have shown that animals from Barrow Island do not begin to use significant amounts of water for evaporative cooling until the ambient temperature rises above this level. Several hides are constructed within a home range of about 8–10 hectares. On Barrow Island, it is a selective feeder, browsing mainly on colonising shrubs and also eating the tips of spinifex leaves which become a major part of the diet in long-unburnt and undisturbed areas. It does not drink, even when water is available. Well adapted to these harsh conditions, it has a low urine production and a water turnover (expressed as a percentage of total body water) of only 5.3 per cent per day, far less than has been measured in any other mammal of comparative size.

The Spectacled Hare-wallaby is solitary but up to three may occasionally be seen feeding together. Vocalisation is limited to a warning hiss and a soft clicking sound made by the male in the presence of an oestrous female and by the female and pouch-young to communicate with each other.

Breeding takes place throughout the year but on Barrow Island there are peaks of births in March and September. Sexual maturity is reached by both male and female at about one year. The period from mating to birth, providing no young is in the pouch, is 29–31 days. The species exhibits embryonic diapause: birth is followed very closely by mating and the resultant embryo ceases development at an early stage. Should the first young be lost prematurely, the quiescent embryo resumes development to be born 28–30 days later. If the pouch-young develops normally, it vacates the pouch at about 150 days and the quiescent embryo then resumes development to be born one to two days after the permanent vacation of the pouch by the first young.

While the Spectacled Hare-wallaby appears to be secure on Barrow Island, in Queensland and in parts of the Northern Territory, concern

D. WATTS

*The Spectacled Hare-wallaby inhabits tropical grass-lands.*

has been expressed that it may be continuing to decline in arid and semi-arid areas on the mainland. It once occurred on Montebello Islands, just to the north of Barrow Island, but became extinct between 1912 and 1950, almost certainly because of predation by the feral Cat. A watch should be kept on mainland populations to ensure these do not suffer the same fate.

A.A. BURBIDGE AND P.M. JOHNSON

## Lagorchestes hirsutus

**Size**
**Bernier and Dorre Islands**
*Head and Body Length*
310–360 (330) mm (males)
360–390 (375) mm (females)
*Tail Length*
260–280 (270) mm (males)
245–305 (275) mm (females)
*Weight*
1245–1800 (1580) g (males)
780–1960 (1740) g (females)
**Tanami Desert**
*Weight*
800–1600 (1220) g (males)
900–1250 (1310) g (females)

**Identification** Rufous above, paler below;
head rufous. Tail brownish-black above, pale ru-
fous below. Animals from Bernier and Dorre
Islands larger, but have a shorter tail and are
grey-brown above, with dark grey head.
Forearms paler than body. Fur long, soft.
**Other Common Names** Western Hare-wal-
laby, Brown Hare-wallaby, Spinifex-rat (desert),
Whistler (south-west). Indigenous names:
Atnukwa (Aranta, eastern); Irlraku, Mala
(Kaititja); Landaa, Landalpartyi (Karatjari);
Mala, Matjirri (Kartutjarra); Mala, Parranti,
Tarnnga, Tiwilpa/Liwilpa, Warku, Witjari/
Witjarri (Kukatja); Mala (Luritja); Mala, Raltatu
(Mangala); Mala, Matjiri/Matjirri, Ngartinpa,
Tarnnga (Manytjilytjarra); Mala, Tarnnga,
Tjiwilpa (Ngaanyatjarra); Mala, Tarnnga,
Tjanpitja, Tjiwilpa/Tiwilpa, Warku
(Ngaanyatjarra); Mala, Tarnnga, Tjanpitja,
Tjiwilpa/Tiwilpa, Warku (Ngaatjatjarra);
Matjirri, Ninngka (Nyamal); Mala, Matjirri
(Nyangamarta); Mala, Matjirri (Nyiyapali); Mala,
Parranti, Tarnnga, Tintinpa, Tjiwilpa/Tiwilpa,
Tjunpu, Warku, Wirrini (Pintupi); Mala,
Tjanpitja, Warku (Pitjantjatjarra); Mala

# Rufous Hare-wallaby

## *Lagorchestes hirsutus*

### Gould, 1844

her-syute'-us: 'hairy dancing-hare'

The Rufous Hare-wallaby was once very common
and widespread throughout most of the arid and
semi-arid parts of central and western Australia,
particularly in the spinifex hummock grasslands
of the sand plain and sand-dune deserts; also in
spinifex on gravelly plains; in areas of tussock
grass; and in shrublands. Its mythology is widely

(Putitjarra); Mala, Witjari (Tjaru); Mala, Malyi,
Witjari (Walmatjari); Liwilpa, Mala, Parranti,
Tarnnga, Witjari (Wangkatjungka); Kunatjinpa,
Mala, Parranti, Tipirri, Witjari (Warlpiri); Mala,
Matjirri (Warnman); Wurrup (south-east).
**Status** Rare.
**Subspecies** *Lagorchestes hirsutus hirsutus*, main-
land Australia.
*Lagorchestes hirsutus bernieri*, Bernier Island,
Western Australia.
*Lagorchestes hirsutus dorreae*, Dorre Island,
Western Australia.
**References**
Bolton, B.L. and P.K. Latz (1978). The Western
    Hare-Wallaby *Lagorchestes hirsutus* (Gould)
    (Macropodidae) in the Tanami Desert. *Aust.
    Wild. Res.* 5: 285–293.
Lundie-Jenkins, G.W., C.M. Phllips and
    P.J. Varman (1993). Ecology of the Rufous
    Hare-wallaby, *Lagorchestes hirsutus* Gould
    (Marsupialia: Macropodidae) in the Tanami
    Desert, N.T. II. Diet and feeding strategy.
    *Wildl. Res.* 20: 477–494.

M. LOCHMAN

*Once widespread on the mainland, the Rufous Hare-wallaby now exists as wild populations only on Bernier and Dorre Islands.*

known by Aborigines who once hunted it for food and early explorers and travellers commented on its abundance.

It persisted in the Great Sandy and Gibson deserts until the 1950s but its Western Australian distribution is now restricted to Bernier and Dorre Islands in Shark Bay. In 1986, an extensive search in the Great Sandy Desert failed to locate it. Two wild populations recently became extinct in the Tanami Desert of the Northern Territory. One was extinguished by the Fox and the influ- ence of drought; the second disappeared in November 1991 when wildfire removed more than 90 per cent of its spinifex habitat.

The Tanami Desert colonies occupied spin- ifex grassland on sand plain interspersed with deflated dunes, salt pans and saline samphire areas. This habitat centres on an ancient infilled drainage line that is characterised by scattered patches of *Melaleuca* shrubs and giant mounds of the termite, *Nasutitermes triodiae*.

In south-western Australia, where John Gilbert obtained the type specimen, the Rufous Hare-wallaby occurred on sand plains with low, woody shrubs. On Bernier and Dorre Islands, it uses this habitat together with spinifex hummock

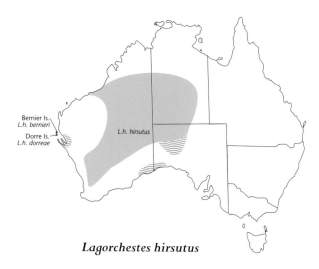

*Lagorchestes hirsutus*

Bernier Is.
*L.h. bernieri*

Dorre Is.
*L.h. dorreae*

*L.h. hirsutus*

grasslands. It shelters by day in a shallow scrape dug under a spinifex hummock or low shrub: this excavation may be developed into a burrow more than 70 centimetres deep, especially during the intense heat of summer. When flushed from a hide it escapes in an explosive zigzag burst of speed, often uttering a high-pitched, nasal squeak.

A thriving, captive-breeding colony has been established from the Tanami Desert colonies and research is proceeding on a population successfully reintroduced into the wild. Significant predation by the Cat is hampering these efforts.

The Rufous Hare-wallaby can cope with a high-fibre diet of spinifex but strongly prefers more nutritious forbs and grasses with a greater content of water. Seed heads are often eaten and insects may be taken during dry periods. It sometimes feeds in burned areas or saline flats where the vegetation is generally more succulent and nutritious than in adjoining mature spinifex or shrubland areas in which it shelters.

On Bernier and Dorre Islands the Rufous Hare-wallaby is less abundant than the Banded Hare-wallaby. Past observations on the ratios of Rufous to Banded Hare-wallabies suggested that the populations of the former fluctuate considerably. Numbers were high in 1906 but less so

in 1910. In 1959, they appeared to be more common and in 1978, numbers had risen dramatically in an area on Dorre Island which had been burned five years previously. However, the observed differences may reflect the localities in which collections were made rather than major shifts in the ratio of the two species over time. In 1988 and 1989 the populations on Dorre and Bernier Island were estimated at 1700 and 2600 respectively.

Reproduction in the mainland subspecies is continuous under favourable seasonal conditions. Pouch life lasts about 124 days. Females become reproductive at 5–18 months and males at about 14 months. Unlike many other macropods, females are slightly larger than males.

In the central deserts, Aborigines hunted the Rufous Hare-wallaby by tracking it to its hide and stamping on it or on the entrance to its burrow, then digging it out. Some hunters used a small leafy branch attached to a spear or long stick which was swished around to imitate the flapping wings of a Wedge-tailed Eagle, leading the wallaby to crouch in its shelter and enabling the hunter to approach closely. Pintupi and Warlpiri people also hunted it by making brush fences and, with much shouting (sometimes with the aid of dogs), driving the animals through gaps where hunters lay in wait with clubs. H.H. Finlayson has described the skilled use of fire by Aborigines in hunting the Rufous Hare-wallaby. This and other Aboriginal burning practices resulted in a tight mosaic of vegetation in various stages of recovery that provided structural and floristic diversity favourable to the species. Extensive summer wildfires eliminate the tall spinifex required for shelter and reduce the long-term availability of food. The Tanami Desert environment into which the wallaby is being reintroduced is being enhanced by the use of a mosaic burning program, designed to create and maintain such diversity.

K.A. JOHNSON AND A.A. BURBIDGE

*Eastern Hare-wallaby. Illustration in John Gould's* The Mammals of Australia.

# Eastern Hare-wallaby

## *Lagorchestes leporides*

### (Gould, 1841)

lep'-or-eed'-ayz: 'hare-like dancing-hare'

*Lagorchestes*, a name from Greek roots meaning 'dancing hare', was coined by John Gould for a genus of macropods characterised by a hairy muzzle and resembling the European Hare in size and in texture and colour of the fur. This description applied particularly to the Eastern

Hare-wallaby, which was the first species to come to his notice. Additionally, in Gould's view, its habits were hare-like; it was solitary, usually sat close to a well-formed 'seat' or scrape under the shelter of a tussock in open country and had a fleet, darting gait. 'Its powers of leaping are also extraordinary ... While out on the plains in South Australia, I started a Hare-Kangaroo before two fleet dogs, after running to a distance of a quarter of a mile (400 metres), it suddenly doubled and came back upon me ... I stood perfectly still and the animal had arrived within twenty feet (6 m) before it observed me,

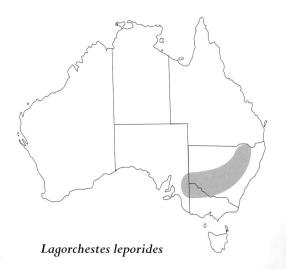

*Lagorchestes leporides*

*Lagorchestes leporides*

**Size**
*Head and Body Length*
450 mm
*Tail Length*
320 mm

**Identification** Grizzled brown above, hairs white tipped, grey below. Indistinct rufous area around eyes; faint hip-stripe; feet and forepaws greyish; tail brown above, white below. Distinguished by black patch on elbow.
**Recent Synonyms** None.
**Other Common Names** Brown Hare-wallaby.
**Status** Extinct.
**Subspecies** None.
**Reference**
Thomas, O. (1888). *Catalogue of the Marsupialia and Monotremata in the collection of the British Museum (Natural History)*. British Museum (N.H.), London. 82–84.

when ... instead of branching off to the right or left; it bounded clear over my head.' J.L.G. Krefft estimated that it could jump to a height of 8 feet (1.8 metres).

The decline to extinction of the Eastern Hare-wallaby was rapid. Gould reported it to be 'tolerably abundant' in 'all the plains' of South Australia in the 1840s, particularly 'between the belts of the Murray and the mountain ranges'. He also recorded it in New South Wales from the Liverpool Plains and the vicinity of the Namoi and Gwydir Rivers. Krefft found it to be common in the 1850s on the plains around the Murray–Darling junction but there is still mention of it thereafter. It disappeared prior to intensive settlement and is very poorly represented among museum specimens. Since 1890, when a specimen was obtained from Booligal, New South Wales, there have been no further indications of its survival. Nothing is known of its biology apart from the notes on its behaviour recorded by Gould and Krefft.

R. STRAHAN

## WALLABIES AND KANGAROOS

### *Macropus*

Species of *Macropus* range in size from the Parma and Tammar Wallabies (3–4 kilograms) to the Red Kangaroo (up to 85 kilograms). Over this more or less continuous gradient it is not possible to make an absolute distinction between wallabies and kangaroos but the species that are called wallabies do not exceed about 20 kilograms. Wallaroos are distinguished from kangaroos by their habitat: they tend to live in steep, hilly country, whereas typical kangaroos prefer flat or undulating land.

Habitats range from wet forest to desert but most species of *Macropus* live in open forest, woodland and grassland where their well-developed grazing adaptations are particularly appropriate. Measured in terms of the number of species, total numbers of individuals, or extent of distribution, *Macropus* must be reckoned to be the most successful of the contemporary macropods and probably the most successful of the Australian marsupial genera.

In view of the range of habitats, it is not surprising that species differ in details of their behaviour, particularly those activities related to conservation of body water. Those from arid regions are physiologically adapted to a low intake of free water and the Tammar Wallaby is able to drink sea water. Differences in patterns of reproduction tend also to be related to habitat: regular breeding is inappropriate to environments where rainfall is unpredictable.

Some species are solitary but most—particularly the larger wallabies, the kangaroos and all but one of the wallaroos—are gregarious. In this, as in other general aspects of their biology, they resemble the antelopes of the African plains but their social organisation is much less complex.

Only one species, the Toolache Wallaby, has become extinct since European settlement. Several species have declined but the larger ones continue to thrive to such an extent that, in many areas, they are regarded as agricultural nuisances or pests. Land that has been cleared to provide pasture for sheep or cattle also offers food for kangaroos and wallabies and this is utilised by the macropods, provided there is adequate cover in which they can rest during the day. Because the food preferences of sheep and kangaroos overlap only slightly, the two can normally coexist without serious competition but, under conditions of drought or overstocking, sheep and kangaroos compete for the last resources of food.

Kangaroos and wallabies are protected by legislation that varies from State to State but, except in Victoria, permits are readily granted to cull local populations that are deemed by wildlife authorities to be excessive. As it is ridiculous to waste the skins and carcasses of more than a million animals killed annually, provision is made for their commercial use.

C. ANDREW HENLEY/LARUS

# Agile Wallaby

## *Macropus agilis*

### (Gould, 1842)

mak'-roh-poos  ah-jil'-is: 'agile long-foot'

The Agile Wallaby is the most common macropo-did in tropical coastal Australia and the southern and eastern lowlands of Papua New Guinea. Its preferred habitat is along rivers and streams in open forest and the adjacent grasslands but in the Northern Territory it is abundant from the coastal sand-dunes to the base of the more rugged inland hills.

### *Macropus agilis*

**Size**

*Head and Body Length*
715–850 (800) mm (males)
593–722 (650) mm (females)
*Tail Length*
692–840 (770) mm (males)
587–700 (640) mm (females)
*Weight*
16–27 (19) kg (males)
9–15 (11) kg (females)

**Identification** Sandy brown above; whitish below. Head may have median dark brown stripe between eyes and ears and faint light buff cheek-stripe. Distinct light stripe on thigh. Edges of ears and tail tip black. Distinguished from *Onychogalea unguifera* by absence of dark dorsal stripe and upright stance when in rapid gait; from *Macropus rufogriseus* by paler colour, gait and paler extremities; from *M. parryi* by shorter tail, held more or less horizontally in rapid gait; from *M. dorsalis* by pale colour and lack of mid-dorsal stripe.
**Recent Synonyms** *Wallabia agilis*, *Protemnodon agilis*.

It appears to eat most native grasses and may dig 30 centimetres or more into the soil to obtain the roots of Ribbon Grass and some other species. In the Northern Territory, it is attracted to areas regenerating after fire. During floods, it retreats to high ground where food is less plentiful but, as the water recedes, it moves into the blacksoil plains where it eats sedges and short grasses. It is also known to browse on Coolabah leaves and to eat the fruits of Leichhardt trees and native figs.

The Agile Wallaby is gregarious, living in groups of up to ten which may aggregate into

*Macropus agilis*

D. WATTS

Right: *During most of the day, the Agile Wallaby rests in deep vegetation, emerging in the late afternoon to graze in open areas.*
Left: *The Agile Wallaby is a common macropod in tropical Australia.*

**Other Common Names** Sandy Wallaby, Kimberley Wallaby, Jungle Wallaby, Grass Wallaby, River Wallaby.

**Status** Abundant.

**Subspecies** The present definition of ranges of subspecies by State boundaries is obviously artificial and must be taken as no more than indicative.

*Macropus agilis agilis*, Northern Territory.

*Macropus agilis nigrescens*, Western Australia.

*Macropus agilis jardinii*, Queensland.

*Macropus agilis papuanus*, southern and south-eastern Papua New Guinea and some adjacent islands.

**References**

Kirkpatrick, T.H. (1970). The agile wallaby in Queensland. *Qld. Agr. J.* 96: 179–180.

Merchant, J.C. (1976). Breeding biology of the Agile Wallaby, *Macropus agilis* (Gould) (Marsupialia: Macropodidae), in captivity. *Aust. Wild. Res.* 3: 93–103.

much larger mobs on feeding areas. It is alert and, in comparison with other macropodids, a nervous animal, a disposition often demonstrated by foot-thumping while the head is held high.

Males are considerably larger than females and have much stouter forelimbs. In captivity, females become sexually mature at an average age of 12 months and males at 14 months but, in the wild, maturity is probably reached several months later. Breeding can occur throughout the year and, after a gestation period of about 30 days, a single young is born. It remains in the pouch for seven to eight months and suckles at foot until it is weaned at the age of 10–12 months. Females usually mate shortly after birth and the resultant blastocyst remains quiescent while the pouch is occupied. A study of the Agile Wallaby in the Northern Territory indicated that there is a high mortality of pouch-young but that lost young are replaced about a month later from dormant blastocysts.

The abundance of the Agile Wallaby and its effect on crops and pastures has led to its being declared a pest in certain areas. Poisoning campaigns have been carried out in Western Australia and the Northern Territory and a bounty system still operates in the cane-growing districts of Queensland.

J.C. MERCHANT

# Antilopine Wallaroo

## *Macropus antilopinus*

(Gould, 1842)

an'-til-oh-peen'-us: 'antelope (-haired) long-foot'

The Antilopine Wallaroo occupies the ecological niche in the monsoonal tropical woodlands that is filled by the Eastern and Western Grey Kangaroos in the forests and woodlands of eastern and southern Australia and by the Red Kangaroo in the arid and semi-arid woodlands. More slender and long-limbed than its relatives in the Euro—wallaroo group, it resembles the Grey and Red Kangaroos in general appearance and behaviour.

It is found in the more open *Eucalyptus* woodlands where perennial grasses are prominent in the understorey. Most of its habitat is on flat or gently undulating terrain but it is sometimes also found in rocky hill country where the kangaroo commonly seen is the Euro. In Arnhem Land it is occasionally observed in the company of the Euro or the Black Wallaroo

where the three species patronise the same waterhole near the base of an escarpment.

Unlike other wallaroos, it is gregarious, occurring in groups of varying size and composition. Solitary individuals or groups of two are most commonly seen but in these cases other animals are usually close by. Groups of three to eight are regularly seen. Larger groups of up to 30 or more are observed occasionally but these appear to be aggregations of smaller groups that have come together because of some disturbances such as the presence of humans or Dingos.

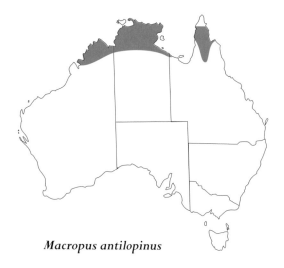

*Macropus antilopinus*

---

### *Macropus antilopinus*

**Size**

(4 males, 4 females)
*Head and Body Length*
965–1200 (1064) mm (males)
778–835 (805) mm (females)
*Tail Length*
780–890 (815) mm (males)
679–700 (692) mm (females)
*Weight*
30–49 (37) kg (males)
16–20 (17.5) kg (females)

**Identification** Males reddish-tan above, paler (almost white) below; limbs similar, tips of paws and hindfeet black; no distinctive facial marks. Females usually have pale grey head and forequarters but may be all grey, or all reddish-tan like the males. (General body colour of male Euro in the same areas is medium to dark greyish-fawn, female paler).

**Recent Synonyms** *Osphranter antilopinus*.

**Other Common Names** Antilopine (or Antelope) Kangaroo. In Northern Territory known locally as 'Red Kangaroo'.

**Status** Usually common, at least in Arnhem Land.

**Subspecies** None.

*The Antilopine Wallaroo is a large tropical kangaroo.*

During the dry season, when the temperature is high, it is inactive during daytime and rests in the shade of trees, bushes or rocks, usually in the vicinity of a waterhole. It becomes active and moves out to graze in the later afternoon. On overcast or rainy days during the wet season, it may be active at any time. It feeds largely if not entirely on grasses. In a study carried out south of Darwin, population densities at two sites, one more favourable than the other, were 11.4 and 30.9 per square kilometre during the wet season and 8.8 and 19.0 during the dry season. One radio-collared male had a home range of 76.1 hectares.

Few data have been recorded on breeding in wild populations and no systematic study has been made. Small, naked, pouch-young have been collected from March to July and females carrying large pouch-young have been observed between September and March. It seems probable that births occur throughout the year but there is a concentration of births towards the end of the wet season and in the early part of the dry season. Observations on two females and a male kept in captivity in Canberra indicate a long oestrous cycle, the lengths of seven cycles being 36–44 (average 40) days. The two gestation periods measured were of approximately 34 days duration and the length of pouch life of young was approximately 270 days. Females appear not to have a post-partum oestrus.

J.H. CALABY

### References

Croft, D.B. (1987). Socio-ecology of the antilopine wallaroo, *Macropus antilopinus*, in the Northern Territory, with observations on sympatric *M. robustus woodwardi* and *M. agilis*. *Aust. Wildl. Res.* 14: 243–255.

Russell, E.M. and B.J. Richardson (1971). Some observations on the breeding, age structure, dispersion and habitat of populations of *Macropus robustus* and *Macropus antilopinus* (Marsupialia). *J. Zool. Lond.* 165: 131–142.

# Black Wallaroo

## *Macropus bernardus*

### Rothschild, 1904

bern-ard'-us: 'Bernard (Woodward)'s long-foot'

L.F. SCHICK

*The tropical Black Wallaroo is the smallest and stockiest of the Euro-wallaroo group of macropods.*

Of the group of kangaroos containing the Euro and wallaroos, the Black Wallaroo is the smallest member. It is the only species among the large wallabies and kangaroos to have a diploid chromosome number of 18, all other species having 16 (or 20 in the case of the Red Kangaroo). Because of its restricted geographical distribution in a remote part of Australia, its extreme shyness and the difficult terrain in which it lives, it is one of the least known of the kangaroos.

It is found in a relatively small area of western and central Arnhem Land, where it inhabits the steep, rocky escarpments and tops of a deeply dissected plateau that rises abruptly from a relatively flat plain. The escarpments and plateau are variously covered with eucalypt woodland with a sparse understorey of grass and shrubs, a sparse community of spinifex and shrubs growing on very shallow soil or cracks or in the rock surface, or small patches of rainforest. There is also much bare rock. The climate is monsoonal with a highly reliable wet season from about December to March.

The Black Wallaroo is solitary; no more than three (an adult male and female and a large

### *Macropus bernardus*

**Size**

(four males, one female*)

*Head and Body Length*
595–725 (683) mm (males)
646 mm (female)

*Tail Length*
545–640 (609) mm(males)
575 mm (female)

*Weight*
19–22 (21) kg (males)
13 kg (female)

\* Females probably grow to a larger size than the measurements indicate.

**Identification** Males dark sooty-brown to black, females pale grey to grey-brown. Paws, feet and apical part of tail black in males, dark brown in females; no face markings. Relatively short ears. (Some small young male Euros occurring in the same area may be dark and could be confused with the Black Wallaroo.)

*Macropus bernardus*

young), are ever seen together. It is not active in daytime except on overcast days during the wet season. Daylight hours are usually spent in the shade of large rocks or trees.

The Black Wallaroo is extremely wary and difficult to approach. When disturbed, it leaps up and moves rapidly up the escarpment until out of sight. When cornered at higher elevations it is also adept at rapid descent and will leap from large rocks or cliffs 3 metres or more in height. It may come down from the escarpments to drink at springs or waterholes near the base or to graze on verges of roads or tracks close to the escarpment. In these situations it is occasionally seen in company with the Euro or Antilopine Wallaroo.

There is little information on breeding but large, furred young have been seen in pouches in the middle of the dry season (June–September). It has been kept in captivity on apparently only one occasion, when a group of five was received at Taronga Zoo in November 1918. The only observation that seems to have been published on them was that 'they were very quiet and docile'.

J.H. CALABY

**Recent Synonyms** None.
**Other Common Names** Black Kangaroo; Northern Black Wallaroo; Bernard's Wallaroo (or Kangaroo).
**Status** Common in its restricted habitat and range.
**Subspecies** None.
**References**
Press, A.J. (1989). The abundance and distribution of black wallaroos *Macropus bernardus* and common wallaroos *Macropus robustus* on the Arnhem Land Plateau, Australia (in) G. Gregg, P. Jarman and I. Hume (eds) *Kangaroos, Wallabies and Rat-kangaroos*. Surrey Beatty and Sons, Sydney. pp. 783–786.
Parker, S.A. (1971). Notes on the small black wallaroo, *Macropus bernardus* (Rothschild, 1904) of Arnhem Land. *Vic. Nat.* 88: 41–43.

# Black-striped Wallaby

## *Macropus dorsalis*

### (Gray, 1837)

dor-sah'-lis: 'notably-backed long-foot'

Although the Black-striped Wallaby is one of the common macropods of eastern Australia, it is among the least known because of its preference for a habitat in which it can stay well hidden and because it is often confused with the Red-necked Wallaby with which it shares much of its range. Nevertheless, there are many features by which the two can be distinguished.

Forested country with a dense shrub layer is the preferred habitat. This includes rainforest margins; brigalow scrub, particularly in a phase

### *Macropus dorsalis*

**Size**
*Head and Body Length*
680–820 (760) mm (males)
530–615 (590) mm (females)
*Tail Length*
740–830 (765) mm (males)
540–615 (595) mm (females)
*Weight*
18–20 (16) kg (males)
6–7.5 (6.5) kg (females)

**Identification** Overall brown above with mid-dorsal dark stripe from neck to rump; sides paler, grading almost to white below. White spot on cheek, behind and below eye; horizontal white stripe on thigh. Distinguished from *Macropus rufogriseus* by back stripe, thigh stripe and distinctive gait—short hop, head held low, body strongly curved with rump tucked under the body, forearms usually extended forward and outward from the body.

**Recent Synonyms** *Wallabia dorsalis.*
**Other Common Names** Scrub Wallaby (confusingly referred to as 'Nailtail' and 'Pademelon' in parts of Queensland).
**Status** Common.
**Subspecies** None.
**References**

Calaby, J.H. (1966). Mammals of the Upper Richmond and Clarence Rivers, New South Wales. *CSIRO Div. Wild. Res. Tech. Pap.* No. 10. Canberra.

Gould, J. (1857). *The Mammals of Australia.* The author, London, text to plates 26 and 27.

*The name of the Black-striped Wallaby refers to the prominent mid-dorsal stripe.* B.J. NOLAN

of regrowth; open forest with a thick acacia or other shrub understorey; and lantana thickets. Most of the daylight hours are spent resting under cover; feeding normally takes place from dusk to dawn, usually on pasture with some tree cover. Even if forced into more open country by dry conditions, the Black-striped Wallaby seldom ventures far from suitable cover.

By day, groups of 20 or more of both sexes and all ages rest under cover in more or less permanent camps. If alarmed, they move off in the same direction, quickly forming a single file and it requires major or continued disturbance to divide and scatter the group. Normal movement is along established routes which are well-marked pads in dense vegetation leading to grazing areas within or along the edge of the forest; or trodden paths in more open country. Although the Black-striped Wallaby is a social species, aged males live as solitary individuals.

Much of the original habitat of the Black-striped Wallaby has been destroyed or modified for agricultural and pastoral purposes but it is still abundant, particularly in central subcoastal Queensland where it is usually regarded as an agricultural pest. It is legally protected throughout its range but, if a landholder can demonstrate that it is responsible for significant losses, it may be taken on pest-destruction permits issued by the wildlife authorities of Queensland or New South Wales.

Only superficial details of the biology are known. The gestation period is 33–35 days and pouch life is 210 days; sexual maturity is reached at about 14 months in the female and 20 months in the male. The female comes into oestrus shortly after giving birth and can carry a quiescent blastocyst. Longevity is 10–15 years.

T.H. KIRKPATRICK

*Macropus dorsalis*

# Tammar Wallaby

## *Macropus eugenii*

### (Desmarest, 1817)

yue-jeen'-ee-ee: 'Eugene (now St Peter) Island long-foot'

With populations on ten or more offshore islands and disjunct mainland populations in Western Australia and South Australia, the Tammar Wallaby was, at one time, distributed in at least a dozen areas, many of them isolated for more than 10 000 years. It is one of the smallest of the species of *Macropus*, reaching its largest size in the population on Kangaroo Island, South Australia. The population on Flinders Island, South Australia, now extinct, differed considerably from the Kangaroo Island form in having a finer, more graceful build and a short, sleek coat.

The Tammar Wallaby requires dense low vegetation for daytime shelter and open grassy areas for feeding. It inhabits coastal scrub, heath, dry sclerophyll forest and thickets in mallee and woodland. During the day it rests in scrub and, although it begins to move at dusk, it does not leave the scrub until after dark, returning to it before dawn.

On the semi-arid islands inhabited by some populations, fresh water may be unavailable for long periods. This is especially so on the islands of Houtman Abrolhos, Western Australia, where rain falls only in June, July and August. Animals taken from Abrolhos to a laboratory in Perth were found to be able not only to drink sea water but also to maintain body weight and even to suckle young on a diet of dry food while drinking only sea water. On Garden Island, Western Australia, the Tammar has been seen drinking from the sea.

Each individual has a defined home range which overlaps the home ranges of others. While several wallabies may feed in the same

## *Macropus eugenii*

**Size**
**Kangaroo Island***
*Head and Body Length*
590–680 (643) mm (males)
520–630 (586) mm (females)
*Tail Length*
380–450 (411) mm (males)
330–440 (379) mm (females)
*Weight*
6–10 (7.5) kg (males)
4–6 (5.5) kg (females)

*Animals from other populations are smaller, often considerably so.

**Identification** Dark, grizzled grey-brown above, becoming rufous on the sides of the body and on the limbs, especially in males. Pale grey-buff below.
**Recent Synonyms** *Thylogale eugenii, Thylogale flindersi, Protemnodon eugenii.*
**Other Common Names** Tammar, Kangaroo Island Wallaby, Dama Wallaby, Dama Pademelon.
**Status** Common, limited.
**Subspecies** Insufficient is known about the many distinct populations to justify the designation of subspecies. The population from Garden Island, Western Australia, shows some genetically-fixed biochemical differences from the Kangaroo Island population.
**Extralimital Distribution** New Zealand.
**References**
Inns, R.W. (1980). *Ecology of the Tammar,* Macropus eugenii *(Desmarest) in Flinders Chase National Park, Kangaroo Island*. Ph.D. Thesis, Univ. Adelaide.
Poole, W.E., J.T. Wood and N.G. Simms (1991). Distribution of the Tammar, *Macropus eugenii* and the relationships of populations as determined by cranial morphometrics. *Wildl. Res.* 18: 625–639.

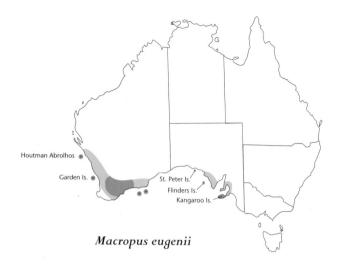

*Macropus eugenii*

Houtman Abrolhos
Garden Is.
St. Peter Is.
Flinders Is.
Kangaroo Is.

pouch for eight to nine months and leaves the pouch in September or October. Females become mature at about nine months while they are still suckling, but males do not become mature until nearly two years old.

The rate of reproduction is high, more than 90 per cent of all females carrying a pouch-young by the end of the breeding season. However, in some years many pouch-young are lost, especially by one-year-old females. In all years mortality is high among juveniles during their first summer and may reach 40 per cent. When the summer has been long, hot and dry, food becomes limited and of poor quality and many

J. LOCHMAN

*Most populations of Tammar Wallabies are on southern offshore islands. It is the small member of the genus* Macropus.

area, no social grouping has been observed except between females and their young at foot.

The breeding cycle is well known only for the Kangaroo Island population. The Tammar Wallaby is one of only two macropodid species which show a strictly seasonal pattern of breeding. Most young are born in late January, very few being born outside January, February and March and (under natural conditions) none from July to December. Within a few hours of giving birth the female mates and the resulting embryo remains quiescent during lactation. However, the quiescent embryo is not reactivated when the young is ready to leave the pouch in September to October. From late June to November the embryo remains quiescent even if the pouch-young is lost and lactation ceases altogether. At this time, females are under the inhibitory influence of day-length which prevents reactivation of the embryo. The embryo formed in September or October in a young female which has mated for the first time also becomes quiescent. Typically, the quiescent embryos of young and older females are reactivated within a few days after December 22 (the summer solstice) and the young are born about 40 days later, 12 months after the mating at which they were conceived. The single young is suckled in the

adults die at the onset of cold, wet weather. While natural predators of the Tammar Wallaby are few, the feral Cat is believed to have made a significant contribution to the extinction of the Flinders Island population. On Kangaroo Island, males may live to at least 11 years and females

A. WILLIAMS

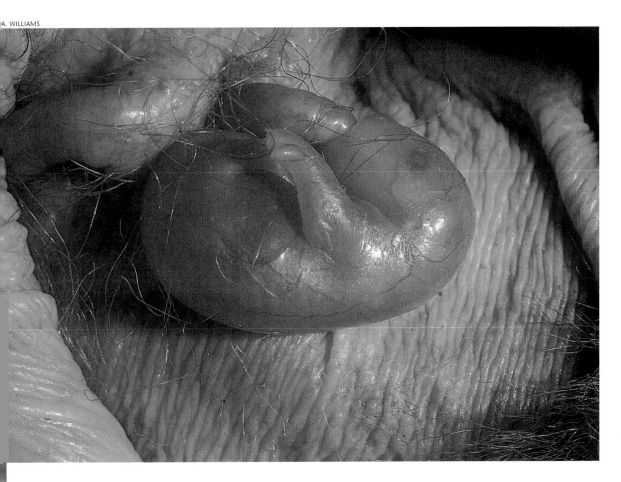

*Reproduction of the Tammar Wallaby has been intensively studied. The illustration is of a very early pouch young.*

to 14 years. A wild female known to be 13 or 14 years old carried a pouch-young.

Although the Tammar Wallaby was probably always restricted to islands and parts of the adjacent mainland of south-western and southern Australia, clearing of scrub has reduced the shelter available for this species and the range of several populations has been reduced since European settlement. It was formerly abundant in some regions of mainland South Australia but, if it survives at all, it is only in a small area near Cleve. Numbers have also been greatly reduced by clearing of the land for wheat growing in south-western Western Australia. The Flinders Island population is almost certainly extinct, its numbers having been severely reduced by loss of habitat, bushfires and predation by the feral Cat. None now survive on St Francis, St Peter, or Thistle Island. Skulls found on Revesby and North Gambier Islands may represent unsuccessful introductions. Introduced populations are thriving on Greenly Island, South Australia and it at least two localities in New Zealand. The species continues to be abundant on Kangaroo Island, despite ongoing persecution outside the Flinders Chase National Park. It is also found on North Twin Peaks Island and Middle Island in the Recherche Archipelago; Garden Island, off Fremantle; and East and West Wallabi Islands, in the Abrolhos.

M.J. SMITH AND L. HINDS

K. GRIFFITHS

# Western Grey Kangaroo

## *Macropus fuliginosus*

(Desmarest, 1817)

fool-idge'-in-oh'-sus: 'sooty long-foot'

*The Western Grey Kangaroo is predominately a grazer, able to eat coarse grasses and also to browse on some selected shrubs.*

While charting the southern coast of Australia in the *Investigator*, Matthew Flinders landed on a large island at the entrance to Spencer Gulf in March 1802. His crew shot many large brown kangaroos which Flinders regarded as similar to the grey kangaroos of the eastern mainland and, in recognition of the supply of fresh meat, he named the place Kangaroo Island. In January 1803, a number of these kangaroos were captured by naturalists from the French research vessel *Géographe* and several were brought alive to the zoological gardens in Paris. One of two specimens which eventually reached the Museum of Natural History in Paris was described in 1817 as the type of *Macropus fuliginosus*, now known as the Western Grey Kangaroo.

Despite a clear account of its origin, this species was mistakenly attributed to Tasmania until 1923 when the Forester Kangaroo of that island was identified as a form of the grey kangaroo of the eastern mainland. Further research demonstrated, in 1971, that the species from Kangaroo Island also exists on the mainland, ranging over the southern part of the continent from the Indian Ocean to western Victoria and central New South Wales. Surveys in 1982 show that its distribution extends into south-western Queensland. Similarities between populations in south-western Australia and on Kangaroo Island in reproduction and in tolerance to fluoroacetate (a poison present in many legumes from south-western Australia but not in those from south-eastern Australia) suggest that the species originated in the south-western part of the mainland. By 1990, morphometric analyses of cranial and

---

### *Macropus fuliginosus*

**Size**
(one year or more old)
*Head and Body Length**
946–2225 mm (males)
971–1746 mm (females)
*Tail* ‡
425–1000 mm (males)
443–815 mm (females)
*Weight*
3.0–53.5 kg (males)
4.5–27.5 kg (females)

\* Measured from nose to tail tip with animal extended so that dorsal surface approximates to a straight line.
‡ Measured on ventral surface. May be subtracted from head to tail measurement to give an approximate measure of the head and body length.

**Identification**  Distinguished from other kangaroos except *Macropus giganteus* by finely haired muzzle. Distinguished from *Macropus giganteus* by brown colour. Males have a strong, characteristic odour.
**Recent Synonyms**  *Macropus major fuliginosus, Macropus major ocydromus, Macropus ocydromus, Macropus melanops.*

M. WRIGHT

**Other Common Names** Black-faced
Kangaroo, Mallee Kangaroo, Sooty Kangaroo,
Stinker.

**Status** Abundant.

**Subspecies** *Macropus fuliginosus fuliginosus*,
Kangaroo Island.

*Macropus fuliginosus ocydromus*, Type-locality,
York, Western Australia.

*Macropus fuliginosus melanops*, Type-locality,
Adelaide district, South Australia.

The two supposed mainland subspecies are not
representative of discrete populations but represent points on a continuous east-west cline.

**References**

Kirsch, J.A.W. and W.E. Poole (1972).
Taxonomy and distribution of the grey kangaroos, *Macropus giganteus* Shaw and *Macropus
fuliginosus* (Desmarest) and their subspecies
(Marsupialia: Macropodidae). *Aust. J. Zool.*
20: 315–339.

Poole, W.E. (1976). Breeding biology and current status of the grey kangaroo, *Macropus
fuliginosus fuliginosus* of Kangaroo Island,
South Australia. *Aust. J. Zool.* 24: 169–187.

Poole, W.E., Carpenter, S.M. and Simms, N.G.
(1990). Subspecific separation in the western
grey kangaroo, *Macropus fuliginosus*: a morphometric study. *Aust. Wild. Res.* 17:
159–168.

body measurements suggested that Western Grey
Kangaroos comprise at least two principal subspecific forms, a uniform group, *M. f. fuliginosus*,
from Kangaroo Island and a composite group on
the mainland that intergrades clinally in its morphological traits along a west–east gradient. *M. f.
melanops* takes precedence as the name for the
mainland complex.

Many aspects of the biology of the Western
Grey Kangaroo are so similar to those of the
Eastern Grey Kangaroo that they need not be described separately. However, the mean lengths of
the oestrous cycle (35 days) and gestation (30.5
days) are a little shorter, the pouch life of 42
weeks is about a fortnight shorter and, in inter-

esting contrast to the Eastern Grey Kangaroo,
the Western Grey does not exhibit embryonic
diapause.

Mixed populations of Eastern and Western
Grey Kangaroos occur on the plains of central
and western New South Wales and in western
Victoria and adjacent south-eastern South Australia. In captivity, western males may produce
hybrids from eastern females (never the reverse)
but no natural hybrids have yet been found in the
field.

In the pastoral area of South Australia where
the Western, but not the Eastern, Grey Kangaroo is found, an aerial survey in the spring of
1978 indicated a population of 295 000 with an
average density of 1.44 kangaroos to the square
kilometre in this small proportion of its range.

A protected species, it is common throughout its distribution and no mainland population is
at risk. Where damage to crops, pastures, or improvements can be demonstrated, State faunal
authorities may issue licences to reduce their
numbers. Permits to export products from
Western Grey Kangaroo are granted by the federal government only to those states with both an
approved species-specific management program
and an associated separate annual application to
take an agreed number of kangaroos.

W.E. POOLE

*Macropus fuliginosus*

# Eastern Grey Kangaroo

## *Macropus giganteus*

### Shaw, 1790

jee'-gahn-tay'-us: 'gigantic long-foot'

M. JONES

*The male Eastern Grey Kangaroo is a powerful animal.*

The first kangaroos collected by Europeans from the east coast of Australia were taken near what is now Cooktown, Queensland, when Cook's expedition spent seven weeks there in 1770. Early descriptions of these kangaroos appear to be composites based on three specimens brought back to Europe and, as convention dictates that the first-used scientific name takes precedence, it has been essential to establish their identity. Unfortunately, little survives of the three animals other than the date of collection and their weights, but diagnostic characters of the skull and teeth, together with the recorded weight, have been used to identify one animal from an original wash drawing of its (lost) skull as an Eastern Wallaroo. Another has been identified from a photograph of a skull (which was subsequently destroyed during World War II) as a Grey Kangaroo. A new specimen specially collected near Cooktown and lodged in the Queensland Museum, is now recognised as the type (neotype) of the latter species.

Grey kangaroos have a wide and almost continuous distribution between the inland plains and the coast where annual rainfall is more than 250 millimetres, in habitats ranging from semi-arid mallee scrub through woodland to forest. Throughout this vast range, grey kangaroos were 'discovered' more than once by various explorers and it is not surprising that taxonomists dealing

*Relations between an infant Eastern Grey Kangaroo and its mother are quite intimate, involving a bond which is not broken until the juvenile is ready to fend for itself.*

with widely separated localities interpreted the considerable range of colour and form to represent as many as five different species. For many years there was much controversy over the status of grey kangaroos but it is now clear that there are only two species; the Eastern and Western Grey Kangaroo.

Distributed throughout most of the eastern States, including Tasmania, the Eastern Grey Kangaroo is predominantly a grazing animal with specific food preferences, restricted to grasses and forbs, which preference is retained even when the number of available plant species is depleted during drought. Its favoured food is grass and, on this diet, it has a lower nitrogen requirement and intake of dry matter than sheep of equivalent weight.

M. JONES

During the hours of daylight, the Eastern Grey Kangaroo usually rests in the shade or shelter of trees and shrubs, moving out to graze from late afternoon to early morning, when animals tend to aggregate in more open country.

Competition between large adult males to mate with receptive females may involve physical struggle. M. JONES

Communications between mother and young and between males and oestrous females involves a series of clucking sounds. Aggressive males and alarmed individuals of both sexes give vent to a guttural cough.

Breeding occurs throughout the year but there is a peak of births in summer. Males, which are notably larger than females, display an increasing interest in a female by remaining near and making occasional close inspections for about four days as she approaches oestrus. Competition between males for a female's attention may occasionally lead to fights. Courtship involves the male sniffing the region of the female's cloaca and pouch opening, together with frequent pawing at her head and clutching at the base of her tail with the forepaws. Attempts to mount are made over a period of several days before successful copulation occurs. This may last up to 50 minutes.

At parturition, the female adopts a crouching posture with the tail behind in its normal position and hindlegs thrust forward with toes in the air, the weight of the body apparently being taken on the heels. The newborn animal climbs from the urogenital opening to the pouch in a few minutes

### *Macropus giganteus*

**Size**

(one year or more old)

*Head and Body Length**

972–2302 mm (males)

958–1857 mm (females)

*Tail ‡*

430–1090 mm (males)

446–842 mm (females)

*Weight*

4–66 kg (males)

3.5–32 kg (females)

\* Nose to tail tip with animal extended so that dorsal surface approximates to a straight line.

‡ Measured on ventral surface. May be subtracted from head to tail measurement to give an approximate measure of the head and body length.

**Identification** Distinguished from other kangaroos in having a hairy muzzle, with fine hairs in the area between the nostrils and upper lip. Differs from *Macropus fuliginosus* in having grey fur.

**Recent Synonyms** *Macropus canguru, Macropus major*.

Following long disputes over the specific name of the grey kangaroos, the International Committee of Zoological Nomenclature suppressed the name *canguru* in 1966. The name *major* was reserved for a possible subspecies from the Sydney area.

and shortly afterwards attaches to one of the four teats. Twin young have been recorded, but usually only one, weighing just over 800 milligrams is produced. The oestrous cycle has an average length of about 46 days, considerably longer than the 36 days required from gestation. Females do not normally come into oestrus immediately after giving birth, as in the Red Kangaroo and some other macropods, but return to oestrus about 11 days after loss of pouch-young. Perhaps in response to a sudden improvement in nutrition, a female carrying a pouch-young more than four months old may occasionally mate again. The fertilised egg from this mating develops to the stage of a blastocyst but ceases further development until the incumbent young vacates the pouch at the age of about 11 months. A young animal begins to leave the pouch for short periods at the age of about nine months but continues to be suckled, from the same teat which it used

when in the pouch, until it is about 18 months old. At this stage, the mother may also have a pouch-young about eight months old.

In areas of low and variable rainfall, populations of kangaroos may vary considerably in response to the availability of food and water. Aerial surveys in 1975–76 indicated that the plains of New South Wales supported between 1 500 000 and 1 600 000 Eastern and Western Grey Kangaroos but further counts in 1977 indicated that, over the intervening two years, the population had increased at an annual rate of 13 per cent despite a legal annual harvesting rate of 4 per cent. Between 1981 to 1984 declining populations—due to drought in the eastern Australian pastoral zones—have been most marked in New South Wales (72 per cent in Western Greys and 54 per cent for Eastern Greys), but average to good rainfall in that State from 1984 to 1987 resulted in pronounced increases in numbers of kangaroos (38 per cent for Western Greys and 90 per cent for Eastern Greys). Droughts with consequent mortality and reduced fecundity are intrinsic to the ecology of the kangaroos. Dry periods may occur locally almost every year and regionally about every ten years, counterbalancing to some extent the rapid increase in numbers between droughts.

For many thousand years, Aborigines killed kangaroos for food and skins and, although they may have been the only significant predator on these animals, they could not have had much effect upon the large populations. Early European settlers valued kangaroos as a source of meat and hides but, as land was cleared, native pastures were changed by the impact of domestic stock and improved by human intervention and, with

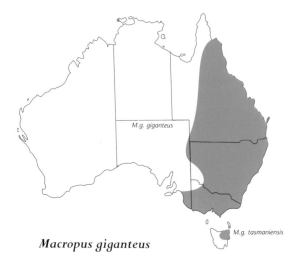

*Macropus giganteus*

the provision of many more watering points, these large, native, grazing animals came to be regarded as agricultural pests. It is indeed likely that development of the pastoral industry led to a marked increase in kangaroo populations. Local reductions in numbers were achieved by a variety of methods: initially bounties were paid on their scalps, then skins were taken for leather manufacture and later both skins and meat were used.

Both species of grey kangaroos are protected by law and are extremely common throughout their extensive range on the mainland. In Tasmania, the Eastern Grey Kangaroo, with an estimated population of less than 20 000, remains vulnerable. Despite active management of the kangaroo within the State's national parks, most of its range is on private land, where populations are subject to pressure from habitat destruction and poaching. Numbers have shown a disturbing decline in recent years. Management of the Eastern Grey Kangaroo is under the control of State faunal authorities which may grant licenses to shoot excess animals where these are deemed to be causing damage to fences, crops, or pastures. Export of kangaroo products taken under such licenses is subject to the approval of federal government authorities.

W.E. POOLE

**Other Common Names** Great Grey Kangaroo, Forester, Scrub Kangaroo, Scrubber.
**Status** Abundant.
**Subspecies** *Macropus giganteus giganteus*, eastern Australia.
*Macropus giganteus tasmaniensis*, Tasmania; may be justified by differences in skull and pelage.
*Macropus giganteus major*, supposedly from Sydney region (has not been demonstrated to be distinct).

**References**

Kirsch, J.A.W. and W.E. Poole (1972). Taxonomy and distribution of the grey kangaroos, *Macropus giganteus* Shaw and *Macropus fuliginosus* (Desmarest) and their subspecies (Marsupialia: Macropodidae). *Aust. J. Zool.* 20: 315–339.

Poole, W.E. (1982). *Macropus giganteus*. *Mamm. Species no. 187*. American Society of Mammalogists.

Fletcher, M, C.J. Southwell, N.W. Sheppard, G. Caughley, D. Grice, G.C. Grigg and L.A. Beard (1990). Kangaroo population trends in the Australian rangelands, 1980–87. *Search*. 21: 28–29.

# Toolache Wallaby

## *Macropus greyi*

### Waterhouse, 1845

gray'-ee: 'Gray's long-foot'

*Toolache Wallaby. Illustration in John Gould's* The Mammals of Australia.

The elegance of its slim, graceful body, and swiftness and erratic gait of the Toolache Wallaby left a lasting impression on those who had seen one. It was an open-country wallaby which lived at the edge of Stringy-bark heath in the grassy habitat developed on land too wet in winter to carry eucalypt cover. These areas are flat or at most gently undulating, becoming swampy in winter and the depressions are filled by matted growth of Black Rush among tussocks of tall coarse Cutting Grass and Kangaroo Grass. A strip of grassland sometimes 50 metres wide between the edge of the heath and the beginning of the tussocks provided an ideal feeding area. Little islands of slightly higher ground carrying sparse stands of She-oak, *Allocasuarina verticillata*, were used for shelter during the day. The heath-land includes frequent clumps of dwarf grass-trees and banksias occurring singly or in clumps, but these were not used as shelter.

The grassland habitat between tussock and heath provided a background for the display of the Toolache Wallaby's striking characteristics of high speed and unusual gait. It was by far the swiftest of the native animals in its region: on grassland it could outrun all but the best dogs and, on reaching the tussock or low scrub, was adroit at changing direction or length of stride among the frequent obstacles while maintaining a high speed for long periods. Because of its erratic

*Macropus greyi*

---

*Macropus greyi*

**Size**
*Head and Body Length*
810 mm (males)
840 mm (females)
*Tail*
730 mm (males)
710 mm (females)

**Identification** Pale greyish-fawn above, fawn below. Tail almost white and with crest on distal third. Dark cheek-stripe from muzzle to ear; back banded with 10–12 light grey bars.
**Recent Synonyms** None.
**Other Common Names** Monkeyface, Onetwo, Grey's Wallaby.
**Status** Presumed extinct.
**Subspecies** None.
**References**
Finlayson, H.H. (1927). Observations on the South Australian members of the subgenus 'Wallabia'. *Trans. Roy. Soc. S. Aust.* 51: 363–377.
Horton, D.R. and P. Murray (1980). The extinct Toolache Wallaby (*Macropus greyi*) from a spring mound in north-western Tasmania. *Rec. Queen Victoria Mus.* Launceston 71: 1–12.

progression, its gait was sometimes interpreted as being of 'two short hops and then a longer one'.

On poorer country it appears to have been a solitary animal but where grassland was developed it was gregarious, grazing and resting in groups. Each group showed a strong attachment to its own area and returned to it, even after repeated disturbance.

The Wedgetailed Eagle was a predator of the young, as evidenced by observed attacks and the presence of skulls in the bones accumulated under their nests. The Fox also preyed on young animals but its effect on populations was probably slight compared with the effect of hunting by man. Its presence in open country and its qualities of speed and endurance made it an ideal test of the prowess of the local residents' hunting dogs, while its exceptionally beautiful skin commanded the attention of shooters. Although once occurring abundantly within its preferred habitat, its total population apparently had declined to a few individuals by the early 1920s and the last positive record of wild individuals was obtained in 1924. An intensive survey conducted by the South Australian National Parks and Wildlife Service in 1975 and 1976 found that the Toolache Wallaby had been seen by reliable naturalists as recently as 1972 but no specimens were obtained.

M.J. SMITH

# Western Brush Wallaby

## *Macropus irma*

(Jourdan, 1837)

erm'-ah: 'Irma's(?) long-foot'

*Macropus irma*

In some respects the Western Brush Wallaby resembles larger kangaroos in its habits. It is a grazer rather than a browser and has an optimum habitat of open forest or woodland, particularly favouring rather open, seasonally wet flats with low grasses and open, scrubby thickets. It is found in some areas of mallee and heathland, is uncommon in wet sclerophyll forest and is absent from the true Karri forests which have dense undergrowth.

*The Western Brush Wallaby lives in the open forest or woodland, usually grazing in open areas.*

Although still relatively common over much of its range, it has not been the subject of detailed study. Little is known of its food preferences but it appears to be able to manage without free water. Activity is greatest in the early morning and late afternoon and it rests during the hotter part of the day, singly or in pairs in the shade of a bush or in small thickets. It is more diurnal in its habits than other macropods in the region. Clearly adapted for life on open ground, it is a speedy animal, able to weave or sidestep

BABS & BERT WELLS

### Macropus irma

**Size**

*Head and Body Length*
(1200) mm
*Tail Length*
540–970 (720) mm
*Weight*
7–9 (8) kg

**Identification** Pale grey with distinct white facial stripe, black and white ears, black hands and feet. Long tail with crest of black hair, particularly towards extremity. Moves fast with head low and tail extended.
**Recent Synonyms** None.
**Other Common Names** Black-gloved Wallaby.
**Status** Possibly declining.
**Subspecies** None.
**Reference**
Gould J. (1852). *The Mammals of Australia, Vol. 2.* The author, London. Plates 20, 21.

with ease as it moves low to the ground with its long tail extended.

Males and females are similar in size and appearance. The breeding season has not been definitely established but young appear to be born in April or May, emerging from the pouch in October and November.

In 1973 and 1974 a dramatic increase in the number of Foxes in south-western Australia appears to have led to a decline in populations of the Western Brush Wallaby. It is thought that juveniles not long out of the pouch may fall victim to this predator. Surveys of the Western Brush Wallaby and Western Grey Kangaroo were conducted in south-eastern forest areas in October of 1970 and 1990. These showed that, while the kangaroo population remained constant, the wallabies decreased by about 80 per cent. The Western Brush Wallaby is now uncommon throughout its range but, like the Brush-tailed Bettong, its numbers increase in response to Fox-baiting.

P. Christensen

# Parma Wallaby

## *Macropus parma*

### Waterhouse, 1845

par'-mah: 'parma (an Aboriginal name for this species) long-foot'

Pine plantations being expanded during the 1960s on Kawau Island, off Auckland, New Zealand, suffered considerable damage from wallabies, the original stock of which had been brought from Australia almost a century previously. Foresters were doing their best to eliminate or control these exotic pests until 1965, when it was discovered that the population consisted of two species: the Tammar Wallaby, fairly widespread in Australia; and the Parma Wallaby, assumed to have become extinct during the nineteenth century. A stay of execution was granted while some of the latter were exported to zoos and research institutions in Australia and other parts of the world to establish breeding colonies from which, it was hoped, the species might perhaps be returned to the wet forests of southern New South Wales where it had first been seen by John Gould around 1840.

*Macropus parma*

K. ATKINSON

*Once thought to be extinct, the Parma Wallaby is now
known to exist in many parts of the Great Dividing
Range. The female is illustrated sitting on its tail in the
birth position.*

It was not long, however, before this enthusi-
asm was found to be misdirected. A live female
was taken near Gosford, New South Wales, in
1967 and subsequent surveys demonstrated that it
still occurs in New South Wales, at least from the
Watagan Mountains north to the Gibraltar
Range—and that numbers appear to be increasing.

The Parma Wallaby lives in wet and dry
forests and occasionally in rainforests but its optimum habitat appears to be wet sclerophyll forest
with a thick, shrubby understorey associated with
grassy patches. Primarily nocturnal, it takes
cover among the shrubs during the day and
emerges at dusk, or shortly before, to feed on
grasses and herbs. When hopping, it remains
close to the ground in an almost horizontal position with its forearms tucked tightly against its
body. At medium pace, the tail is curved upwards
in a shallow U-shape.

The male is larger than the female and has a
more robust chest and forelimbs. On Kawau
Island, there is a sharply defined breeding season
from March to July, controlled by the nutritional
state of the female. Available evidence from wild
populations in Australia suggests that most births
occur between February and June. In Australia,
females become sexually mature when one year
old but on Kawau Island, maturity is not attained
until two, occasionally three, years of age. Males
appear to reach maturity at 20–24 months of age
in both regions.

The single young is born after a gestation of
about 35 days. Its first excursions from the pouch
are made at the age of 23–25 weeks and it quits
the pouch when about 30 weeks old and about
750 grams in weight. Weaning is not completed
for another 10–14 weeks, during which time the
female may give birth to a second young.

The Parma Wallaby is normally solitary
under natural conditions although feeding aggregations of two (rarely three) are sometimes
observed in Australia. Larger aggregations occur
on Kawau Island where the population density is

## Macropus parma

**Size**

**Australian populations**

*Head and Body Length*

482–528 mm (males)

447–527 mm (females)

*Tail Length*

489–544 mm (males)

405–507 mm (females)

*Weight*

4.1–5.9 kg (males)

3.2–4.8 kg (females)

**Identification** Uniform greyish-brown back and shoulders, white throat and chest; dark stripe along spine ending mid-back, white stripe on upper cheek. About 50 per cent of animals with white tip to tail. Tail about same length as head plus body. Distinctive faecal pellets are flattened, square to slightly rectangular.

**Recent Synonyms** *Thylogale parma, Wallabia parma.*

**Other Common Names** White-throated Pademelon, White-throated Wallaby.

**Status** Rare, scattered.

**Subspecies** None.

**References**

Maynes, G.M. (1979). Distribution and aspects of the biology of the Parma Wallaby, *Macropus parma*, in New South Wales. *Aust. J. Wildl. Res.* 4: 109–205.

Wodzicki, K. and J.E.C. Flux (1967). Rediscovery of the White-throated Wallaby, *Macropus parma* Waterhouse 1846, on Kawau Island, New Zealand. *Aust. J. Sci.* 29: 429–430.

much higher. It appears that the effects of nutritional stress and crowding on Kawau Island have exercised some selective pressures: females are smaller than those in Australia and their captive-reared female progeny remain small even when uncrowded and well fed.

G. MAYNES

A. DENNIS

# Whiptail Wallaby

## *Macropus parryi*

Bennet, 1835

pa'-ree-ee: 'Parry's long-foot'

Once considered a contender for the title of Captain Cook's Kangaroo (see page 336), the Whiptail Wallaby has a distribution which barely reaches Cooktown, Queensland, its densest populations being in southern Queensland and

C. ANDREW HENLEY/LARUS

Left: *Quite appropriately, the Whiptail Wallaby was known in some regions as the 'Pretty-faced Wallaby'.*

Above: *Female Whiptail Wallaby with young-in-pouch and at foot.*

northern New South Wales. It is an inhabitant of undulating or hilly country with open forest and a grass understorey.

It is a grazer, feeding primarily on grasses and other herbaceous plants, including ferns. Except during drought, it seldom drinks, apparently getting sufficient moisture under normal circumstances from its food and from dew. It is unusually diurnal for a macropod. Feeding activity is at a maximum at dawn and continues into the early morning with increased periods of rest during the middle of the day. Feeding is resumed in the late afternoon and continues into the night.

*Macropus parryi*

345

## Macropus parryi

**Size**

*Head and Body Length*
up to 924 mm (males)
up to 755 mm (females)
*Tail Length*
861–1045 (941) mm (males)
728–858 (781) mm (females)
*Weight*
14–26 (16) kg (males)
7–15 (11) kg (females)

**Identification** Light grey (winter), brownish-grey (summer) above; white below. Forehead and base of ears dark brown. White stripe on upper lip; light brown stripe down neck to shoulder; white hip-stripe. Tail long, slender, with dark tip.
**Recent Synonyms** *Wallabia elegans*.
**Other Common Names** Pretty-face Wallaby, Grey-faced Wallaby, Grey or Blue Flier.
**Status** Common.
**Subspecies** None.
**References**

Kaufmann, J.H. (1974). Social ethology of the whiptail wallaby, *Macropus parryi* in north-eastern New South Wales. *Anim. Behav.* 22: 281–369.

Maynes, G.M. (1973). Aspects of reproduction in the Whiptail Wallaby, *Macropus parryi*. *Aust. Zool.* 18: 43–46.

During the hotter part of the day, the Whiptail Wallaby seeks patches of shade and, when the air temperature rises above 30°C, animals may lick their forearms to increase evaporative cooling.

A social species, it lives in groups of up to 50 individuals, comprised of subgroups of ten or less which typically include adults and subadults of both sexes and young at foot.

A subgroup occupies a home range which may partially or completely overlap those of other subgroups. Vocal communication includes a soft cough, indicating fear or submission; a sound intermediate between a hiss and a growl which is used by females as a defensive threat; and a soft clucking made by courting males. When alarmed, an individual thumps the ground with its hindfeet and all animals in its group hop away, taking a zig-zag course which is probably confusing to a predator. Under natural conditions, females do not breed until they are older than 18–24 months and males do not have the opportunity to mate before the age of two to three years. Courtship of a female in oestrus is characterised by the dominant male and a number of subordinate males following the female. A dominant male keeps others at a distance by chasing and by a ritualised gesture of aggression which involves pulling up clumps of grass with the forepaws while directing its head towards the threatened individual. Copulation occurs on only one day per oestrous cycle of 41–44 days.

Gestation takes 34–38 days, the young being born while the female sits on the base of its tail with its back supported by a firm object and the tail extended between the outstretched hindlegs. The young becomes detached from the teat when about 23 weeks old, vacates the pouch at about 37 weeks but continues suckling until about 15 months old. Throughout this period, the mother grooms the young, decreasing as it approaches independence.

Towards the end of the pouch life of its young, the female comes into oestrus and mates again, producing a quiescent blactocyst which does not develop further until the pouch is vacated.

Being a grazer, the Whiptail Wallaby probably benefited from the early agricultural practice of ringbarking, which permitted more grass to grow in forests while retaining adequate shelter. Total clearing of forest has had a detrimental effect on the species in some areas but it is sufficiently widespread and represented in national parks and reserves to be in no danger.

P.M. JOHNSON

# Common Wallaroo

## *Macropus robustus*

### Gould, 1841

roh-bus'-tus: 'robust long-foot'

E. BEATON/TERRA AUSTRALIS

*The Common Wallaroo is a large, rather shaggy-haired Kangaroo.*

Discovery of a wash drawing of a skull among the papers of Sir Joseph Banks demonstrated that a Common Wallaroo was one of the three macropods collected in 1770 near the Endeavour River, Queensland by Cook's expedition. The species was subsequently described by Gould from two animals collected in 1839 in the Liverpool Range, New South Wales.

Its habitat is varied but usually features steep escarpments, rocky hills or stony rises; areas where caves, overhanging rocks and ledges pro-

---

### *Macropus robustus*

**Size**

*Head to tail*
1138–1986 mm (males)
1107–1580 mm (females)
Measured from nose to tail tip with animal extended so that dorsal surface approximates a straight line.

*Tail*
551–901 mm (males)
534–749 mm (females)
Measured on ventral surface. May be subtracted from head to tail measurement to give an approximate measure of the head and body length.

*Weight*
7.25–46.5 kg (males)
6.25–25 kg (females)

**Identification** Dark grey to reddish above, paler below. Fur shaggy, coarse, varying from long to short among the subspecies. Bare black rhinarium, as in other wallaroos.

---

vide shelter and relief from extreme heat in areas experiencing prolonged periods of high temperature. It leaves this shelter in the cool of the evening to graze, primarily upon grasses and shrubs, usually within a limited home range which may include lower slopes and surrounding plains. Well adapted to aridity, it can maintain itself and even breed successfully on pastures of low protein content and can survive without frequent access to free water if it has access to refuges from solar radiation and to food plants of sufficient water content.

In common with the Antilopine and Black Wallaroos, to which species it is closely related, the Common Wallaroo has a large, bare, black rhinarium. It also shares with other wallaroos a distinctive stance; shoulders thrown back, elbows tucked into the sides and wrists raised. When disturbed, it utters a loud hiss, in association with exhalation. It also has a characteristic 'cch-cch' vocalisation.

The Common Wallaroo tends to spend most of the day in rocky areas on hillsides, moving into open grasslands (and crops) in the late afternoon or evening.

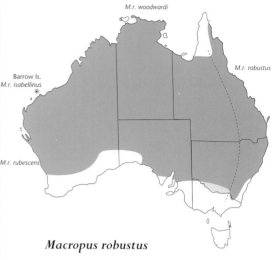

*Macropus robustus*

There is marked sexual dimorphism: mature males attain twice the weight of mature females and usually are darker in colour. The two best known subspecies differ so markedly that they have distinct common names. *Macropus robustus robustus*, which inhabits the eastern and western slopes of the Great Dividing Range, has shaggy, dark grey fur and is known as the Wallaroo or Eastern Wallaroo. The reddish and shorter-haired *Macropus robustus erubescens*, which occupies most of the continent westward of this region, is known as the Euro. Two smaller subspecies have more limited distributions.

Sexual maturity is attained at 18–24 months. Breeding may occur throughout the year but reproduction is reduced in times of drought and may cease if drought is prolonged. There is a post-partum oestrus and embryonic diapause. Females of the Eastern Wallaroo and Euro differ somewhat in their reproductive patterns. It has been established that the oestrous cycle of the former has a mean length of 34 days but data from female Euros indicate a mean of about 45 days in this subspecies. In the Eastern Wallaroo,

**Recent Synonyms** *Osphranter robustus*, *Osphranter erubescens*, *Osphranter isabellinus*, *Osphranter reginae*, *Osphranter antilopinus woodwardi*, *Osphranter antilopinus cervinus*.

**Other Common Names** Euro, Eastern Grey Wallaroo, Red Wallaroo, Roan Wallaroo, Barrow Island Wallaroo, Hill Kangaroo. Indigenous name:  Biggada.

**Status** Abundant.

**Subspecies** *Macropus robustus robustus*, eastern non-arid, temperate to tropical mainland.
*Macropus robustus erubescens*, remainder of mainland except extreme northern and southern parts.
*Macropus robustus woodwardi*, Kimberley, Western Australia; north-western Northern Territory.
*Macropus robustus isabellinus*, Barrow Island, Western Australia.

**References**

Ealey, E.H.M. (1967). Ecology of the euro, *Macropus robustus* (Gould) in north-western Australia. *CSIRO Wildl. Res.* 12: 9–80.

Poole, W.E. and J.C. Merchant (1987). Reproduction in captive wallaroos: the eastern wallaroo, *Macropus robustus robustus*, the euro, *M. r. erubescens* and the antilopine wallaroo, *M. anthlopinus*. *Aust. Wild. Res.* 14: 225–242.

J. LOCHMAN

Above: *Euros at a waterhole.*
Right: *Like other wallaroos, the Common Wallaroo has a bare, black rhinarium (muzzle).*

the mean length of gestation is 34 days and pouch life is about 261 days; the respective periods are about 34 and 244 days in the Euro.

The Common Wallaroo is essentially solitary. No long-term studies of population dynamics have been undertaken but it appears that the density of populations is governed both by the amount of available shelter and by proximity to food and water. Seasonal conditions, competition from other grazing animals and human predation may reduce populations to less than the apparent carrying capacity of an area. Densities as high as 13 per square kilometre have been recorded and, following a population crash, as low as 0.04 per square kilometre.

Although the Common Wallaroo is a protected species, permits may be granted by State wildlife authorities to reduce numbers in prescribed areas.

W.E. POOLE

R. & A. WILLIAMS

# Red-necked Wallaby

## *Macropus rufogriseus*

### (Desmarest, 1817)

rue'-foh-griz-ay'-us: 'red-grey long-foot'

This is the common large wallaby of the forests of south-eastern Australia, including Tasmania and the larger islands around the Tasmanian coast and in Bass Strait. The original picture of its distribution on these islands is somewhat confused because of extinctions and introductions in historical times. It is found in the eucalypt forests where there is at least a moderate shrub stratum with open areas nearby and also in tall coastal heath communities. It is essentially a grazing animal and subsists largely on grasses and herbs. Forest clearing has undoubtedly reduced its numbers in some places, but its present geographic distribution is probably little different to what it was at the beginning of European settlement. Indeed, partial clearing for stock grazing, which creates more forest edge and pasture, appears to have had a beneficial effect on its numbers. It persists in many areas which are largely cleared provided that some patches of forest are left on hill tops or along creeks to serve as shelter. It is common over most of its range and especially abundant in Queensland, north-eastern New South Wales and Tasmania.

The Red-necked Wallaby is essentially solitary but it may occur in such high density that 30 or more may be seen grazing together at night on a few hectares of lucerne. However, when such a group is disturbed it does not behave as a unit; individuals or pairs may run off in any direction. During the day the Red-necked Wallaby rests among dense patches of shrubs inside the forest. Only individuals or females with large young at foot (occasionally accompanied by an adult male) are seen together in these shelters. Individuals emerge from shelter to graze in the late afternoon. They may be abroad much earlier on wet or dull days, especially in the cooler months. Home ranges

*Macropus rufogriseus*

**Size**
**Mainland**
*Head and Body Length*
770–888 (823) mm (males)
708–837 (772) mm (females)
*Tail Length*
703–876 (797) mm (males)
664–790 (720) mm (females)
*Weight*
15–23.7 (18.6) kg (males)
12–15.5 (13.8) kg (females)
**Tasmania**
*Head and Body Length*
712–923 (782) mm (males)
659–741 (706) mm (females)
*Tail Length*
691–862 (768) mm (males)
623–778 (695) mm (females)
*Weight*
15–26.8 (19.7) kg (males)
11–15.5 (14) kg (females)

**Identification** Mainland: Male grizzled medium grey to reddish above, with pronounced reddish-brown neck, white or pale grey below; female somewhat paler. In both sexes, muzzle, paws and largest toe black; white stripe on upper lip. Tasmania and Bass Strait islands: Darker grey; more brownish neck. Distinguished from the 13 other macropodids that are found in some part of the range of this wallaby by size or colouration. Similar to Black-striped Wallaby in Queensland and northern New South Wales but lacks pronounced, narrow, dark dorsal stripe from the neck to centre of back and white hip-stripe.
**Recent Synonyms** *Wallabia rufogrisea*.

are relatively small and stable throughout the year, those of males being larger than those of females.

The species is protected by law in all States in which it occurs, but it may be killed under licence as a pest of crops or pastures in Queensland and Tasmania, or during open seasons in Tasmania. A small number may also be killed as crop pests in

D. WATTS

D. WATTS

Above: *Female Red-necked Wallaby indicating disinterest in an almost independent, non-suckling juvenile.* Left: *Female Red-necked Wallaby with large pouch-young.*

**Other Common Names** Bennett's Wallaby (Tasmania), Brush Wallaby, Eastern Brush Wallaby, Brush Kangaroo, Brusher, Red Wallaby (mainland).

**Status** Common to abundant in most parts of its range.

**Subspecies** *Macropus rufogriseus rufogriseus*, Tasmania and Bass Strait islands. (Until recently the name *M. r. rufogriseus* was restricted to animals from King Island or the Bass Strait islands and the Tasmanian form was named *M. r. fruticus*).

*Macropus rufogriseus banksianus*, Australian mainland.

**References**

Merchant, J.C. and J.H. Calaby (1981). Reproductive biology of the red-necked wallaby (*Macropus rufogriseus banksianus*) and Bennett's wallaby (*M. r. rufogriseus*) in captivity. *J. Zool. Lond.* 194: 203–217.

Johnson, C. (1987). Macropod studies at Wallaby Creek. IV. Home range and movements of the red-necked wallaby. *Aust. Wildl. Res.* 14: 125–132.

New South Wales if the landholder can demonstrate that there is a pest problem. Most of those killed are not utilised, but there is a small export trade in skins from Queensland and Tasmania: the Tasmanian subspecies has longer and denser fur and is more attractive to skin buyers.

The Red-necked Wallaby is of great interest to students of reproduction because the breeding patterns of its two subspecies are quite different, a situation unique among kangaroos. On the mainland, females give birth in all months, with perhaps a slight increase in relative number of births during summer. In Tasmania, however,

D. WATTS

there is a well-defined breeding season, with births occurring between late January and July. The great majority of young are born in February and March. The rate of reproduction is high and virtually all females of breeding age carry pouch-young. The only other wallaby having the breeding pattern of Tasmanian animals is the Tammar of Kangaroo Island, South Australia. The different breeding patterns of the two subspecies of the Red-necked Wallaby are not affected by captivity.

*The range of the Red-necked Wallaby extends into the colder parts of Tasmania.*

The length of the oestrous cycle is approximately 33 days and the gestation period about 30 days. There is a post partum oestrus and embryonic diapause. In the Tasmanian form, females without pouch-young that mate at the end of the breeding season, do not give birth until the following breeding season, up to eight months later. In females whose young survive their full pouch life, the interval from post-partum mating to the birth of the resulting young may be as long as 12 months. In the mainland form in all months and in the Tasmanian subspecies during the breeding season, a new young resulting from a post-partum mating is not born until 16–29 days after permanent emergence of the large pouch-young.

The pouch life is about 280 days duration although the young continue to be suckled until 12–17 months old. In captivity, females begin to breed at about 14 months of age and males are capable of breeding about the age of 19 months.

J.H. Calaby

M.r. banksianus

King Is.

Furneaux Group

M.r. rufogriseus

*Macropus rufogriseus*

# Red Kangaroo

## *Macropus rufus*

(Desmarest, 1842)

rue'-fus: 'red long-foot'

*Despite its adaptation to arid conditions the Red Kangaroo must drink. Its distribution and numbers have increased in response to the establishment of stock watering points by graziers.* C. ANDREW HENLEY/LARUS

The only kangaroo truly characteristic of the arid zone, this species is found mainly in the better-watered plains country and low open woodlands, but subsists sparsely in the deserts. Hills and ranges are the preserve of the Euro, *M. robustus*, which is not restricted to the inland.

Surprisingly for such a large and arid-adapted herbivore, it depends upon green herbage, the abundance of which determines the proportion of females that breed and the survival of their young. As the supply of green herbage dwindles

during droughts, breeding diminishes. In Central Australia, half the mothers cease to breed after three months of summer without rain (after five months in winter). Survival of pouch-young is even more sensitive to drought, for half die (in the pouch) after only two to three months of drought. Maturity in both sexes is usually reached at two to three years but females can commence breeding up to six months earlier in good years.

Why is there such sensitivity to drought? With pregnancy lasting only 33 days and young weighing only about 2 grams, the metabolic cost of producing a neonate is low; the cost of maintaining a growing young for eight months in a drought is considerably higher. It is therefore a better strategy to tolerate frequent pregnancies

C. ANDREW HENLEY/LARUS

Left: *Female Red Kangaroo with advanced pouch-young.*

Right: *The Red Kangaroo is social, moving in groups led by a dominant male.*

with early losses and replacement of young, than to function, for example, like ruminant cattle. In the Red Kangaroo, it is the breeders which survive and continue to produce new young into a drought. In cattle, suckling is such a load that both mother and calf may perish. Even if the mother survives the load of lactation, it may fail to breed for a year or more afterwards.

The Red Kangaroo has another reproductive adaptation to long droughts. After the final pregnancy, females fail to prepare for the usual oestrus following birth: they become anoestrous, but continue to suckle. The physiological mechanisms for this apparent paradox are unknown. About 20 per cent of young that survive a drought are suckled by anoestrous females which tend to be the older, better-conditioned mothers. With the breaking of the drought, full breeding potential is regained within days of green herbage sprouting. Both the lactating and non-lactating anoestrous females mate.

Not all males remain reproductive during a drought. Some fail to produce sperm, the proportion doing so being related to duration of bouts of hot weather. Resumption of spermatogenesis is slow, so some oestrous females fail to be impregnated after droughts break.

Males eventually grow to about twice the mass of females but, as sizes diverge, there are fewer old males: this is probably a natural effect of continual growth in a difficult environment. Legal harvesting for hides and/or meat today targets larger kangaroos and also reduces the proportion of older males. Such harvesting is possible and regulated in the relatively abundant populations of

## Macropus rufus

**Size**

*Head and Body Length*
935–1400 (1150) mm (males)
745–1100 (1000) mm (females)
*Tail Length*
710–1000 (880) mm (males)
645–900 (820) mm (females)
*Weight*
22–85 (66) kg (males)
17–35 (26.5) kg (females)

**Identification** Red to blue-grey above, distinctly white below. Adults distinguished from other species of kangaroos by white underparts, black and white patch at side of muzzle and broad white stripe from corner of mouth to base of ear. Rhinarium naked, sharply delineated, dusky.

**Recent Synonyms** *Megaleia rufa.*

**Other Common Names** Plains Kangaroo, Blue-flier (female).

Indigenous name: Marloo.

**Status** Abundant.

**Subspecies** Several subspecies have been described but their validity is not well established.

**References**

Caughley, G., N. Shepherd and I. Short (1987). *Kangaroos and their Ecology and Management in the Sheep Rangelands of Australia.* Cambridge Univ. Press.

Corbett, L. and A. Newsome (1987). The feeding ecology of the dingo. III. Dietary relationships with widely hypothesis of alternation of predation. *Oecologia* 74: 215–227.

Grigg, G., P. Jarman and I. Hume (eds) (1989). *Kangaroos, Wallabies and Rat-kangaroos,* 2 vols. Surrey Beatty and Sons, Sydney.

inland Queensland, New South Wales, South Australia and Western Australia. Numbers were probably much lower before European settlement of inland Australia. The scarcity of Red Kangaroos beyond the Dingo Barrier Fence is associated, however, with poorer habitat and increasing aridity. Dingo populations increase when the Rabbit is abundant but the Red Kangaroo is preyed upon when Rabbit populations crash in drought. At such times, at least, the Dingo probably regulates Red Kangaroo populations (unless it is itself regulated by human endeavours, as in sheeplands). High

*The Red Kangaroo, largest of the macropods, inhabits quite arid country.*

populations of the Red Kangaroo in semi-arid areas of inland Australia are in part due to eradication of the Dingo.

It is difficult to estimate the extent of competition for food between the Red Kangaroos and sheep and cattle. Although normal diets are sufficiently different for it to be suggested that both the native and introduced herbivores could be run together profitably, there is also strong evidence of competition in dry times.

A. Newsome

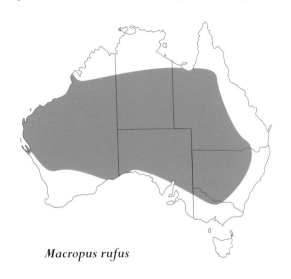

**Macropus rufus**

## NAILTAIL WALLABIES

### *Onychogalea*

The three small and distinctively marked species of the genus *Onychogalea* have a small, horny spur at the end of the tail, the function of which is quite unknown. Additionally they are characterised by very slender upper incisors which are inclined forward and decrease evenly in size from the central pair backwards.

The claws of the forefoot, which are long and strongly developed, are used in excavating a scrape under vegetation in which they rest during the day. When moving at speed, a nailtail wallaby extends its forelimbs outwards and downwards from the axis of the body. The apparently circular motion described by these limbs is responsible for the old vernacular name, 'organ-grinder'.

Nailtail wallabies have fared very badly since European occupation. One is extinct and one is now rare; the third is common locally.

# Bridled Nailtail Wallaby

## *Onychogalea fraenata*

### (Gould, 1841)

on'-i-koh-gah -lay'-ah freen-ah'-tah: 'bridled nailed-weasel'

In the mid-nineteenth century, the Bridled Nailtail Wallaby ranged from the Murray River in the south to Charters Towers in the north and was common over at least part of the range. Even around the turn of the century it was present in sufficient numbers to be shot for its pelts. Today, it is restricted to a small area near Dingo in central Queensland and is locally common on an area of 11 470 hectares set aside as Taunton Scientific Reserve. Since cattle were excluded from the habitat of the Bridled Nailtail Wallaby, its numbers have increased from a few hundred to about 1500.

It is an animal of the semi-arid inland and earlier lived mostly on the slopes and plains to the west of the Great Dividing Range in a mixture of tall shrubland and grassy woodland. The surviving population exists on the more fertile soils of the district, which support open eucalypt forest and woodland and brigalow scrub. It has a preference for scrub edges and adjacent vegetation, grazing and sheltering in the shrubland and grazing in the grassy woodland. Home ranges overlap within and between sexes and vary in size from 20–90 hectares. Male ranges tend to be larger. The Bridled Nailtail Wallaby is usually solitary but females may have young at foot and feeding aggregations of four to five animals are not uncommon. Unlike many other macropods, it is not known to give an alarm thump with its hindfeet.

During the day, it rests under a shelter bush, where it may scratch a shallow depression: several such bushes may be used. When disturbed in the daytime it may slip quietly away, unseen, or hop along an indirect path to another shelter bush. When chased, it may seek refuge in hollow

logs and one was observed by Gould to climb within the trunk of a hollow tree. A Bridled Nailtail Wallaby may avoid detection by lying prone in long grass, or by crawling out of sight under low shrubs where it remains even when closely approached. Such reactions are all appropriate to a life in grassy woodlands and indicate that the species has had a long association with this environment. At Taunton, it is preyed upon by the Dingo and Cat.

About an hour before dusk, the Bridled Nailtail Wallaby begins to make its way to the edge of the scrub, often grazing in small clearings on the way. It rarely ventures more than 200 metres from the edge of the scrub except during drier months. In winter it often basks in late afternoon sunlight at the scrub edge before moving out to feed. The diet consists of mixed forbs, grass and browse, the latter becoming dominant during the drier seasons. Chenopod forbs are favoured, together with soft-leaved grasses such as species of *Chloris*, *Sporobolus* and *Bothriochloa* and

*The Bridled Nailtail Wallaby is so called in respect of a white 'bridle' from the neck to behind the forearm.*

D. WATTS

C. ANDREW HENLEY/LARUS

*Female Bridled Nailtail Wallaby with young in pouch.*

groups of four to five animals are sometimes seen. Mating may occur at night, when in the open, or under a shelter bush during the day.

At least in central and northern Queensland, the decline of the Bridled Nailtail Wallaby appears to be associated with effects of the pastoral industry. It may be ascribed either to competition with stock for food or to disturbance of the ground cover. The Fox is rare or absent from this habitat and the wallaby vanished prior to widespread clearing of habitat. Other disturbing factors may have affected it in the southern part of its range, including the Rabbit and Fox.

M. EVANS AND G. GORDON

malvaceous forbs. Most browse is taken from the shrub, *Eremophila mitchelli*. The forepaws are used to rake aside dry material in tussocks and prostrate forbs to expose greener leaves and shoots. Surface water is drunk during drier periods.

Breeding appears to occur throughout the year, although pouch-young and young-at-foot are seen more frequently during late spring and summer, when conditions are more favourable for plant growth. Females in oestrus are usually accompanied by a single male, but oestrous

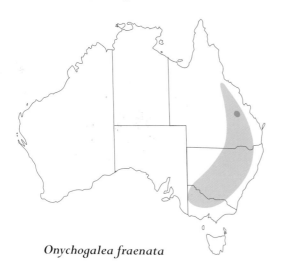

*Onychogalea fraenata*

## Onychogalea fraenata

**Size**
*Head and Body Length*
510–700 mm (males)
430–540 mm (females)
*Tail Length*
380–540 mm (males)
360–440 mm (females)
*Weight*
5–8 kg (males)
4–5 kg (females)

**Identification** A white 'bridle' line running from the centre of the neck down behind the forearm on each side of the body; horny pointed 'nail' on tail tip.
**Other Common Names** Flashjack, Pademelon, Merrin, Waistcoat Wallaby.
**Status** Endangered.
**Subspecies** None.
**Reference**
G. Gordon and B.C. Lawrie (1980). The rediscovery of the bridled nailtail wallaby (*Onychogalea fraenata* Gould) in Queensland. *Aust. Wildl. Res.* 7: 339–345.

# Crescent Nailtail Wallaby

## *Onychogalea lunata*

### (Gould, 1841)

lune-ah'-tah: 'crescent (-marked) nailed-weasel'

John Gilbert, who collected the specimens upon which John Gould erected this species in 1841, noted that it 'is found in the gum forests of the interior of Western Australia, where there are patches of thick scrub and dense thickets, in the open glades intervening between which it is occasionally seen sunning itself, but at the slightest alarm immediately takes itself to the shelter of thick scrub...'. Gilbert's 'forests' would today be termed woodlands. He stated that it made no nest but formed a hollow in soft ground beneath a shrub.

Early naturalists recorded that the Crescent Nailtail Wallaby, or Wurrung, would sometimes run into a hollow log when chased. B.W. Leake, an early settler of the Kellerberrin district,

*The Crescent Nailtail Wallaby. Illustration from John Gould's* The Mammals of Australia.

Western Australia, noted that it would 'make for a hollow tree with a hole in the bottom. Into this it would go and clamber up the sides, until it got some distance up inside the tree. To procure Wurrungs for food the Aborigines used to light a fire and smoke them out'. Leake said it lived in open timbered country.

It was apparently quite plentiful in the agricultural districts of south-western Western Australia until just after 1900. G.C. Shortridge collected 23 specimens for the British Museum from near Pingelly and Wagin between 1904 and 1907 but thereafter it declined rapidly, the last recorded specimen coming from Cranbrook in 1908. Leake noted its disappearance from Kellerberrin by 1899.

The Crescent Nailtail Wallaby also occurred in Central Australia. The Elder Expedition collected it in the Everard Range in north-western South Australia in 1891. Previously, in 1888, three South Australian specimens had been sent to the British Museum by Sir George Grey and Wood Jones later noted that it had been reported 'from the centre' in 1884. The Horn Expedition of 1894 obtained two specimens from Alice

## Onychogalea lunata

**Size**

*Head and Body Length*

371–508 mm

*Tail Length*

153–330 mm

*Weight*

approx. 3.5 kg

**Identification**  Generally ashy grey above, pale grey below, with defined crescent-like white shoulder stripe from chest to scapular regions. Ill-defined pale hip stripe. Horny nail at tip of tail.

**Recent Synonyms**  None.

**Other Common Names**  Lunated Nailtail Wallaby.

Indigenous names: Wurrung (south-western Western Australia); Urnda (Adyamathana); Kurnda (Njadjuri, Barngarla, Narunga, Kaurna); Warlpartu (Kartutjarra); Tjawalpa (Luritja); Tjawalpa, Warlpartu (Manytjilytjarra); Tjawalpa (Ngaanyatjarra); Tjawalpa, Warlpatju/Walpatju (Ngaatjatjarra); Tjawalpa, Warlpatju/Walpatju (Ngaatjatjarra); Tjawalpa, Wamarru, Warlpatju/Walpatju (Pintupi); Tjawalpa (Pitjantjatjarra); Pitarri, Tjawalpa/Tawalpa (Warlpiri).

**Status**  Presumed extinct.

**Subspecies**  None.

**References**

Burbidge, A.A., K.A. Johnson, P.J. Fuller and R. Southgate (1988). Aboriginal knowledge of the mammals of the central deserts of Australia. *Aust. Wildl. Res.* 15: 9–39.

Leake, B.W. (1962). *Eastern Wheatbelt Wildlife.* The author, Perth (printed by Docket Book Co., Perth).

Tunbridge, D. (1991). *The story of the Flinders Ranges mammals.* Kangaroo Press, Sydney.

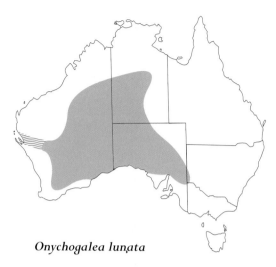

*Onychogalea lunata*

Springs, extending the known range northward. W.A. Wills collected one near Rawlinna on the Nullarbor Plain in 1927 or 1928 and it has been recorded frequently from the surfaces of subfossil deposits throughout the Nullarbor. In the 1930s, H.H. Finlayson found it still occurring near the Everard Ranges as well as near the Musgrave Ranges, South Australia and the Cavenagh Range, Western Australia. Writing in 1961, he stated that it was still extant in some areas and that one had been killed between the Tarlton and Jervois Ranges, Northern Territory, as late as 1956. The report in Ride's 'A Guide to the Native Mammals of Australia' of a 1964 specimen from near the Warburton range, Western Australia, cannot now be substantiated.

Aborigines from the central deserts of Western Australia, western Northern Territory and north-western South Australia recall that the Crescent Nailtail Wallaby inhabited most types of country, including stony hills and was particularly abundant in mulga country. They say that it ate grass and sheltered, lying on its side, under a low shady tree or shrub, or sometimes under a spinifex hummock. It was hunted by building brush fences and enclosures and driving the animals into these or through a gap behind which people waited with clubs. They also say that it had disappeared from the western desert by the 1940s. Aborigines from the Flinders Ranges area of South Australia recall that it lived throughout the ranges and on the adjacent plains but had disappeared from this area by the 1890s.

The Crescent Nailtail Wallaby was once quite common in a variety of habitats throughout much of central, southern and south-western Australia, but was unable to withstand the changes wrought by European settlement.

A.A. BURBIDGE

# Northern Nailtail Wallaby

## *Onychogalea unguifera*

### (Gould, 1841)

ung-wif'-er-ah: 'nail-bearing nailed-weasel'

Like several other genera of native mammals, *Onychogalea* is represented by a south-western, an eastern-south-eastern and a tropical northern species. The tropical representative is the Northern Nailtail Wallaby, first collected near Swan Point on the Dampier Peninsula, Western Australia, in 1838 during the post-Darwinian voyage of HMS *Beagle*. Most of its range lies between the northern Australian coast and the 500 millimetre isohyet, but it extends several hundred kilometres further inland in north-western Australia and tends to avoid areas of higher rainfall in Arnhem Land and the Kimberley.

The Northern Nailtail Wallaby inhabits open woodlands with a tussock grass understorey, tall shrublands and grasslands with scattered trees or shrubs. It particularly favours the edges of black-soil plains (tussock grasslands) such as in the Barkly Tablelands area of the Northern Territory, the gulf region of western Queensland and the East and West Kimberley regions of Western Australia. It also occupies coastal plains interspersed with *Melaleuca* thickets. In these habitats, it shelters in an adjacent shrubland or woodland during the day and moves onto the grassland to feed at night. In the central part of its range in the Northern Territory it commonly occupies Coolabah, *Eucalyptus microtheca*, open woodlands and tall shrublands of Lancewood, *Acacia shirleyi* and Bullwaddy, *Macropteranthes keckwickii*.

Herb cover is a major factor determining sites that constitute suitable habitat. It is a very selective feeder, choosing mostly dicotyledonous herb foliage, succulents, fruits and green grass shoots. Grass is only eaten in substantial amounts

*The Northern Nailtail Wallaby is reasonably common in its woodland habitat.*

P.M. JOHNSON

## *Onychogalea unguifera*

**Size**

*Head and Body Length*

540–690 (600) mm (males)

490–600 (570) mm (females)

*Tail Length*

600–730 (660) mm (males)

600–650 (630) mm (females)

*Weight*

6–9 (7.5) kg (males)

4.5–7 (5.8) kg (females)

**Identification**  Small sandy wallaby; lower mid-dorsal dark stripe continuing onto tail; tuft of dark hairs towards end of tail, nail on tail tip.

**Recent Synonyms**  None.

**Other Common Names**  Organ-grinder, Sandy Nailtail.

Indigenous names: Karrabul, Mangarra (Kimberley, Western Australia); Kurrurung, Kurrurungku (central desert); Tjunma (Barkly region, Northern Territory); Ulala (Arnhem Land, Northern Territory).

**Status**  Common.

**Subspecies**  *Onychogalea unguifera unguifera*, north-western Australia.

*Onychogalea unguifera annulicauda*, north-eastern Australia.

**References**

Ingleby, S. (1991). Distribution and status of the Northern Nailtail Wallaby, *Onychogalea unguifera* (Gould, 1841). *Wildl. Res.* 18: 655–676.

Ingleby, S., M. Westoby and P.K. Latz (1989). Habitat requirements of the Northern Nailtail Wallaby, *Onychogalea unguifera* (Marsupialia: Macropodidae) in the Northern Territory and Western Australia (in) G. Grigg, P. Jarman and I. Hume (eds) *Kangaroos, Wallabies and Rat-kangaroos*. Aust. Mammal. Soc. and Surrey Beatty and Sons, Sydney. pp. 767–782.

*Onychogalea unguifera*

when herbs are not available and shrub foliage is usually avoided. There is little or no seasonal change in the pattern of diet selection.

The Northern Nailtail Wallaby is solitary but may form feeding aggregations of up to four animals. It is active between dusk and dawn, sheltering during the day in a shallow scrape under low, dense shrubs (commonly Conker-berry, *Carissa lanceolata*), grass tussocks, spinifex hummocks, or the multi-stemmed base of tall, woody shrubs usually in open areas such as the edges of blacksoil plains, or other run-on areas, including floodplains, claypans, swamps and gilgais.

The Northern Nailtail Wallaby has survived European settlement much better than other nailtail wallabies. There is little evidence of a decline in its geographical distribution or abundance during the last century, although numbers may have declined locally in parts of the West Kimberley region of Western Australia. It appears to be under little immediate threat of severe population reduction. Most suitable habitats are under pastoral lease and there is little evidence that it is adversely affected by judicious cattle grazing. However, its preferred habitats are poorly represented in national parks or other conservation reserves throughout northern Australia and this situation should be rectified.

S. Ingleby and G. Gordon

## ROCK-WALLABIES

### *Petrogale, Peradorcas*

As their name implies, rock-wallabies live in rocky habitats. Among the most beautiful of the macropods, they are frequently brightly coloured and distinctly adorned. The hindfoot differs from that of all other macropods in that the claw of the fourth toe projects only slightly, if at all, beyond the large toe-pad. Locomotory thrust is provided by friction between the extensively granulated sole and the surface of rocks. The soles of the hindfeet have a more obvious fringe of stiff hairs than in other kangaroos and the tail, which performs an essential function in balancing the animal when it is hopping in a three-dimensional environment, is less tapered and carried arched over the back.

All rock-wallabies are terrestrial and predominantly bipedal but they readily ascend trees with sloping trunks. They are usually found as localised colonies of up to several hundred animals but individual outcrops may be occupied by no more than ten animals. The optimum habitat consists of extensive rock outcrops with deep fissures and caves, associated with feeding areas at the base, on top, or along terraces on the rock face. Vertical or near-vertical rock faces are rarely used except where there are breaks in the rock or rock falls which provide upward and downward escape routes. Rock-wallabies may occasionally occupy steep slopes with a cover of 1 metre or more of grass or other vegetation between boulders, in which case animals camp next to a boulder, under the vegetation.

In many areas, suitable rocky outcrops are not more than 2–10 kilometres apart. In others, a distance of 100 kilometres or more of unsuitable flat or undulating country may separate one outcrop from the next. That rock-wallabies have crossed these long distances in the past is demonstrated by current distribution or fossil remains. Moreover, the present distribution of unique chromosomes and genetic markers, is evidence of their spread over long distances following an origin as single events within individual animals.

The degree of restriction of populations to individual rocky outcrops is unknown. Movement between widely distributed areas of suitable habitat is probably dependent on intervening refuges in the form of tall grass, low scrub and isolated trees affording temporary shelter to migrating animals. Such long-distance movements may now be severely restricted due to the presence of introduced herbivores which have reduced the vegetation cover and introduced carnivores which increase the hazards associated with migration.

The preferred food is grass but herbs and some leaves and fruits are also eaten. As in other macropodids the large forestomach is adapted for microbial fermentation of cellulose. Some rock-wallabies that occupy extremely arid areas appear to survive without drinking, water turnover being reduced by sheltering during the day in caves where the relative humidity is considerably higher than outside and the air temperature 10–15°C cooler. They usually emerge in the late afternoon or early evening to feed on adjacent herbage during the night. After a cold night, animals may bask in the sun during the early morning. In winter, they often bask in the morning and again in the evening, before moving off to feed.

There are no pronounced colour differences between males and females of a particular form but females may be up to 30 per cent smaller than males of the same age. Breeding in females is potentially continuous after the onset of sexual maturity at one to two years of age but may be influenced by seasonal factors. Embryonic diapause is a feature of reproduction. Observations on captive animals suggest that after a young animal has permanently left the pouch but is not yet weaned, it is left in a sheltered position to which its mother returns for suckling. This difference from most other kangaroos, in which the newly emerged young follows its mother at foot for several months, appears to be an adaptation to the more difficult terrain.

Rock-wallabies are found only on mainland Australia and offshore islands and are not represented in the many suitable rocky habitats in Tasmania and Papua New Guinea. Their greatest diversity is found on the Australian Shield (the older, Pre-Cambrian, part of Australia) and they may have originated there as the western equivalent of the pademelons (Thylogale). Assuming a western origin of the group, it must also be assumed that two subgroups crossed the artesian basins (which have little suitable rock-wallaby habitat) and invaded the Great Dividing Range where they have undergone comparatively recent and rapid speciation.

G.B. Sharman and G.M. Maynes

# Allied Rock-wallaby

## *Petrogale assimilis*

### Ramsay, 1877

pet'-roh-gah'-lay ass-sim'-il-is:
'similar rock-weasel'

B.J. NOLAN

*Allied Rock-wallaby.*

The first specimen of the Allied Rock-wallaby known to science was collected from Palm Island late last century, during the voyage of the *Chevert* and was formally described in 1877 by E.P. Ramsay, then Curator of the Australian Museum, Sydney. Like the related Unadorned Rock-wallaby, it generally lacks distinct markings. However, the two species are clearly distinguishable by the shape and number of their chromosomes, as well as characteristic blood and tissue proteins. The Allied Rock-wallaby is widespread in north-eastern Queensland, being found from Home Hill (where it contacts *P. inornata*), north-west to Croydon and south-west to Hughenden. Populations of *P. assimilis* from near Hughenden, in central-western Queensland, are most unusual having high levels of genetic diversity. *P. assimilis* is also found on Magnetic and Palm Islands.

*Petrogale assimilis*

**Size**

*Head and Body Length*
470–590 (512) mm (males)
445–550 (496) mm (females)

*Tail Length*
409–545 (497) mm (males)
445–550 (487) mm (females)

*Weight*
4.7 kg (males)
4.3 kg (females)

**Identification** Similar to *P. sharmani*, *P. inornata* and *P. mareeba*. General colouration varies according to type of rock on which it lives. Mostly grey-brown above (but can be dark brown), paler sandy brown on underparts, forearms, hindlegs and at base of tail. Some specimens tend to russet on the rump and base of tail. Pale cheek-stripe, slight axillary patch and indistinct dorsal headstripe occasionally present. Paws and feet darker than limbs. Tail darkens to almost black towards end, with a slight brush at tip. Moult as for *P. inornata*. Most readily identified by the shape and number of its chromosomes.

**Recent Synonyms** *Petrogale puella*.

**Other Common Names** Torrens Creek Rockwallaby (part).

**Status** Generally common, tends to be rarer on the western edge of its range.

**Subspecies** None.

**References**

Barker, S.C. 1990. Behaviour and social organisation of the allied rock-wallaby *Petrogale assimilis*, Ramsay 1877 (Marsupialia: Macropodoidea). *Aust. Wildl. Res.* 17: 301–311.

Close, R.L. and J.N. Bell (1990). Age estimation of pouch-young of the allied rock-wallaby, *Petrogale assimilis*, in captivity. *Aust Wildl. Res.* 17: 359–367.

Horsup, A. and H. Marsh (1992). The Diet of the allied rock-wallaby, *Petrogale assimilis*, in the wet-dry tropics. *Wildl. Res.* 19: 17–33.

The Allied Rock-wallaby is an opportunistic and generalist feeder, enabling it to survive in the harsh and unpredictable wet-dry tropics. Forbs and browse form the major part of its diet, but grass can also be an important food source, particularly after rain when newly sprouted grass shoots become available. Fruits, seeds and flowers are also eaten.

There appears to be no distinct breeding season, births being recorded in all months of the year. Studies in captivity have shown that the gestation period and the oestrous cycle are 30–32 days, pouch life extending for six to seven months. Males and females reach sexual maturity at around 18 months and can live for up to 13 years.

Within a colony, basically linear dominance hierarchies are established and stable relationships form between mature males and females. Pair-bonding between an adult male and female is characterised by regular social grooming, sharing of daytime resting places and other parts of their

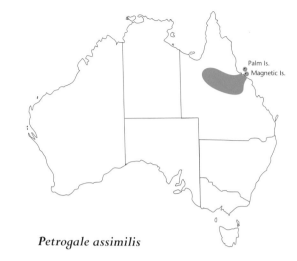

*Petrogale assimilis*

exclusive home range. Pairs forage together, usually at night.

The Mount Claro population (now *P. sharmani*) and the Mareeba population (now *P. mareeba*) were previously listed as chromosome races of *P. assimilis*.

*Allied Rock-wallaby.*

M.D.B. ELDRIDGE AND R.L. CLOSE

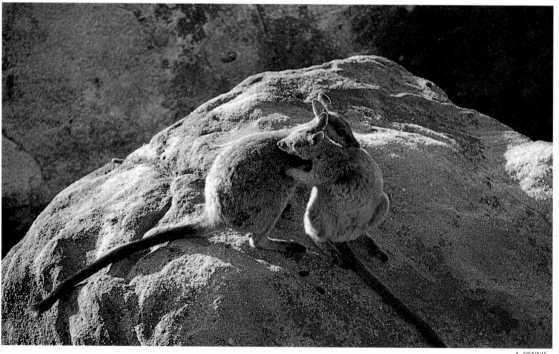

A. DENNIS

# Short-eared Rock-wallaby

## *Petrogale brachyotis*

### (Gould, 1841)

brak'-ee-oh'-tis: 'short-eared rock-weasel'

The type specimen of the Short-eared Rock-wallaby was collected in the vicinity of Hanover Bay, Western Australia. Few specimens have since been taken close to this locality, but the species is found from the Kimberley through Arnhem Land and thence eastward along the Gulf of Carpentaria near the eastern edge of the Australian Shield. The southerly range corresponds roughly to the 700 millimetre median isohyet and thus to the monsoonal area of the

*Short-eared Rock-wallaby.*

J. LOCHMAN

---

*Petrogale brachyotis*

**Size**

*Head and Body Length*
435–550 (492) mm (males)
405–485 (456) mm (females)

*Tail Length*
387–550 (479) mm (males)
320–520 (441) mm (females)

*Weight*
3.2–5.6 (4.4) kg (males)
2.2–4.7 (3.7) kg (females)

**Identification**  Ears usually less than half length of head. Uniform in colour on back, with variable whitish margins. Fur short and fine. Tail darkens distally with a distinct brush in some specimens. Distinguished from *P. penicillata* and *P. lateralis* by short, uniformly coloured ears. Distinguished from *P. concinna* (with which it is sympatric over part of its range) by distinct markings and much greater size.

Kimberley Race: Light grey above, white to greyish white below. Dark brown neck-stripe to above level of shoulder. Side-stripe absent, represented by extension of white axillary patch. Faint suggestion of hip-stripe on some specimens.

Victoria River Race: Pale grey above. Markings almost absent. Short white neck-stripe (30–40 mm), axillary patch smaller and paler than in other races. Similar in size to nominate race.

Arnhem Land Race: Usually dark grey or brown above. Distinct white or buff side-stripe. Distinct dark brown to black neck-stripe and dorsal stripe. A well developed black axillary patch and pale hip-stripe are sometimes present. Forearms and legs pale to bright cinnamon in some populations.

**Recent Synonyms** *Petrogale venustula* (part), *Petrogale wilkinsi* (part), *Petrogale longmani* (part).

**Other Common Names** Roper River Rock-wallaby (part), Longman's Rock-wallaby (part), Brush-tailed Rock-wallaby.

**Status** Mostly common, but locally rare and perhaps extinct on extreme west of range. May also be declining in the Gulf region, on the eastern edge of range.

**Subspecies** Three geographically distinct populations are given interim recognition. Kimberley Race: Kimberley region, Western Australia. The name *P. b. brachyotis* is applicable to this population.

Victoria River Race: Victoria River District, Northern Territory.

Arnhem Land Race: Alligator Rivers region, Arnhem Land, western Gulf of Carpentaria, Groote Eylandt, Northern Territory. (This highly variable population includes forms previously described as *P. venustula*, *P. longmani*, or *P. wilkinsi*.)

**References**

Briscoe, D.A., J.H. Calaby, R.L. Close, G.M. Maynes, C.E. Murtagh and G.B. Sharman (1982). Isolation, introgression and genetic variation in rock-wallabies (in) R.H. Groves and W.D.L. Ride (eds) *Species at Risk*. Research in Australia. Australian Academy of Science, Canberra. pp. 73–87.

Eldridge, M.D.B., P.G. Johnston and P.S. Lowry (1992). Chromosomal rearrangements in Rock-wallabies, *Petrogale* (Marsupialia: Macropodidae). VII. G-banding analysis of *P. brachyotis* and *P. concinna*: Species with dramatically altered karyotypes. *Cytogenet and Cell Genet.* 61: 34–39.

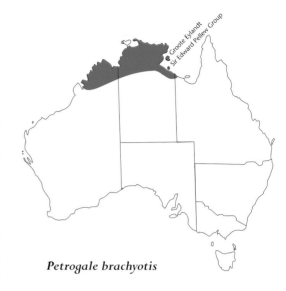

*Petrogale brachyotis*

Australian Shield. Within its preferred range of savanna grassland, it occurs on low rocky hills, cliffs and gorges. A study at Mount Borradaile in Arnhem Land has shown that browse and seeds are significant dietary items.

Although the Short-eared Rock-wallaby is widespread and abundant throughout much of northern Australia, it remains a little-known and poorly studied species. Despite its larger size, genetic studies have shown it is most closely related to *P. concinna* and *P. burbidgei*. While it is very unusual to find different species of rock-wallaby living together, *P. brachyotis* is sympatric with both *P. concinna* and *P. burbidgei* in different parts of its range.

The Short-eared Rock-wallaby is exceedingly variable in size and colour pattern. Specimens from Arnhem Land, Roper River and Groote Eylandt have been described as separate species and grouped with the Black-footed and Brush-tailed Rock-wallabies. The subspecies *P. brachyotis signata* is of doubtful validity; it is possible that *P. brachyotis* is divisible into three races, from the Kimberley, Victoria River District and Arnhem Land.

G.B. Sharman, G.M. Maynes, M.D.B. Eldridge and R.L. Close

# Monjon

## *Petrogale burbidgei*

### Kitchener and Sanson, 1978

ber'-bid-jee: 'Burbidge's rock-weasel'

J. LOCHMAN

*Monjon (once known as the Warabi), smallest of the Rock-wallabies.*

Smallest of the rock-wallabies, the Monjon had, until recently, escaped discovery because it is restricted to the rugged, inhospitable parts of the Kimberley region, Western Australia, where it is easily confused in the field with the Nabarlek or Little Rock-wallaby.

### *Petrogale burbidgei*

**Size**

*Head and Body Length*
3-6–353 (322) mm

*Tail Length*
264–290 (276) mm

*Weight*
960–1430 (1258) g

**Identification** Back olive, with a tawny and blackish marbling. Face predominantly clay-coloured with light horizontal stripe from snout through the eye to base of ear. Indistinct light greyish-olive stripe along midline of head and neck. Tail light greyish-olive. Flanks deep olive buff; undersurface ivory yellow. Paws and feet light greyish-olive with a fuscous black undersurface. Distinguished from most rock-wallabies by smaller size and from the similar-sized *Petrogale concinna* by shorter ears (less than 35 millimetres long).

**Recent Synonyms** None.

**Other Common Names** Warabi.

**Status** Common, limited.

**Subspecies** None.

**Reference**

Kitchener, D.J. and G. Sanson (1978). *Petrogale burbidgei* (Marsupialia, Macropodidae), a new rock wallaby from Kimberley, Western Australia. *Rec. West. Aust. Mus.* 6: 269–285.

It inhabits the highly fractured King Leopold sandstone country along the coastline and some of the offshore islands in the Bonaparte Archipelago. The associated vegetation is characterised by a low open woodland of eucalypts, acacias, figs, *Terminalia* and *Owenia*. During the day it can be seen resting, sometimes squeezed beneath horizontal crevices in the sandstone. It is easily disturbed from such situations and at Mitchell Plateau and Bigge Island large numbers of aroused animals have been seen moving swiftly among rocks.

The Monjon differs from the similar-sized Nabarlek in habitat preference: the two species have not been recorded from the same mainland locality or the same islands in the Bonaparte Archipelago. Little is known of its general biology although the presence of small similar-sized pouch-young in August and October suggests that the period of births is protracted.

D.J. KITCHENER

Bigge Is.
Katers Is.
Boongaree Is.

*Petrogale burbidgei*

# Cape York Rock-wallaby

## *Petrogale coenensis*

### Eldridge and Close, 1992

koh'-en-en'-sis: 'Coen rock-weasel'

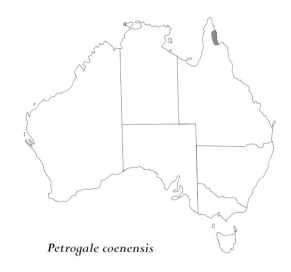

*Petrogale coenensis*

The Cape York Rock-wallaby is a rare and little-known species, found on eastern Cape York Peninsula from Musgrave to the Pascoe River. It is known only from six adult specimens collected between 1981 and 1987. Originally referred to as the 'Cape York Race', this chromosomally distinct population was discovered in 1981 and described as a new species in 1992. The specific name coenensis refers to the north Queensland town of Coen, which is located at the centre of its known distribution and is also near the type locality.

Although *P. coenensis* has blood and tissue proteins very similar to those of *P. godmani*, it differs from *P. godmani* by four chromosome rearrangements and clearly represents a distinct species. (*P. coenensis* was earlier placed within *P. godmani*).

The Cape York Rock-wallaby appears to be rare within its known range and only small scattered colonies have been identified. Future research is required to assess the extent of its populations.

M.D.B. Eldridge and R.L. Close

R.L. CLOSE

## Petrogale coenensis

### Size

(three specimens of each set)
*Head and Body Length*
540–565 mm (males)
440–510 mm (females)
*Tail Length*
485–540 mm (males)
470–500 mm (females)
*Weight*
4.3–5.0 kg (males)
4.0–4.2 kg (females)

**Identification** Similar to *P. godmani*. Grey-brown above; paler (sandy brown to buff) underparts, forearms, hindlegs and base of tail. Pale to buff cheek-stripe. Mid-dorsal head-stripe extending down neck to upper back. Tail darker than body towards the base, with a slight brush at tip. Increasing density of dirty white hairs towards tip of tail results in a distinct silvery tail tip in most specimens. (In one individual the fur on the underparts was very short, dense and pale, almost white, with mottled darker patches on belly and towards the rump and flanks; another had purple pigmentation over the chest and throat, distal two-thirds of the tail being silvery white. Most readily identified by the shape and number of chromosomes.

**Recent Synonyms** None.
**Other Common Names** None.
**Status** Rare within its limited distribution.
**Subspecies** None.
**Reference**
Eldridge, M.D.B. and R.L. Close (1992). Taxonomy of rock wallabies *Petrogale* (Marsupialia: Macropodidae). I. A revision of the eastern *Petrogale* with the description of three new species. *Aust. J. Zool.* 40: 605–625.

*Cape York Rock-wallaby.*

G.D. SANSON

*Nabarlek.*

# Nabarlek
## *Petrogale concinna*
### Gould, 1842

kon-sin'-ah: 'elegant pouched-gazelle'

Until the recent discovery of the marginally smaller Monjon, the Nabarlek was the smallest known macropodid, hence its other common name, Little Rock-wallaby.

The Nabarlek is unique among all marsupials in apparently producing an unlimited number of supernumerary molars. These erupt at the posterior margin of the jaws and migrate forward in the same way as in the large kangaroos, with early loss of the premolars. The number of molars, all of the same size, varies from four to six in one quadrant

## Petrogale concinna

**Size**

*Head and Body Length*
290–350 (319) mm
*Tail Length*
220–310 (297) mm
*Weight*
1.05–1.5 (1.3) kg (dry season)
1.07–1.7 (1.4) kg (wet season)

**Identification** Dull rufous above, marbled with light grey and black (brilliant rust-red in *P. c. concinna*); greyish-white below. Indistinct dark grey to black shoulder-stripe sometimes present. Tail with black brush tip. Similar to *Petrogale burbidgei* but has longer ears (more than 35 mm) and supernumerary molars.
**Recent Synonyms** *Petrogale concinna.*
**Other Common Names** Little Rock-wallaby.
**Status** Rare, limited.
**Subspecies** *Petrogale concinna concinna*, Mary and Victoria Rivers district, Northern Territory. Probably rare.
*Petrogale concinna monastria*, Kimberley region, Western Australia.
*Petrogale concinna canescens*, eastern Arnhem Land, Northern Territory.
**Reference**
Sanson, G.D., J.E. Nelson and P. Fell (1985). Ecology of *Peradorcas concinna* in Arnhem Land in a wet and dry season. *Proc. Ecol. Soc. Aust.* 13: 69–72.

at one time with another tooth erupting behind.

Studies on a population from Arnhem Land indicate that, in the wet season, the diet includes the grasses, *Cyperus cuspidatus*, *Eriachne sp.* and *Fimbristylis sp.* which, at this time of the year, cover the margins of the blacksoil plains that are not inundated by floodwaters. In the dry season, the Nabarlek shelters in sandstone crevices and caves during the day and emerges at night to forage for a fern, *Marsilea crenata*, which grows on the edges of billabongs. Individuals may move several hundred metres into the blacksoil plains

in search of this food, in contrast to the sympatric Short-eared Rock-wallaby, which has not been observed to move away from the shelter of rocks.

The silica content of *Marsilea* may reach 26 per cent (dry weight) and in the grasses referred to above it is as high as 15 per cent. It is probable that this extremely abrasive diet is associated with the Nabarlek's peculiar system of continual replacement of molar teeth.

During the wet season it is partly diurnal, basking on rocks for up to three hours after dawn and feeding for several hours before dusk. In the dry season, when the harsher conditions are reflected in a diminished body weight, it is much more secretive and seldom leaves the shelter of rocks until after dark. Irrespective of the season, the diurnal activity of the Nabarlek is greatly reduced by the presence of the White-breasted Sea Eagle, which is a known predator.

Breeding probably occurs throughout the year, but a greater number of large pouch-young and young-at-heel have been observed in February (wet season) than August (dry season). Observations on a captive colony have demonstrated embryonic diapause. The oestrous cycle is 32–35 days and the length of gestation is 30–32 days. Young leave the pouch at about 180 days and sexual maturity is attained in the second year of life.

G.D. SANSON

*Peradorcas concinna*

# Godman's Rock-wallaby

## *Petrogale godmani*

### Thomas 1923

god'-man-ee: 'Godman's rock-weasel'

*Petrogale godmani*

The first specimen of Godman's Rock-wallaby was collected at Black Mountain, south-west of Cooktown in 1922 by T.V. Sherrin during an expedition funded by the Godman Trust. Oldfield Thomas described the species in 1923.

It is found from the Mitchell River and near Mount Carbine, west to 'Pinnacles' and north to Bathurst Head. Although Godman's Rock-wallaby is still relatively common throughout its range it is now rarely seen on Black Mountain, where it was once abundant.

*Godman's Rock-wallaby.*

Genetic studies of *P. godmani* have yielded conflicting results. Its chromosomes indicate that it is most closely related to *P. assimilis*, but its blood and tissue proteins suggest that it is also very closely related to *P. coenensis*.

On its southern border, *P. godmani* forms a narrow hybrid zone with *P. mareeba*. Despite major genetic differences, some hybrids have

R.L. CLOSE

## Petrogale godmani

**Size**

*Head and Body Length*

505–570 (533) mm (males)

495–528 (508) mm (females)

*Tail Length*

520–640 (566) mm (males)

480–540 (505) mm (females)

*Weight*

5.2 kg (males)

4.3 kg (females)

**Identification** Similar in colouration to *P. inornata*, *P. sharmani*, *P. mareeba* and *P. assimilis*. Generally more distinctly buff or pale cinnamon on foreparts, distal one-third to half of tail frequently dirty white. Most readily identified by the shape and number of chromosomes.

**Recent Synonyms** None.

**Other Common Names** None.

**Status** Common within its restricted distribution.

**Subspecies** None.

**Reference**

Eldridge, M.D.B. and R.L. Close (1992). Taxonomy of rock-wallabies *Petrogale* (Marsupialia: Macropodidae). I. A revision of the eastern *Petrogale* with the description of three new species. *Aust. J. Zool.* 40: 605–625.

proved fertile and a limited exchange of genetic markers has occurred. The presence of some typical *P. godmani* genes and characteristics south of the hybrid zone, in specimens of *P. mareeba*, may indicate either that *P. mareeba* is advancing northwards and replacing *P. godmani* or that some *P. godmani* genes are passing through the hybrid zone and entering northern populations of *P. mareeba*.

M.D.B. ELDRIDGE AND R.L. CLOSE

G.B. BAKER

# Herbert's Rock-wallaby

## *Petrogale herberti*

Thomas, 1926

her'-bert-ee: 'Herbert's rock-weasel'

Herbert's Rock-wallaby is widespread and common in south-eastern Queensland, being found from Nanango, northward to the south bank of the Fitzroy River at Rockhampton and westward to beyond Rubyvale and Clermont. In size and colouration it is intermediate between the Brush-tailed Rock-wallaby, to the south and the Unadorned Rock-wallaby, found to the north. Although described as a species in 1926,

*Herbert's Rock-wallaby.*

*P. herberti* was long regarded as a subspecies of *P. inornata* or *P. penicillata*. It has recently been reinstated as a full species.

On its southern border *P. herberti* contacts *P. penicillata* and a narrow hybrid zone is formed. An analysis of the lice parasitic on rock-wallabies indicates that *P. penicillata* and *P. herberti* have only recently come back into contact and begun to hybridise. Studies have shown that two partially differentiated forms of a single louse species are associated with *P. penicillata* and *P. herberti*. These differences probably arose when the populations of the lice (and therefore the rock-wallabies) were isolated from each other. Now that they and their hosts have regained contact, the lice populations are interbreeding.

Although *P. penicillata* and *P. herberti* differ in their chromosomes, as well as blood and tissue proteins, female hybrids are fertile, allowing limited genetic exchange to occur between these species. However, each species is maintaining its genetic identity. Since the hybrid zone is of great biological interest, it is unfortunate that the only known hybrid colony is threatened by a residential subdivision.

M.D.B. Eldridge and R.L. Close

*Petrogale herberti*

## *Petrogale herberti*

**Size**
*Head and Body Length*
502–615 (562) mm (males)
470–566 (517) mm (females)
*Tail Length*
510–660 (572) mm (males)
510–571 (530) mm (females)
*Weight*
5.0–6.7 (6.0) kg (males)
3.7–4.9 (4.3) kg (females)

**Identification**  Grey-brown above, darker on the face and shoulders, tending to tawny on the rump. Chest and belly buff to white. Indistinct pale cheek-stripe, distinct black dorsal stripe from forehead to beyond shoulders. Exterior of ears black towards the base, inside of ears buff. Blackish axillary patch. White side-stripe, more distinct than in *P. penicillata*, from axillary patch to thighs. Forearms and legs brown; forepaws and feet dark brown to black. Tail darkens distally, brush less prominent than in *P. penicillata*. Most readily identified by shape and number of chromosomes.
**Recent Synonyms**  None.
**Other Common Names**  None.
**Status**  Common throughout most of its range.
**Subspecies**  None.
**References**
Barker, S.C., R.L. Close and D.A. Briscoe (1991). Genetic divergence and speciation in *Heterodoxus octoseriatus*. *Int. J. Parasitol.* 21: 479–482.
Eldridge, M.D.B. and R.L. Close (1992). Taxonomy of Rock-wallabies *Petrogale* (Marsupialia: Macropodidae). I. A revision of the eastern *Petrogale* with the description of three new species. *Aust. J. Zool.* 40: 605–625.

# Unadorned Rock-wallaby

## *Petrogale inornata*

### Gould, 1842

in'-or-nah'-tah: 'unadorned rock-weasel'

*Petrogale inornata*

The type specimen of the Unadorned Rock-wallaby was collected in June 1839, from Cape Upstart, south of Townsville, during a survey voyage of HMS *Beagle*. John Gould identified the specimen as a new species, *P. inornata*, in 1842. The species name refers to a lack of distinct markings.

It is found in the ranges of central coastal Queensland from Home Hill southwards to the north bank of the Fitzroy River at Rockhampton. It is also found on some islands of the Whitsunday group.

There appears to be no distinct breeding season, births being recorded in all months of the year. Studies in captivity show that the gestation period and the oestrous cycle are 30–32 days, with pouch life extending for six to seven months. Both sexes reach maturity at about 18 months.

M.D.B. ELDRIDGE AND R.L. CLOSE

*Unadorned Rock-wallaby.*

D. WHITFORD

### *Petrogale inornata*

**Size**

*Head and Body Length*
507–570 (535) mm (males)
454–560 (507) mm (females)
*Tail Length*
490–640 (548) mm (males)
430–560 (504) mm (females)
*Weight*
3.4–5.6 (5.0) kg (males)
3.1–5.0 (4.2) kg (females)

**Identification** Similar to *P. sharmani*, *P. assimilis*, *P. mareeba*. General colouration varies according to type of rock on which populations live. Mostly grey-brown above; paler sandy brown on underparts, forearms and hindlegs; almost buff at base of tail. Some individuals darker, tending towards dark grey or dark brown. Pale cheek-stripe and slight mid-dorsal stripe occasionally present. Degree of ornamentation varies with latitude, animals in the south of range having more noticeable markings (including a light side-stripe). Tail darkens to almost black towards tip, with slight brush. Moults in autumn to predominantly grey on back and flanks and becomes progressively more sandy through the year. Most readily identified by the shape and number of chromosomes.

**Recent Synonyms** None.

**Other Common Names** Plain Rock-wallaby.

**Status** Common throughout most of its range.

**Subspecies** None.

**References**

Johnson, P.M. (1979). Reproduction in the plain rock-wallaby *Petrogale penicillata inornata* Gould, in captivity, with age estimation of the pouch-young. *Aust. Wildl. Res.* 6: 1–4.

Eldridge, M.D.B. and R.L. Close (1992). Taxonomy of rock wallabies *Petrogale* (Marsupialia: Macropodidae). I. A revision of the eastern *Petrogale* with the description of three new species. *Aust. J. Zool.* 40: 605–625.

# Black-footed Rock-wallaby

## *Petrogale lateralis*

### Gould, 1842

lat-er-ahl'-is: 'notable-sided rock-weasel'

The Black-footed Rock-wallaby is a widespread and genetically diverse species consisting of six chromosomally distinct forms; four named subspecies; and two (as yet unnamed) forms, currently designated as chromosome races. *P. l. lateralis* has 11 pairs of chromosomes of similar shape to those of the presumed ancestor of all modern rock-wallabies.

The nominate subspecies, *Petrogale l. lateralis*, was described by John Gould in 1842 from specimens collected by John Gilbert in Western Australia. Gould reported that it was 'very abundant in all rocky districts of Swan River, particularly the Toodyay' but the species is not found in this locality today. Gilbert described *P. l. lateralis* as a 'remarkably shy and wary animal, feeding only at night in little open patches of grass and never…going more than two or three hundred yards from its rock retreats'.

Hackett's Rock-wallaby, *Petrogale lateralis hacketti* was formally described in 1905 but had, in fact, first been 'discovered' by the explorer Matthew Flinders more than a century earlier. On 13 January 1802, Flinders reported that a party returned from Mondrain Island with a 'few small kanguroos [sic] of a species different from any I had before seen'. The specimens apparently did not survive and were not described, but some fine colour drawings of the animals by Ferdinand Bauer, the artist aboard the *Investigator*, have survived and are now in the Museum of Natural History, London. *P. l. hacketti* is restricted to three islands of the Recherche

## *Petrogale lateralis*

### Size

#### *Petrogale lateralis lateralis*
*Head and Body Length*
497–529 (512) mm (males)
446–486 (472) mm (females)
*Tail Length*
483–605 (551) mm (males)
407–516 (490) mm (females)
*Weight*
4.1–5.0 (4.5) kg (males)
3.1–3.8 (3.5) kg (females)

#### *Petrogale lateralis hacketti*
*Head and Body Length*
506–570 (533) mm
*Tail Length*
460–574 (532) mm
*Weight*
4.8–5.3 (5.0) kg

#### *Petrogale lateralis pearsoni*
*Head and Body Length*
480–570 (539) mm
*Tail Length*
320–510 (400) mm
*Weight*
2.3–4.5 (3.4) kg

#### *Petrogale lateralis purpureicollis*
*Head and Body Length*
521–610 (548) mm (males)
490–570 (524) mm (females)
*Tail Length*
450–606 (562) mm (males)
470–532 (507) mm (females)
*Weight*
6.0–7.1 (6.5) kg (males)
4.7–5.7 (5.0) kg (females)

### West Kimberley Race (one adult male)
*Head and Body Length*
475 mm
*Tail Length*
527 mm
*Weight*
3.5 kg

### MacDonnell Ranges Race
*Head and Body Length*
450–521 (491) mm
*Tail Length*
507–597 (558) mm
*Weight*
2.8–4.5 (4.1) kg

### Identification

*Petrogale lateralis lateralis*
Dark grey-brown above; paler on chest passing to
dark brown on belly. Face dark with distinct white
to sandy cheek-stripe; light brown patches at base of
ears. Dark brown to black dorsal stripe from between ears to beyond shoulders. Feet sandy, digits
black. Forearms sandy dorsally, darker ventrally;
paws dark brown to black. White side-stripe with
wider dark brown stripe immediately ventral, extending from axillary patch to thighs. Coat thick and
'woolly', particularly about the rump, flanks and
base of tail. Tail brownish-grey proximally, passing
to black distally with a slight terminal brush.
Changes to a lighter and browner pelage in summer.

*Petrogale lateralis hacketti*
Similar to *P.l.lateralis* but larger.

*Petrogale lateralis pearsoni*
Dark grizzled grey-brown above, passing to almost
silver on shoulders and neck. Chest and belly pale
yellow to buff, paler distally. Face dark grey-brown
with wide, light yellow to buff cheek-stripe, paler
below eye. Chin almost white; base of ears buff.
Distinct dark dorsal stripe from between ears to beyond shoulders. Forearms white proximally, passing
to buff distally; paws dark brown to black. White
side-stripe with wider rich dark brown stripe immediately ventral, extending from axillary patch to
thighs. Coat thicker and more 'woolly' than in *P. l.
lateralis*. Tail light brown, passing to black distally
with slight terminal brush.

*Petrogale lateralis purpureicollis*
Light grey-brown above, greyer on shoulders; light
sandy-brown on chest and belly; occasionally tending

to light rufous on rump. Pale cheek-stripe; chin whitish. Brown dorsal stripe from forehead to between the ears. Characteristic purple pigmentation, of variable intensity, can extend from face to the neck and shoulders. Feet and forearms sandy with dark brown digits. Dark brown axillary patch. Occasionally, dark brown markings extend from axillary patch to thigh but no distinct side stripe. Fur generally short but dense. Tail sandy-brown, darker proximally. Terminal brush less distinct than in *P. l. lateralis*. Changes to a predominantly sandy-orange pelage in summer.

*West Kimberley Race*

Similar to *P. l. lateralis* but smaller.

*MacDonnell Ranges Race*

Dark grizzled brown above, passing to grey on shoulders. Paler on chest; belly buff. Sandy cheek-stripe; ears dark brown, paler smoky brown at base. Dark brown to black dorsal stripe from between ears to beyond the shoulders. Feet grey-brown; digits black. Forearms sandy dorsally and darker ventrally; paws dark brown to black. White side-stripe with wider dark brown stripe immediately ventral, extending from axillary patch to thighs. Coat shorter and less dense than in *P. l. lateralis*. Tail dark grey proximally becoming browner distally; terminal brush less distinct. Changes to a predominantly sandy-brown pelage in summer.

## Recent Synonyms

*Petrogale pearsoni* (part), *Petrogale hacketti* (part), *Petrogale purpureicollis* (part).

## Other Common Names

Side-striped Rock-wallaby, Black-flanked Rock-wallaby, West Australian Rock-wallaby, Recherche Rock-wallaby (part), Hackett's Rock-wallaby (part), Pearson Island Rock-wallaby (part), Purple-necked Rock-wallaby (part), Purple-collared Rock-wallaby (part).

## Status

*Petrogale lateralis lateralis*

Has declined seriously on the mainland with many local extinctions. Remaining populations are vulnerable and most require active management to ensure survival.

*Petrogale lateralis hacketti*

Remains common within its restricted distribution.

*Petrogale lateralis pearsoni*

Remains common within its restricted distribution.

*Petrogale lateralis purpureicollis*

Common throughout most of its range.

*West Kimberley Race*

Poorly known, but appears to be rare within its restricted distribution.

*MacDonnell Ranges Race*

Common in some parts of range. Elsewhere populations declining with many local extinctions.

## Subspecies

*Petrogale lateralis lateralis*

Formerly widespread throughout much of Western Australia, now known only from North-west Cape, Barrow Island, Salisbury Island, Calvert Ranges in the Little Sandy Desert and scattered remnant colonies in the Wheatbelt.

*Petrogale lateralis hacketti*

Mondrain, Wilson and Combe Islands, Recherche Archipelago, Western Australia.

*Petrogale lateralis pearsoni*

Pearson Islands (Investigator Group), Thistle Island, Wedge Island (Gambier Islands), South Australia.

*Petrogale lateralis purpureicollis*

North-western Queensland from Dajarra and west of Mount Isa to eastern end of the Selwyn Ranges, extending northwards to Lawn Hill and possibly into Northern Territory.

*West Kimberley Race*

Known only from the Edgar Range, West Kimberley district, south of the Fitzroy River.

*MacDonnell Ranges Race*

Central Australia, extending to eastern Western Australia and northern South Australia.

## References

Kinnear, J.E., M.L. Onus and R.N. Bromilow (1988). Fox control and rock-wallaby population dynamics. *Aust. Wildl. Res.* 15: 435–450.

Eldridge, M.D.B., R.L. Close and P.G. Johnston (1991). Chromosomal rearrangements in rock-wallabies, *Petrogale* (Marsupialia: Macropodidae). IV. G-banding analysis of the *Petrogale lateralis* complex. *Aust. J. Zool.* 39: 621–627.

B.J. THOMSON

*Black-footed Wallaby (MacDonnell Ranges race).*

Archipelago, Western Australia and can be found in the rock outcrops of the island's interior, as well as along the rocky shore, where it feeds on succulents and grasses. Although similar in appearance to *P. l. lateralis*, *P. l. hacketti* has only ten pairs of chromosomes. Like *P. l. hacketti*, the West Kimberley Race has ten pairs of chromosomes but one differs in shape.

An insular South Australian form, the Pearson Island Rock-wallaby, *Petrogale lateralis pearsoni*, was described as a separate species in 1922. Like *P. l. lateralis*, it has 11 pairs of chromosomes but one pair is of a slightly different shape. Originally found only on North Pearson Island, it was accidentally introduced to South and Middle Pearson Islands in 1960 and now seems to be well established. Animals deliberately introduced to Thistle Island in 1974 and Wedge Island in 1975 have apparently thrived.

The large and distinctly coloured Purple-necked Rock-wallaby *P. l. purpureicollis* of western Queensland was originally described as a separate species by A.S. Le Souef in 1924. Although currently included in *P. lateralis*, recent studies have shown that it does in fact represent a distinct species. Like *P. l. lateralis*, it has 11

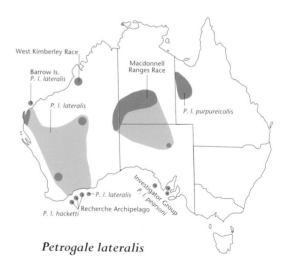

*Petrogale lateralis*

pairs of chromosomes, but two pairs are of a different shape. The characteristic purple colouration of *P. l. purpureicollis* varies considerably in intensity sometimes appearing as a faint pink wash through the fur of the head and neck; at other times of the year, the face, head, neck and shoulders can be a brilliant red to purple.

The MacDonnell Ranges Race, from Central Australia, has 11 pairs of chromosomes, distinctly different from those of *P. l. lateralis* and more similar to those of *P. l. purpureicollis*. Populations

Western Australia but is now found only in small numbers in several widely scattered localities. In the south-west, colonies are largely confined to scattered outcrops of granite boulders in remnants of mallee scrub surrounded by cleared agricultural land. Recent studies of these populations have clearly implicated the introduced Fox in its massive decline and show that, without adequate control, Fox predation will continue to suppress the remnant rock-wallaby populations, making them increasingly vulnerable to extinction. It is encouraging to note that, where introduced predators are controlled, populations of the Black-footed Rock-wallaby are thriving.

This continuing decline gives urgency to the task of identifying remnant populations of *P. lateralis* and mapping the distribution of each subspecies. Rock-wallaby populations attributable to *P. lateralis* have been reported from the Tanami Desert, north-western South Australia and the Warburton region of Western Australia. While these populations remain unidentified, priorities for conservation and management will be difficult to assess.

M.D.B. ELDRIDGE AND R.L. CLOSE

P.M. JOHNSON

Above and right: *Black-footed Wallaby,* P. l. purpureicollis.

of the MacDonnell Ranges Race have declined greatly in some areas, but remain common in others. This major decline commenced in the 1930s and is still continuing; populations at Ayers Rock (Uluru) and the Olgas (Kata Tjuta) have recently disappeared.

The decline of the Black-footed Rock-wallaby on the Western Australian mainland has been even more dramatic. It was once abundant and distributed over much of

P.M. JOHNSON

# Mareeba Rock-wallaby

## *Petrogale mareeba*

### Eldridge and Close, 1992

mar-eeb'-ah: 'Mareeba rock-weasel'

The existence of this chromosomally distinct rock-wallaby has been known since 1974 when Hayman and Martin published a diagram of the chromosomes of a rock-wallaby labelled *P. pencillata inornata*. However, the location and distribution of this population remained largely unknown until 1979, when rock-wallabies collected from around the North Queensland town of Mareeba were found to have similar chromosomes. This population became known as the Mareeba Race, until it was described as a species in 1992. *P. mareeba* had formerly been grouped with *P. sharmani* (then called the Mount Claro Race) in *P. assimilis* and, earlier, in *P. inornata*. *P. mareeba* is found from the Mitchell River and near Mount Carbine (where it forms a hybrid zone with *P. godmani*), west to Mungana and south to the Burdekin River near Mount Garnet.

*Petrogale mareeba*

### *Petrogale mareeba*

**Size**

*Head and Body Length*
425–548 (485) mm (males)
425–500 (471) mm (females)
*Tail Length*
420–530 (486) mm (males)
415–467 (441) mm (females)
*Weight*
4.5 kg (males)
3.8 kg (females)

**Identification** Similar to *P. inornata*, *P. assimilis* and *P. sharmani*. General colouration varies according to type of rock on which it lives. Mostly grey-brown above; paler sandy brown-buff on underparts, forearms, hindlegs and base of tail. Some specimens considerably darker. Pale cheek-stripe. Mid-dorsal head-stripe occasionally present. Tail darkens distally, with slight brush. Dirty white tail-tip of variable length occasionally present in some northern populations. Moult as in *P. sharmani*. Most readily identified by the shape and number of chromosomes.

**Recent Synonyms** None.

**Other Common Names** None.

**Status** Common within its restricted distribution.

**Subspecies** None.

**Reference**

Eldridge, M.D.B. and R.L. Close (1992). Taxonomy of rock-wallabies *Petrogale* (Marsupialia: Macropodidae). I. A revision of the eastern *Petrogale* with the description of three new species. *Aust J. Zool.* 40: 605–625.

The Mareeba Rock-wallaby is a cryptic species which cannot be reliably distinguished from *P. sharmani* and *P. assimilis* on morphological criteria. Nevertheless, it is genetically distinct. The chromosomes of *P. mareeba* can be readily derived from those of *P. sharmani* and these two

HANS & JUDY BESTE

*Mareeba Rock-wallaby.*

# Brush-tailed Rock-wallaby

## *Petrogale penicillata*

(Gray, 1825)

pen'-iss-il-ah'-tah: 'brush (-tailed) rock-
weasel'

This was the first rock-wallaby to be named by European scientists. Referred to as the Tuft-tailed or Mountain Kangaroo, it was originally described as *Kangurus penicillatus* in 1825, becoming *Petrogale penicillata* in 1837.

Originally an abundant and widespread species, it was found in suitable rocky areas in a wide variety of habitats, including rainforest gullies, wet and dry sclerophyll forest, open woodland and rocky outcrops in semi-arid country. Most commonly, the sites occupied by *P. penicillata* have a northerly aspect, allowing the animals to sun themselves in the morning and evening. Sites with numerous ledges, caves and crevices are favoured, these providing numerous daytime rest areas with multiple escape routes. Studies in central and southern New South Wales

species must share a recent common ancestor.

As in *P. inornata* and *P. assimilis*, the colouration of *P. mareeba* is highly variable and tends to match the colour of the rock on which the animals live. For example, individuals from basaltic outcrops are almost black, while those from lighter coloured rock types can be light grey-brown to sandy brown. What little is known indicates that the biology of *P. mareeba* is similar to that of *P. assimilis*.

M.D.B. Eldridge and R.L. Close

*Petrogale penicillata*

C.A. HENLEY

## *Petrogale penicillata*

**Size**

*Head and Body Length*
529–586 (557) mm (males)
510–570 (536) mm (females)
*Tail Length*
510–700 (611) mm (males)
500–630 (563) mm (females)
*Weight*
5.5–10.9 (7.9) kg (males)
4.9–8.2 (6.3) kg (females)

**Identification** Brown above, tending to rufous on the rump and grey on the shoulders. Chest and belly paler (in some animals a white blaze on the chest). White to buff cheek-stripe; black dorsal stripe from forehead to back of head. Exterior of ears black, inside of ears buff. Black axillary patch often extending as a dark stripe to the margin of the hindlegs. Pale grey side-stripe sometimes present. Feet and paws dark brown to black. Tail darkens distally with a prominent brush. Pelage long and thick, particularly about the rump, flanks and base of tail. Animals from north of range tend to be lighter and have a less prominent tail brush.

*Brush-tailed Rock-wallaby.*

**Recent Synonyms** None.
**Other Common Names** None.
**Status** Common in south-eastern Queensland and northern New South Wales. Populations in the southern and western parts of the range have declined and must be considered vulnerable. Remnant populations in Victoria and western New South Wales endangered.
**Subspecies** None.
**References**

Short, J. (1982). Habitat requirements of the brush-tailed rock-wallaby, *Petrogale penicillata* in New South Wales. *Aust. Wildl. Res.* 9: 239–246.

Short J. (1989). The diet of the brush-tailed rock-wallaby in New South Wales. *Aust. Wildl. Res.* 16: 11–18.

Short, J. and G. Milkovits (1990). Distribution and status of the brush-tailed rock-wallaby in south-eastern Australia. *Aust. Wildl. Res.* 17: 169–179.

have shown that the Brush-tailed Rock-wallaby feeds mainly on grasses and forbs but also eats a significant amount of browse. Seeds, fruit and flowers are eaten opportunistically.

Since European settlement, the Brush-tailed Rock-wallaby has declined in Victoria, as well as in southern and western New South Wales. Its former abundance can be gauged from the thousands that were shot for the skin trade and as supposed agricultural pests. Bounties were paid on over half a million rock-wallabies killed in New South Wales between 1884 and 1914. In 1908, 92 590 rock-wallaby skins were marketed through a single Sydney wool-broking company.

G. WHITMONT

The species declined dramatically in Victoria in the early 1900s and was thought to be extinct until 1953 when a few colonies were rediscovered near the Snowy River, East Gippsland. In 1970, additional remnant populations were located in the Grampians. In New South Wales, it has declined west of the watershed of the Great Divide where sheep grazing is the major land use. A well-known colony around Jenolan Caves, New South Wales, has all but disappeared in the 1990s. Introduced predators (primarily the Fox) and competition with introduced herbivores (such as the Goat and Rabbit) have been widely suggested as precipitating its decline.

The Brush-tailed Rock-wallaby was successfully introduced to Hawaii and New Zealand. On the island of Oahu, a small population of rock-wallabies, descended from two animals,

*Brush-tailed Rock-wallaby.*

has existed since 1916. In New Zealand, where the species was introduced in the 1870s, the Brush-tailed Rock-wallaby is found on Kawau, Rangitoto and Motutapu Islands. On some of these islands it has reached pest proportions and is regularly culled.

In the north of its range, *P. penicillata* contacts *P. herberti* and an apparently narrow hybrid zone has formed. In this zone, some female hybrids are fertile, allowing limited exchange of genes between the two populations. This exchange is not widespread and each species appears to be retaining its genetic identity.

M.D.B. ELDRIDGE AND R.L. CLOSE

# Proserpine Rock-wallaby

## *Petrogale persephone*

### Maynes, 1982

per-sef'-oh-nee: 'Proserpine rock-weasel'
(Greek Persephone was the same goddess as
the Roman Proserpine)

B.J. NOLAN

The Proserpine Rock-wallaby is found in a rather intensively settled area but was not brought to scientific attention until 1976. It was, however, known to members of the Proserpine branch of the Wildlife Preservation Society of Queensland and undoubtedly to other local people. It has been found in a small number of localities near Proserpine and on several islands in the Whitsunday group. It usually inhabits rocky outcrops in closed canopy forests, surrounded by open woodland with a grass understorey. The rock outcrops provide shelter and home sites and the grassy understorey of the surrounding forest provides food.

The Proserpine Rock-wallaby is a relict—the remnant of an apparently more widely distributed species that occupied a greater range before the smaller and more successful Unadorned Rock-wallaby became common in adjacent areas. Different kinds of rock-wallabies are seldom found together and, during the evolution of the group, newly evolved, more successful forms have probably repeatedly replaced older, less well-adapted forms. However, the distribution of *P. persephone* remains perplexing. On the mainland, its distribution is within that of *P. inornata*, although both species are also found on different islands in the Whitsunday group. At no known locality do both species occur together, although not all sites where *P. persephone* might occur have been surveyed.

Since the Proserpine Rock-wallaby has the smallest known distribution of any rock-wallaby, it is extremely vulnerable to any habitat loss or disturbance. It is presently endangered by the rapid rate of human occupation of central coastal Queensland and the associated activities. Specimens are frequently found as road kills near Airlie Beach.

*Petrogale persephone*

*Proserpine Rock-wallaby.*

There are no pronounced colour differences between the sexes but males can be up to 50 per cent heavier than females. Little is known about reproduction but the few available observations indicate a prolonged (perhaps continuous) breeding season and the occurrence of embryonic diapause.

Like *P. xanthopus*, *P. persephone* has 11 pairs of chromosomes of similar shape to those of the ancestor of all modern rock-wallabies. Unlike most other rock-wallabies, *P. persephone* and *P. xanthopus* have developed major morphological and genetic differences without any significant associated changes in the shape and number of their chromosomes. Despite major morphological, ecological and genetic differences, *P. persephone* is able, in captivity, to produce fertile hybrids and backcross hybrids with *P. xanthopus*.

G.B. SHARMAN, G.M. MAYNES, M.D.B. ELDRIDGE
AND R.L. CLOSE

## *Petrogale persephone*

**Size** (adult)
*Head and Body Length*
501–640 (581) mm (males)
526–630 (575) mm (females)
*Tail Length*
580–676 (623) mm (males)
515–624 (582) mm (females)
*Weight*
5.6–8.8 (7.2) kg (males)
4.1–6.4 (5.2) kg (females)

**Identification** Ears with brownish-orange hair outside, black internally. White to buff lateral stripe on upper lip, passing below eye to about level of ear with platinum or pale pear grey band below, arising at corner of mouth and passing back to ear region. Shoulders raw sienna, passing to dark brown in axillary region with rest of dorsal surface light brown, lightly pencilled with white hairs. Chin white, throat varying from white to alabaster; chest and general ventral surface light yellow to butter yellow. Toes of fore- and hindlimbs and about half of hindfeet black. Recently moulted individuals may be grey rather than brown in overall colour. Overall impression of a basically dark coloured animal is heightened by black feet and black dorsal surface of the tail. White to yellowish-white tail tip of variable length in most individuals.

**Recent Synonyms** None.

**Other Common Names** None.

**Status** Vulnerable due to restricted distribution. Currently threatened by clearing for the rapidly expanding tourist industry.

**Subspecies** None.

**Reference**

Maynes, G.M. (1982). A new species of rock-wallaby, *Petrogale persephone* (Marsupialia: Macropodidae) from Proserpine central Queensland. *Aust. Mammal.* 5: 47–58.

# Rothschild's Rock-wallaby

## *Petrogale rothschildi*

### Thomas, 1904

roths'-chile-dee: 'Rothschild's rock-weasel'

---

### *Petrogale rothschildi*

**Size**
**Mainland**
*Head and Body Length*
532–592 (559) mm (males)
463–526 (500) mm (females)
*Tail Length*
580–704 (636) mm (males)
539–613 (570) mm (females)
*Weight*
5.0–6.6 (6.1) kg (males)
3.7–5.3 (4.1) kg (females)
**Dampier Archipelago**
*Head and Body Length*
426–470 (447) mm
*Tail Length*
412–487 (458) mm
*Weight*
2.6–3.9 (3.4) kg

**Identification** Greyish-brown above; dull brown below. Upper surface of head and ears rich dark brown, contrasting with lighter cheeks and throat and with grey neck and shoulders. No neck-stripe or dorsal stripe. Distinguished from the *Petrogale lateralis* by these characters, absence of side-stripe and shorter, uniformly brown ears. At times, fur may be suffused with purplish pigmentation, especially around back of neck and shoulders.
**Recent Synonyms** None.
**Other Common Names** Roebourne Rock-wallaby.

---

Rothschild's Rock-wallaby, one of the larger species of the genus *Petrogale*, is found in the Hamersley Range region of Western Australia and extends to some offshore islands in the Dampier Archipelago. The Hamersley Range is a dissected plateau rising from the Australian Shield and is characterised, geologically, by the presence of igneous rocks and, botanically, by grass-steppe and shrub vegetation. Since this vegetation extends beyond the range of Rothschild's Rock-wallaby, it appears that its distribution is limited by the characteristic granite rock piles and outcrops of its habitat. Air temperature in the deep caves between the rocks may be as much as 15°C cooler than outside the rock pile.

The first specimen was collected in 1901 by J.T. Tunney and sent to Oldfield Thomas, Curator of Mammals, British Museum (Natural History), who named the species after the Hon. Walter Rothschild, who had supported Tunney's expedition.

Rothschild's Rock-wallaby was known from Lewis Island (Dampier Archipelago) but this population is now extinct. Stock from Enderby Island has been successfully reintroduced to West Lewis Island.

It would appear that, like *P. lateralis*, populations of *P. rothschildi* are being suppressed by

Dampier Archipelago
Rosemary Is.
Enderby Is.
Dolphin Is.

*Petrogale rothschildi*

C. JENNER

*Rothschild's Rock-wallaby.*

exotic predators. Recent studies have implicated the Fox in the decline of *P. rothschildi* in Western Australia. While *P. rothschildi* has remained conspicuously abundant on the Fox-free outer islands of the Dampier Archipelago, populations on the mainland and inner islands (where the Fox is present) have declined. Where the Fox is controlled, the numbers of Rothschild's Rock-wallabies have increased substantially. Effective predator control will therefore be essential in ensuring the long-term survival of this species.

G.B. SHARMAN, G.M. MAYNES, M.D.B. ELDRIDGE
AND R. CLOSE

**Status** Relatively common in areas of suitable habitat, but appears to have generally declined on the mainland; remains abundant on outer islands of the Dampier Archipelago (where Fox is not present).

**Subspecies** None described, but animals from Enderby, Rosemary and West Lewis Islands, Dampier Archipelago, are markedly smaller and may warrant subspecific recognition.

**Reference**

Briscoe, D.A., J.H. Calaby, R.L. Close, G.M. Maynes, C.E. Murtagh and G.B. Sharman (1982). Isolation, introgression and genetic variation in rock-wallabies (in) R.H. Groves and W.D.L. Ride (eds) *Species at risk: Research in Australia*. Australian Academy of Science, Canberra. pp. 73–87.

# Sharman's Rock-wallaby

## *Petrogale sharmani*

### Eldridge and Close, 1992

shar'-man-ee: 'Sharman's rock-weasel'

*Petrogale sharmani*

Sharman's Rock-wallaby is one of a number of physically similar species found in north-eastern Queensland. Discovered at Mount Claro, it is found in the Seaview and Coane Ranges, west of Ingham. Its known distribution totals 200 000 hectares, being confined by *P. assimilis* to the south and west and by *P. mareeba* to the north. Morphologically, it is indistinguishable from *P. mareeba* and *P. assimilis*, but genetic studies have shown it to be a distinct species. Studies of its chromosomes show that it is most closely related to *P. mareeba*, although differing from it in chromosome shape and number.

The existence of this chromosomally distinct population (originally called the 'Mount Claro Race') has been known since 1976. It was not described as a species until 1992. *P. sharmani* had previously been grouped with *P. mareeba* (then known as the 'Mareeba Race') in *P. assimilis* and earlier in *P. inornata*. Sharman's Rock-wallaby is named after Professor G.B. Sharman in recognition of his enormous contribution to marsupial biology and to *Petrogale* in particular.

*Sharman's Rock-wallaby.*

R.L. CLOSE

## Petrogale sharmani

**Size**

*Head and Body Length*
490–530 (511) mm (males)
455–475 (465) mm (females)
*Tail Length*
500–532 (515) mm (males)
435–482 (465) mm (females)
*Weight*
4.4 kg (males)
4.1 kg (females)

**Identification** Similar to *P. inornata*, *P. assimilis* and *P. mareeba*. Generally grey-brown above; paler sandy brown on underparts, forearms, hind-legs and almost buff at base of tail. Pale cheek-stripe and pale patch on face between the eyes. Slight mid-dorsal head-stripe occasionally present. Tail darkening to almost black distally, with a slight brush at tip. Moults as in *P. inornata*. Most readily identified by the shape and number of chromosomes.
**Recent Synonyms** None.
**Other Common Names** Mount Claro Rock-wallaby.
**Status** Common within known range but restricted distribution makes it particularly vulnerable to any habitat loss or disturbance.
**Subspecies** None.
**Reference**
Eldridge, M.D.B. and R.L. Close (1992). Taxonomy of rock-wallabies *Petrogale* (Marsupialia: Macropodidae). I. A revision of the eastern *Petrogale* with the description of three new species. *Aust. J. Zool.* 40: 605–625.

Following the autumn moult, specimens of *P. sharmani*, like those of *P. inornata*, *P. assimilis* and *P. mareeba*, become progressively more sandy throughout the year, until their colouration matches the mature wallaby grass, which grows around the rock outcrops where it lives.

M.D.B. ELDRIDGE AND R.L. CLOSE

HANS & JUDY BESTE

*Yellow-footed Rock-wallaby.*

# Yellow-footed Rock-wallaby

## Petrogale xanthopus

### Gray, 1855

ksan'-thoh-poos: 'yellow-footed rock-weasel'

This brightly coloured and distinctly ornamented rock-wallaby is most commonly associated with the Flinders Ranges. However, an equally large population is now known from the ranges of the Adavale Basin in south-west Queensland. This distinctive subspecies remained largely undetected until the mid-1980s but extensive survey work has shown it to be relatively abundant in suitable habitat.

Like *P. persephone*, the Yellow-footed Rock-wallaby has 11 pairs of chromosomes of similar

## *Petrogale xanthopus*

### Size
#### *Petrogale xanthopus xanthopus*
*Head and Body Length*
480–650 (600) mm
*Tail Length*
570–700 (690) mm
*Weight*
6–11 kg
#### *Petrogale xanthopus celeris*
*Head and Body Length*
560–600 (580) mm
*Tail Length*
565–675 (520) mm
*Weight*
6–12 kg

### Identification
*Petrogale xanthopus xanthopus*
Fawn-grey above; white below. Distinct white cheek-stripe; ears orange; rich brown mid-dorsal stripe from crown of head to centre of back; reddish-brown axillary patch. Buffy-white side-stripe and brown and white hip-stripe. Forearms, hindlegs and feet rich orange to bright yellow. Tail orange-brown with irregular dark brown annulations, tail tip dark brown or white. Brightness of ornamentation and distinctiveness of tail annulations variable.

*Petrogale xanthopus celeris*
Similar to *P. x. xanthopus* but generally paler in colouration. Dorsal stripe black; light orange-brown above eyes; ears grey-brown; forearms, legs and base of tail light orange-brown; hip-stripe indistinct; tail annulations less pronounced.

**Recent Synonyms** *Petrogale celeris* (part).

**Other Common Names** Ring-tailed Rock-wallaby.

**Status** Has declined throughout range since European settlement. Within restricted distribution in South Australia (Flinders Ranges) and Queensland (Adavale Basin), remains relatively common in areas of suitable habitat. Smaller

shape to the presumed ancestor of all modern rock-wallabies. On the reasonable hypothesis that rock-wallabies originated somewhere on the Australian Shield, animals sharing a common ancestry with the Yellow-footed Rock-wallaby once extended as far as the Great Dividing Range.

The Yellow-footed Rock-wallaby inhabits semi-arid country. Some populations are found in association with permanent fresh water which may be restricted to mere soaks at the edges of

outlying populations in the Gawler Ranges and Olary Hills, South Australia, the Gap and Coturaundee Ranges, New South Wales, and Cavaway Range, Queensland, may be endangered.

### Subspecies
*Petrogale xanthopus xanthopus*
Flinders Ranges, Gawler Ranges and Olary Hills, South Australia; Gap and Coturaundee Ranges, New South Wales.
*Petrogale xanthopus celeris*
Gowan, Grey, Cheviot, Yangang and Macedon Ranges, bounded by Adavale, Blackall and Stonehenge, south-western Queensland.

### References
Copley, P.B. and A.C. Robinson (1983). Studies on the Yellow-footed Rock-wallaby *Petrogale xanthopus* Gray (Marsupialia: Macropodidae). II. Diet of the Yellow-footed Rock-wallaby at Middle Gorge, South Australia. *Aust. Wildl. Res.* 10: 63–77.

Poole, W.E., J.C. Merchant, S.M. Carpenter and J.H. Calaby (1985). Reproduction, growth and age determination in the Yellow-footed Rock-wallaby *Petrogale xanthopus* Gray, in captivity. *Aust. Wildl. Res.* 12: 127–136.

Lim, L., A.C. Robinson, P.B. Copley, G. Gordon, P.D. Canty and D. Reimer (1987). *The conservation and management of the Yellow-footed Rock-wallaby* Petrogale xanthopus *Gray 1854*. Dept. Env. Plan. S. Aust. Publ. 4: 1–94.

rock faces. Others exist where even this scanty source appears to be unavailable. It lives in colonies of up to 100 individuals in suitable rocky areas, where there is frequently a greater diversity of plants than on the surrounding arid plains. Individual animals occupy home ranges of 150–200 hectares centred on the rocky areas and overlapping with other members of the colony. When conditions are favourable it can breed continuously. Studies on captive animals indicate a gestation period of 31–32 days and a pouch life of about 194 days. During summer, when ambient daytime temperatures are often above 40°C, it is strictly nocturnal, but some activity occurs, by day and night, in winter.

Like most rock-wallabies, it appears to be regularly hunted by the Wedge-tailed Eagle, *Aquila audax*. On one occasion, seven eagles were observed circling over a colony and making periodic swoops at the rock-wallabies as they ran across the rock face. Predation by man has also been a significant factor. Late last century, hundreds of Yellow-footed Rock-wallaby skins were exported annually to London from Adelaide. This period of intensive hunting led to a significant reduction of numbers in some areas, particularly the Flinders Ranges.

Although the Yellow-footed Rock-wallaby is now strictly protected, it must compete for food with introduced herbivores (Goat and Rabbit)

S. PERKINS

*Yellow-footed Rock-wallaby.*

and survive in an environment that has been degraded by these animals and domestic stock. Fox predation may also have become an important factor in limiting populations.

In most areas it has declined significantly since European settlement. Surviving colonies in New South Wales appear to be the remnants of larger and more widespread populations. Throughout its range there has been considerable alteration of vegetation due to grazing and browsing by domestic stock and feral herbivores, as well as the removal of timber for the mining industry and fencing. Studies of the diet of the Yellow-footed Rock-wallaby indicate that grass, forbs and browse are the dominant components, browse becoming more important in dry seasons. The effect of drought is therefore compounded by increased competition for food with introduced browsers such as the Goat. During a moderate drought in New South Wales in 1982–83, numbers of Yellow-footed Rock-wallabies dropped by 60 per cent. After the drought broke, numbers soon recovered where the Goat was controlled but, where there was no control, rock-wallaby numbers remained depressed.

G.B. SHARMAN, G.M. MAYNES, M.D.B. ELDRIDGE
AND R.L. CLOSE

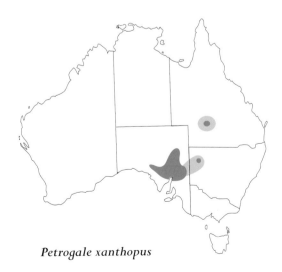

**Petrogale xanthopus**

## PADEMELONS

### *Thylogale*

Pademelons are small, compact-bodied and relatively short-tailed macropods inhabiting rainforest or wet sclerophyll forest with a dense understorey. Typically, they make runways through the ground vegetation. They feed on rather succulent grasses and also browse on shrubs.

In common with most marsupials from rainforest and wet sclerophyll forests, pademelons have suffered a reduction in range due to land clearance, but they are still common where the habitat remains. They are not seriously disturbed by selective logging.

# Tasmanian Pademelon

## *Thylogale billardierii*

### (Desmarest, 1822)

thie'-loh-gah'-lay  bil-ard'-ee-air'-ee-ee:
'Billardière's pouched-weasel'

The Tasmanian Pademelon, which formerly occurred in south-eastern South Australia and Victoria, is extinct on the mainland but still common and widely distributed throughout Tasmania and the larger islands of Bass Strait. It favours the dense vegetation of wet sclerophyll forests, rainforests and tea-tree scrubs but also occurs in those parts of open, grassy, dry sclerophyll forests where moist gullies, patches of sedge, or similar dense vegetation provide suitable cover for daytime shelter.

The diet consists mainly of short green grasses and herbs, occasionally supplemented by browse from taller woody plants. The forepaws are used to hold food and for crude manipulation of herbage in the mouth. In the Tasmanian high country, where deep snowfalls are common, the forelimbs are important in uncovering buried vegetation: diggings are often concentrated beneath spreading trees, the branches of which catch much of the snow, leaving a shallower covering below.

Like many wallabies of similar size, the Tasmanian Pademelon is most numerous where grassy clearings adjoin shelter vegetation. At

## Thylogale billardierii

**Size**

*Head and Body Length*

630 mm (males)

560 mm (females)

*Tail Length*

345–483 (417) mm (males)

320 mm (females)

*Weight*

3.8–12 (7.0) kg (males)

2.4–10 (3.9) kg (females)

**Identification** Rich dark brown to dark grey-brown above. Buff below with a rufous tinge, especially in the lower abdomen. Tail about two-thirds length of head and body.

**Recent Synonyms** None.

**Other Common Names** Red-bellied Pademelon, Rufous Wallaby.

**Status** Abundant.

night it travels to the pasture to feed, rarely venturing more than 100 metres from the forest edge. It usually returns to shelter by morning twilight, daytime grazing in clearings being very uncommon. The home range may be as large as 170 hectares and individuals sometimes travel more than 2 kilometres through forest from the daytime rest site to reach a regularly used feeding area.

Feeding aggregations of ten or more individuals may form at night but there is no evidence of a colonial structure or of persisting bonds between individuals. When disturbed, members of groups scatter rapidly for cover. Vocalisations are uncommon but occasionally a guttural hiss is uttered during squabbles and males produce a clucking sound when following oestrous females.

*Once found in South Australia and Victoria, the Tasmanian Pademelon is now restricted to Tasmania.*

D. WATTS

**Subspecies** None.
**References**

McCartney, D.J. (1978). *Reproductive biology of the Tasmanian Pademelon,* Thylogale billardierii. Hons. Thesis, Univ. Tasmania.

Mollison, B.C. (1960). Progress report of the ecology and control of marsupials in the Florentine Valley. *Appita* 14: 21–36.

Rose, R.W. and D.J. McCartney (1982). Reproduction of the red-bellied pademelon *Thylogale billardierii* (Marsupialia). *Aust. Wildl. Res.* 9: 27–32.

King Is.    Furneaux Group

*Thylogale billardierii*

Males are much larger than females, sometimes reaching more than twice the weight of the average adult female and always having a disproportionately greater musculature in the forelimbs and chest.

Gestation occupies 30 days and pouch life is approximately 200 days. Breeding is continuous but over 70 per cent of young are born in April, May and June. Young become sexually mature before 14–15 months, males when achieving a weight of about 4 kilograms and females 3.5 kilograms.

The flesh is palatable and has been a favoured table item since early European settlement. The fur is dense and long and the leather fine and soft, which has led to sustained trade in its pelts. In some areas where the Tasmanian Pademelon causes substantial losses to agricultural crops, it is necessary for a proportion to be destroyed.

K.A. JOHNSON AND R. ROSE

*The Tasmanian Pademelon is primarily a grazer but also browses on woody plants.*

D. GREIG

# Red-legged Pademelon

## *Thylogale stigmatica*

### (Gould, 1860)

stig-mah'-tik-ah: 'pricked (pattern) pouched-weasel'

*Thylogale stigmatica*

Although rainforest is the preferred habitat of the Red-legged Pademelon, it also occurs in wet sclerophyll forests and occasionally in deciduous vine thickets. In New Guinea, it is reported to occur in lowland rainforest as well as low mixed savanna near swamps. It is occasionally found in lantana thickets on land originally covered by rainforest.

Its diet includes leaves and fruit from a wide variety of rainforest plants as well as the grasses which grow at the forest edge, two of these being *Paspalum notatum* and *Cyrtococcum oxyphyllum*.

*The Red-legged Pademelon inhabits rainforest and subtropical to tropical wet sclerophyll forests, eating a variety of plant material.*

Commercial crops such as oats and ryegrass are also grazed. Leaves are either browsed directly from standing vegetation or picked up from the forest floor, often in a quite old and brittle state. Diet varies considerably between northern Queensland and north-eastern New South Wales, the two areas where dietary studies have been conducted. This difference is reflected in the use of habitat. In the southern part of its range, the Red-legged Pademelon rarely leaves the forest and its diet is composed almost entirely of leaf and fruit material, with the Moreton Bay Fig, *Ficus macrophylla*, being one of the possibly favoured food

M. TRENERRY

## Thylogale stigmatica

**Size**

*Head and Body Length*
470–536 (492) mm (males)
386–520 (463) mm (females)
*Tail Length*
372–473 (443) mm (males)
301–445 (357) mm (females)
*Weight*
3.7–6.8 (5.1) kg (males)
2.5–4.2 (4.1) kg (females)

**Identification** Fur soft, thick, grey-brown above (dark in rainforest forms, pale in forms from open country); cream below (pale grey in southern population). Cheeks, forearms, outside and inside of hindlegs rufous brown. Tail short, thick.

**Recent Synonyms** None.

**Other Common Names** Pademelon, Northern Red-legged Pademelon.

**Status** Common.

**Subspecies** *Thylogale stigmatica stigmatica*, Cairns region. *Thylogale stigmatica coxenii*, Cape York. *Thylogale stigmatica wilcoxi*, southern Queensland, New South Wales. *Thylogale stigmatica oriomo*, south central Papua New Guinea lowlands.

species. In northern Queensland, the forest edge is utilised to a greater degree and a high proportion of grass is incorporated in the diet.

Substantial amounts of dicotyledonous plants, ferns and non-grassy monocotyledonous material is also consumed, with favoured species including leaves and fruit of rainforest figs, *Ficus* spp.; fruit of the Burdekin Plum, *Pleiogynium timorense* and climbing monocotyledonous plants such as *Calamus* spp. and *Smilax* spp. Seeds of the Pink Ash, *Alphitonia whitei*, are also eaten. An individual has been seen to eat the seeds of a rainforest plum, *Prunus turnerana,* out of a Cassowary dropping. The reason for the great latitudinal variation in diet is not yet explained but it may be related to competition with the Red-necked Pademelon,

which is sympatric with the Red-legged Pademelon in the south.

The Red-legged Pademelon is active throughout most of the 24-hour cycle but activity is lowest between midday and early afternoon and around midnight. During daylight hours, activity is within the forest, where much time is spent searching for food. When feeding, it typically moves in a slow quadrupedal gait with nose close to the forest floor. In northern Queensland, animals proceed rapidly to the rainforest edge after dusk and remain grazing there until just prior to dawn. Peaks in the rate of movement occur during these journeys to and from the forest edge. When resting, the tail is swung between the extended hindlegs and, sitting on the base of its tail, the animal leans back against a tree or rock. As it goes to sleep, the head droops forward to rest on the tail or on the ground beside it.

In northern Queensland, the home range is 1–4 hectares and is partitioned into spatially distinct nocturnal and diurnal ranges. The diurnal range is located well within the forest and is often more than twice the size of the nocturnal range, which is located primarily on pasture. When grazing, the Red-legged Pademelon rarely ventures more than 70 metres from the forest edge. If disturbed on pasture, it dashes back into the forest, often along well-defined runways. In northern Queensland, it forms an important component of the Dingo's diet. Other predators include the Tiger Quoll and large Amethystine Pythons.

Several vocalisations are used in social behaviour. A harsh rasping sound is uttered in hostile interactions and by a female rejecting a male during courtship. The courting male makes a soft clucking sound and a similar sound is used by a female calling her young. A sound commonly heard is the 'alarm thump', made with the hindfeet as an animal flees a disturbance.

Reproduction is characterised by a gestation period of 28–30 days and an oestrous cycle of 29–32 days. Post-partum oestrus and mating occurs 2–12 hours after the birth of the young. The sex of pouch-young is distinguishable at three to four weeks; vibrissae appear at eight to ten weeks; teat detachment occurs at 13–18 weeks; ears become erect at 15–18 weeks; eyes open at 16–18 weeks; hair becomes visible at 19–21 weeks; feet become dirty (indicating short excursions from the pouch) at 22–26 weeks; and young leave the pouch at 26–28 weeks. Age at weaning is approximately 66 days after pouch emergence. Females become sexually mature at about 48 weeks, males at about 66 weeks.

The Red-legged Pademelon is solitary but individuals may form feeding aggregations beneath fruiting trees or when grazing on pasture. Distribution within the range is discontinuous, particularly in the north, where it appears to be limited by the availability of vegetation providing adequate cover. Extensive clearing of rainforest has reduced the available habitat but, in northern Queensland, this may have benefited the populations in the remaining forest by increasing the availability of grass. Thus, in the rainforest fragments that remain, Red-legged Pademelons usually exist in good numbers, although such fragments often comprise only a few hectares. Sufficient parks and reserves exist throughout its range to secure its status.

P.M. JOHNSON AND K.A. VERNES

### References

Cooke, B.N. (1979). *Field observations of the behaviour of the macropod marsupial* Thylogale stigmatica *(Gould).* M.Sc. Thesis, University of Queensland, Brisbane.

Johnson, P.M. and K. Vernes (1994). Reproduction in the Red-legged Pademelon, *Thylogale stigmatica* Gould (Marsupialia: Macropodidae) and age and development estimation of the Pouch-young. *Wildl. Res.* 21: 553–558.

Vernes, K. 1994. *Life on the edge: the ecology of the Red-legged Pademelon* Thylogale stigmatica *(Gould) (Marsupialia: Macropodidae) in fragmented rainforest in the North Queensland Wet Tropics.* M.Sc. Thesis, James Cook University, Townsville.

K. GRIFFITHS

*The Red-necked Pademelon inhabits dense forest but feeds on the grassy edges.*

# Red-necked Pademelon

## *Thylogale thetis*

### (Lesson, 1837)

thet'-iss: 'Thetis (Bougainville's ship) pouched-weasel'

The Red-necked Pademelon is one of three similar species which inhabit the dense rainforest and eucalypt forests of eastern Australia. Populations within a forest are usually sparse but, where the forest edge adjoins grassy areas or pasture, animals may be numerous. Animals travel routinely to feeding grounds in the evening along well-defined runways where they graze on short green grasses and shrubs until shortly before dawn. Movement from and to the forest occurs rapidly and in a seemingly purposeful manner, the timing of movement being remarkably consistent from day to day and varying only with seasonal changes in day length. Between periods of sleeping, it moves about in the forest, grazing or browsing and, on chilly winter mornings, may be seen basking in small open areas where the sun penetrates to the forest floor.

The Red-necked Pademelon travels quadrupedally when moving slowly: the tail drags behind and is never supportive. The forepaws are often used to hold food items and manipulate material protruding from the mouth. No nest is made, resting sites being no more than shallow depressions in the leaf litter. It is a very timid animal and rarely moves more than 100 metres from the forest edge into a clearing: the home range is 5–30 hectares. Small aggregations of feeding animals often form in clearings but these rapidly disrupt when disturbed and the pademelons scatter back to the forest. The Fox and Dingo are known predators, as probably are large birds of prey.

Captive females may breed before they are 12 months old but, in natural conditions, they rarely reach maturity before 17 months of age. Females breed continuously but, at least in north-eastern New South Wales, there is a peak of births in autumn and in spring. More than 70 per cent of the young survive pouch life but mortality may be very high after the young are weaned.

Mothers call to their young with a repetitive 'click' and young use a high-pitched sibilant squeak during pouch life and until a few months after independence. A guttural growl is made by

## Thylogale thetis

**Size**

*Head and Body Length*
300–620 (520) mm (males)
290–500 (420) mm (females)
*Tail Length*
270–510 (430) mm (males)
270–370 (350) mm (females)
*Weight*
2.5–9.1 (7.0) kg (males)
1.8–4.3 (3.8) kg (females)

**Identification**  Brownish-grey above; whitish below. Reddish neck and shoulders. Distinguished from *Thylogale stigmatica* by distribution of reddish fur. Distinguished from *Macropus parma* by colouration and by shorter tail, which is held rod-like when hopping.
**Recent Synonyms**  None.
**Other Common Names**  Pademelon Wallaby.
**Status**  Common.
**Subspecies**  None.
**Reference**
Johnson, K.A. (1980). Spatial and temporal use of habitat by the Red-necked Pademelon, *Thylogale thetis*. *Aust. Wildl. Res.* 7: 157–166.

males and females when threatening and a repetitive 'cluck' is made by males when showing sexual interest in females. One or two loud thumps are often made with the hindfeet when animals are alarmed.

The Red-necked Pademelon is a creature of the forest edge and the density of a population is related to the length of such an edge. Where land has been cleared in a mosaic for agriculture or forestry, populations may reach pest proportions; appropriate management in such circumstances calls for forested areas that are more or less circular in shape.

K.A. Johnson

*Thylogale thetis*

## QUOKKA AND SWAMP WALLABY

### *Setonix, Wallabia*

Apart from its very short tail, the Quokka looks much like any other wallaby but details of its dentition, skull structure, chromosomes and blood proteins indicate that it cannot be included, with the typical wallabies, in the genus *Macropus*. Nor do these characters link it with any other species of macropod but it may be related to the rock-wallabies. The Swamp Wallaby also appears to be 'ordinary' but, in respect of the above characters, it is even more anomalous. For these reasons and others based on the skull and dentition, each species has been placed in a genus of its own.

Both the Quokka and Swamp Wallaby are browsers and it seems likely that each is a relict of a separate line of evolution from early browsing macropods. There are some indications that the Swamp Wallaby shares a distant ancestry with *Macropus*.

# Quokka

### *Setonix brachyurus*

## (Quoy and Gaimard, 1830)

see'-ton-ix brak'-ee-yue'-rus:
'bristle-footed short-tail'

When the Dutch navigator Willem de Vlamingh saw the Quokka in 1696 on an island off the mouth of the Swan River, Western Australia, he described it as 'a kind of rat as big as a common cat' and, in its honour, named the island Rotte-nest ('rat nest', now spelled Rottnest). It had earlier been observed there, in 1658, by his fellow countryman Samuel Volckertzoon who regarded it as 'a wild cat resembling a civet-cat but with browner hair'. Although his description is very misleading (and seems more appropriate to a quoll), it constitutes the second definite European reference to an Australian marsupial.

The suitability of the Quokka as a laboratory animal and the ease with which it can be studied on this almost flat, poorly vegetated island has put it in the forefront of marsupial studies in Australia. Much of the pioneer work on macropod nutrition, temperature regulation, reproduction, immunology, ecology and behaviour was conducted on Rottnest Island populations of this small wallaby.

*Setonix brachyurus*

At the time of European settlement the Quokka was common in south-western Western Australia. Fossil evidence suggests that it has always been restricted to this region and to Rottnest and Bald Islands but, while numbers on Rottnest Island remained high, mainland populations declined so severely in the twentieth century that, by 1960, it was known only from a few isolated populations in swamps near Perth. The last decade has seen some recovery and it is now common in the moister parts of the extreme south-western part of the State.

Despite its apparent preference for densely vegetated, moist conditions, the Quokka survives in large numbers on Rottnest Island in a harsh, seasonally arid habitat where the low vegetation affords scant cover and potable water is available in summer only from around salt lakes and on some beaches. Plants grow during the winter and progressively decline in water and nitrogen content through the summer to a low point in March. By the end of summer, animals become anaemic and many die, those populations farthest from sources of fresh water suffering the highest mortality. Seasonal debility arises from insufficient drinking water for metabolic and thermoregulatory needs which, in turn, leads animals to eat more succulent but less nutritious plants, thus adding the effects of nitrogen deficiency to those of dehydration.

The Quokka has an excellent thermoregulatory ability at ambient temperatures up to 44°C

*Setonix brachyurus*

**Size**
*Head and Body Length*
435–540 (487) mm (males)
400–500 (468) mm (females)
*Tail Length*
260–310 (289) mm (males)
245–285 (265) mm (females)
*Weight*
2.7–4.2 (3.6) kg (males)
2.7–3.5 (2.9) kg (females)

**Identification**  Grizzled grey brown above with tinge of rufous. Fur long, thick, coarse. Ears very short, rounded. Tail short. No definite body markings.
**Recent Synonyms**  None.
**Other Common Names**  Short-tailed Wallaby, Short-tailed Pademelon.
**Status**  Common, limited.
**Subspecies**  None.
**References**
Tyndale-Biscoe, H. (1973). *Life of Marsupials.* Edward Arnold, London.
Sharman, G.B. (1975). Studies on marsupial reproduction III. Normal and delayed pregnancy in *Setonix brachyurus. Aust. J. Zool.* 3: 56–70.

but, although such temperature is seldom encountered on the island, animals around soaks fight for the available shelter on hot summer days.

Local populations are widely dispersed during winter but, as the days begin to become hotter in November, they begin to converge at night, sometimes from as far afield as 2 kilometres, around soaks that provide the only permanent fresh water. Each soak is used exclusively by a group of animals from the surrounding area and the group has a well developed social organisation. Adult males form a linear hierarchy based on age and are dominant to females and juveniles, which themselves have no ranking. Males defend an individual space and this defence is most

K. GRIFFITHS

*The very short-tailed Quokka is now largely restricted to Rottnest Island.*

marked in the vicinity of their resting sites. Populations living in areas distant from soaks form groups of 25–150 adults which occupy group territories. Very few individuals move outside their group territories, the boundaries of which are generally coincident with topographic features.

On the mainland, the Quokka appears to be able to breed throughout the year but on Rottnest Island the breeding season is brief. Females come into oestrus in January if the year is mild but, in hot years, do not do so until March. The single young is carried in the pouch until August and is suckled until about October. Most females carry a quiescent blastocyst resulting from a mating shortly after the first birth of the year but few blastocysts resume development after that young has left the pouch.

D. J. KITCHENER

# Swamp Wallaby

## *Wallabia bicolor*

### (Desmarest, 1804)

wol-ah'-bee-ah bie'-kol-or: 'two-coloured wallaby'

A combination of genetic, reproductive, dental and behavioural characteristics set the Swamp Wallaby so far apart from other wallabies that it is classified as the sole living member of the genus *Wallabia*. For example, while wallabies in the genus *Macropus* have 16 chromosomes, the Swamp Wallaby has 11 in the male and ten in the female.

The Swamp Wallaby lives in thick undergrowth in forest, woodland and heath in eastern and southern Australia from Cape York to south-western Victoria. Its presence was reported in south-eastern South Australia in the early twentieth century but if it occurs there today it is quite rare. Areas of dense grass or ferns, sometimes in wet spots on hillsides of open eucalypt forest, provide daytime shelter from which it emerges to feed at night. In Queensland, brigalow scrub

*Wallabia bicolor*

**Size**

*Head and Body Length*
723–847 (756) mm (males)
665–750 (697) mm (females)

*Tail Length*
690–862 (761) mm (males)
640–728 (692) mm (females)

*Weight*
12.3–20.5 (17) kg (males)
10.3–15.4 (13) kg females)

**Identification** Dark brown to black above; light yellow to strong rufous-orange below. Light yellow to light brown cheek-stripe (conspicuous in northern part of range, slight in the south). Extremities usually darker but tip of tail is occasionally white, especially in Queensland. Distinguished from other wallabies within its range by very dark colour. Gait different from other wallabies: head low, with tail straight out behind.

**Recent Synonyms** *Protemnodon bicolor*.

**Other Common Names** Black Wallaby, Black-tailed Wallaby, Fern Wallaby, Black Pademelon, Stinker (Queensland), Black Stinker (New South Wales).

**Status** Common.

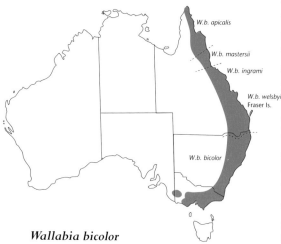

*Wallabia bicolor*

is particularly favoured and in the brigalow belt of southern inland Queensland it is a common species. Local distribution over its range appears to be determined by the availability of adequate dense vegetation for shelter. Although solitary, it may aggregate when feeding.

Limited information suggests that food plants include both pasture and shrub species. In Queensland, agricultural crops adjacent to suitable shelter are readily grazed especially during winter and in dry times. In Victoria, the Swamp Wallaby eats a wide range of native and exotic vegetation including pine tree seedlings. Preference is shown for the coarse browse supplied by shrubs and bushes rather than grass. Among

G. WHITMONT

*The relationships of the Swamp Wallaby to other macropods remains puzzling: it is primarily a browser.*

months old. After a gestation of 33–38 days, a single young is born. The Swamp Wallaby exhibits embryonic diapause but the mating which results in a quiescent blastocyst takes place up to eight days before the birth of an established foetus. The Swamp Wallaby thus differs from most, if not all, marsupials in having a gestation period longer than the oestrous cycle. The pouch life of the young is complete by eight to nine months but it continues to suckle as a young-at-foot until about 15 months old.

The coarse fur and small body size make the Swamp Wallaby unattractive to commercial shooters.

J.C. MERCHANT

**Subspecies** The species needs revision at infraspecific level and no great confidence can be placed in the supposed subspecies listed below.
*Wallabia bicolor bicolor*, New South Wales, Victoria.
*Wallabia bicolor apicalis*, Cape York, Queensland.
*Wallabia bicolor mastersii*, Cairns region, Queensland.
*Wallabia bicolor ingrami*, Queensland, south of Cairns.
*Wallabia bicolor welsbyi*, Stradbroke Island, Queensland.

**References**

Edwards, G.P. and E.H.M. Ealey (1975). Aspects of the ecology of the Swamp Wallaby *Wallabia bicolor* (Marsupialia: Macropodidae). *Aust. Mammal.* 1: 307–317.

Kirkpatrick, T.H. (1970). The Swamp Wallaby in Queensland. *Qld. Agr. J.* 96: 335–336.

Hollis, C.J., J.D. Robertshaw and R.H. Harden (1986). Ecology of the swamp wallaby in north-eastern New South Wales, Australia, I. Diet. Aust. *Wildl. Res.* 13: 355–66.

the plants eaten are Bracken Fern (believed to be poisonous to cattle) and the introduced hemlock (which is highly poisonous to man and stock). It had been suggested that the browsing habit may be reflected in the shape of the molars, which differs from that of other wallabies. Like the Agile Wallaby, the Swamp Wallaby has a broad fourth premolar which is never shed and is used for cutting coarse plant material.

Breeding, which may occur throughout the year, begins in both sexes when they are 15–18

## Subfamily Sthenurinae

The Banded Hare-wallaby appears to be the sole survivor of a once large group of macropods. Unlike any other living macropods, its lower incisors bite against the upper incisors and the fur over the rump is transversely barred. While this opinion is favoured by the Editor, it must be mentioned that there is a school of thought that regards this species as a macropodine that, by convergent evolution, has come to resemble a sthenurine.

# Banded Hare-wallaby

## *Lagostrophus fasciatus*

(Pèron and Lesueur, 1807)

lag'-oh-stroh'-fus fas'-ee-ah'-tus: 'banded turning-hare'

First figured enchantingly, if not very accurately by C.A. Lesueur from specimens taken at Shark Bay, Western Australia, in 1801, the Banded Hare-wallaby was one of the earliest macropodids to be described. Although once more widespread, it is now restricted to Bernier and Dorre Islands in Shark Bay.

On these islands it is commonly found among dense *Acacia ligulata* scrub, beneath which it forms runs and in which it shelters during the day. Other low, spreading shrubs and bushes (such as *Heterodendron oleifolium*, *Acacia coriacea* and *Diplolaena dampieri*) may be utilised and several individuals are often found within one patch of scrub. Although adults of each sex appear to live within well-defined individual home ranges or territories, relations between adult females and between juveniles and adults appear to be largely unaggressive. Interactions between

*Lagostrophus fasciatus*

**Size**
*Head and Body Length*
400–450 (430) mm
*Tail Length*
350–400 (370) mm
*Weight*
1.3–2.1 kg
Occasional males up to 2.5 kg, occasional females up to 3.0 kg

**Identification** Dark grizzled grey above, with transverse dark bands across lower back and rump; greyish-white below. Distinguished from *Lagorchestes hirsutus* and *Bettongia lesueur* (the only other macropods on Bernier and Dorre Islands) by much darker colour.
**Recent Synonyms** None.
**Other Common Names** Indigenous names: Marnine, Merrnine, Munning.
**Status** Common in suitable areas on Bernier and Dorre Islands. Vulnerable to further habitat disturbance or introduction of predators to these islands.
**Subspecies** *Lagostrophus fasciatus fasciatus*, Bernier and Dorre islands.
*Lagostrophus fasciatus albipilis*, south-western Western Australia (extinct).

M. LOCHMAN

*The Banded Hare-wallaby appears to be the sole surviver of an otherwise extinct subfamily of macropods.*

males, on the other hand, are characterised by a high level of aggression, the intensity of fighting apparently being related to competition for food. Since the mortality rate of adult males (30 per cent per annum) is higher than that of females (25 per cent), local groups probably include more females than males, each male defending a territory which overlaps those of several females.

Surveys in 1988/89 showed that the total insular population was about 7700, divided al-most equally between the two islands. On each island, it was associated with thick scrub shelter; the *Triodia* grassland which had developed on Dorre Island after an extensive fire in 1973 being little used. Historical data suggest that it has long been more abundant than the Rufous Hare-wallaby on the islands but it is likely that populations fluctuate considerably in response to varying annual rainfall.

Feeding takes place at night in open areas with scattered shelter. Grasses normally make

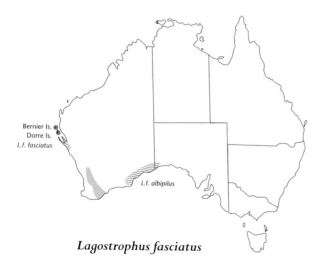

*Lagostrophus fasciatus*

January or February may give birth later in the first half of the year. Some females also mate shortly after giving birth and retain a quiescent blastocyst which begins to develop following the premature loss of the preceding pouch-young, or just prior to its normal vacation of the pouch; others return to oestrus at this time. This variability generally leads to births of some young from December to September, but females usually are reproductively inactive in October and November. Severe summer drought can disrupt the above pattern, breeding being resumed after the drought has broken.

Young spend about six months in the pouch and are weaned about three months later. Both males and females are capable of breeding in the first year of life, but usually do not do so until the second year. Although females appear capable of producing two young per year, they normally manage to rear only one.

R.I.T. PRINCE

up less than half the dietary intake, the remainder being made up by malvaceous and leguminous shrubs and other dicotyledonous plants. Free water is seldom available but heavy dews are frequent on the islands and moisture from this source, together with water in the food, appears to meet the animal's needs.

Fossil remains show that the range of the Banded Hare-wallaby extended across the southern Nullarbor Plain into the lower Murray River region of South Australia within the past 3000–5000 years. Comments by Gould and Sturt suggest that it may have been found in South Australia within the past 150 years but, since European settlement, no live animals have been taken on the mainland except in south-western Australia.

A peak of births occur in the latter half of summer, but females that are unsuccessful in

### References

Tyndale-Biscoe, C.H. (1965). The female urogenital system and reproduction of the marsupial *Lagostrophus fasciatus*. *Aust. J. Zool.* 13: 225–267.

Short, J. and B. Turner (1992). The distribution and abundance of the banded and rufous hare-wallabies, *Lagostrophus fasciatus* and *Lagorchestes hirsutus*. *Biol. Conserv.* 60: 157–166.

# ORDER NOTORYCTEMORPHIA
## *Marsupial Mole*

## FAMILY NOTORYCTIDAE
### *Marsupial Mole*

The remarkable resemblance of the Marsupial Mole to the eutherian Golden Mole of Africa is a striking example of convergent evolution. The two are almost identical in size and shape and each is covered with fine, silky fur. Both are blind and lack external ears; each has a horny shield on the snout and has only the stub of a tail. In each, the claws of the feet have been transformed into spade-like structures. Derived from vastly different ancestors, each has arrived at the same solution to problems posed by a burrowing existence in sandy deserts.

The Marsupial Mole's degenerate dentition is not diprotodont but some authorities argue that neither is it polyprotodont (as in dasyuroids and perameloids). Whether or not the hindfoot shows signs of syndactly (indicating a relationship with the perameloids or diprotodonts) is also in doubt. A chromosome complement of 2n=20 has been cited as evidence of a relationship to the dasyuroids and/or perameloids but biochemical studies do not demonstrate a close relationship with *any* other marsupials.

## Marsupial Mole

### *Notoryctes typhlops*

### (Stirling, 1889)

noh'-toh-rik'-tayz  tif'-lops:
'blind southern-digger'

The Marsupial Mole spends most of its life underground and is blind, the eyes being reduced to vestigial subcutaneous lenses. The ears are represented externally only by small holes surrounded by dense hair. The snout is protected by a horny shield and the tail is reduced to a leathery stub. It is widely distributed over the deserts of Australia, particularly in sand-dunes, interdunal flats and sandy soils along river flats.

The moles of Eurasia and America use all five toes of the forelimb to construct more or less permanent burrows but the Marsupial Mole, like the Golden Mole, digs with the aid of two flattened claws. As it progresses, the burrow fills in behind, so that it could almost be said to 'swim' through the desert sand which is its home. In firm sand, its path is marked by a barely perceptible tube, oval in cross-section and filled with looser material. Often it burrows horizontally about 10–20 centimetres below the surface and then, for no apparent reason, dips vertically to depths exceeding 2.5 metres. A shaft sunk in an unsuccessful attempt to locate an animal at the bottom of such a burrow was criss-crossed with as many as six older sand-filled tubes.

The Marsupial Mole occasionally comes to the surface and seems more inclined to do so after rain. Above the ground, it moves in a sinu-ous fashion, dragging its legs to leave parallel furrows on either side of the central depression made by the body and downwardly turned tail. Captive animals often feed while above ground then retire below to sleep. Occasionally they fall asleep above the ground for several hours. Ant pupae, sawfly larvae, beetles, scarabaeid and longicorn beetle larvae and the larvae of cossid moths have been eaten by wild and captive an-imals.

The male has no visible scrotum but, con-trary to some claims, the testes do not lie in the abdominal cavity: they are situated between the skin and the abdominal wall. The backwardly opening pouch of the female encloses two teats and a rudimentary pouch may be present in the male. Little is known of reproduction and only a few of the numerous specimens which have reached museums have pouch-young. Still less is known of the social organisation of the species but the burrowing habit and the impermanent

## Notoryctes typhlops

**Size**

*Head and Body Length*
121–159 mm
*Tail Length*
21–26 mm
*Weight*
40–70 g

**Identification** Mole-like: cannot be confused with any other Australian mammal.
**Recent Synonyms** *Notoryctes caurinus*.
**Other Common Names** Indigenous names: Yitjarritjarri (Luritja, Ngaatjatjarra, Pintupi, Pitjantjatjara), Kakarratulpa (Manytjilytjarra, Ngaatjatjarra, Nyangamarta, Pintupi, Putitjarra, Warnman), Yirtarrutju (Ngaanyatjarra, Ngaatjatjarra, Pintupi, Pitjantjatjara), Putjurrputjurrpa (Warlpiri).
**Status** Common (but rarely observed).
**Subspecies** None.
**References**

Stirling, E.C. (1891). Description of a new genus and species of Marsupialia, '*Notoryctes typhlops*'. *Trans. Roy. Soc. S. Aust.* 14: 154–187.

Johnson, K.A. and D.W. Walton (1989). Notoryctidae (in) D.W. Walton and B.J. Richardson (eds) *Fauna of Australia. Volume 1B Mammalia.* Aust. Govt. Publ. Serv., Canberra. pp. 591–602.

*Notoryctes typhlops*

nature of the burrows is suggestive of a solitary existence. Just how a young animal fends for it-self after leaving the pouch remains one of the many mysteries surrounding this fascinating but little-understood animal.

The relationship between the Marsupial Mole and other mammals has been a matter of conjecture ever since Stirling erroneously placed it with the monotremes in 1888. It is undoubt-edly a marsupial but Calaby and others point out

D. ROFF

that it has characteristics in common with almost every other living marsupial family. On the basis of chromosome studies, they placed it with the polyprotodonts, but Baverstock was unable to find any molecular basis for this affiliation and concluded that the Marsupial Mole represents a distinct and unique lineage at least 50 million years old.

A second species, *N. caurinus* was described in 1920 from specimens collected near the Eighty Mile Beach in north-western Western Australia. It differs from the better-known form in weight and in the morphology of the nose-shield and tail. It is not sufficiently known to warrant separate treatment here.

*The Marsupial Mole is so different from all other marsupials that it cannot be classified with confidence.*

Because of its great scientific interest, the survival of the Marsupial Mole is a matter of concern. Its range does not seem to have contracted since its discovery and the rate at which dead specimens reach museums (5–15 per decade) has remained fairly constant through the twentieth century. However, some anecdotal reports indicate a decline in abundance, indicating the need for studies to monitor its abundance.

K.A. JOHNSON

# Subclass Eutheria
## *Eutherian or Placental Mammals*

**ORDER CHIROPTERA**
*Bats*

**ORDER RODENTIA**
*Rodents*

**ORDER SIRENIA**
*Dugong and Manatees*

**ORDER CARNIVORA**
*Carnivorous Eutherian Mammals*

**ORDER LAGOMORPHA**
*Rabbit and Hare*

**ORDER PERISSODACTYLA**
*Odd-toed Ungulates*

**ORDER ARTIODACTYLA**
*Even-toed Ungulates*

The eutherians include humans and many other mammals that are familiar for a variety of reasons. Many are large; some have been domesticated; many are active by day; they are the *only* mammals living in The Old World; and they are much more numerous than monotremes and marsupials. The immense evolutionary radiation of the eutherian mammals (extending from terrestrial, arboreal, burrowing and flying species to those that are entirely aquatic) is unmatched by any other group of living vertebrates but was equalled, in the past, by reptiles. While concepts such as 'superiority' are dangerous in biology, it is reasonable to say that the eutherian mammals are now the dominant vertebrates. In general, where eutherians have been in competition with marsupials, the former have been more successful—but not universally so—and the specialisations of the three living monotremes are such that they do not come into competition with eutherians.

The success of eutherians is often attributed to their method of nurturing a foetus by means of a very efficient placenta, enabling a female eutherian to produce very large and active young (as in horses, deer and elephants). On the other hand, many eutherians give birth to extremely dependent young that are blind and helpless, some of which (like marsupials) attach themselves to their mothers' teats shortly after birth.

It is not generally recognised that some marsupials, such as bandicoots and the Koala, also have well-developed placentas but nevertheless give birth to young that lack fully developed hindlimbs. The widespread use of the term 'placental mammals' for the Eutheria is misleading.

The assumption that the eutherian placenta is a significant 'improvement' upon the marsupial condition has been offered as an explanation for the success of the eutherians but there is no evidence that, in competition between the two types of mammal, the outcome has been determined by efficiency of reproduction. On the contrary, many studies indicate that the different strategies of reproduction in marsupials and eutherians involve a comparable input of energy and have similar outcomes.

Australian eutherians comprise seven orders of living mammals. The Chiroptera (bats) preceded human occupation of the continent, as did the Sirenia (Dugong) and marine Carnivora (fur-seals and seals) in coastal waters. Human activities are responsible for the introduction of some other rodents, terrestrial Carnivora (Cat, Dog, Fox), Lagomorpha (Rabbit, Hare), Perissodactyla (Horse, Donkey) and Artiodactyla (Camel, Pig, Goat, deer and cattle).

Previous Page: *Grey-headed Flying-fox.* K. GRIFFITHS
Inset: *Plains Rat.* L.F. SCHICK

# ORDER CHIROPTERA

## *Bats*

Bats are the only mammals capable of sustained flight, an ability otherwise restricted among living animals to birds and insects. The wing consists of a thin membrane with a leading edge extending from the shoulder to the base of the thumb and thence to the tips of the second and third toes and a trailing edge extending from between the tips of the third, fourth and fifth toes to the ankle. The membrane is usually continued from the ankle to the tail to form a tail-membrane which varies considerably in relative area among the various families of bats. It may reach the tip of a long tail (thereby contributing quite significantly to the total flight surface); incorporate only part of the tail; or be an insignificant ridge of skin extending to the base of the tail. In bats that lack a tail, a membrane may or may not extend between the hindlegs.

Although the skeleton of the forearm of a bat has the same basic structure as that of a human, its proportions are very different. The upper arm is shorter, the forearm much longer and all the toes except the thumb are extraordinarily elongated. The tip of the thumb is free of the wing membrane and bears a claw, used in crawling, for support and for grooming. The fingers (except for the second finger of fruit-bats) are clawless. Flight is powered mainly by muscles that pull the forearm towards the chest. To provide a surface for attachment of these large muscles, the sternum, or breastbone, is keel-shaped, as in flying birds.

The hindlimb extends outwards and backwards from the body, acting as a strut that supports the rear end of the wing. The foot has five clawed toes, usually all of much the same length and arranged parallel to each other. Typically, a long, slender bone, the calcaneum or calcar, extends inwards from the ankle, providing a support for the rear edge of the tail-membrane. Some bats use the hindfeet to catch small fishes near the surface of water: the Large-footed Myotis catches aquatic insects in this manner.

The structure of the hip joint is such that the hindlimb cannot be brought into a vertical supporting position under the body as in most mammals, but in many bats it can be flexed to an intermediate angle (with the axis of the foot at a right angle to that of the body). Despite this ungainly posture, all bats can climb on vertical surfaces, supporting themselves with their thumbs and feet and are capable of a scurrying gait—sometimes quite fast—on a level surface. When resting, a bat hangs by the claws of one or both of its forefeet and the arrangement of ligaments in the foot is such that this does not require any muscular effort: an individual that has died while roosting may therefore remain suspended. Although an inactive bat is usually upside-down, it reverses itself, supporting itself by its thumbs, in order to void urine or faeces. Most bats fold their wings along the side of the body when roosting, but fruit-bats wrap their wings around the body, like a blanket.

Present and past distribution indicate that bats originated in tropical forests. Indeed, it seems that they could hardly have begun to evolve anywhere but in hot, humid areas, for their structure and habits involve very great problems of heat and water loss. Most bats are small and therefore have a relatively large body surface from which heat is lost by radiation: the extended wings act as immense radiators which vastly increase this surface. Much more energy is expended in flight than in other modes of locomotion, so a bat must ingest larger quantities of food than a terrestrial mammal of similar size. It is engaged in considerable effort to get enough food to provide energy to hunt for more and it cannot afford a further drain of heat energy from its warm body to cold surroundings. However, warm surroundings reduce this loss. Animals of relatively large surface area also tend to lose considerable amounts of water by evaporation, but this does not present a problem to those with ready access to drinking water and which live in humid conditions.

To reduce the loss of energy through radiated heat, a small mammal may become torpid when at rest, 'switching off its thermostat' and allowing the temperature of its body to fall to about that of its surroundings (by becoming cooler, it also reduces its rate of energy consumption). It may also reduce the rate of heat loss by taking shelter in a warm or insulated area such as a cave or tree hollow. Bats may employ one or both of these strategies. Loss of water by evaporation can be minimised by roosting in restricted areas, as in humid caves.

Bats appear to have evolved from primitive mammals similar to shrews and, like them, to have been originally insectivorous. Most remain insectivorous and, outside the tropics, their problems of heat loss are exacerbated by a great reduction in the amount of food available in winter. Hibernation—an almost complete cessation of bodily activity through the winter months—is a means by which these difficulties can be evaded. When hibernating, a bat may have a temperature only a few degrees above that of its surroundings. Body temperatures close to $0°C$ have been recorded on occasions in some species but temperatures above zero are more usual. The rate of metabolism of a hibernating bat may be as little as 1 per cent of the rate when flying and the heart rate may be correspondingly slower. However, even this state of barely 'ticking over' requires energy, provided by a store of fat laid down in the body of a bat in the period just before hibernation. This resource is seldom much in excess of probable needs, so disturbance of a hibernating bat colony—returning the animals, even briefly, to maximum metabolic activity—can lead to premature exhaustion of food reserves and consequent death of animals before the return of warm weather.

Because most of the research on bats has been conducted in northern Europe and North America, hibernation has come to be seen as a 'normal' activity of bats but it is largely restricted to the minority of species that have managed to establish themselves in regions where the winter is cold. Some species in warm and cool climates cope with seasonal shortages of food by migration.

Associated with hibernation are two alternative reproductive strategies that are uncommon among mammals. One is storage of sperm in the female reproductive tract from autumn to spring, permitting a female to fertilise an ovum as soon as hibernation comes to an end. Another is for an ovum, usually fertilised in autumn, to develop to an early embryonic stage and then remain quiescent until spring, when development resumes. (The phenomenon of embryonic quiescence or diapause is also characteristic of kangaroos but operates with quite different triggering mechanisms.)

The period of gestation, ranging from 50 to 240 days, is relatively long for small mammals, as is the period of maternal dependence (three to ten weeks). Usually, one young is born in each breeding season. Although it may be one-quarter to one-third of the weight of the mother, it is blind and (except in fruit-bats) hairless. The hindlimbs are well developed, with strongly clawed toes, but the forelimbs are proportionately much smaller than in the adult. The milk teeth, which are usually present at birth, are sharp and recurved and used by the young bat, together with the claws of its thumbs and toes, to cling to the mother. A female bat suckles its young from two teats, one under each armpit, but in some groups there is a false (non-lactating) teat in each groin onto which the young may attach itself by its teeth. Females of some species carry their very young infants with them when foraging but, more commonly, these are left behind, often in nursery colonies in a cave. Such a colony, comprising from a few score to tens of thousands of individuals, generates a considerable amount of heat and moisture, creating a protective environment for the young.

No bats are blind but many insectivorous bats have very small eyes. By emitting ultrasonic calls and assessing the position of objects that reflect these vibrations, many bats are able to navigate and to locate prey without using sight. Very short pulses of ultrasound (or barely audible frequencies in some species) are produced rapidly (up to 200 times per second) by the larynx and are projected either through the mouth or through the nostrils. Reflections are collected by the pinnae, or ear-flaps, which are usually large and often complex in shape. A prominent lobe, the tragus, which covers part of the aperture of the pinna in many bats, increases the directional sensitivity of the ear. Horseshoe-bats and leafnosed-bats, which lack a tragus, have a fold of skin at the base of the pinna (the antitragus), which probably has a similar function. They also have projecting structures on the snout which serve to direct ultrasonic cries emitted through the nostrils. Members of the family that includes the Ghost Bat use sight as well as echolocation when hunting: they have a nose-leaf, a tragus and large eyes. The process of echolocation is complex and involves far more than the simple determination of the direction and distance of a sound-reflecting object. Emissions may vary in intensity, in frequency and in pulse rate and these components themselves are varied in respect of a bat's interest in its surroundings—whether 'cruising', 'searching', or 'homing-in' on prey. Experimental research has yet to provide full explanations of how the various kinds of bat calls are used to locate a *single* object and it must be recognised that, at any given moment, an individual is receiving information from every sound-reflecting object in the beam (and range) of its emitted vibrations. A considerable part of the mid-brain is devoted to interpretation of the ultrasonic input to the ears and it seems likely that signals are integrated into a 'coarse-grained picture' of the surroundings, including trees, foliage, rocks, holes, flying insects and other bats. Most fruit-bats have large eyes and no capacity for echolocation, but members of the African and Asian genus *Rousettus*, which roost in dark caves, have a simple echolocatory system based on audible clicks produced within the mouth rather than by the larynx.

Bats are a very successful mammalian group, second only in number of species to the rodents: about 40 per cent of the species of living mammals are rodents and about 20 per cent are bats. Of the 15 or so families of bats currently recognised, six are represented in Australia; these also occur in other continents, particularly in the tropics, and one large family, the Vespertilionidae, has a worldwide distribution. Significantly, all the families occurring in Australia and New Guinea are well represented in south-eastern Asia and, perhaps even more significantly, only one of the families occurring in Asia (mouse-tailed bats, of the family Rhinopomatidae, which extends no further eastward than central Sumatra) does *not* also occur in Australia. This biogeographic pattern shows that the Australian bats are derived from the Asian fauna.

Most of the genera of Australian bats also occur in New Guinea and Asia and, in each genus, there are usually more New Guinean than Australian species. Tropical New Guinea offers more ecological niches for bats than Australia but the greater species diversity in this region is probably a reflection of a 'filtering' effect on bat colonisation from Indonesia into New Guinea and thence to Australia. Two-thirds of the Australian bat species are found on Cape York, which appears to have been the main region of entry from New Guinea. As elsewhere in the world, the diversity of bats diminishes with increasing latitude. Only six species are known in Tasmania.

Bats have undergone less evolutionary radiation in Australia than rodents, which seem to have entered the continent at much the same time. Only two genera, *Rhinonycteris* and *Macroderma*, are restricted to Australia and neither of these is remarkably different from its nearest Asian relative. About half of the species are endemic.

Bats fall into two suborders, so distinct that some authors postulate that they may not share a common ancestor. The Megachiroptera, with some species weighing more than 1 kilogram and with a wingspan up to 1.6 metres, includes the flying-foxes, fruit-bats and blossom-bats. The Microchiroptera is a very diverse group of smaller and mainly insectivorous bats.

# SUBORDER MEGACHIROPTERA: MEGABATS

These bats tend to be large (up to 1 kilogram or more and with a wingspan up to 1.6 metres). They have claws on the first and second toes of the forelimb (only on the first in the Microchiroptera) and the tail, if present at all, is short. When roosting, a megachiropteran hangs upside-down by a hindfoot with the wings wrapped around its body and the head held perpendicular to the chest. At birth, it is somewhat more advanced than a newborn microchiropteran, often with a fine covering of hair. For some weeks it is carried by its mother, clinging to the fur with its feet and to a teat with its teeth.

The seemingly more 'simple' structure of megachiropterans has led to the belief that they are descendants of a group of primitive bats ancestral to the microchiropterans. There is little evidence to support this view but it does seem that they represent an early branch from the main line of bat evolution, retaining some primitive features while specialising to feed on plants. Most herbivorous mammals are characterised by a number of special masticatory and digestive mechanisms required to deal with vegetable fibres, but the megachiropterans, by concentrating on plant juices, avoid this challenge. One consequence of their specialisation is that they are largely restricted to tropical forests where succulent fruits can be found throughout the year. Those living in temperate regions usually engage in considerable seasonal migrations in search of food. Megachiropterans do not hibernate. The Megachiroptera includes only one family, the Pteropodidae.

# FAMILY PTEROPODIDAE

## *Flying-foxes, fruit-bats, blossom-bats*

*The Northern Blossom-bat feeds mainly on nectar from eucalypts and rainforest trees.*

The most familiar members of this family are the large flying-foxes and fruit-bats of Africa, Asia and the Australian region. Most are tropical but a few species extend into warm-temperate regions. The family also includes some quite small species that feed mainly on nectar and pollen.

The family is divided into four subfamilies, three of which are represented in Australia. The Macroglossinae comprises blossom-bats that are characterised by small size and a long, flexible tongue with a brush-like tip. The Nyctimeninae, with only two species in Australia, have tubular nostrils. The Pteropodinae comprises the large fruit-eating bats.

## Subfamily Macroglossinae

The two genera of blossom-bats that comprise this group have a long, flexible, brush-tipped tongue that is used to lap nectar. The nostrils are simple.

# Northern Blossom-bat

## *Macroglossus minimus*

(Geoffroy, 1810)

mak'-roh-glos'-us  min'-im-us:

'smallest long-tongue'

Widespread through South-East Asia and Melanesia, the Northern Blossom-bat extends across northern Australia where it roosts in palms, bamboos, mangroves, monsoon- and rainforest; individuals and small groups shelter under loose bark, within hollows, or among leafy crowns and leaf bases. Elsewhere in its range, it may roost in buildings or adjacent gardens.

Between dusk and dawn, it forages actively in environments ranging from sclerophyll woodland to rainforest. Flowers of *Eucalyptus*, *Melaleuca*, mangrove, *Barringtonia* and bottle-brushes are common sources of food. In settled areas, the flowers of the introduced Sausage Tree and Century Plant are

*The Northern Blossom-Bat is characterised by flap-like interfemoral membranes ('pyjamas') on each leg.*

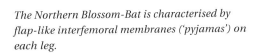

B.G. THOMSON

## Macroglossus minimus

**Size**

*Head and Body Length*
59–64 (61) mm
*Tail Length*
minute
*Forearm Length*
39–42 (40) mm (males)
38–40 (39) mm (females)
*Weight*
12–17 (14) g (males)
10–14 (12) g (females)

**Identification** Small, fawn to reddish; paler on chest. Ears broad, rounded. Distinguished from *Synconycteris* by presence of flap-like inter-femoral membranes ('pyjamas') on each leg. Tail a tiny, black point. Mature males have V-shaped gland on neck or chest, producing a distinctive odour.
**Recent Synonyms** *Odontonycteris lagochilus*, *M. lagochilus*.
**Other Common Names** None.
**Status** Common, unlimited.
**Subspecies** Four subspecies, of which the Australian population is *Macroglossus minimus pygmaeus*.
**Extralimital Distribution** South-East Asia, Melanesia.
**References** None relevant.

also exploited. Durian trees in north-eastern Queensland appear to be pollinated by Northern Blossom-bats, which are attracted in considerable numbers to flowering trees. Native cauliflorous *Syzygiums* are also a major attractant. Together with *Syconycteris australis*, where their ranges overlap, *Macroglossus minimus* appears to be a major pollinator of a wide variety of native species. A foraging bat lands on or hovers near a flower, lapping nectar with its long papillated tongue and exchanging pollen deposited on its fur.

Social activity such as chasing, wing-clapping and vocalisations are often heard near feeding trees.

In north-eastern Queensland, *M. minimus* is also frugivorous, eating the female fruits of several species of dioecious fig as well as *Timonius* sp. In contrast with a flying-fox, this bat is able to swallow relatively large seeds in comparison with its body size and, for these plants, it is an important disperser. There is also evidence of it eating leaves.

The teeth are extraordinarily reduced and a wide range of dental abnormalities have been noted, in particular the variable reduction of incisors and the presence of additional molars, indicative of the lack of selective significance of teeth in this fluid-feeder.

No females with young have ever been netted, but lactating females in north-eastern Queensland appear in December, with some evidence of a second weaker peak of births in February. Breeding appears to become aseasonal in the far northern part of its distribution.

The ranges of *Macroglossus minimus* and *Syconycteris australis* overlap in north-eastern Queensland, where the two bats appear to be virtually indistinguishable in size and colour and to have virtually the same feeding preferences and breeding patterns.

J.L. McKean, M. McCoy and H.J. Spencer

*Macroglossus minimus*

# Common Blossom-bat

## *Syconycteris australis*

(Peters, 1867)

sike'-oh-nik'-ter-is ost-rah'-lis:
'southern fig-bat'

M. TRENERRY

*The mouse-sized Common Blossom-bat is one of the smallest of the pteropodids.*

Although also referred to as the Queensland Blossom-bat, this species is widely distributed in New Guinea and, in Australia, extends well into New South Wales. It is one of the smallest pteropodids: the size of a mouse. In areas where the Common Blossom-bat coexists with the very similar *Macroglossus minimus*, it is often very difficult to differentiate between the two species, the most useful character being the presence or absence of an interfemoral flap on the inside of each leg: *Syconycteris* lacks these.

With large eyes and a keen sense of smell, it seeks out blossoms loaded with nectar and pollen. It may hover in front of a flower, but appears incapable of feeding in flight: instead it

*Syconycteris australis*

crash-lands onto a promising blossom. The pointed snout and long brush-tipped tongue are well adapted to gathering nectar and the body hairs are covered with minute scale-like projections in which pollen lodges. Interestingly, pollen is not consumed directly from a flower but by grooming the fur and wings after a bout of nectar feeding. Pollen passes rapidly through the gut, usually beginning to appear in the faeces within 45 minutes of ingestion. Pollen provides almost all the protein in the diet; sugars in the nectar meet the energy requirements.

While it is considered to be a specialist blossom-feeder, there appears to be considerable geographic variation in the involvement of fruit and leaf material in the diet. Bats in the southern extension of the range are mostly strict nectarivores, whereas in the north-east Queensland area they are facultative frugivores and, to a lesser extent, folivores. Fig seeds, particularly from the female fruit of *Ficus variegata*, *F. congesta*, *F. racemosa* and other species, as well as the seeds of *Timonius* sp.—a common member of the Rubiceae—are frequently found in the faeces.

In tropical Australia, the Common Blossom-bat roosts among rainforest foliage and feeds on nectar and pollen from a variety of rainforest trees, including the cauliflorous *Syzigium* species. It is an important pollinator of *Syzigium*. At the southern limits of the range, heathland and melaleuca swamps are critical feeding habitats: *Banksia*, *Melaleuca*, *Callistemon* and certain eucalypts are favoured sources of food in these

areas. Subtropical rainforest and coastal eucalypt forests are visited when heaths and paperbarks are not flowering.

Coastal rainforest is a particularly important habitat in New South Wales, providing a moist microclimate for roosting in the subcanopy layer. Such areas are usually adjacent to heathland feeding areas. Roosts here are changed frequently but the distance between consecutive roosts is usually less than 50 metres. In the wet tropics where there is no heathland, the bats often visit fruit orchards to feed on blossoms of flowers such as Wax-jambu and Durian.

The Blossom-bat has been recorded as commuting up to 4 kilometres to feed from a roost, such mobility being necessary to exploit spatially and temporally patchy food resources. In coastal New South Wales some species, such as

---

## Syconycteris australis

### Size
*Head and Body Length*
*circa* 60 mm
*Tail Length*
Nil
*Forearm Length*
38–44 (41) mm (males)
39–45 (43) mm (females)
*Weight\**
16–22 (19) g (males)
16–26 (19) g (females)
\* Northern population slightly lighter.

**Identification** Fur fawn to reddish and very soft. Nostrils raised above surface of muzzle. Long, brush-like tongue; incisors slender and weak. Differs from *Macroglossus minimus* in having obvious upper incisors and lacking interfemoral flaps (pyjamas) on legs, there being a strip of long fur in this location. Muzzle is slightly shorter and eyes

relatively large. Males lack scent gland.
**Recent Synonyms** None
**Other Common Names** Queensland Blossom-bat, Eastern Blossom-bat.
**Status** Common in limited habitat.
**Subspecies** None.
**Extralimital Distribution** New Guinea.
**References**

Nelson, J.E. (1964). Notes on *Syconycteris australis*, Peters, 1867 (Megachiroptera). *Mammalia* 28: 429–432.

Law, B.S. (1992). Physiological factors affecting pollen use by the Queensland Blossom Bat (*Syconycteris australis*). *Functional Ecol.* 6: 257–264.

Law, B.S. (1993). Roosting and foraging ecology of the Queensland Blossom-bat (*Syconycteris australis*) in north-eastern New South Wales: flexibility in response to seasonal variation. *Wildl. Res.* 20: 419–31.

HANS & JUDY BESTE

*Common Blossom-bat feeding on a banana flower.*

*Banksia integrifolia*, flower consistently for up to six months and blossom-bats show great fidelity to these feeding areas. In the wet tropical areas, it is an extremely opportunistic feeder and can appear in large numbers at a briefly flowering food tree. The extent of seasonal movements is not known, but some marked bats return to the same feeding sites over several years. An individual appears to defend its food source from other bats by chasing the intruder, vocalising and clapping its wingtips together.

In New South Wales, one young is usually born in October or November and another between February and April. Lactation lasts up to three months. It remains active during the relatively cold winter. In northern Australia and the New Guinean part of its range, breeding appears to occur throughout the year.

The Common Blossom-bat enters torpor rapidly, and arouses equally rapidly. In captivity, torpor occurs when food is withheld and the daytime temperature is below 26°C: the minimum temperature of torpid individuals is about 20°C.

Blossom-bat abundance in a given area is related to the abundance of food flowers. In New South Wales, artificial supplementation of nectar in winter leads to an increase in numbers. This indicates that the capacity of blossom-bats to meet the energy costs of flight and thermoregulation limits their population size, rather than (as has been suggested for flying-foxes) the availability of protein in the diet.

B.S. LAW AND H.J. SPENCER

## Subfamily Nyctimeninae

The members of this family have a short snout, simple ears, and prominent tubular nostrils. The lower incisors are absent and the lower canines bite against the upper incisors. The second digit of the forelimb lacks a claw.

# Eastern Tube-nosed Bat

## *Nyctimene robinsoni*

### Thomas, 1904

nik'-tee-may'-nay  rob'-in-sun-ee:
'Robinson's moonlight (-bat)'

One of the smaller Australian megachiropterans, the Eastern Tube-nosed Bat is distinguished by its protuberant nostrils and the bright yellow spots on its wings, ears and nostrils. The wings are brown in colour, very variably spotted and together with a dark stripe along the midline of the back, ensure that when the animal is roosting with its wings wrapped around it, it closely resembles a dead leaf. It appears to be primarily solitary, but groups of up to five bats in one tree have been observed relatively frequently in the Queensland coastal rainforest.

In the northern part of its Australian range, it appears to be a fruit specialist, favouring figs,

*The Eastern Tube-nosed Bat flies slowly and is able to hover while feeding.*

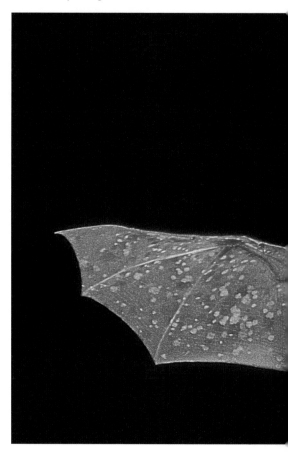

*Nyctimene robinsoni*

particularly *Ficus variegata*, the common cluster fig, as well as the Blue Quandong *Eleaocarpus grandis*, Burdekin Plum *Pleiogynium timorense* and other fleshy fruit. It is reported to feed on *Banksia* blossoms in heathland communities but, unlike the blossom bats, *Nyctimene* has quite a short tongue and, in this case, may be capitalising on an abundant resource. Guavas are a favourite food and individuals have been seen feeding in these trees only centimetres from the ground and thus becoming the favoured victim of cats in the area. In north-east Queensland, *Nyctimene* is in danger of becoming an orchard pest through its interest in exotic fruit, especially Soursop (*Annona muricata*).

It has been suggested that the tubular nostrils are an adaptation to feeding on mushy fruit, functioning as miniature 'snorkels' while the snout is immersed in pulp. However, the tube-

M. TRENERRY

*The Eastern Tube-nosed Bat is distinguished by pale (usually yellow) spots on the wings and ears.*

nosed bat normally eats fruit while hanging from its feet and thumb-claws, resting the fruit on its belly, or simply draped around the fruit: hence the nostrils are above the fruit. The dentition is peculiar in that there are no incisors on the lower jaw, their function being assumed by the lower canines which are almost in contact with each other and bite against the two upper incisors. The combination of the upper and lower canines, however, make a very effective mechanism for tearing off a very large fruit and holding it while in flight.

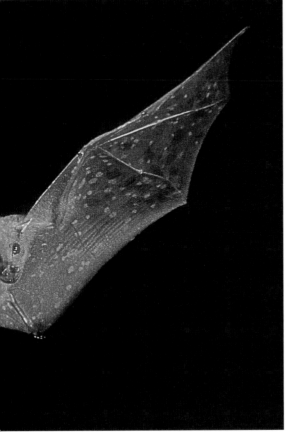

J. PETTIGREW & L.S. HALL

## *Nyctimene robinsoni*

### Size
*Head and Body Length*
100–110 mm
*Tail Length*
20–25 mm
*Forearm Length*
60–70 mm
*Weight*
30–50 g

**Identification**  Distinguished from all other Australian bats by prominent tubular nostrils and clear yellow to green spots (varying from brown to whitish) on wings and ears. Prominent mid-dorsal stripe.

**Recent Synonyms**  *Nyctimene tryoni*, *Nyctimene albiventer* (part).

**Other Common Names**  None.

**Status**  Common.

**Subspecies**  None. Specimens identified as subspecies of *Nyctimene albiventer* from Australia are now regarded as being *N. robinsoni*.

### References

Richards, G.C. (1986). Notes on the natural history of the Queensland Tube-nosed bat. *Macroderma* 2: 64–67.

Spencer, H.J. and T.H. Fleming (1989). Roosting and foraging behaviour of the Queensland tube-nosed bat, *Nyctimene robinsoni*, Pteropodidae: preliminary radio-tracking observations. *Aust. Wildl. Res.* 16: 413–420.

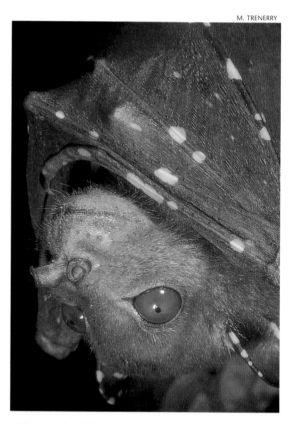

M. TRENERRY

*The nostrils of the Eastern Tube-nosed Bat lie on tubular prominences.*

It is a highly manoeuvrable flyer and can hover with ease. The presence of an active Eastern Tube-nosed Bat is readily detected by its characteristic bleat-like call, even though this appears to be seasonal in nature (through summer) and may be restricted to males. Day roosts are usually within 1 kilometre of the feeding grounds and some studies indicate that individuals have considerable roost site fidelity, the same roost being used for months to years. Other studies show that individuals tend to roost by day in the last food tree used on the previous night. Tagging studies in coastal north Queensland indicate that some bats may be resident in an area for three years or more. Females give birth to one young which is carried by the mother until quite large. (Recent DNA studies suggest that a second species of *Nyctimene* is present in north-eastern Queensland, distinguished by different dentition, a narrower head and a more uniform fur colouration than in *N. robinsoni*.)

L.S. HALL, G.C. RICHARDS AND H.J. SPENCER

# Torresian Tube-nosed Bat

## *Nyctimene vizcaccia*

Thomas, 1914

viz-karch'-ee-ah:
'Vizcaccia's moonlight (-bat)'

G. RICHARDS

*The Torresian Tube-nosed Bat resembles the Eastern Tube-nosed Bat but is smaller.*

● Moa Is.

*Nyctimene vizcaccia*

This small fruit bat and the Torresian Flying-fox are found only on Moa Island in Torres Strait, centred upon the rainforest there. The more ubiquitous Eastern Tube-nosed Bat, *Nyctimene robinsoni*, also occurs on the island, but is less restricted to rainforest.

Very little is known of the biology of the Torresian Tube-nosed Bat and its relationship with extralimital populations.

G.C. RICHARDS

---

### *Nyctimene vizcaccia*

**Size**
*Head and Body Length*
90–100 mm
*Tail Length*
15–20 mm
*Forearm Length*
60–65 mm
*Weight*
approximately 40 g

**Identification** Similar to *Nyctimene robinsoni* but less robust, fawn-brown instead of grey-brown with a less pronounced dorsal stripe and smaller in size.

**Recent Synonyms** *Nyctimene cephalotes vizcaccia*, *Nyctimene bougainville*.

**Other Common Names** None.

**Status** Endangered.

**Subspecies** None.

**Extralimital Distribution** New Guinea, Solomons.

**References** None relevant.

## Subfamily Pteropodinae

This very large group includes the fruit-bats and flying-foxes, characterised by large eyes; simple external ears; and a simple tongue that cannot be extended beyond the lips. Only two genera are represented in Australia: *Pteropus*, including all the flying-foxes; and *Dobsonia*, the Bare-backed Fruit-bat, in which the wings are inserted near to the middle of the back.

# Bare-backed Fruit-bat

## *Dobsonia moluccensis*

(Quoy and Gaimard, 1830)

dob-sohn'-ee-ah   mol'-uk-en'-sis:
'Moluccan Dobson's (bat)'

This bat is immediately recognisable by its apparently hairless back. In most other genera of bats, the wings join the body along the flanks but, in species of *Dobsonia*, the bases of the wings meet along the centre of the back and, since the wings are virtually hairless, the back appears to be naked. Between the undersurface of the wing and the actual back is a large pocket, well furred on its undersurface.

The unusual relationship between the wings and the body gives the Bare-backed Fruit-bat a higher than usual ratio between wing area and body weight, permitting it to fly at low speed without stalling. Apart from the obvious benefit that this confers on the bat as it searches for food in dense vegetation, it seems also to assist it to hover when approaching a roosting site within a cave or similar dark place. In Indonesia and Melanesia, it may be found in caves in colonies

*Dobsonia moluccensis*

**Size**
*Head and Body Length*
circa 300 mm
*Tail Length*
20–30 mm
*Forearm Length*
135–155 mm
*Weight*
380–500 g (males heavier than females)

**Identification** Dark brown to almost black; face rather like a dog. Wing membranes meet in mid-dorsal line. Only a single pair of upper and lower incisors.
**Recent Synonyms** *Dobsonia magna*, *Dobsonia moluccense*.
**Other Common Names** Spinal-winged Fruit-bat, Spinal-winged Bat.
**Status** Rare and limited in Australia, more abundant extralimitally.
**Subspecies** There are four subspecies of which the Australian population constitutes *Dobsonia moluccensis magna*.
**Extralimital Distribution** Ambonia, Ceram, Aru Islands, Papua New Guinea, the Bismarck, dEntrecasteaux and Louisiade Archipelagos.
**Reference**
Dwyer, P.D. (1975). Notes on *Dobsonia moluccense* (Chiroptera) in the New Guinea Highlands. *Mammalia* 39: 113–318.

H. MILLEN

*The Bare-backed Fruit-bat is a slow flier in dense vegetation.*

of thousands but Australian groups are much smaller. Colonies of a hundred or so have been found scattered in dark areas under piles of large boulders and roosting sites have also been found in old mines, abandoned houses and dense vegetation, always in near-darkness. It is the only Australian megachiropteran to regularly roost in caves.

The Bare-backed Fruit-bat has a distinctive 'flapping' flight as it approaches and hovers before landing among the native fruit and blossoms on which it feeds. The Northern Bloodwood is one of its favourite food trees and it is also attracted to flowering bananas.

The structure of social groupings is not known, but the bats tend to be more vocal in their daytime retreats than when feeding at night, when they are very wary. Males mature at about two years of age and copulation occurs at the end of the wet season in May and June. Between September and November each female gives birth to a single young which is carried for about one month and nursed for a further four to five months.

**Dobsonia moluccensis**

Two pairs of scent glands probably have some function in mating. One extends from the base of the ear downward and forward to the angle of the mouth; the other is near the junction of the wing-membrane with the body. These glands produce a characteristic musty odour but do not inhibit the appetites of the New Guineans who frequently eat this bat. Australian Aborigines do not eat the Bare-backed Fruit-bat, possibly because it is far less abundant than flying-foxes.

L.S. HALL

# Black Flying-fox

## *Pteropus alecto*

### Temminck, 1837

te'-roh-poos  ah-lek'-toh: 'fury wing-foot'

In coastal northern Australia, it is not uncommon to witness the evening exodus of flying-foxes from their camps. These are often in mangrove swamps and from a nearby beach or vantage point one can see thousands of bats heading towards their nocturnal feeding grounds. Camps often contain several species.

Mangrove islands in the estuaries of most northern rivers usually contain camps of the Black Flying-fox, often consisting of hundreds of thousands of individuals. Smaller camps of several hundred to several thousand may be found in mangroves, paperbark swamps or occasionally in patches of rainforest. It is a high-roosting species and seeks rather dense leaf cover. The perimeter of a camp is usually guarded by old males which observe intruders and, if these come too close, fly into the camp to raise an alarm.

Under natural conditions, the preferred food includes the blossoms of eucalypts, paperbarks and turpentines. Groups are known to travel as far as 50 kilometres from their camp to feed on

*Pteropus alecto*

## *Pteropus alecto*

**Size**
*Head and Body Length*
240–260 mm
*Forearm Length*
150–182 mm
*Weight*
500–700 g

**Identification** Largest of the flying-foxes. Short black fur over whole of body, sometimes frosted below. Reddish collar around back of neck; brown eye-ring may be present.
**Recent Synonyms** *Pteropus gouldii.*
**Other Common Names** Gould's Fruit-bat, Black or Blackish Fruit-bat.
**Status** Common in Australia and extralimitally.
**Subspecies** This widespread species has at least four subspecies, of which the Australian representative is *Pteropus alecto gouldii.*
**Extralimital Distribution** Celebes, Lombok, Sumba and New Guinea.
**References**
Ratcliffe, F.N. (1931). The flying-fox (*Pteropus*) in Australia. *Bull. Counc. Sc. Ind. Res. Aust.* 53: 1–80.
Nelson, J.E. (1965). Behaviour of Australian Pteropodidae (Megachiroptera). *Anim. Behav.* 13: 544–547.

such trees. Other native and introduced blossoms and fruits particularly mangoes, are also eaten.

Despite its apparently laboured flight, the Black Flying-fox may travel at 25–35 kilometres per hour with an average wing-beat of 120 per minute. Occasionally an animal may glide for 20–30 metres and it is usual to glide into the roost. This is sometimes an elegant performance as the bat stalls, twists its body and grasps a branch in a smooth sequence—but it is not uncommon for one to land among the foliage in an uncontrolled crash.

There is much activity in the camp during the mating season which lasts from March to April. Large males establish a territory on a branch where each defends a circle about 1 metre in diameter. Here they spend a lot of time grooming and displaying their external genitalia. (The testes, which normally reside in a tight scrotum, descend into a prominent position at this time.)

*The Black Flying-fox is very social and roosts in camps containing as many as hundreds of thousands of individuals.*

# Dusky Flying-fox

## *Pteropus brunneus*

Dobson, 1878

broon-ay'-us: 'dark brown wing-foot'

The only known specimen of this animal is an adult male held in the Museum of Natural History, London. According to its label, it was

Erection of the penis while grooming is frequent. A small amount of courtship takes place and is usually in the form of neck-biting and licking of the female's vulva. Intromission occurs for 1–1.5 minutes. Up to four copulations between the same pair can occur in half an hour.

Females give birth to a single young in October but newborn have been seen from August until the end of November. In the northern part of the range, young have been recorded during all months of the year. Although well developed, a newborn bat is incapable of flight and is carried by the mother for the first month or so. During this period it grips the mother's fur and teats with its claws and recurved milk teeth. The teats are located towards the junction of the wing and body, and the young gives the appearance of hiding under its mother's wing. At the age of one month the young is left behind in the camp at night. At two months it can fly and spends much time flapping its wings, but does not leave the camp until about three months old.

L.S. HALL

### *Pteropus brunneus*

**Size**
*Head and Body Length*
Approximately 210 mm
*Forearm Length*
118 mm
*Weight*
Probably 200 g

**Identification** Overall mid-brown; reddish-gold tinge due to pale tips on hairs. Mantle across shoulder area slightly lighter brown, with partially concealed buff-coloured glandular tufts at sides of neck (male). Head paler golden brown. Ventral surface scarcely lighter than dorsal. (Colours may have faded due to age of specimen). Dorsal surface of tibia furred along proximal three-quarters of length.
**Recent Synonyms** None.
**Other Common Names** Percy Island Flying-fox.
**Status** Probably extinct.
**Subspecies** None.
**References**
Andersen, K. (1912). *Catalogue of the Chiroptera in the collection of the British Museum, Second edition. Vol. 1: Megachiroptera*. British Museum (Natural History), London.
Koopman, K.F. (1984). Taxonomic and Distributional Notes on Tropical Australian Bats. *Amer. Mus. Novit.* 2778: 1–48.

*Pteropus brunneus*

collected on Percy Island by Captain Denham of HMS *Herald*. The Museum register indicates that the skin and skull were purchased in 1874 from Stevens' Sale Rooms, erroneously labelled *Pteropus gouldi*.

Little if anything has been added to our knowledge of *P. brunneus* since it was described by Dobson in 1878. Late in the nineteenth century it was said to be plentiful on Percy Island and had been seen flying over the nearby Australian coast. Reference was also made to the odour of its glandular secretions, 'resembling musk'.

The specimen is clearly distinct from the four recognised *Pteropus* species occurring in Australia. It is smaller (forearm length 118 millimetres), with glandular tufts partially concealed by fur on the shoulders. The tibias are furred dorsally. It appears to belong to the *Pteropus hypomelanus* taxonomic grouping within the flying-foxes, also represented by a number of forms around the Coral Sea. Karl Koopman regards *P. admiralitatum solomonis* as the form most closely resembling *P. brunneus*.

It seems likely that the Dusky Flying-fox was represented by a small population in the Percy Isles, vulnerable to habitat alteration and destruction of food sources within a suitable range.

P. CONDER

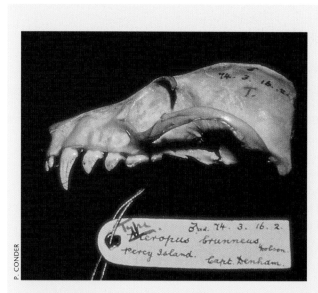

Above: *Skull of the Dusky Flying-fox.*
Right: *Museum specimen of the Dusky Flying-fox.*

# Spectacled Flying-fox

## *Pteropus conspicillatus*

### Gould, 1850

kon-spis'-il-ah'-tus: 'spectacled wing-foot

HANS & JUDY BESTE

*The Spectacled Flying-Fox is distinguished by a prominent ruff of pale fur.*

This species is unique among Australian flying-foxes in being the only rainforest specialist on the mainland. It is primarily frugivorous and appears to have an integral role in rainforest regeneration. Camps of Spectacled Flying-foxes are usually located within rainforest and those that are not are rarely further than 6 kilometres away from this habitat. Consequently, the distribution pattern of this species, restricted in Australia to north-eastern Queensland, follows almost exactly that of the rainforest in this region.

Many rainforest trees need animals to transport their seeds: the further away from the parent, the greater their chance of survival to maturity, so long-distance dispersal contributes to rainforest regeneration. It seems that the Spectacled Flying-fox is a

*Pteropus conspicillatus*

major dispersal agent over short distances of fine seeds—such as those of figs, which pass quickly through the digestive tract. Long-distance dispersal is mainly attributable to competition between individuals. After dusk, groups fill their feed trees, well-defined feeding territories being established after a few squabbles. Later in the night, new arrivals invade the territories of the residents, leading to confrontations after which the raider is evicted from the tree. However, a raider usually steals a fruit and carries this for some distance before alighting to feed. Thus, the needs of the tree in the final and crucial stage of its reproduction are met in a process that is an

## *Pteropus conspicillatus*

**Size**

*Head and Body Length*
220–240 mm
*Forearm Length*
160–180 mm (males)
155–175 mm (females)
*Weight*
580–850 g (males)
500–650 g (females)

**Identification** Almost black, with prominent straw-coloured fur surrounding the eyes and along the muzzle. Prominent yellow neck ruff of distinctive hair. Ruff and head silver-blonde in some individuals. Body fur may have a frosted appearance. Distinguished from other Australian flying-foxes by the yellowish spectacles.
**Recent Synonyms** None.
**Other Common Names** Spectacled Fruit-bat.
**Status** Vulnerable in parts of its limited range in Australia.
**Subspecies** Three subspecies are recognised: *Pteropus conspicillatus conspicillatus* is the only form in Australia.
**Extralimital Distribution** New Guinea and offshore islands, including Woodlark, Alcester, Kiriwina and Halmahera.
**References**
Richards, G.C. (1990) The Spectacled Flying-fox, *Pteropus conspicillatus* (Chiroptera: Pteropodidae), in north Queensland. 1. Roost sites and distribution patterns. *Aust. Mammal.* 13: 17–24.
Richards, G.C. (1990). The Spectacled Flying-fox, *Pteropus conspicillatus* (Chiroptera: Pteropodidae), in north Queensland. 2. Diet, seed dispersal and feeding ecology. *Aust. Mammal.* 13: 25–31.

integral link in the regeneration of the entire rainforest ecosystem.

As in the other three flying-foxes, males anoint their neck ruffs with a strongly smelling liquid secreted through their erect penis as part of their grooming activities. This liquid appears to contain fatty acids and together with the secretions of the sebaceous glands of the neck, is used as a marking scent by the males who rub it onto branches. In the case of the Spectacled Flying-fox, the secretion is cranberry red, which imparts a bloody appearance to the neck ruff.

The Spectacled Flying-fox has recently been demonstrated to have the greatest tolerance to

*Rainforest camps of the Spectacled Flying-Fox may contain thousands of individuals.*

G.A. HOYE

*The Spectacled Flying-Fox has pale rings around its eyes.*

ranges of ambient temperature of any mammal: air temperatures from freezing to almost 40°C cause almost no change in metabolic rate. Their wings and their ability to control heat loss through them, seems a major factor in this ability.

Conception occurs in April to May, but sexual activity is continuous from about January to June. As with other flying foxes, the females give birth to one young a year, in the October to December period. Juveniles are nursed for over five months and, on weaning, congregate in nursery trees in the colony. The juveniles fly out for increasing distances with the colony at night and are 'parked' in nursery trees, often kilometres distant from the colony and are brought back to the colony in the morning.

The Spectacled Flying-fox has been nominated as vulnerable, particularly in the southern part of its range in the Cooktown to Ingham region. It is known to raid orchards but its natural food supply has diminished, concomitant with the loss of rainforest for agricultural and development purposes. Clearing has also exposed the species to heavy mortality from the native paralysis tick when the bat is feeding on low shrubs of the introduced Wild Tobacco, *Solanum mauritanium*, growing in degraded pasture land. Lethal control measures such as aerial electric fences and shooting have also exerted significant pressure on the population.

G.C. RICHARDS AND H.J. SPENCER

# Large-eared Flying-fox

## *Pteropus macrotis*

### Peters, 1867

mak-roh'-tis: 'large-eared wing-foot'

*Pteropus macrotis*

Long known from New Guinea, this species has only recently been recorded from Australian territory in Torres Strait. From a camp of mangrove-covered mud banks near Boigu Islands, large numbers fly each night to the New Guinea coast about 12 kilometres away. The Large-eared Flying-fox is known to eat the flowers of the Coconut Palm but it also forages in the dry monsoon scrubs around the coast, probably feeding

*Primarily New Guinean, the Large-eared Flying-fox has only recently been recorded from Australian islands in Torres Strait.*

on a variety of fruits and blossoms. Bats return to the camp during night or around dawn. Locating a camp on a smaller island probably poses a much easier navigational challenge than finding one on the relatively featureless coast and an island refuge has the advantage of having fewer predators than a mainland site.

In a camp, bats hang at all levels from the canopy to branches that are partly submerged at high tide. It is interesting that the island camp

L. HALL

## *Pteropus macrotis*

**Size**

*Head and Body Length*
197–217 mm
*Forearm Length*
136–142 mm
*Weight*
315–415 g
*Ear*
31–37 mm

**Identification** Ears long and pointed relative to other species of *Pteropus*. Brownish fur all over head and body. Small size.

**Recent Synonyms** *Pteropus epularis.*

**Other Common Names** Sappur, Epauletted Flying-fox.

**Status** Restricted to northern islands of Torres Strait in Australia, but widespread around coastal New Guinea.

**Subspecies** *Pteropus macrotis macrotis*, Indonesia and New Guinea.
*Pteropus macrotis epularis*, Torres Strait Islands, coastal New Guinea.

**Extralimital Distribution** New Guinea, Aru Islands.

**References**

Flannery, T.F. (1995). *Mammals of New Guinea*, 2nd edition. Reed Books, Sydney.

Hall, L.S. and G.C. Richards (1991). Flying-fox camps. *Wildlife Australia* 28: 19–22.

apparently included no Black Flying-foxes, by far the most abundant pteropid in this region.

Nothing is known of its reproduction. The subspecific name, *epularis*, refers to a tuft of hairs emerging from glandular tissue on the shoulder. An increase in size and secretory activity of this gland during the breeding season has been observed in other flying-foxes.

L.S. HALL

# Grey-headed Flying-fox

## *Pteropus poliocephalus*

Temminck, 1825

poh'-lee-oh-sef'-ah-lus:
'grey-headed wing-foot'

More research has been devoted to the study of this species than any other bat in Australia. Some of this effort has been motivated by the fact that the Grey-headed Flying-fox is an occasional pest of cultivated fruit crops; some relates to its important role in the dispersal of seeds and pollen of trees and other plants. It occurs in a coastal belt from Rockhampton to Melbourne and occasional individuals are found on Bass Strait islands. It is infrequently found west of the Great Dividing Range.

Bats commute daily to foraging areas, usually within 15 kilometres of the day roost; a few individuals may travel up to 50 kilometres. The Grey-headed Flying-fox feeds on a wide variety of flowering and fruiting plants and it is responsible for the seed dispersal of many rainforest

*Pteropus poliocephalus*

## Pteropus poliocephalus

**Size**

*Head and Body Length*
230–289 (255) mm
*Forearm Length*
138–180 mm
*Weight*
600–1000 (700) g

**Identification** Large flying-fox with reddish-yellow mantle completely encircling neck. Head grey or whitish grey; other body fur dark brown, long and shaggy. Thick leg fur extends to ankle, in contrast to other *Pteropus* in which it reaches only to the knee. Some individuals can be difficult to distinguish from *P. alecto*, with which it sometimes hybridises.

**Recent Synonyms** None.

**Other Common Names** Grey-headed Fruit-bat.

**Status** Common in limited habitat; numbers in camps may vary widely with season and year.

**Subspecies** None.

**References**

Eby, P. (1991). 'Finger-winged night workers': managing forests to conserve the role of Grey-headed Flying-foxes as pollinators and seed dispersers (in) D. Lunney (ed.). *Conservation of Australia's forest fauna*. Roy. Zool. Soc. of NSW, Mosman. pp. 91–100.

Martin, L., P.A. Towers, M.A. McGuckin, L. Little, H. Luckhoff and A.W. Blackshaw (1987). Reproductive biology of flying-foxes (Chiroptera: Pteropodidae). *Aust. Mammal.* 10: 115–118.

Parry-Jones, K. and M.L. Augee (1991). Food selection by grey-headed flying-foxes (*Pteropus poliocephalus*) occupying a summer colony site near Gosford New South Wales. *Wildl. Res.* 18: 111–124.

K. GRIFFITHS

*The Grey-headed Flying-fox has a varied diet of fruits and blossoms.*

trees, such as native figs and palms. It also feeds extensively on the blossoms of eucalypts, angophoras, tea-trees and banksias and it is probably an important pollinator of these trees. In suitable winds it can cruise at speeds in excess of 35 kilometres per hour for extended periods. One animal was found electrocuted on a power-line 1000 kilometres from where it had been banded six months earlier.

The social organisation is complex and, as in most other megachiropterans, is organised around roost sites, known as camps. Camps are commonly formed in gullies, typically not far from water and usually in vegetation with a dense canopy. Roost sites are a pivotal resource, for these are places where mating, birth and the rearing of young occur, in addition to being day-to-day refuges from predators. Camp sites are traditional, in the sense that, once established, they are used repeatedly, but not all camps are used with the same intensity. Some are used at the same time every year by hundreds of thousands of flying-foxes; others only sporadically by a few hundred animals. Few camps are occupied on a year-round basis, the one in the Sydney suburb of Gordon being an exception. Other

*The Grey-headed Flying-fox has the most southern distribution of any pteropid. This colony is in a Sydney suburb.*

camps are occupied when suitable food supplies are available. Camps of the Grey-headed Flying-fox often also contain Black and Little Red Flying-foxes.

Most conceptions occur in March or April, but mating may take place at almost any time of the year. Gestation is about six months, with most births in October; a few as late as February. Twins are extremely rare and may not survive in the wild. Young animals are carried on the ventral surface of their foraging mothers for four to five weeks after birth. They are then left in the camp at night and suckled when the female returns. Young are sometimes abandoned by their mothers, particularly when food is short in early summer. Vocal communication is highly sophisticated, over 20 different situation-specific calls being used.

The species thrives in captivity and, given suitable food, breeds freely.

C.R. TIDEMANN

G.B. BAKER

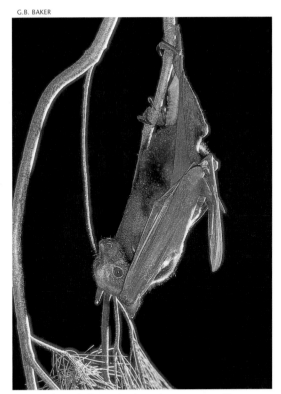

*The Little Red Flying-fox is an important pollinator and seed-distributor in forests.*

# Little Red Flying-fox

## *Pteropus scapulatus*

Peters, 1862

skap'-yue-lah'-tus:
'notably shouldered wing-foot'

Few megachiropterans are as nomadic as the Little Red Flying-fox, or as tolerant of environmental differences: ranging between temperate and subequatorial climates and from moist coastal to the dry inland of the continent. Regional populations fluctuate widely between years, particularly in the inland where the blossoming of primary food resources is sporadic.

K. ATKINSON

B.G. THOMPSON

*The Little Red Flying-fox has reddish-brown fur over most of the body.*

Nomadism is probably a response to patchy food resources.

Foraging flying-foxes are key species in forest ecosystems where they spread pollen and seeds within and between forests, maintaining their health and diversity. All dominant plant species within the geographic range are included in the diet—nectar and pollen from eucalypt flowers being predominant. Fruits, leaves, growing shoots, bark, sap and insects are also eaten. When preferred foods are scarce, the Little Red Flying-fox may raid orchards and damage fruit crops. Although a nuisance, it is not a serious pest.

During the day, groups congregate in tall vegetation—sclerophyll woodland and forest, paperbark, bamboo and mangrove communities—usually near water. Roosting sites are often congested, with 5000 to one million individuals, hundreds of which may roost on the

same tree, for periods ranging from a day to two months. Roosts sometimes abut those of the more sedentary Black Flying-fox or Grey-headed Flying-fox, both of which occupy each site for several months. Noisy roosting activities such as courting, mating and conflict during early morning and late afternoon contrast markedly with the period of rest and sleep from mid-morning to late afternoon. When disturbed and before taking flight from roosts as low as

## *Pteropus scapulatus*

**Size**
*Head and Body Length*
195–235 (215) mm
*Forearm Length*
125–156 (138) mm (males)
125–148 (137) mm (females)
*Weight*
350–604 (461) g (males)
310–560 (434) g (females)

**Identification** Reddish-brown all over. Pelage varies from brown to yellow on the neck and shoulders, around the eyes and under the wings, or grey on the crown. Distinguished from related *Pteropus* species by small size; naked or sparsely furred legs; and reddish-brown wing membranes which appear transparent in flight.
**Recent Synonyms** *Pteropus elseyii.*
**Other Common Names** Collared Flying-fox, Collared Fruit-bat, Little Red or Reddish Fruit-bat.
**Status** Common, limited.
**Subspecies** None.
**References**
Ratcliffe, F.N. (1932). Notes on the fruit bats (*Pteropus spp.*) of Australia. *J. Anim. Ecol.* 1: 32–57.
Nelson, J.E. (1965). Movements of Australian flying foxes (Pteropodidae: Megachiroptera). *Aust. J. Zool.* 13: 53–73.

*Pteropus scapulatus*

# Torresian Flying-fox
## *Pteropus* sp.

At the time of writing, this species is awaiting the publication of its formal scientific description. When discovered on Moa Island in Torres Strait, it was a great surprise to learn that such a large and obvious animal had not been recorded before. Well known to the residents of St Paul's Mission on the Island, who call it 'sapur', it provides a striking example of how little attention

1 metre above ground, individuals generally scramble to high branches. At dusk, columns leave the site after flying in spiral formation and probably taking a bearing on likely feeding areas.

Reproduction is similar to, but out of phase with, related species other than the Black Flying-fox in northern Australia, where there is a marginal overlap in the cycle. Mating takes place between November and January and parturition in April and May. Mothers carry suckling young for one month, after which they are left at the roost and suckled periodically through the night until able to fly. Parental care, including suckling and grooming, continues for several months of semi-independence during which juveniles develop basic skills: finding and choosing suitable food is part of this developmental process. The Red Flying-fox makes maternity sites in suitable forests in national parks and reserves but none have been specifically established for their protection.

Predators including humans, raptors, arboreal reptiles, aquatic reptiles and large fishes catch flying-foxes when roosting, in the air near roosting sites, or when drinking from the water surface. The extent to which populations are controlled by predation is not known, but humans have the greatest impact by clearing forests and removing food and shelter.

M. McCoy

*The Torresian Flying-fox resembles the Black Flying-fox but is smaller and has much stouter molars.*

N. CHOPPING

Moa Is.

**Pteropus sp.**

**Pteropus sp.**

**Size**
*Head and Body Length*
160–200 mm
*Forearm Length*
128–141 mm
*Weight*
210–240 g

**Identification** Similar in appearance to *Pteropus alecto* but distinguishable by its much smaller size, its much heavier dentition, particularly its larger molars.
**Recent Synonyms** None.
**Other Common Names** Sapur.
**Status** Vulnerable and rare.
**Subspecies** None.
**Extralimital Distribution** Unknown.
**Reference**
Richards, G.C., L.S. Hall and J.D. Pettigrew.
*A new fruit-bat of the genus Pteropus (Chiroptera: Pteropodidae) from Torres Strait, Australia (ms submitted).*

has been given to Torres Strait by mammalogists.

Moa Island lies halfway between the tip of Cape York Peninsula and the New Guinean mainland. It is part of a chain of continental islands connecting these two land masses and has one of the largest tracts of rainforest of all 40 or so islands in the Strait, many of which are coral atolls. Even during past ice ages, when sea-levels were lowered and Australia and New Guinea were connected, its rainforest would probably have been an 'island' of green within a 'sea' of open eucalypt savanna. These conditions—and those between the ice ages and the present day—have apparently permitted this bat to evolve as a distinct species. DNA studies indicate its close relationship with the Black Flying-fox, *Pteropus alecto*, a species that migrates seasonally through the region.

Its dentition includes heavy molars that distinguish fruit-eating species from those that primarily feed on nectar. It is probably dependent upon rainforest fruits and undoubtedly plays an integral part in the ecology and dynamics of the island's rainforest, being the only disperser of large seeds.

Nothing is known of its breeding apart from the observation that large young are carried by females in December. Since only one colony is known at present, the conservation of its camp is essential.

G.C. RICHARDS

# SUBORDER MICROCHIROPTERA: MICROBATS

Microchiropterans comprise a very large and diverse group of small bats (up to 170 grams and wingspan up to 30 centimetres, but usually much smaller), most of which feed on insects. Their evolution has led to such diversity that the group is not readily diagnosed by external characters but it may be noted that the forelimb bears only one claw (two in the Megachiroptera); that, when an animal is roosting, its head hangs downwards or is raised perpendicular to the back (perpendicular to the chest in megachiropterans); and that the wings are usually folded against the sides of the body when an animal is roosting (wrapped around the body in megachiropterans). All microchiropterans are capable of echolocation.

Most are insectivorous but a few are predators on terrestrial mammals and even on other bats; some catch fishes or aquatic invertebrates; some (the true vampires) feed on mammalian blood; and some have secondarily become adapted to a diet of fruit, nectar and pollen. They may be solitary or gregarious and may roost in caves, crevices, tree holes, under bark, or in the open. Many take advantage of human habitations and excavations. With the exception of the Ghost Bat, which is a large predator on other mammals, all the Australian microchiropterans are insectivorous. Most microchiropterans live in the tropics but some (particularly members of the family Vespertilionidae) have successfully invaded cool-temperate regions where winter temperatures may fall below 0°C.

Six families are represented in Australia: the Megadermatidae, of which the Ghost Bat is the only Australian species; the Rhinolophidae, or horseshoe-bats; the Hipposideridae, or leafnosed-bats; the Emballonuridae, or sheathtail-bats; the Molossidae, or freetail-bats; and the Vespertilionidae, a large and cosmopolitan family of bats that have no inclusive common name.

## FAMILY MEGADERMATIDAE
### *False vampires*

This family comprises only three genera and five species, found in central Africa, southern Asia and Australia. They are characterised by possession of large eyes, a noseleaf and a tragus on the ear. The large ears are joined above their bases and a tail is lacking. False vampires are predators on large insects and other invertebrates and on a wide variety of small vertebrates, including terrestrial mammals (caught on the ground) and bats (taken in flight). The Ghost Bat, largest member of the family and largest microchiropteran in the world, is restricted to Australia.

# Ghost Bat

## *Macroderma gigas*

### (Dobson, 1880)

mak'-roh-derm'-ah jee'-gas: 'giant large-skin'

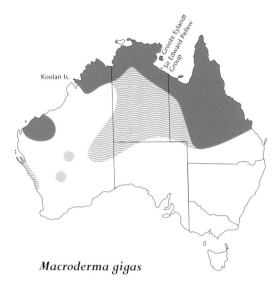

*Macroderma gigas*

This is Australia's only carnivorous bat, a predator on large insects, frogs, lizards, birds, small mammals and even other bats (including bentwinged, horseshoe, leafnosed and sheathtailed bats and the Little Cave Bat). Much of its prey is

---

### *Macroderma gigas*

**Size**
*Head and Body Length*
100–130 (115) mm
*Forearm Length*
102–112 (105) mm
*Weight*
140–165 (150) g

**Identification** Light to dark grey above; paler below. Long ears joined together; large eyes; simple noseleaf; no tail. Largest microchiropteran bat in Australia.
**Recent Synonyms** *Megaderma gigas*.
**Other Common Names** False Vampire, False Vampire Bat, Australian False Vampire Bat.
**Status** Sparse to rare.
**Subspecies** None.
**References**
Douglas, A.M. (1967). The natural history of the Ghost Bat *Macroderma gigas* (Microchiroptera, Megadermatidae), in Western Australia. *West. Aust. Nat.* 10: 125–138.
Tidemann, C.R., D.M. Priddel, J.E. Nelson and J.D. Pettigrew (1985). Foraging behaviour of the Australian Ghost Bat, *Macroderma gigas* (Microchiroptera: Megadermatidae). *Aust. J. Zool.* 33: 705–713.

---

captured on the ground. It appears to scan the area with its large eyes and ears and, unlike other microchiropterans, it does not make its echolocatory call continuously when in flight. It swoops on its prey, envelops it in its wings and kills with powerful bites. The victim is then taken to an established feeding site, usually a rock overhang or small cave, which is easily recognised by the accumulation below it of discarded parts of prey. During the day, it rests in large caves, mines, or deep rock fissures.

Ghost bats have a long fossil record in Australia, with a number of Tertiary species of *Macroderma* from the rich Oligo-Miocene deposits of Riversleigh Station, north-western Queensland and another from Pliocene sediments in Wellington Caves, New South Wales. Subfossil remains of the Ghost Bat in south-western Australia, into the ranges of South Australia and in eastern New South Wales, testify to an earlier, more widespread distribution. Contraction of its range may be coincident with the onset of more arid conditions about 10 000 years ago, leading to a reduction in food resources for all the fauna, but particularly the predators. However, there is evidence that small populations survived in south-western Western Australia and the Flinders Ranges until not less than 200 years ago.

The present patchy and widespread distribution of the Ghost Bat in northern Australia includes habitats as diverse as the arid Pilbara region and the lush north Queensland rainforests,

G.B. BAKER

J. LOCHMAN

Above: *The Ghost Bat is a predator on prey ranging from bats to terrestrial mammals.*
Right: *The Ghost Bat is by far the largest of Australia's microbats. Some individuals are very pale.*

so the exact reasons for its decreased range are still open to speculation. It is certain, however, that this endemic species is sensitive to disturbance and the few remaining large populations are at risk. One of the largest colonies known is in a series of gold mine adits at Pine Creek, Northern Territory. Smaller colonies range in size from a few to well over 400; groups in excess of 1000 individuals are unusual.

Mating probably takes place in July or August. Gestation occupies about three months and females bear a single young between September and November each year. Mothers form nursery colonies, which include males, and the young are weaned on prey which mothers bring to the roost. Juveniles hunt with their mothers until they become completely independent.

G.C. RICHARDS AND S. HAND

447

# FAMILY RHINOLOPHIDAE
## *Horseshoe-bats*

D. WHITFORD

*The Eastern Horseshoe-bat roosts mainly in caves but also in many equivalent places.*

I n the first edition of this work, the Rhinolophidae was interpreted in a broad sense to include the horseshoe-bats as well as the leafnosed-bats, here referred to the family Hipposideridae. Although sharing many characters, the two families differ in the form of the noseleaf, the structure of the foot and details of the dentition.

Rhinolophids have an elaborate noseleaf with three components. The lower, horseshoe-shaped part covers the upper lip and surrounds the nostrils. A pointed vertical component, the lancet, lies between the nostrils. The third part (sella), lying between the horseshoe and the lancet, is flattened and folded. The large ears lack a tragus.

Two species of *Rhinolophus* occur in Australia.

# Eastern Horseshoe-bat

## *Rhinolophus megaphyllus*

### Gray, 1834

rine'-oh-loh-fus meg'-ah-fil'-us:
'great (nose-) leafed crested-snout'

The Eastern Horseshoe-bat roosts in a variety of locations including caves, disused mine tunnels, roadside culverts, wartime bunkers, boulder piles and occasionally houses. High temperatures and humidities are characteristic of most roosts, but complete darkness is not essential since individuals often roost in the twilight zone of caves and mines. It is a common bat, more so in the north of its range, but it does not form huge colonies like those of the bentwing-bats. The largest known colonies rarely contain more than 2000 animals, with most colonies between five and 50 individuals. Roosts are located in a range of habitats including rainforest, eucalypt open forest and woodland.

The Eastern Horseshoe-bat forages among dense vegetation along gullies, over roads and creeks in rainforest and within open forest and woodland. Hunting does not occur over open

*Rhinolophus megaphyllus*

### *Rhinolophus megaphyllus*

**Size**
*Head and Body Length*
42–58 (44) mm
*Tail Length*
38–43 (40) mm
*Forearm Length*
44–51 mm
*Weight*
7–14 g

**Identification** Greyish-brown fur, slightly lighter underneath. Queensland populations have a rufous coloured form. Bats with grey fur retain this colour throughout life but rufous individuals undergo colour changes according to age and sex. The noseleaf is complex, with a horseshoe-shaped outer margin and fleshy projection from within.

**Recent Synonyms** *Rhinophyllotis megaphyllus*.

**Other Common Names** None.

**Status** Common in a wide variety of habitats throughout much of its range, uncommon in southern Australia but range may be expanding.

**Subspecies** Four subspecies recognised; all Australian populations are *Rhinolophus megaphyllus megaphyllus*.

**Extralimital Distribution** Eastern New Guinea, Bismarck Archipelago, St Aignan's Island in Louisiade Archipelago.

areas such as pastures and it rarely moves outside stands of forest or woodland when foraging or commuting to foraging areas. The full height of the forest is utilised during hunting, from within centimetres of the ground to the top of the canopy, but never above it. It has been observed to take spiders on the ground.

Flight is normally slow and fluttery and is interspersed with sudden changes of direction and foraging height. It often flies very close to the canopy, hovering momentarily within the foliage. Tracking of individuals carrying radio-tags has revealed that some bats fly up to 2 kilometres away from the roost when feeding. Where a number of

*The Eastern Horseshoe-bat has a slow, fluttering flight.*

roosts are available, bats may periodically move between them.

A wide range of insects are taken, all of which appear to be detected and captured while the bat is flying. Moths are the most frequent prey. Other prey includes beetles, bugs, cockroaches, dragonflies, flies, grasshoppers and wasps. A regular habit of this bat is to return at night to the daytime roost with prey which is then dismembered and eaten there.

The males store sperm for up to four months prior to copulation and for a further four months after conception. Mating takes place in late autumn and winter, with ovulation and fertilisation occurring soon after—the females do not store sperm prior to fertilisation. The only described instance of copulation was inside a cave at about midday, with both the male and female hanging from the roof by their toes.

*The Eastern Horseshoe-bat is common over most of its range and may be spreading.*

Adult females are present at maternity sites, varying in size from 15 to several thousand bats, between September and March. These are usually not more than 30 kilometres from non-maternity roosts. The maternity colonies are distinct from the females' winter roosts, which are occupied by males all year. Males may also be present at the maternity sites. Although the Eastern Horseshoe-bat now roosts in many human-made structures, nearly all known maternity colonies are in natural caves.

Females give birth to a single young in November or December after a gestation of four to four-and-one-half months. Times of birth are similar in north-eastern New South Wales and on Cape York, indicating that the reproductive cycle does not vary with latitude.

Nursing lasts for about eight weeks and the young are weaned by the end of January. Adult size is reached at five to six weeks. When being nursed or carried in flight by its mother, the young is attached to its mother's underside, with its head

D. WHITFORD

G. ANDERSON

# Large-eared Horseshoe-bat

## *Rhinolophus philippinensis*

Waterhouse, 1843

fil'-ip-een-en'-sis:
'Philippines crested-snout'

facing the female's tail and feet grasping her fur. Males become sexually mature early in their second year, females in their second or third year.

Seasonal changes in weight are normally slight, but the activity level of bats at the day roosts changes seasonally. During the colder months of May to September, many individuals in southern Australian localities enter torpor. Torpid bats are also observed at high-altitude roosts in northern Australia, such as the Atherton Tableland. Such inactivity is for short periods, many bats being torpid during the day but active by dusk and leaving the roost to forage. Foraging flights have been observed at temperatures as low as 5°C in south-eastern Queensland.

C.R. PAVEY AND R.A. YOUNG

### References

Hall, L.S., R.A. Young and A.P. Spate (1975). Roost selection of the Eastern Horseshoe bat *Rhinolophus megaphyllus. Proc. 10th Biennial Conf. Aust. Speleol. Fedn.*: 47–56.

Krutzsch, P.H., R.A. Young and E.G. Crichton (1992). Observations on the reproductive biology and anatomy of *Rhinolophus megaphyllus* (Chiroptera: Rhinolophidae) in eastern Australia. *Aust. J. Zool.* 40: 533–549.

Young, R.A. (1975). Aging criteria, pelage colour polymorphism and moulting in *Rhinolophus megaphyllus* (Chiroptera) from south-eastern Queensland, Australia. *Mammalia* 39: 75–111.

The Large-eared Horseshoe-bat, with its long ears and an intricate horseshoe-shaped noseleaf topped off with a prominent projection, is one of the most striking bats to be found in an Australian cave. Its common name is very apt, since the size of its ears are an important feature in its identification.

*The Large-eared Horseshoe-bat is aptly named.*

G.A. HOYE

## Rhinolophus philippinensis

**Size**

*Head and Body Length*
62–65 mm
*Tail Length*
25–35 mm
*Forearm Length*
50–60 mm
*Weight*
8–15 g

**Identification**  A bat with characteristic rhinolophid noseleaf; grey-brown above, slightly lighter below. Differs from *R. megaphyllus* in having longer ears (more than 20 mm) and forearms (more than 50 mm).
**Recent Synonyms**  *Rhinolophus maros*.
**Other Common Names**  None.
**Status**  Rare in Australia.
**Subspecies**  There are six recognised subspecies, the Australian population being *Rhinolophus philippinensis robertsi*.
**Extralimital Distribution**
Borneo, Philippine Islands, Kei Islands, Sulawesi, New Guinea and Timor.

occur together, individuals often roost within groups of Eastern Horseshoe-bats. Since the latter species is much more common, it is likely that the presence of the Large-eared Horseshoe-bat is often overlooked in mixed colonies.

There are no known maternity sites in Australia. Pregnant females have been captured during October at Mount Molloy (west of Cairns) and at Coen (Cape York Peninsula) and during late November at Iron Range (also on Cape York). This suggests a similar reproductive period to the Eastern Horseshoe-bat, with young being born in late October or November.

This bat hunts in rainforest and woodland where it catches insects during continuous flight. Insects taken include beetles, moths and grasshoppers. Hunting sites within rainforest are often within gaps such as roads and tracks and it tends to fly within the lower half of the canopy. A radio-tagged individual at Iron Range frequently flew out of the rainforest across cleared land to catch insects attracted to a generator-powered light beside a building.

It has one of the lowest frequency calls of any rhinolophid bat. The smaller form has a maximum constant frequency pulse of 28–32 kilohertz (about 40 kilohertz in the larger form). Low-frequency calls, which travel

Two forms of this bat have been observed within its range in northern Queensland. They differ in size, characteristics of the echolocation call and structure of the noseleaf. One form has been variously considered to be an undescribed species, a population of a horseshoe-bat from Sulawesi, or as a hybrid between the Large-eared and Eastern Horseshoe-bats. It is unclear whether what we call the Large-eared Horseshoe-bat is one or two species.

It does not appear to form large colonies. In Australia, scattered individuals are found in caves and mines, often in association with the Eastern Horseshoe-bat. Where the two species

*Rhinolophus philippinensis*

**References**

Tate, G.H.H. (1952). Results of the Archibold Expeditions No. 67. A new *Rhinolophus* from Queensland (Mammalia, Chiroptera). *Amer. Mus. Novit.* 1578: 1–3.

Hall, L.S. (1989). Rhinolophidae (in) D.W. Walton and B.J. Richardson (eds). *Fauna of Australia 1B Mammalia*. Bureau of Flora and Fauna, Canberra. pp. 857–863.

further in a humid atmosphere, are advantageous to a bat that hunts in rainforest.

The mines and caves in which it roosts during the day are found in rainforests and also in open eucalypt forest and woodland. Location of a day roost in a particular habitat does not necessarily mean that feeding occurs in the same area. Low numbers of roosts in some areas may mean that feeding sites are a considerable distance from day roosts. In open forest and woodland, foraging habitats may be among thicker vegetation along creeks and gullies.

The biology of the Large-eared Horseshoe-bat is largely unknown.

C.R. PAVEY

*The noseleaf of the Large-eared Horseshoe-bat is an elaborate structure.*

G.B. BAKER

# FAMILY HIPPOSIDERIDAE
## *Leafnosed-bats*

G.B. BAKER

*The Fawn Leafnosed-bat has a slow, fluttering flight, often close to the forest floor.*

It is confusing that, although *Hipposideros* means 'horseshoe,' members of this family are known as leafnosed-bats (distinct from the horseshoe-bats of the family Rhinolophidae).

Superficially, hipposiderids resemble rhinolophids but lack a lancet in the noseleaf; what seems to be an equivalent structure is usually in the form of three small components, sometimes with pointed tips. The digits of the hindfoot have only two phalanges (three in rhinolophids). There are only two lower premolars (three in rhinolophids).

Two genera occur in Australia: *Hipposideros* and the endemic, monotypic *Rhinonicteris*.

# Dusky Leafnosed-bat

## *Hipposideros ater*

Templeton, 1848

hip'-oh-sid'-e-ros ah'-ter: 'black horseshoe(-bat)'

Widely distributed and delicate in appearance, the Dusky Leafnosed-bat is the smallest Australian member of its family. Its day-roosts include caves in limestone and sandstone, old mine adits and tree-hollows. When roosting in caves, it shows a strong preference for areas that are pitch-dark and have a high temperature and humidity. Such caves are small and the colony size is correspondingly low—less than 30 or 40 animals. Colonies in mines are larger, one famous location at Bramston Beach in north-eastern Queensland having supported about 100 individuals since at least 1964. Roosts are shared with a variety of other species including the Eastern Horseshoe-bat, Orange Leafnosed-bat, Little Cave Eptesicus and Common and Little Bentwing-bats.

*Hipposideros ater*

### *Hipposideros ater*

**Size**

*Head and Body Length*

45 mm

*Tail Length*

23–28 mm

*Forearm Length*

37–43 mm

*Weight*

4.5–10 g

**Identification** Light grey, mottled grey or gingerish above; slightly paler below. Ears rounded and broadly triangular, bluntly pointed. Noseleaf squarish in outline and without additional leaflets.

**Recent Synonyms** *Hipposideros bicolor.*

**Other Common Names** Dusky Horseshoe-bat.

**Status** Uncommon over a wide range in northern Australia.

**Subspecies** Seven subspecies recognised, two from Australia:

*Hipposideros ater aruensis*, north-eastern Queensland.

*Hipposideros ater gilberti*, Kimberley regions and the Northern Territory.

Subspecies of other populations not yet determined.

**Extralimital Distribution** India, Sri Lanka, Philippine Islands, Little Nicobar Island through the East Indies to New Guinea.

**References**

Hill, J.E. (1963). A revision of the genus *Hipposideros. Bull. Brit. Mus. (Nat. Hist.) Zool.* 11: 1–129.

Schulz, M. and K. Menkhorst (1986). Roost preferences of cave-dwelling bats at Pine Creek, Northern Territory. *Macroderma* 2: 2–7.

Northern Territory roost-sites are regularly inhabited by the Ghost Bat, which preys upon it. The roosting requirements of these two species are very similar and the same chamber within a

G.B. BAKER

*The Dusky Leafnosed-bat roosts in caves and tree-hollows.*

manoeuvrable flight enable it to move through the understorey of rainforests with apparent ease. The diet is not well known but moths, particularly members of the family Noctuidae, form a considerable portion of the diet. This is significant, because these moths possess tympanic organs that can pick up the ultrasound of bats and thereby avoid them. However, moths are most sensitive to bats producing calls at frequencies between 30 and 70 kilohertz and, since this bat has a very high frequency call of 154–164 kilohertz, it is unlikely to be detected by noctuid moths.

Maternity roosts include both caves and mines, usually the same locations that are used as non-maternity roosts in winter. The single young is born in October or November. Some young can fly short distances by the middle of December and weaning takes place within a couple of weeks of this stage of development.

Its known range has been greatly extended inland over the past decade. During the 1980s it was sighted for the first time at Chillagoe (150 kilometres west of Cairns), Cloncurry (north-western Queensland) and Blackdown Tableland (central Queensland). It may perhaps have been overlooked in the past at these places. Recently, it has been seen in the Dulcie Ranges in the southern Northern Territory, by far the furthest inland locality of any Australian leafnosed- or horseshoe-bat. Given its preference for foraging in dense vegetation, its inland occurrence is surprising. Suitable hunting sites in inland areas would be pockets of dry vine scrub and tree-lined creeks. Since this habitat is limited, the low number observed away from coastal locations is not surprising. However, since it has also been observed to hunt in eucalypt woodland at Chillagoe, the available data is insufficient to make a generalisation.

mine is sometimes occupied by both. This may indicate that roosting places are a limited resource for cave- and mine-inhabiting bats, for it might be expected that the Dusky Leafnosed-bat would roost away from a predator if it had the choice. The Dusky Leafnosed-bat behaves cautiously when leaving its roost at night if the Ghost Bat is present. At such times it has been reported to fly into very small passages where the larger species cannot manoeuvre.

Dense vegetation is the favoured hunting habitat. This includes lowland and highland rainforest in coastal Queensland and mangroves in coastal Western Australia. Flight is slow, low to the ground and very manoeuvrable, often with changes in direction. It frequently hovers while hunting. When seen at close range in flight, it resembles a large moth. Its small size and slow,

C.R. PAVEY AND C.J. BURWELL

# Fawn Leafnosed-bat

## *Hipposideros cervinus*

### (Gould, 1854)

ser-veen'-us: 'tawny horseshoe (-bat)'

The Fawn Leafnosed-bat is widespread from Malaysia through Melanesia but, in Australia, is the most restricted of all members of its family, being found only on Cape York Peninsula, north of about Coen. Although it has been found at only a small number of Australian localities, it is reasonably common where it does occur. A colony may contain up to 900 bats. It utilises caves and mines as roosting places, often in association with the Eastern Horseshoe-bat and (less frequently) the Diadem Leafnosed-bat. It is similar in appearance to the Eastern Horseshoe-bat, but that species is slightly larger, with an elaborately structured noseleaf.

The orange form of this bat is uncommon but at least one colony, at Iron Range on Cape

C. ANDREW HENLEY/LARUS

*The Fawn Leafnosed-bat roosts in caves and abandoned mines.*

---

### *Hipposideros cervinus*

**Size**

*Head and Body Length*
50–55 mm

*Tail Length*
21–31 mm

*Forearm Length*
42–48 mm

*Weight*
6–9 g

**Identification**  Can be a variety of colours, including grey, greyish-brown and orange. Distinguished from other leafnosed-bats and horseshoe-bats by shape of noseleaf, lower portion wider than upper, narrowing below nostrils to expose leaflets.

---

York Peninsula, consists almost entirely of orange individuals. It is likely that the orange fur results from bleaching of the hair pigment by high concentrations of ammonia combined with high humidity in the mine. Such bleaching of bat fur has been experimentally documented for some North American bats. All three species of leafnosed-bats that roost in the Iron Range mine consist of orange or rufous individuals, whereas this colour is rare in adjacent colonies (within 1 kilometre) roosting in mines with low ammonia concentrations.

It hunts in a variety of habitats including rainforest, especially in the vicinity of creeks and open eucalypt forest. There is anecdotal

**Recent Synonyms** *Hipposideros galeritus*.
**Other Common Names** Fawn Horseshoe-bat.
**Status** Restricted and uncommon; more common extralimitally.
**Subspecies** There are four subspecies, the Australian form being *Hipposideros cervinus cervinus*.
**Extralimital Distribution** Widely spread from Malaysia, Sumatra, Java, Borneo, Sulawesi, New Guinea, Solomon Islands to Philippine Islands in the north and New Hebrides in the east.

**References**
Jenkins, P.D. and J.E. Hill (1981). The status of *Hipposideros galeritus* Cantor, 1846 and *Hipposideros cervinus* (Gould 1854) (Chiroptera: Hipposideridae). *Bull. Brit. Mus. (Nat. Hist.) Zool.* 41: 279–294.
Flannery, T.F. (1995). *Mammals of New Guinea*, 2nd edn. Reed Books, Sydney.

evidence that it feeds on insects attracted to lights in the vicinity of settlements and individuals have been observed to fly into buildings at night. When bats leave a roost at dusk, they follow established pathways, often along creeks and gullies, for some distance. As hunting commences, single animals or small groups separate from the main group and fly into the forest. This movement pattern is reversed at dawn. An observer positioned above the entrance of a large roost at first light will notice three or four 'streams' of bats flying through the forest, converging on the opening to their daytime roost. On such occasions the flight is fast and direct, with slight changes in direction taken to swerve around obstacles such as branches or tree trunks.

Foraging flight is slow and fluttering, often low to the ground. A bat may fly very close to the vegetation, hovering momentarily as it slowly moves around the edge of a sapling or shrub.

When hunting, it can fly within 20 centimetres or less of the ground or over water. Foraging bats sometimes perch for short periods on thin twigs of trees. Known prey includes beetles, bugs, moths, cockroaches, flies and parasitic wasps.

The single young is born in November or December. Apart from this, little is understood of the reproductive cycle and few maternity sites are known. It appears that the maternity roost is the same as the winter roost, being located in a cave or mine. Maternity sites can be shared with the Eastern Horseshoe-bat, whose young are born at a similar time.

The ultrasonic calls have the typical leafnosed-bat pattern, with a constant frequency pulse of 144–145 kilohertz and a terminal frequency modulation dropping to 120 kilohertz. Within the leafnosed- and horseshoe-bats there is a strong relationship between the body size of a species and the constant frequency component of its call: as body size increases, the call frequency decreases. The Fawn Leafnosed-bat, a small species, has a very high call frequency while the lowest frequency among Australian species is found in the large Diadem Leafnosed-bat.

C.R. Pavey and C.J. Burwell

*Hipposideros cervinus*

G.B. BAKER

*Some Diadem Leafnosed-bats have orange fur.*

# Diadem Leafnosed-bat

## *Hipposideros diadema*

### (Geoffroy, 1813)

die'-ah-dem'-ah: 'diadem horseshoe(-bat)'

To deal with the large variety of insects that it eats—scarabs, weevils, click beetles, moths, ants and plant bugs—the Diadem Leafnosed-bat has a rather fearsome set of teeth with which it can deliver a painful bite to humans. Almost as varied as its diet is the range of sites where individuals roost during the day: it is not uncommon to find one hanging from the ceiling of a cave or disused mine, in an old shed and even in a large culvert under a road. In its extralimital distribution, large hollow trees and even the tree canopy itself, are used as daytime roosting sites. It is very wary and, in exposed roosts such as culverts, one seldom gets a second look at it before it makes a swift and silent exit.

Like most leafnosed-bats it favours warm, humid locations, often of small dimensions and is frequently found with other species, particularly the Eastern Horseshoe-bat. The largest numbers are found in caves, where females apparently congregate to form maternity groups towards the end of the year. Females pregnant with a single embryo have been collected in September and lactating females in January.

G.A. HOYE

*Many Diadem Leafnosed-bats have pale patches on the shoulder and the belly.*

body is continually twisted from side to side and its ears are in constant motion, ultrasonically scanning the approaches. As an appropriate insect comes into range, the bat releases its grip, flies towards it, snatches it and returns to the branch. If a swarm of insects suddenly appears, it catches a number, killing them and storing them in its cheekpouches for later consumption.

Its ultrasonic calls are similar to those of the other horseshoe-bats studied so far, having a constant frequency with a terminal modulation: 58 kilohertz for the long pulse, dropping to 47 kilohertz.

It is the only Australian bat which regularly has patches of light-coloured fur, giving it a spectacular appearance in a spotlight. No specimens from the Northern Territory have yet been recorded with these light patches. Bright orange individuals have been seen on Cape York Peninsula. It is highly probable that what is now called *H. d. inornatus* is a valid species.

L.S. HALL AND G. RICHARDS

A Diadem Leafnosed-bat hunts in a specific area which may be up to 10 kilometres from its daytime roost. In rainforest, it flies up and down a track, frequently landing on a branch that overlooks the track while it chews its prey. In eucalypt forest, it may hang from the branch of a tree over a creek bed. As it waits in ambush, its

H.d. inornatus

H.d. reginae

**Hipposideros diadema**

C. ANDREW HENLEY/LARUS

## Hipposideros diadema

**Size**

*Head and Body Length*
75–85 mm
*Tail Length*
30–40 mm
*Forearm Length*
78–82 mm (*H. d. reginae*)
68–73 mm (*H. d. inornatus*)
*Weight*
30–50 g

**Identification** Fur grey to yellowish-brown, usually with darker tips, often with prominent lighter patches on each shoulder and on belly. Upper section of noseleaf divided into four depressions by ridges, wider and narrower than lower section. Three supplementary leaflets under each side of lower noseleaf.

**Recent Synonyms** None.

**Other Common Names** Large Horseshoe-bat, Diadem Bat, Diadem Horseshoe-bat.

**Status** Limited to sparse; common extralimitally.

**Subspecies** There are approximately 17 subspecies of *Hipposideros diadema*, two of which occur in Australia:

*Hipposideros diadema reginae*, northern coastal Queensland.

*Hipposideros diadema inornatus*, northern parts of the Northern Territory.

**Extralimital Distribution** Widespread in Burma, Malaysia, Vietnam, Philippines, Sumatra, Java, Borneo, Celebes, New Guinea and Solomon Islands.

**Reference**

McKean J.L. (1970). A new subspecies of the horseshoe-bat *Hipposideros diadema*, from the Northern Territory. *West. Aust. Nat.* 11: 138–140.

*The Diadem Leafnosed-bat roosts in caves and abandoned mines (here in a disused goldmine at Iron Range, Cape York).*

# Semon's Leafnosed-bat

## *Hipposideros semoni*

### Matschie, 1903

sem'-on-ee: Semon's horseshoe(-bat)'

The small number of records of this bat include an intriguing variety of daytime roosting sites: the door handle of a car, a clothes closet, an oven and a picture rail in an unoccupied house. More commonly it roosts in caves, mines or rocky overhangs and cracks where light is much reduced. It is not known whether roosts are changed seasonally nor whether colonies are formed. It has been found roosting only as solitary individuals, but it has been observed to feed in a group of at least eight.

Semon's Leafnosed-bat has been caught in mist-nets placed along tracks in rainforest. Its short, broad wings give it the capacity for slow, fluttering flight as it searches for food in and around dense vegetation and it is seldom seen more than 2 metres above the forest floor. It is also found in drier open woodlands of Cape York, commonly between June and September, the driest part of the year.

*Hipposideros semoni*

461

The conspicuous noseleaf is rather square in outline, consisting of an upper portion, mostly above the level of the eyes and a lower portion which contains a depression for the nostrils. There is a central fleshy protuberance on the lower part, just above the nostrils and another

*Little is known of the biology of Semon's Leafnosed-bat.*

on the top edge of the upper portion of the noseleaf. The function of the protuberance and leaflets is unknown.

Females have two false teats in the pubic region and two milk-bearing teats in the thoracic area. Nothing is known of its breeding biology or social behaviour.

In the first edition of this book, this species was called the Greater Wart-nosed Horseshoe-bat but, despite an aversion to patronymics, the author and editor now reject this clumsy name.

L.S. HALL

## *Hipposideros semoni*

**Size**

*Head and Body Length*
40–50 mm
*Tail Length*
10–20 mm
*Forearm Length*
42–46 mm
*Weight*
12–16 g

**Identification** Brown above, lighter below. Wing-membrane near body covered with whitish-brown hair. Noseleaf with club-shaped protuberance 6–8 mm long, projecting vertically from centre of lower portion; small wart on centre of upper noseleaf. Upper noseleaf divided into four depressions with two supplementary leaflets under each side of lower portion. Distinguished from *Hipposideros stenotis* by longer central wart.

**Recent Synonyms** *Hipposideros muscinus.*
**Other Common Names** Wart-nosed Horseshoe-bat, Semon's bat, Greater Wart-nosed Horseshoe bat.
**Status** Rare.
**Subspecies** None.
**Extralimital Distribution** New Guinea.
**Reference**
Van Deusen, H.M. (1975). History of Semon's horseshoe bat in Australia. *N. Qld. Nat.* 42: 4–5.

# Northern Leafnosed-bat

## *Hipposideros stenotis*

Thomas, 1913

sten-oh'-tis: 'narrow-eared horseshoe(-bat)'

Most specimens of this little-known bat have been found in cracks and caves along the western escarpment of the Arnhem Land plateau, an area typified by sandstone cliffs, gorges and waterholes bordered by paperbark trees. Others have come from Derby, Western Australia, and from abandoned mines in the gulf country of the Northern Territory near the Queensland border and the vicinity of Mount Isa, Queensland. The wide separation of these localities indicates that it can tolerate a wide range of environmental conditions. Intensive searching in the small caves and almost inaccessible rock piles that they favour would be necessary to establish its distribution more definitely.

As in Semon's Leafnosed-bat, the noseleaf carries two warty protuberances, one at about the level of the eyes and another on the top edge. The lower section of the noseleaf has a

shallow parabolic portion surrounding the nostrils which has been shown, in other horseshoe-bats, to beam echolocating calls forwards while shielding these from the bat's own ears. The cochlea of the inner ear is particularly large and presumably is able to analyse fine changes in the frequency and intensity of its call. Among leafnosed-bats, the frequency of the call is predominantly constant and inversely related to the size of the bat; higher frequencies being required to detect smaller prey. It is a reasonable postulate that the as-yet-unrecorded, ultrasonic cry of this species will be of a constant and high frequency.

## *Hipposideros stenotis*

**Size**

*Head and Body Length*
42–45 mm
*Tail Length*
25–30 mm
*Forearm Length*
42–45 mm
*Weight*
6–10 g

**Identification** Brown above; slightly paler below. Ears haired for one-third of length, acutely tapered at the tip. Noseleaf complex, covering much of the face, with two supplementary leaflets and box-like cavities in the upper section. A protuberant 'wart' on centre of the top edge of upper portion. Similar to *Hipposideros semoni*, but wart in centre of noseleaf is smaller (up to 8 mm long in *H. semoni*).
**Recent Synonyms** None.
**Other Common Names** Dahl's Horseshoe Bat, Lesser Wart-nosed Horseshoe-bat.
**Status** Apparently rare, endemic. May be underestimated.
**Subspecies** None.
**Reference**
Hill, J.E. (1963). A revision of the genus *Hipposideros*. *Bull. Brit. Mus. (Nat. Hist.) Zool.* 11: 1–129.

Boulanger Is.
Bathurst Is.
Koolan Is.

*Hipposideros stenotis*

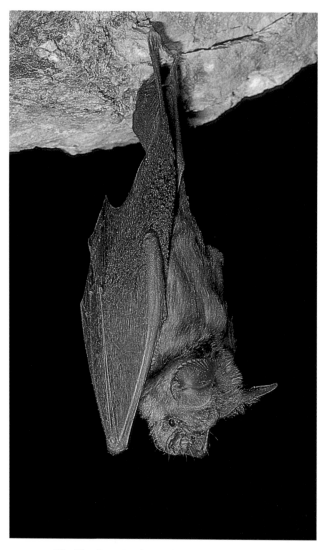

*The Northern Leaf-nosed Bat does not cluster when roosting in a cave.* G.B. BAKER

# Orange Leafnosed-bat

## *Rhinonicteris aurantius*

(Gray, 1845)

or-ant'-ee-us; 'golden nose-bat'

The beautiful golden fur and complexity of the face are striking features of the Orange Leafnosed-bat but it has largely escaped the attention of naturalists and very little is known of its biology. It occurs in the Kimberley region of Western Australia and the 'top end' of the Northern Territory and there are records from Camooweal and Lawn Hill, Queensland.

Gould reported in 1863 that it roosts during the day in the hollow trunks of eucalypt trees

Like Semon's Leafnosed-bat, it has a fluttering flight, rather like that of a butterfly, characterised by frequent changes of direction as it hunts moths, beetles, mosquitos and flies. When roosting, this species hangs pendulous-like from the ceiling. It does not cluster and usually only one or two are found near the entrance. The breeding season could be long, as pregnant females have been caught in July and a female with attached young was caught in late January.

L.S. HALL

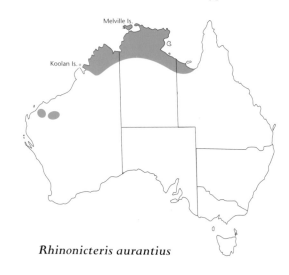

*Rhinonicteris aurantius*

*The Orange Leaf-nosed Bat preys mostly upon moths and beetles, taken in flight.* B.G. THOMSON

*The Orange Leafnosed-bat roosts in caves.* G. ANDERSON

but all subsequent observations indicate that it roosts in caves, in colonies ranging from 20 to several thousand individuals. Probably because it is unable to maintain a high body temperature when resting, it usually chooses sites that are very warm and humid, often deep caves in limestone. In view of its poor temperature regulation it is interesting that it does not cluster for warmth: individuals hang from the roof of the cave about 12 centimetres apart. The Dusky Leafnosed-bat has similar roosting requirements and both species occasionally mingle in a roost.

At dusk the Orange Leafnosed-bat emerges to feed, mostly on moths but also on beetles, shield-bugs, parasitic wasps, ants, chafers and weevils. It frequently returns to the cave to groom and to digest its food, behaviour which suggests that most of its feeding takes place in the open woodland close to the home cave. Some individuals have been observed flying further afield onto open blacksoil plains.

The Orange Leafnosed-bat is very sensitive to human interference and quickly takes to the wing at a slight disturbance, often retiring to a more distant chamber in the cave to avoid an intruder. Like many other cave-dwelling species, it may completely abandon a roost if subjected to continual human disturbance: this occurred with a colony at Katherine, Northern Territory. Nothing is known of its breeding but the changing of caves on a regular seasonal basis may be an indication of a seasonal breeding cycle.

Close relatives of this species were common inhabitants of limestone caves in the Riversleigh area, north-western Queensland during the Oligo-Miocene (25–15 million years ago) and Pliocene (5–3 million years ago).

S. JOLLY AND S. HAND

## *Rhinonicteris aurantius*

### Size
*Head and Body Length*
45–53 (50) mm
*Tail Length*
24–28 (25) mm
*Forearm Length*
47–50 (48) mm
*Weight*
8–10 (9) g

**Identification** Orange fur, occasionally darkened by brown-tipped hairs; darker fur around eyes. Noseleaf complex; lower part broad with central gap at the front; upper part scalloped. Nasal pits deep. Ears small and acutely pointed.
**Recent Synonyms** None.
**Other Common Names** Golden Horseshoe-bat.
**Status** Sparse. This endemic species is the only member of its genus.
**Subspecies** None.
**Reference**
Gray, J.E. (1845). Description of some new Australian animals (in) E.J. Eyre, *Journals of Expeditions of Discovery into Central Australia and Overland from Adelaide to King George's Sound in the years 1840 and 1841.* Appendix, Vol. 1, London.

# FAMILY EMBALLONURIDAE
## *Sheathtail-bats*

*The Common Sheathtail-bat roosts communally in the twilight zone of a cave.*

The tail of a bat is usually firmly attached to the tail-membrane and acts as a central strut. In sheathtail-bats, however, the tail appears to pierce the tail-membrane from below and to protrude freely above its surface. It is, in fact, in organic connection with the tail-membrane but by means of an extremely elastic sheath that, in effect, permits the tail-membrane to slide along the tail, from the base to the tip. This arrangement increases the mobility of the hindlimbs, thus making quadrupedal locomotion easier. The ability to vary the area of the tail-membrane may also contribute to the fine control of flight. Sheathtail-bats are fast fliers and their narrow wings are so long that, when an individual is at rest, the tips are folded back over the rest of the membrane. The face of a sheathtail-bat is more pointed than in most microchiropterans and bears prominent, forwardly directed nostrils. There is no noseleaf and the pinna of the ear has a tragus.

Thirteen genera are distributed between South America, Africa, southern Asia and the Australia–New Guinean region.

# Yellow-bellied Sheathtail-bat

## *Saccolaimus flaviventris*

### (Peters, 1867)

sak'-oh-lay'-mus flah'-vee-vent'-ris: 'yellow-bellied throat-pouch'

The contrast between the black dorsal fur and the white to yellowish belly of this bat distinguishes it from most other species in southern Australia.

It is widespread across Australia and its apparent rarity is probably due to its flying so high and fast that it is seldom collected. In eucalypt forests it feeds above the canopy but in mallee or open country it comes lower to the ground.

Usually solitary, but occasionally occurring in colonies of less than ten individuals, the Yellow-bellied Sheathtail-bat roosts in tree hollows and has been found in the abandoned nests

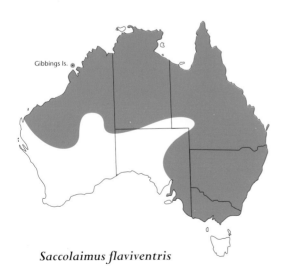

*Saccolaimus flaviventris*

*Saccolaimus flaviventris*

**Size**
*Head and Body Length*
76–87 mm
*Tail Length*
20–35 mm
*Forearm Length*
74–80 mm
*Weight*
30–60 g

**Identification** Glossy jet black on back; white to yellow underneath. Head flattened; muzzle sharply-pointed.
**Recent Synonyms** *Taphozous flaviventris.*
**Other Common Names** White-bellied Sheathtail-bat, Yellow-bellowed Freetail-bat.
**Status** Rare in widespread habitat.
**Subspecies** None.
**References**
Hall, L.S. and G. Gordon (1982). The throat pouch of the Yellow-bellied Bat, *Taphozous flaviventris*. *Mammalia* 46: 247–252.
Chimimba, C.T. and D.J. Kitchener (1987). Breeding in the Australian Yellow-bellied Sheathtail-bat, *Saccolaimus flaviventris* (Peters, 1867) (Chiroptera: Emballonuridae). *Rec. West. Aust. Mus.* 13: 241–248.

of Sugar Gliders. Occasionally it is found resting on the walls of buildings in broad daylight, and one such individual, caught at Queanbeyan, New South Wales, appeared to be so exhausted that it made no effort to escape. Similar reports suggest that it is migratory in southern Australia and that individuals found resting in the open are in the course of a winter migration from cooler to warmer areas. The Common Bentwing-bat, which is only half as

*The Yellow-bellied Sheathtail-bat is distinctively patterned in black and white (or yellow).*

# Papuan Sheathtail-bat

## *Saccolaimus mixtus*

Troughton, 1925

mix'-tus: 'intermediate throat-pouch'

This species was described in 1925 on the basis of specimens from New Guinea but three specimens were taken near the Pascoe River near the tip of Cape York Peninsula in 1948, another seven in 1981 and three in 1982 from the vicinity of Weipa, Queensland.

It emerges at dusk and, like species of *Taphozous*, forages over the canopy of open forest, venturing closer to the ground in open areas or along unobstructed flyways. The three earlier Australian specimens were 'near a gully filled with fringe forest' while those collected in 1982 were caught as they flew within 4 metres of the ground along a bush track in closed *Eucalyptus tetrodonta* forest.

large, is known to migrate long distances to hibernating or maternity sites at certain times of the year.

Males have a prominent throat-pouch which is devoid of glandular tissue but a subcutaneous gland lies behind it. The throat-pouch is represented by a rudimentary fold of skin in the female.

G.C. RICHARDS

*The range of the Papuan Sheathtail-bat extends into northern Australia.*

B.G. THOMSON

## *Saccolaimus mixtus*

**Size**

*Head and Body Length*
72–77 (75) mm

*Tail Length*
22–26 (24) mm

*Forearm Length*
62–68 (65) mm

**Identification** Brown above, darker on head and shoulders; pale buff brown below. Hairs not bicoloured as in *T. georgianus* and *T. australis*. Throat-pouch well developed in male, less so in female. Wing pockets of both sexes lined with pale hairs.
**Recent Synonyms** *Taphozous mixtus*.
**Other Common Names** New Guinea Sheathtail-bat, Wing-pouched Saccolaimus, Troughton's Sheathtail-bat, Allied Freetail-bat.
**Status** Apparently rare in Australia.
**Subspecies** None.
**Extralimital Distribution** New Guinea.
**Reference**
Troughton, E. le G. (1925). A revision of the genera *Taphozous* and *Saccolaimus* (Chiroptera) in Australia and New Guinea, including a new species and a note on two Malayan forms. *Rec. Aust. Mus.* 14: 313–341.

Data on the specimens from western Papua indicates that the Papuan Sheathtail-bat roosts in limestone caves. There is, however, a strong possibility that the species may also utilise tree hollows.

G.C. RICHARDS AND B. THOMSON

*Saccolaimus mixtus*

# Bare-Rumped Sheathtail-bat

## *Saccolaimus saccolaimus*

### (Temminck, 1838)

sak'-oh'-lay'-mus:
'throat-pouched throat-pouch'

In India and Malaysia where this species is also found, it is called the Tomb Bat because of its habit of roosting in such structures. All records in Australia indicate that it roosts only in trees, where it has a close association with *Eucalyptus alba*. This frequently has hollow limbs and spouts where three to four individuals have been found roosting in one tree-hole.

Its most characteristic feature is the naked rump: the fur on the back does not extend

### *Saccolaimus saccolaimus*

**Size**

*Head and Body Length*
90–100 mm

*Tail Length*
25–35 mm

*Forearm Length*
72–76 mm

*Weight*
40–50 g

**Identification** Reddish-brown to dark brown above, irregularly flecked with patches of white fur. Naked rump. Throat patch present in males, rudimentary in females. Wing-pouches absent.
**Recent Synonyms** *Saccolaimus nudicluniatus*, *Taphozous nudicluniatus*.
**Other Common Names** Naked-rumped Freetail Bat, Naked-tailed Saccolaimus, Tomb Bat. Naked-rumped Sheathtail-bat.
**Status** Rare, scattered (in Australia); common extralimitally.

A. COMPTON

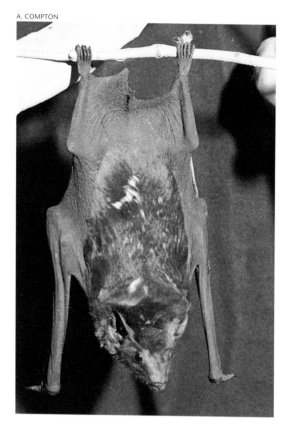

*The fur of the Bare-rumped Sheathtail-bat is flecked with white.*

beyond the hips and, even for some distance below this, is very sparse. It is also distinguished by a curious flecking of white patches in the fur of the upper part of the body. As in most Australian members of the genus *Saccolaimus*, the male has a throat-pouch, a structure which is rudimentary in the female, being represented by a well-defined, almost naked area, bordered behind by a semicircular ridge.

In Queensland, the Bare-rumped Sheathtail-bat is restricted to well-timbered coastal areas from Townsville to Cooktown: it is also found in the Alligator Rivers region of the Northern Territory. Over its entire distribution it uses a variety of habitats ranging from dry woodland in India to dense rainforest in Malaysia and it has been found roosting in the eaves of houses, in hollow trees, between boulders, in caves and, as mentioned, in old monumental structures. This behaviour is in contrast to that of most other emballonurid bats, which usually roost in caves or similar rocky areas. The bat has a characteristic posture as it sits on a surface, with the forearms spread and the body supported on its four limbs. When disturbed, it scuttles across the surface in a spider-like manner.

Although gregarious, it does not form tight clusters. It is generally very silent but, when caught, it has a loud cry and can deliver a painful bite. The wings are long and narrow and it flies high, straight and fast—possibly the main reason why so little is known about it. Females are lactating during the tropical wet season and one young is born, but the exact periods of mating and parturition in Australia are unknown.

L.S. HALL

**Subspecies** The Australian population may possibly be a distinct subspecies of the Asian *Taphozous saccolaimus* but more studies are required to determine this.

**Extralimital Distribution** India, Burma, Malaysia, Sumatra, Java, Borneo, West Irian, Papua New Guinea, Timor.

**Reference**

Goodwin, R.E. (1979). The bats of Timor: systematics and ecology. *Bull. Amer. Mus. Nat. Hist.* 163: 73–122.

***Saccolaimus saccolaimus***

# Coastal Sheathtail-bat

## *Taphozous australis*

### Gould, 1854

taf'-oh-zoh'-us ost-rah'-lis:
'southern throat-pouched tomb-dweller'

B.G. THOMSON

*The Coastal Sheathtail-bat is seldom found more than a few kilometres from the ocean.*

This species is rarely found further than a few kilometres from the ocean, where it roosts in sea caves, rock fissures, boulder piles and abandoned structures, such as old concrete military bunkers. Large colonies exist in more extensive caves, the first specimen to be described having been found in a sea cave containing 'great numbers'. Individuals usually roost separately but, in winter, clusters of two to five bats may form. Within a large roost site, a colony occupies a

main area near the entrance and, when disturbed, it quickly takes flight, moving to a more inaccessible area deeper in the roost. In smaller roosts, particularly shallow sea caves, bats will fly out in the daylight to a similar roost nearby.

Beginning at nightfall, beetles and other insects are hunted, often to be eaten inside the daytime roost or at a nearby feeding station. Ecological studies show that it only uses habitats that are within a kilometre or so from the ocean, such as dune scrubs and coastal paperbark swamps, where it forages just above the canopy. This has important implications for the conservation of this rare species and the effects of habitat modification such as sand-mining should be monitored closely.

---

### *Taphozous australis*

**Size**
*Head and Body Length*
80–90 mm
*Tail Length*
20–25 mm
*Forearm Length*
63–67 mm
*Weight*
30–50 g

**Identification** Similar to *Taphozous georgianus* but distinguished by smaller size, paler colour and presence of throat-pouch (functional in male, vestigial in female). Two colour variants may occur in the same colony, but fawn individuals more common than light grey ones.
**Recent Synonyms** *Taphozous fumosus*, *Saccolaimus australis*.

*Taphozous australis*

# Common Sheathtail-bat

## *Taphozous georgianus*

### Thomas, 1915

jorge'-ee-ah'-nus: '(King) George (Sound) tomb-dweller'

Very little is known of its reproductive biology but groups of breeding females have been found in September and most births probably occur in October or November. The testes of males are retracted into the abdomen in September but descend into a tight scrotum in April.

As with most members of this genus, there is a build-up of body fat in autumn. Being a primarily tropical species, it is unlikely to undergo hibernation but, in winter in the southern part of its range (just north of Rockhampton), individuals may go into temporary torpor.

G.C. RICHARDS

This bat roosts in the twilight zone near the entrance of a cave in the rocky country of northern Australia. It does not hang from the roof but clings to the wall, propped on its forearms. It is alert and, when disturbed, scurries about, crabwise and retreats into a crevice. Individuals roost singly, well separated, and some have been seen to defend territories. As many as 100 bats have been found in a cave but this is exceptional; fewer than 20 and often only one or two are typical. An individual may use many caves, regularly moving from one to another.

The Common Sheathtail-bat emerges at dusk and flies high and fast, catching insects and eating them in flight. It is not known to drink. Substantial fat reserves are accumulated in autumn and utilised in winter. In Central Queensland,

**Other Common Names** Neck-pouched Taphozous, Little Sheathtail-bat, North-eastern Sheathtail-bat.
**Status** Uncommon in limited habitat. Very rare extralimitally.
**Subspecies** None.
**Extralimital Distribution** New Guinea.
**Reference**
Troughton, E. le G. (1926). A revision of the genera *Taphozous* and *Saccolaimus* (Chiroptera) in Australia and New Guinea, including a new species and a note on two Malayan forms. *Rec. Aust. Mus.* 14: 313–341.

Bigge Is.
Augustus Is.
Koolan Is.
Cockatoo Is.
Barrow Is.

*Taphozous georgianus*

B.G. BAKER

*The Common Sheathtail-bat flies high and fast, catching insects in flight.*

### *Taphozous georgianus*

**Size**

*Head and Body Length*
75–89 (82) mm

*Tail Length*
21–32 (27) mm

*Forearm Length*
66–74 (71) mm

*Weight*
April: 30–51 (41) g
September: 19–31 (26) g
Whole year: 19–51 (32) g

**Identification** Dark brown above, base of hairs creamy brown. Lighter brown below, base of hairs grey. Fine hairs along undersurface of arm grizzled grey. Sparse yellow-brown hairs on undersurface of base of tail. Similar to *Taphozous hilli* but slightly different colour, males distinguished by absence of pouched neck-gland.

average body weight reaches a peak of 41 g in April but this declines to 26 g in September. Energy is conserved in times of food scarcity by allowing the body temperature to fall during the day. Bats are often quite sluggish when first encountered in a roost in mid-winter.

Mating usually occurs in August or September. In December, most females give birth to a single young which is well developed, has open eyes and is as much as one-quarter of its mother's weight. It is carried by its mother until it is about half her weight and is able to fly at the age of three to four weeks. Adult size is reached at three months and sexual maturity at about nine months. Females mate for the first time at about this age but males do not do so until about 21 months old.

# Hill's Sheathtail-bat

## *Taphozous hilli*

### Kitchener, 1980

hil'-ee: 'Hill's tomb-dweller'

**Recent Synonyms** *Saccolaimus georgianus*.
(Confused until recently, with *Taphozous hilli*.)
**Other Common Names** Unpouched Freetail
Bat, Sharp-nosed Bat.
**Status** Common.
**Subspecies** None, but tail may be shorter in
Northern Territory population.
**References**
Kitchener, D.J. (1973). Reproduction in the
Common Sheathtail Bat, *Taphozous georgianus*
(Thomas) (Microchiroptera: Emballonuridae)
in Western Australia. *Aust. J. Zool.* 21:
375–389.
Jolly, S. (1990). The biology of the common
sheathtail bat, *Taphozous georgianus* (Chiroptera:
Emballonuridae), in Central Queensland. *Aust.
J. Zool.* 38: 65–77.

Over much of its range, Hill's Sheathtail-bat
roosts alongside the Common Sheathtail-bat, a
species from which it is distinguished with difficulty. Although it has been well represented in
collections for many years, it was only recently
recognised as a distinct species.

It appears to roost exclusively in caves,
mines or adits. In Western Australia, it is widely

J. LOCHMAN

Testes remain in the abdominal cavity for
most of the year but descend into the scrotum
for a few months in summer. Sperm is produced
in summer and autumn and stored in the tail of
the epidymis until mating in the following
spring. Storage of sperm is unusual in a tropical
bat and it is interesting to note that whereas in
most mammals the tail of the epididymis is
closely adherent to its testes, in sheathtail-bats
it is permanently located in the cooler environment of the scrotum. Only the right ovary is
functional and foetuses develop only in the right
horn of the uterus.

Only about 10 per cent of females and less
than 3 per cent of males survive for more than
four years. It has been suggested that the higher
male mortality may reflect mating competition
between them. In some other members of this
family, breeding behaviour is based on harems
(leading to intense competition between males)
but we do not yet know whether this is the case
in the Common Sheathtail-bat.

S. JOLLY

distributed in the semi-arid Pilbara and Murchison regions and the Gibson Desert, and it has
been found as far eastwards as the vicinity of
Tennant Creek, Northern Territory. It is uncommon in the sandy deserts, specimens having been
recorded only from a single breakaway area in
the Great Sandy Desert, and it does not penetrate the Kimberley region. Extensive mining
operations in Western Australia have probably

G.B. BAKER

Above: *Hill's Sheathtail-bat is a fast-flying predator on insects.*

Left: *When roosting, Hill's Sheathtail-bat clings to the walls of a cave with all four limbs.*

led to an extension of its range and may continue to do so because adits and mines appear to be utilised soon after they have been abandoned.

Like the Common Sheathtail-bat, Hill's Sheathtail-bat is insectivorous. As the two species are of similar size and occupy the same roosts, it would be interesting to know how they partition available resources; differences in the size of canine teeth suggest that they may eat different prey.

Females give birth to a single young between early summer and mid-autumn, after which the female reproductive organs become relatively quiescent until early winter: no oestrous female has been collected earlier than August. Males appear to be fertile throughout the year but the testes are scrotal in summer, inguinal from late autumn to winter and abdominal from mid-winter to spring. Despite the extended fertility of males, copulation appears to be restricted to a brief period in the year, since no sperm has been recorded in the reproductive tracts of females.

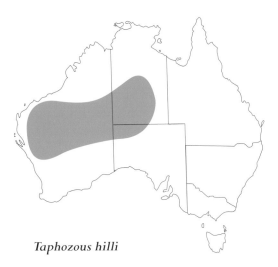

*Taphozous hilli*

D. MATTHEWS

## *Taphozous hilli*

**Size**

*Head and Body Length*
63–75 (71) mm
*Tail Length*
26–35 (30) mm
*Forearm Length*
63–72 (68) mm
*Weight*
20–25 (22) g

**Identification** Rich brown above, grading to lighter brown on rump; light buff to orange-buff tipped with olive-brown below. Wings greyish-brown. Similar to *Taphozous georgianus* but slightly different colour; males distinguished by pouched neck-gland.

**Recent Synonyms** Until 1980 confused with *Taphozous georgianus*.

**Other Common Names** Slender-toothed Sheathtail-bat.

**Status** Probably common to uncommon, habitat not limited.

**Subspecies** None.

**References**

Kitchener, D.J. (1980). *Taphozous hilli* sp. nov. (Chiroptera: Emballonuridae), a new sheath-tail-bat from Western Australia and Northern Territory. *Rec. West. Aust. Mus.* 8: 161–169.

Kitchener, D.J. (1976). Further observations on reproduction in the common sheathtail-bat, *Taphozous georgianus* Thomas, 1915 in Western Australia, with notes on the gular pouch. *Rec. West. Aust. Mus.* 4: 335–347.

The depth of the pouch surrounding the neck-gland of males is correlated with seasonal enlargement of the seminiferous tubules of the testes and of the accessory male glands. This suggests that the neck-gland plays a role in social behaviour, particularly mating, but this has not yet been demonstrated.

D.J. KITCHENER

# Arnhem Sheathtail-bat

## *Taphozous kapalgenis*

### McKean and Friend, 1979

kap'-al-gen-sis: 'Kapalga tomb-dweller'

The relatively recent (1979) discovery of this distinctive species in northern Australia suggests that its apparent rarity may simply be due to the

*The Arnhem Sheathtail-bat roosts in caves and rock-clefts.*

lack of bat specialists working on faunal surveys in tropical Australia. The Arnhem Sheathtail-bat is quite different from any other Australian emballonurid, its nearest relative being found in Flores in the Lesser Sunda group of islands. This interesting zoogeographical link parallels that of the endemic Arnhem Land Black-banded Pigeon, *Ptilinopus alligator*, the closest related species of which, *P. cinctus*, also occurs in the Lesser Sundas.

Although so far only known from the Arnhem Land region, it is quite possible that the species occurs across the Top End of the Northern Territory, particularly since some Aboriginal people who have been shown a specimen claim that it occurs in the Rose and Roper Rivers regions and roosts in the base of pandanus leaves. Around the CSIRO Division of Wildlife Research study area at Kapalga on the South Alligator River, Northern Territory, the habitat of the Arnhem Sheathtail-bat

***Taphozous kapalgenis***

## *Taphozous kapalgenis*

**Size**
*Head and Body Length*
75–85 (80) mm
*Tail Length*
20–23 (22) mm
*Forearm Length*
58–63 (60) mm
*Weight*
26 g (one specimen)

**Identification** Pale brown pelage often tinged orange; two broad, white, ventral flank stripes.
**Recent Synonyms** None.
**Other Common Names** White-striped Sheathtail-bat.
**Status** Rare, scattered (see text).
**Subspecies** None.
**Reference**
McKean, J.L. and G.R. Friend (1979). *Taphozous kapalgenis*, a new species of sheathtail-bat from the Northern Territory, Australia. *Victorian Nat.* 96: 239–241.

includes the margins of blacksoil plain swamps, open woodland dominated by *Eucalyptus papuana* and *Pandanus spiralis* and tropical layered woodland with a mixture of *Eucalyptus tectifica*, *E. clavigera* and *E. papuana*. When feeding, the bats fly rapidly with abrupt deviations and emit a succession of loud, shrill notes.

Most fly and feed above the tree canopy but come lower to the ground in open areas or along unobstructed flyways. In 1980, one individual was captured in Kakadu National Park as it flew along a narrow road corridor in dense riverine vegetation.

J. McKean and B. Thomson

# FAMILY MOLOSSIDAE
## *Freetail-bats*

*The flattened head of the Southern Freetail-bat enables it to roost in narrow cavities.* P. GERMAN

The common name of members of this family refers to the length of tail projecting from the tail-membrane. They are also known as mastiff-bats, in reference to the wrinkled, hound-like jowls of most species. There is no noseleaf. The ear lacks a tragus. A glandular patch of no known function is present on the neck of some species.

With long, narrow wings, they are fast fliers, usually hunting above the forest canopy but also taking insects on the ground. Because of their effective quadrupedal gait, they are sometimes known as 'scurrying bats.'

The family, which contains 12 genera and about 80 species, is represented on every continent. Three genera *Chaerephon*, *Mormopterus* and *Nyctinomus*, are represented in Australia by six species.

# Northern Freetail-bat

## *Chaerephon jobensis*

(Miller, 1902)

keer'-e-fon jobe'-en-sis:
'Jobi (Island) chaerophon' (a proper name)

*Chaerephon jobensis*

Found only in the tropics, the Northern Freetail-bat obtains food and shelter in open forest rather than in rainforest. Its fast and direct flight is appropriate to tropical woodlands but it does not have sufficient manoeuvrability to forage in dense vegetation. The closely related White-striped Freetail-bat has similar restrictions but, for reasons not yet understood, the two species have quite distinct ranges, the border of which approximates the Tropic of Capricorn.

When in flight, most bats use their sonar calls continuously to navigate, avoid obstacles and catch prey. This seems not to be the case in the Northern Freetail-bat, which always flies at a considerable distance above the ground or tree-tops. When flying from roost to a foraging area, or when flying between foraging areas, it uses only one sonar pulse every 10 or 20 seconds

*The Northern Freetail-bat roosts in hollow tree spouts.*

B.G. THOMSON

## Chaerephon jobensis

**Size**

*Head and Body Length*
80–90 (84) mm
*Tail Length*
35–45 (41) mm
*Forearm Length*
46–52 (49) mm
*Weight*
20–30 (26) g

**Identification** Chocolate to grey-brown above; slightly greyer below. Lips prominently wrinkled. No neck-pouch in either sex.
**Recent Synonyms** *Nyctinomus jobensis, Nyctinomus plicatus, Chaerephon plicatus, Tadarida jobensis.*
**Other Common Names** Northern Mastiff-bat.
**Status** Common, widespread.
**Subspecies** Australian population referable to *Chaerephon jobensis colonicus.*
**Extralimital Distribution** Papua New Guinea, Solomons, New Hebrides.
**References**
Lestang, A. de (1929). A bat colony (*Chaerephon plicatus colonicus*) in North Queensland. *Aust. Zool.* 6: 106–7.
Begg, R.L. and J.L. McKean (1982). Cave dwelling in the molossid bat *Tadarida jobensis colonicus. Northern Territory Naturalist* 5: 12.

G.B. BAKER

# Beccari's Freetail-bat

## *Mormopterus beccarii*

### Peters, 1881

morm-op'-ter-us  bek-ar'-ee-ee:
'Beccari's winged-hobgoblin'

(compared with five to ten per second in all other species). Presumably it does so merely to ensure that it maintains correct height.

It usually roosts in hollow tree spouts, which may contain large colonies. Colonies of 300–500 have been found in the ceilings of buildings. A cave-roosting colony has been recorded from the Northern Territory, as well as in the Solomon Islands.

The timing of reproductive events is unknown, but juveniles have been found in colonies in December and January, so the breeding season may occur in summer as with many other species.

G.C. RICHARDS

First described in 1865 from the Moluccan Islands, Indonesia, this tropical bat occurs through New Guinea and across northern Australia. It inhabits desert, semi-arid and mesic regions of tropical Australia, where it forages over rainforests, floodplain margins and various eucalypt communities found across the region (open forest, woodland and savanna). Most have been captured along watercourses fringed with River Red Gum and, in wetter regions, also with paperbark and pandanus.

It is robust and muscular with thick, elastic flight membranes. The wings are short, narrow and pointed, with high wing-loading and aspect ratio, so flight is swift, direct and agile, but not

*Beccari's Freetail-bat inhabits Indonesia and New Guinea, extending into northern Australia.*

manoeuvrable. It hunts flying insects above the tree canopy and along the unobstructed corridors of river courses, only approaching ground level over pools. Its echolocation calls are well suited to foraging in uncluttered habitat, with lowest frequency of 23.5 kilohertz and call duration of 13 milliseconds.

Beccari's Freetail-bat usually roosts in tree hollows, but colonies of up to 50 have been found under roofs in urban areas of Queensland. Like other freetail-bats, it can scurry about on surfaces, which may explain why its stomach contents sometimes include flightless insects as well as beetles and moths that are caught in flight.

Throughout its Australian range, females give birth to a single young during the summer (wet season). Pregnant females have been recorded in November, December and January; lactating females have been captured in January.

It is common in suitable habitat. Populations occur in the following National Parks: Drysdale River, Purnululu, Karijini Range and Rudall River (Western Australia), Kakadu (Northern Territory) and Lakefield (Queensland).

N.L. McKenzie

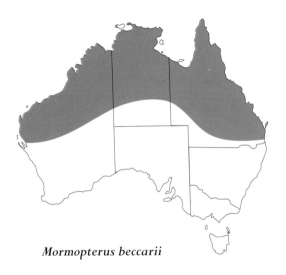

*Mormopterus beccarii*

## Mormopterus beccarii

**Size**

*Head and Body Length*
57–65 (60) mm

*Tail Length*
33–38 (35) mm

*Forearm Length*
36–40 (38) mm

*Digit Length*
40–47 (43) mm

*Weight*
10–18 (14) g

**Identification** Skin on snout and wings mid- to dark brown. Fur on head and back dark brown to greyish with pale bases. Underparts distinctly lighter. Distinguished from *Tadarida australis* and *Chaerephon jobensis* by triangular rather than rounded ears. Significantly larger than all other Australian *Mormopterus*, although *M. beccarii* specimens from the Moluccas and New Guinea have short forearms (34–36 mm).

**Recent Synonyms** *Tadarida beccarii*, *T. loriae*, *Mormopterus loriae*.

**Other Common Names** Northern Scurrying-bat, Beccari's Scurrying-bat.

**Status** Cryptic, but common and widespread.

**Subspecies** *Mormopterus beccarii beccarii*, Australia and Ambon.

*Mormopterus beccarii astrolabiensis*, New Guinea.

**Extralimital Distribution** Ambon and New Guinea.

**References**

Adams, M., T.R. Reardon, P.R. Baverstock and C.H.S. Watts (1988). Electrophoretic resolution of species boundaries in Australian Microchiroptera. IV. The Molossidae (Chiroptera). *Aust. J. Biol. Sci.* 41: 315–326.

Chrome, F.H.J. and G.C. Richards (1988). Bats and gaps: microchiropteran community structure in a Queensland rain forest. *Ecology* 69, 1960–1969.

# Little Northern Freetail-bat

## *Mormopterus loriae*

(Thomas, 1897)

lor'-ee-ee: 'Loria's winged-hobgoblin'

This small freetail-bat was first described from the mouth of Kemp Welsh River in south-eastern New Guinea and it has been recorded in rainforest and mangroves through near-coastal areas of northern Australia and New Guinea. In Western Australia, it is confined to mangrove forests between Point Torment (near Derby) and Exmouth Gulf.

The wings are short, narrow and pointed; flight is swift, direct and agile. Poorly adapted to hunting in confined spaces where manoeuvrability is preferable to speed, it preys upon flying insects in the open air above and beside the forest canopy and is often seen flying through gaps where creeks or roads provide unobstructed corridors through the forest. Like other freetail-bats, it can scurry about on surfaces, where it eats some flightless insects.

In Queensland, colonies have been reported from roofs, tree-hollows and cracks in posts. In

Western Australian mangrove communities, individuals have been found roosting in small spouts and crevices in dead upper branches of the tree, *Avicennia marina*. It emerges at dusk and can be captured at first dark in mist nets set across shadowed glades at the edge of a forest. Swarms of up to a hundred individuals are often seen flying above the mangrove canopy soon after sunset, before dispersing to forage singly or in pairs. The first Northern Territory specimen, collected in September 1948 by the American-Australian Scientific Expedition to Arnhem Land, was taken

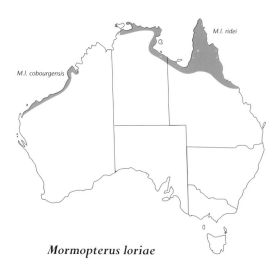

*Mormopterus loriae*

M.l. ridei

M.l. cobourgensis

---

### *Mormopterus loriae*

**Size**

*Head and Body Length*
47–55 (50.5) mm
*Tail Length*
30–36 (33) mm
*Forearm Length*
32.5–35 (34) mm
*Digit 5 Length*
37.5–46 (41.3) mm
*Weight*
6.2–9.0 (7.3) g

**Identification** Smaller than *Mormopterus beccarii* (head to vent less than 56 mm). Forearm shorter than *M. norfolkensis* (unlike the *M. planiceps* species complex). Length of fifth digit divided by head-vent length exceeds 0.75 mm. Unlike a small *Mormopterus* of arid central Australia, the distance across its upper canine teeth exceeds 3.7 mm. Skin of the snout and wings mid-brown. Fur on head and back greyish-brown to mid-brown with white bases. Fur on underside distinctly paler; greyish-lemon to greyish-buff on abdomen and chest; lemon on throat and chin, hairs whiter towards base. Darker individuals reported from north-eastern Australia.

**Recent Synonyms** *Tadarida loriae*, *Mormopterus planiceps ridei*, *M. planiceps cobourgiana*.

G.A. HOYE

*The Little Northern Freetail-bat roosts in small tree spouts and crevices.*

**Other Common Names** Little Northern Scurrying Bat, Little Scurrying Bat, New Guinea Scurrying Bat.
**Status** Common and widespread in suitable habitats.
**Subspecies** *Mormopterus loriae loriae*, New Guinea.
*Mormopterus loriae ridei*, Queensland.
*Mormopterus loriae coburgensis*, Northern Territory.
**Extralimital Distribution** New Guinea.

**References**

Johnson, D.H. (1964). Mammals (in) *Records of the American–Australian Scientific Expedition to Arnhemland*, Vol. 4 (Zoology). Melbourne University Press.

McKenzie, N.L. and A.N. Start (1989). Structure of bat guilds in mangroves: disturbance and determinism (in) D.W. Morris, Z. Abramski, B.J. Fox and M.R. Willig (eds). *Patterns in the structure of mammalian communities*. Spec. Publ. Texas Tech. Univ., Lubbock.

from a group of about ten bats seen flying about a 'meadow-like opening' in the monsoon forest shortly after sunset. The group dispersed soon after it was discovered.

Females give birth to a single young. In Western Australia, young are born during the summer (wet season) and most females are still lactating in March. Juveniles have been captured in mist nets from March to May and have body weights of 3.7–5.3 grams. By early June, there are few subadults in the population; most individuals weigh more than 7 grams and their wing-bones have stopped growing. A similar pattern of summer births is reported in Queensland.

The species is common in suitable habitat. Populations occur in Gurig National Park, Northern Territory, and in Daintree National Park, Queensland.

N.L. McKenzie

# Eastern Freetail-bat

## *Mormopterus norfolkensis*

### (Gray, 1839)

nor'-foke-en'-sis: 'Norfolk (Island) winged-hobgoblin' (erroneous locality)

Often referred to as the Norfolk Island Mastiff-bat, this species is not, however, known from Norfolk Island. The specimen in the British Museum, from which the species was described in 1839, was purchased from a dealer in the previous year and, although it was assumed for some reason to have come from Norfolk Island, no documentary evidence remains to support this assumption. The firm statement by Troughton in his *Furred Animals of Australia* that the type specimen was sent from Sydney by W.S. Macleay, is quite mistaken but it may nevertheless have come from the vicinity of Sydney, for its present range includes that city.

Although its habitat preferences are unclear, most records are from dry eucalypt forest and woodland east of the Great Dividing Range,

| *Mormopterus norfolkensis* |
| --- |
| **Size** |
| *Head and Body Length* |
| 50–55 mm |
| *Tail Length* |
| 35–45 mm |
| *Forearm Length* |
| 36–40 (38) mm |
| *Weight* |
| 7–10 (8.5) g |

**Identification** Considerably smaller than *Tadarida australis* and *Chaerephon jobensis*. Distinguished from *Mormopterus beccarii* by longer fur, translucent, wrinkly wing and tail membranes and lighter build. Greater forearm length than other Australian species of *Mormopterus*.

**Recent Synonyms** *Micronomus norfolkensis*, *Nyctinomus norfolkensis*, *Tadarida norfolkensis*.

**Other Common Names** Norfolk Island Mastiff-bat, Norfolk Island Scurrying Bat, Norfolk Island Freetail-bat, Eastern Wrinkle-lipped Bat, Eastern Micronomus and Eastern Little Mastiff-bat.

**Status** Rare.

**Subspecies** None.

**Reference**
Peters, W. (1881). Uber die Chiropteren-gattung *Mormopterus* und die dahin gehörigen. *Arten. Mber. Preuss. Akad. Wiss*: 482–485.

*Mormopterus norfolkensis*

from southern New South Wales to south-east Queensland. Nevertheless, a number of individuals have been caught at one site flying low over a rocky river through rainforest and wet sclerophyll forest. Seldom has more than one individual been recorded from the 15 or so known locations.

Little is known of its reproductive cycle but the capture of a number of females and no males at one site suggests that the sexes separate at certain times of the year, perhaps for birth and raising of young.

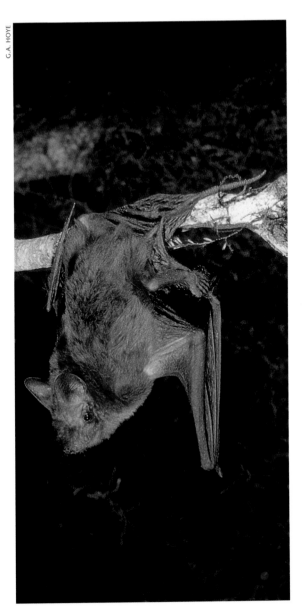

G.A. HOYE

*Little is known of the Eastern Freetail-bat but it appears to roost in trees.*

The Eastern Freetail-bat tends to be predominantly tree-dwelling, but one individual was recorded roosting in the roof of a hut, together with an Eastern Broadnosed-bat and a number of Gould's Wattled Bats.

F.R. ALLISON AND G.A. HOYE

# Southern Freetail-bat

## *Mormopterus planiceps*

(Peters, 1866)

plahn'-ee-seps':
'flat-headed winged-hobgoblin'

The very flat head and body of the Southern Freetail-bat enables it to enter small hollows and crevices to roost: a colony of seven living in the hollow centre of a dead *Casuarina* tree entered and left through a hole the thickness of a man's thumb. Flatness also permits it to roost in the spaces between corrugated iron and roof beams.

Colonies usually include less than ten individuals but may occasionally contain almost 100.

---

### *Mormopterus planiceps*

**Size**
*Head and Body Length*
50–65 (57) mm
*Tail Length*
30–40 (33) mm
*Forearm Length*
31–37 (34) mm
*Weight*
10–14 (11) g

**Identification** Charcoal grey above; notably paler below. Ears triangular, not joined across forehead. Upper lip overhangs lower lip and bears fringe of stiff hairs. No neck-pouch in either sex.

**Recent Synonyms** *Nyctinomus petersi, Nyctinomus planiceps, Micronomus planiceps, Tadarida planiceps.*

**Other Common Names** Little Freetail-bat, Little Flat Bat, Western Mastiff-bat, Western Micronomus, Western Scurrying Bat.

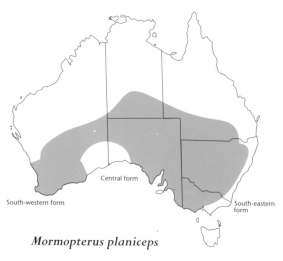

Central form

South-western form

South-eastern form

0

*Mormopterus planiceps*

The Chocolate Wattled Bat, Western Broad-nosed-bat and King River Eptesicus have been recorded as sharing its roosts.

The Southern Freetail-bat has limited manoeuvrability. Its flight is swift and direct as it forages over the forest canopy, over waterholes or along the borders of tree-lined creeks. It may also scurry along the ground or over the trunk of a tree in search of insects, using the thumbs and the hindfeet in a 'rowing' action. By taking its food high in the air and on the ground, it utilises insect resources that are not greatly exploited by other bats.

**Status** Common.
**Subspecies** None described but three forms can be distinguished.
**Reference**
Hill, J.E. (1961). Indo-Australian bats of the genus *Tadarida. Mammalia* 25: 29–56.

It is quiet and gentle to handle but is extremely aggressive towards bats of other species in captivity or in the wild. Aggressive interactions have been observed between aerially foraging bats and between them and individuals of other species, indicating that feeding territories probably exist.

Adult females are in an advanced stage of pregnancy in late November, and the single young is probably born in December. The actual timing of births is yet to be confirmed.

Three forms that may represent distinct species are separable on the basis of fur and penis characters.

G.C. RICHARDS

*The Southern Freetail-bat flies fast and directly over the forest canopy, preying on insects.*

N. SPEECHLY

G.B. BAKER

# White-striped Freetail-bat

## *Nyctinomus australis*

## (Gray, 1838)

nik'-tee-noh'-mus  ost-rah'-lis:
'southern night-dweller'

The grotesque, savage appearance of this bat belies its nature, for it is a very docile animal to handle. Largest of all the Australian

*The White-striped Freetail-bat is distinctively coloured.*

freetail-bats, it occupies a diverse range of habitats but is absent from the tropics where it is replaced by the closely related Northern Freetail-bat.

The foraging behaviour of the White-striped Freetail-bat gives it access to two sources of food, one high, one low, that are not exploited to a great extent by most other bats. Whereas most bats tend to hunt for insects around or underneath trees, the White-striped Freetail-bat flies a fast, direct path above the tree canopy, especially favouring the lines of Red Gums along dry inland creeks, in search

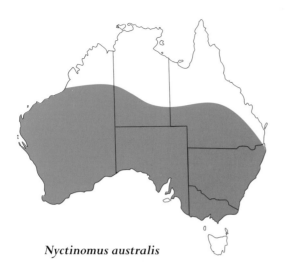

*Nyctinomus australis*

## *Nyctinomus australis*

**Size**
*Head and Body Length*
85–100 (92) mm
*Tail Length*
40–55 (43) mm
*Forearm Length*
57–63 (61) mm
*Weight*
25–40 (36) g

**Identification** Chocolate brown above; slightly lighter below. Distinct stripes of white fur along the junction of the undersurface of the wings and body; additional patches of white fur may occur. Large, fleshy, forward-pointing ears. Upper lip deeply wrinkled. Throat-pouch present in both sexes.
**Recent Synonyms** *Nyctinomus albidus, Austronomus australis, Tadarida australis.*
**Other Common Names** White-striped Bat, White-striped Mastiff-bat, White-striped Scurrying Bat.
**Status** Common to uncommon in widespread habitat.
**Subspecies** *Nyctinomus australis australis,* most of range.
*Nyctinomus australis atratus,* central desert.
**References**
Hall, L.S. and G.C. Richards (1972). Notes on *Tadarida australis* (Chiroptera: Molossidae). *Aust. Mammal.* 1: 46–47.
Thomas, O. (1924). A new species of *Nyctinomus australis. Ann. Mag. Nat. Hist.* Ser. 9, 14: 455–456.

of high-flying moths. Occasionally it descends to the ground where, with its delicate tail-membrane retracted close to the body to avoid injury, it is surprisingly agile in its search for terrestrial insects (for which reason freetail-bats are sometimes called 'scurrying bats'). Some molossid bats are unable to fly from the ground but the White-striped Freetail-bat can launch itself directly into the air, albeit with some difficulty since the long narrow wings, adapted for fast flight, do not provide the uplift necessary for rapid take-off.

Small groups, seldom more than ten, roost in tree hollows or similar retreats but maternity colonies may number several hundred bats. Solitary individuals have been found under loose tree bark, in dead stumps and in ceilings of buildings. There is a record from the early twentieth century of one being found in a rock cavern in Central Australia.

The scant information on the reproductive cycle suggests that females are pregnant during spring and early summer and give birth to a single young towards the end of the year. It is not known whether sperm is stored or foetal development is delayed during the winter, nor is there any information on the expected differ-ence in timing of reproductive events between northern and southern populations or between populations from arid and wet climates.

G.C. RICHARDS

# FAMILY VESPERTILIONIDAE
## *Vespertilionid bats*

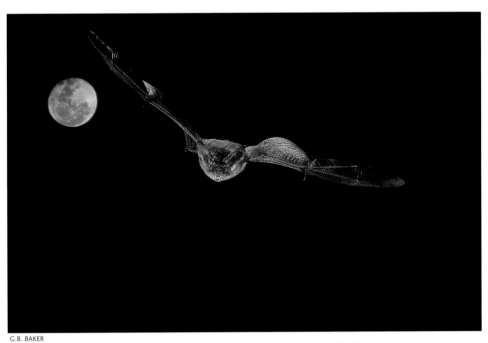

G.B. BAKER

*Gould's Long-eared Bat is a slow, manoeuvrable flier.*

Vespertilionid bats share certain skeletal and dental characteristics but the family is so diverse that there are few external features that serve to distinguish all of them from members of other bat families. For example, while the majority lack a noseleaf, this structure is weakly developed in the long-eared bats of the genus *Nyctophilus*. The tail is relatively long and the tail-membrane extends to, or almost to, its tip. This, in itself, is not diagnostic but it serves to distinguish vespertilionids from other Australian bats.

A notable feature of the family is its success: of all the living species of bats, about one in three is a vespertilionid. Almost half of the Australian bats are in this group. They are found in every continent and, thanks to superior physiological adaptations—particularly hibernation—they extend much further into the cold parts of the world than any other group.

The Australian species fall into four subfamilies. The Kerivoulinae (known outside Australia as 'woolly' or 'painted' bats) has only one Australian representative, the Golden-tipped Bat. The Miniopterinae includes 11 species of bentwing-bats, two of which occur in Australia. The Murininae, has 14 species, one of which occurs in Australia. The Nyctophilinae has seven Australian species. All the other Australian microbats are members of the immense and widespread Vespertilioninae.

## Subfamily Kerivoulinae

Members of this small subfamily have woolly fur and funnel-like ears. Only one species occurs in Australia.

# Golden-tipped Bat

## *Kerivoula papuensis*

### Dobson, 1878

ke'-ree-vule'-ah pah'-pue-en'-sis:
'Papuan kerivoula' (Sri Lankan name for a related species)

The story of the Golden-tipped bat in Australia encapsulates many of the problems that bat conservationists face when developing management prescriptions for a species about which so little is known.

First described in 1878 from a specimen collected in New Guinea, its presence in Australia was confirmed by five specimens collected near Rockhampton in 1884 and a further three speci-

*Kerivoula papuensis*

mens collected from Cape York in 1897. Nearly 84 years passed before the species was recorded again in Australia. Within two years, ten further specimens had been trapped, extending its known range to southern New South Wales. That number has now grown to nearly 50 individual captures at coastal sites ranging in altitude from 50 to 150 metres above sea-level from Cape York to the border of New South Wales and Victoria. This is considerably further south than might be expected from its predominantly tropical distribution in Indonesia, New Guinea and the Philippines.

As trap-effort intensifies and techniques improve, so does our understanding of the distribution of this species. It seems that in coastal forests, bat-traps are essential to capturing this unusual species. For example, a team of workers conducting the North-east Forest Biodiversity Study in New South Wales made an intensive survey of all bat species in the region. They placed traps at some 500 sites and exerted an accumulated trap-effort of over 1100 trap-nights. Relative to some other species, the Golden-tipped bat was well represented, with 31 individual captures at 21 of these sites. Some sites yielded more than one individual, indicating that these were not aberrant findings.

Computer modelling of presumed habitat parameters predicts a distribution in a narrow band down the eastern coast of Australia in areas with high summer rainfall, equable thermal conditions and moist closed lowland forest. These are conditions vulnerable to the effects of climate change. For the most part, these predictions have been borne out by recent trapping results. Most trap sites are in coastal forests, near to where wet and dry forests meet and often in the vicinity of creeks. Captures are generally in dense or tangled

P. GERMAN

*The Golden-tipped Bat has woolly fur with orange tips to the hairs.*

### Kerivoula papuensis

**Size**
*Head and Body Length*
50–60 mm
*Tail Length*
40–50 mm
*Forearm Length*
35–39 mm
*Weight*
circa 6 g

**Identification**  Dark brown above; fur tipped with gold. Golden fur covers forearms and thumbs. Very short, pointed muzzle overhangs lower jaw. Long upper canine teeth fit pockets in lower lip.
**Recent Synonyms**  *Phoniscus papuensis.*
**Other Common Names**  Dome-headed Bat.
**Status**  Rare.
**Subspecies**  None.
**Extralimital Distribution**  New Guinea.
**Reference**  None relevant to biology of species.

vegetation or along creek lines, possibly reflecting both a preference for ecotonal habitats for feeding and passage and an ability to manoeuvre in dense vegetation.

Observations of captive individuals support the idea that it feeds by gleaning, being able to fly slowly in very dense vegetation, hovering for short periods and plucking prey from surfaces while on the wing. The prey collected in this way includes spiders in their webs. Investigation of faecal material and stomach contents of over 20 individuals shows a very high and sometimes exclusive diet of spiders. The echolocation calls are suitable for such foraging and include a short, quiet call with broad frequency sweeps initiated at a very high frequency. The wings are broad and enable slow flight. This combination of foraging behaviour and echolocation make it a difficult species to detect with remote listening devices unless encountered on a flight through the upper forest canopy, when it uses a louder navigational call.

It has been suggested that the Golden-tipped Bat is suited to life in cloud forests of New Guinea where the species is best known. Its current distribution and roosting habits in Australia may reflect this and lead to an association with moist, dense vegetation. The dense long and grizzly coloured coat would provide camouflage among vegetation and protection from the elements. Its known roost sites in New Guinea range from dead foliage to roofs of houses.

The pattern of reproduction can only be surmised. Most recent captures in Australia have occurred in the spring and summer and revealed very little about their reproductive status. No males showed descended testes during that time and several females were in post-lactation condition. It is possible that it breeds during the wet season in the northern part of its range and during the early spring elsewhere, but this has yet to be verified. Males are slightly smaller than females, a common feature of species that form relatively small seasonal maternity colonies and males are transient members.

D.P. WOODSIDE

## Subfamily Miniopterinae

Members of this group have a very long third finger and a narrow wingtip, which is folded back when the bat is at rest.

# Little Bentwing-bat

## *Miniopterus australis*

### (Tomes, 1858)

min'-ee-op-te-rus ost-rah'-lis: 'southern small-wing'

Like the Common Bentwing-bat, the Little Bentwing-bat depends upon specific nursery sites to rear its young. It occupies caves and tunnels during the day and at night forages for small insects beneath the canopy of well-timbered habitats including rainforest, *Melaleuca* swamps and dry sclerophyll forests. Its distribution within Australia becomes increasingly coastal towards the southern limit of its range in New South Wales.

It frequently shares roosting sites with the Common Bentwing-bat and, in winter, the two species may form mixed clusters. In fact, the southernmost breeding population of the Little Bentwing-bat, found in the Macleay River watershed, seems to depend upon the larger nursery colony of Common Bentwing-bats to provide the high temperatures needed to rear its young. This population of approximately 5000 individuals moves between roosts to satisfy different seasonal needs but the distances travelled are less than those recorded for its larger relative. A colony may be biased towards either adult or younger individuals.

In the north of its range, association with the Common Bentwing-bat is less pronounced.

### *Miniopterus australis*

**Size**

*Head and Body Length*
43–48 mm
*Tail Length*
43–48 mm
*Forearm Length*
36–40 mm
*Weight*
7–8 g

**Identification** Chocolate brown above; paler below. Fur with lighter, more subtle shades than in the larger *Miniopterus schreibersii*, which has forearm length greater than 44 mm.
**Recent Synonyms** None.
**Other Common Names** Tomes' Bat.
**Status** Abundant.
**Subspecies** Three subspecies: Australian form is *Miniopterus australis australis*.
**Extralinital Distribution** New Caledonia, New Guinea, Indo–Malayan archipelago, Philippines.
**Reference**
Dwyer, P.D. (1968). The biology origin and adaption of *Minopterus australis* (Chiroptera) in New South Wales. *Aust. J. Zool.* 16: 49–68.

A huge nursery colony found at Mount Etna, near Rockhampton, Queensland, is dominated by Little Bentwing-bats; less than 1 per cent of the approximately 100 000 adult females that assemble here each spring are of the larger species. The evening emergence of this vast colony from a jagged cleft among limestone,

*The Little Bentwing-bat is an abundant species.*

P. GERMAN

with some individuals being plucked from the air by pythons and others dived upon by predatory Ghost Bats, is a spectacular sight.

The very young bats are abandoned by their mothers through the night and remain clinging to the cave walls, packed together in huge pink masses, at temperatures that may reach 37°C. Many fall to the floor of the cave where they become victims to millions of beetles which live there. The temperature deep inside the cave is uncomfortable for adult bats and after suckling their young in the early morning the mothers regularly move to cooler situations near the cave entrance.

The Little Bentwing-bat resembles its larger relative in that females do not store sperm, fertilisation coinciding with mating. Births occur in December. Males are sexually active during the winter, copulatory activity occurs through late July and August and fertilisation takes place in the latter month. Mating times do not vary significantly with latitude nor is there a lengthy period before the embryo implants upon the wall of the uterus.

*Miniopterus australis*

Banded individuals have been recaptured when they were at least five and a half years old but the life span is probably longer than this. Green Tree Frogs, pythons, owls and the Fox are known predators. Conservation of the species requires minimising disturbance at overwintering and nursery roosts. In Australia, only five nursery sites have been reported. Although others undoubtedly occur, it is of concern that two of these known sites are in areas where limestone mining takes place.

P.D. DWYER

G.D. ANDERSON

# Common Bentwing-bat

## *Miniopterus schreibersii*

### (Kuhl, 1817)

shribe'-er-zee-ee: 'Schreibers' small-wing'

Above: *The Common Bentwing-bat is also found in Europe, Africa and Asia.*

Right: *Larger than the Little Bentwing-bat, the Common Bentwing-bat has similar behaviour.*

By day, the Common Bentwing-bat rests in caves, old mines, stormwater channels and comparable structures including occasional buildings. Typically, it is found in well-timbered valleys where it forages, above the tree canopy, on small insects. Its flight is level and fast, punctuated by swift, shallow dives upon its prey. Where conditions are favourable, colonies are often large: some dominated by males, others by females; some including mostly adults, others mostly young.

Large distances are travelled between different roosts according to changing seasonal needs and the dictates of age and reproductive status. The pattern of movement varies in response to local climatic conditions and the dispersion of suitable roosting sites. In south-eastern Australia cold roosts are sought during the winter to allow hibernation at a time when insect food is scarce.

With the onset of spring, adult females move from numerous widely scattered roosts to specific nursery caves that provide high temperature and humidity throughout the year or—in the southern part of the range—have an internal conformation that retains air that has been warmed by the bats' activities. Scattered colonies located within a single large watershed often use a particular nursery cave year after year.

In a nursery cave, young bats, massed upon the ceiling at densities up to 3000 per square metre, are nursed and reared to independence. A single naked young is born to each female, usually in December. While the timing of birth is relatively fixed, mating occurs earlier at higher latitudes. At 28° south in north-eastern New South Wales, mating and fertilisation occur from late May to early June, prior to hibernation. Development of the fertilised ovum, which is always derived from the left ovary, proceeds slowly and implantation does not take place until shortly before the females come out of hibernation late in August. Thereafter, with warmer weather, an abundance of insect food and the selection of warmer roosts by the females, the embryo develops at a normal rate. No

---

### *Miniopterus schreibersii*

**Size**

*Head and Body Length*
52–58 mm

*Tail Length*
52–58 mm

*Forearm Length*
45–49 mm

*Weight*
13–17 g (pregnant females to 20 g). In southern latitudes individuals may enter hibernation weighing 20 g.

**Identification** Blackish to reddish-brown above; paler below. Mantle of contrasting brown in some moulting females; rufescent forms present in some northern Queensland populations. Short muzzle; high-crowned head. Last joint of third finger four times length of preceding joint and folded back along lie of first joint when at rest. Larger than *Miniopterus australis*.

P. GERMAN

other hibernating bat is known to possess this capacity for delayed implantation. In the tropics (15° south), delayed implantation does not occur and, although mating takes place about September, the young are born in December.

Nursery colonies disband between February and March, adults and juveniles going separate ways. At this time some young individuals disperse many hundreds of kilometres. Sexual maturity is reached in the second year of life and longevity may be in excess of 17 years.

The Common Bentwing-bat is preyed upon by owls, pythons, the feral Cat and, occasionally, the Fox. Frequent disturbance of roosts used for hibernation seriously increases winter mortality. Because of its dependence upon relatively few nursery caves, threats to the existence or structural integrity of any of these may place the survival of widespread populations in jeopardy.

P.D. DWYER

**Recent Synonyms** *Miniopterus blepotis*.

**Other Common Names** Schreibers' Long-tailed Bat, Schreibers' Bat, Eastern Bent-winged Bat, Bent-winged Bat, Long-fingered Bat (in Europe).

**Status** Abundant.

**Subspecies** Two subspecies are currently recognised in Australia:
*Miniopterus schreibersii blepotis*, eastern Australia.
*Miniopterus schreibersii orianae*, north-western Australia.

**Extralimital Distribution** New Guinea, Indo–Malayan archipelago, Africa and Eurasia to 45° north.

**References**

Dwyer, P.D. (1966). The population pattern of *Miniopterus schreibersii* (Chiroptera) in north-eastern New South Wales. *Aust. J. Zool.* 14: 1073–1137.

Richardson, E.G. (1977). The biology and evolution of the reproductive cycle of *Miniopterus schreibersii* and *M. australis* (Chiroptera: Vespertilionidae). *J. Zool.* 183: 353–375.

M.s. orianae

M.s. blepotis

*Miniopterus schreibersii*

## Subfamily Murininae

This essentially Asian group has some 14 species, one of which extends into New Guinea and Australia. Members are characterised by tubular nostrils, similar to those of the pteropodid *Nyctimene* but, while this is a fruit-eater, *Murina* is insectivorous.

# Tube-nosed Insect Bat

## *Murina florium*

### Thomas, 1908

myue-reen'-ah flor'-ee-um: 'mouse-like (bat) from Flores'

***Murina florium***

In 1981 a single specimen of this extremely rare bat was captured for the first time in Australia on Mount Baldy (Atherton Tableland, Queensland). Subsequently 11 others have been caught at adjacent localities, two in tropical rainforest near Cedar Bay and another at Iron Range.

The first Australian specimen was kept alive for ten days, permitting observations on its behaviour. Unlike typical vespertilionid bats it does not rest with its wings folded at the sides of the body but in a unique posture. The wings are wrapped around the body, as in fruit-bats, but held away from the chest to form an umbrella. The tail-membrane, sparsely covered with woolly hair, is curled over the posterior edges of the wings, leaving the tips of the wings protruding. When it was resting in this posture and sprayed with a fine mist of water, droplets formed on the hairs of the tail-membrane, ran down the wings and flowed off the wing tips—much as raindrops run off the drip-tips of rainforest leaves. Inasmuch as the habitat in Australia is shrouded in mist and cloud for much of the year, it seemed likely that the usual posture is an adaption which prevents the animal's body from becoming unduly wet. However, subsequent finds of the species in lowland forests cast some doubt on this hypothesis.

### *Murina florium*

**Size**

*Head and Body Length*
47–57 mm
*Tail Length*
31–37 mm
*Forearm Length*
33–36 mm
*Weight*
6–8 g

**Identification** Distinctive, laterally diverging tubular nostrils at point of muzzle. Long hairs on tail-membrane, feet and forearms. Lower outer edge of pinna flap-like, producing a notch. Body hair long, brown or grey, paler tipped.

*The Tube-nosed Insect Bat has been found increasingly in eastern Australia since 1981.*

Another intriguing aspect of its biology is the method of eating insects. When given a moth, the bat holds it in its mouth and, suspended by both its feet and its long thumbs, quickly chews it. Wings, legs and other fragments fall into the tail-membrane and, after eating any edible remnants among these, the bat releases the grip of its feet and, hanging by its thumbs alone, shakes the discarded parts away. This posture is also employed when eating faeces. Such coprophagy, which has not been reported for any other Australian bat, may offer some nutritional advantage for a species living in cool altitudes where insects are irregularly abundant. It is, however, possible that the behaviour is a response to stress: one of the lowland specimens was seen to eat its faeces.

Yet another surprising feature is the appetite of the captive animal for papaw juice, even after a meal of moths and with abundant drinking water. Although far from proven, it seems possible that the Tube-nosed Insect Bat may also feed on nectar.

The soft wide-band echolocation call of short duration (1.5–2 milliseconds) and the slow, butterfly-like flight are both indications that it detects its targets at very short distances (as do gleaners such as species of *Nyctophilus* and *Phoniscus*). The limited evidence available suggests that this species has managed to find a way of life in the worst of tropical habitats and that, by foraging in the canopy of flowering trees, it obtains sufficient insects and/or nectar to ensure its survival.

A recent capture of a *Murina* in Iron Range lowland coastal rainforest is suspected to represent a second species, but this is awaiting confirmation. If proven, this distribution pattern will be similar to that of the extralimital range, where *M. florium* is found in upland habitats and a smaller species is found in nearby coastal regions.

G.C. RICHARDS, R.B. COLES AND H.J. SPENCER

**Recent Synonyms** *Murina lanosa*, *M. toxopeusi*.
**Other Common Names** Flute-nosed Bat.
**Status** Very rare in Australia.
**Subspecies** None.
**Extralimital Distribution** Eastern Indonesia and New Guinea.
**References**

Richards, G.C., L.S. Hall, P. Helman and S.K. Churchill (1982). First discovery of a species of the rare tube-nosed insectivorous bat (*Murina*) in Australia. *Aust. Mammal.* 5: 149–51.

Spencer, H.J., N. Schedvin and B.M. Flick (1992). Re-discovery of Australia's rarest bat, *Murina florium*, the insectivorous tube-nosed bat, in lowland forest in far-north Queensland. *Bat Res. News* 33: 76.

## Subfamily Nyctophilinae

The long-eared bats assigned to this group are separated from other vespertilionids mainly on penis morphology. Some authors extend this essentially Australian group to include genera from northern and central America but there is no strong evidence of this. The status of the subfamily is uncertain.

G.B. BAKER

*The Northern Long-eared Bat is not demanding in its choice of roosts: it will shelter under foliage.*

# Northern Long-eared Bat

## *Nyctophilus arnhemensis*

Johnson, 1959

nik'-toh-fil'-us arn'-em-en'-sis:
'Arnhem (Land) night-lover'

Although the type specimen of the Northern Long-eared Bat was collected at Rocky Bay on Cape Arnhem Peninsula in 1948, a specimen of this species, misidentified as *Nyctophilus timoriensis*, had been collected earlier near Broome, Western Australia, in 1895.

It occurs in suitable habitat throughout the tropical areas of the Northern Territory and Kimberley where average annual rainfall exceeds 500 millimetres, and extends southwards along the coast of Western Australia to Exmouth Gulf. In subcoastal parts of the Northern Territory, it has been recorded in forests of Cajeput, *Melaleuca leucadendra*, open forests of *Eucalyptus miniata*, isolated patches of rainforest, riverine fringes of paperback trees

### *Nyctophilus arnhemensis*

**Size**

*Head and Body Length*
43–52 (49) mm
*Tail Length*
35–43 (39) mm
*Forearm Length*
36–40 (38) mm
*Weight*
6–8 (7) g

**Identification**  Brown all over. Hairs of upper surfaces olive-brown to light olive-brown with dark bases; hairs on undersurfaces very pale brown with dark bases. Pilbara populations darker, some individuals with very dark brown back fur and yellowish-brown tips on undersurface fur. Intermediate in size between *Nyctophilus bifax* and *N. walkeri*. Distinguished from *N. geoffroyi* by small ears, brown colouration and low ridge forming posterior element of noseleaf.

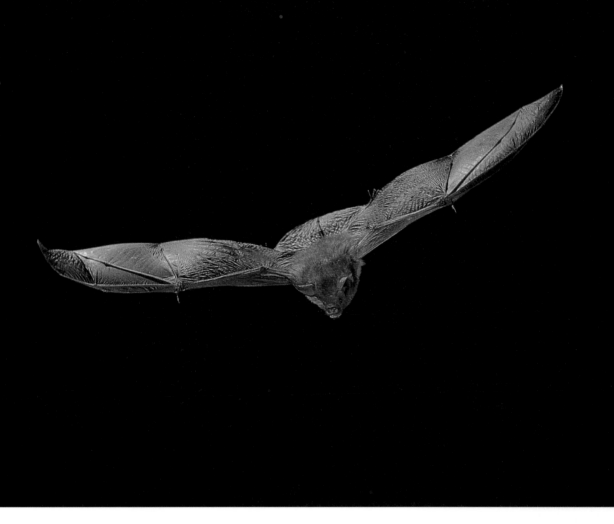

and over freshwater lagoons. It has also been found in mangroves on the islands of the Sir Edward Pellew Group and over waterholes in sandstone country.

Specimens collected in the Kimberley were foraging in rainforest patches, in groves of Cajeput and *Melaleuca argentia* forest that fringe watercourses such as the Drysdale River and in woodlands of *Tristania grandiflora* and *Melaleuca acacioides* around freshwater swamps. The species is common in mangrove forests along Western Australia's northern and western coast, as far south as Gales Bay in Exmouth Gulf.

It hunts insects under, or close to, the foliage of dense vegetation at late dusk and at night. The slow fluttering flight of its short, wide wings is well adapted to confined spaces where manoeuvrability is preferable to speed.

Roost sites appear to be variable. Two of the three Rocky Bay specimens were taken

*The Northern Long-eared Bat flies slowly with great manoeuvrability in search of insects.* G.B. BAKER

**Recent Synonyms** None (some early confusion with *Nyctophilus timoriensis*; not satisfactorily distinguished from *Nyctophilus microtis* from New Guinea).

**Other Common Names** None.

**Status** Common in suitable habitat; habitat limited.

**Subspecies** None.

**References**

Johnson, D.H. (1964). Mammals (in) *Records of the American Australian Scientific Expedition to Arnhem Land*. Vol. 4 (Zoology). Melbourne University Press.

Johnson, D.H. (1959). Four new mammals from the Northern Territory of Australia. *Proc. Biol. Soc. Wash.* 72: 183–188.

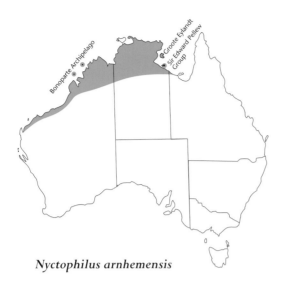

*Nyctophilus arnhemensis*

# Eastern Long-eared Bat

## *Nyctophilus bifax*

### Thomas, 1915

bie'-fax: 'bifurcated (penis bone) night-lover'

This species occurs across northern Australia, where it is locally common in habitats ranging from rainforest to riparian woodland. Ongoing studies suggest that the western subspecies, *N. bifax daedalus*, could prove to be a distinct species and that past records from New Guinea are unlikely to represent this species.

Little is known of its biology or habitat requirements. Radio-tracking studies in northern New South Wales indicate that individuals frequently roost communally in tree-hollows or among dense foliage and that retention of a range of roost sites is essential to its survival.

Studies in New South Wales show that there are seasonal changes in roost-selection from the

from their daytime refuge sites about 2.5 metres from the ground under hanging flaps of the soft, flexible bark of Cajeput trees. Another Northern Territory specimen was taken from the roof of a house at Port Langdon on Groote Eylandt. A Kimberley specimen was found roosting in rainforest foliage on Dampier Peninsula.

Young are born from late in the dry season (October) to the end of the wet season (February). Litters of two have been recorded.

Cryptic and probably more common than records suggest, the Northern Long-eared Bat occurs as secure populations in the pristine mangrove communities of the Prince Regent Nature Reserve and elsewhere on the remote north-western Kimberley coast. Other Kimberley populations occur in the Drysdale River National Park and Coulomb Point Nature Reserve. In the Northern Territory, populations occur in Kakadu National Park and the Keep River Reserve.

N.L. MᴄKᴇɴᴢɪᴇ

*The broad wings of the Eastern Long-eared Bat indicate that it is a slow, manoeuvrable flier.*

G.B. BAKER

## Nyctophilus bifax

### Size

*Head and Body Length*

45–55 mm

*Tail Length*

40–45 mm

*Forearm Length*

38–44 mm

*Weight*

8–12 g

**Identification** Fur brownish to tan above; grey-brown below. Bump on snout behind nose-leaf low and rounded. Distinguished from *N. gouldi* by brownish-tan (rather than greyish) fur; *N. bifax* has less developed snout bump; ears are shorter on average than in *N. gouldi*. External appearance of *N. bifax bifax* similar to *N. arnhemensis* but slightly larger: forearm length of adult females usually greater than 40 mm; adult males usually greater than 38 mm. *N. bifax daedalus* differs from *N. arnhemensis* in lighter fur colour.

**Recent Synonyms** *Nyctophilus gouldi bifax*.

**Other Common Names** North Queensland Long-eared Bat.

**Status** Common, widespread in the coastal tropics, localised in subtropical regions.

**Subspecies** *Nyctophilus bifax bifax*, north-eastern Australia.

*Nyctophilus bifax daedalus*, north-western Australia (possibly a distinct species).

**References**

Lunney, D., J. Barker, T. Leary, D. Priddel, R. Wheeler, P. O'Connor and B. Law. 'Roost selection by the north Queensland Long-eared Bat *Nyctophilus bifax* in the Iluka World Heritage Area, New South Wales.' *Aust. J. Ecol*. 20: 532–537.

Parnaby, H.E. (1987). 'Distribution and taxonomy of *Nyctophilus gouldi* Tomes, 1858 and *Nyctophilus bifax* Thomas, 1915 in eastern Australia'. *Proc. Linn. Soc. NSW*. 109: 153–174.

L. LUMSDEN

*The Eastern Long-eared Bat roosts in tree-hollows or in dense foliage.*

edge of rainforest remnant areas in summer to the centre of rainforest patches in winter.

Females frequently produce twins and carry these at least until their combined weight is equal to that of their mother; thereafter they are left in the maternity roost while the mother is foraging.

The relatively broad wings and short, low-intensity echolocation calls supports limited observations that it is a slow and manoeuvrable flier.

H. PARNABY

*Nyctophilus bifax*

P. GERMAN

# Lesser Long-eared Bat

## *Nyctophilus geoffroyi*

### Leach, 1821

zhef'-roy'-ee: 'Geoffroy's night-lover'

The distribution of the Lesser Long-eared Bat is rivalled only by Gould's Wattled Bat. It is found from semi-arid areas with only sporadic trees, through mallee, woodland and wet forest to alpine areas below the tree-line and tropical northern Australia. Except in some of the wetter temperate forests, it is a common species. In keeping with its ability to exploit such a huge range of habitat types it is rather unselective about roost sites but seems to prefer those with warm, humid conditions. It tolerates extreme heat and has been found in places where the temperature exceeds 40°C. It is basically a crevice-rooster, being often found beneath the exfoliating bark of trees such as acacias and she-oaks, but tree-hollows are also used. Solitary males have been found roosting in such unlikely places as the nests of Fairy Martins, piles of bricks, clothing left hanging outdoors, rolled-up blinds, under a

*The Lesser Long-eared Bat roosts in a variety of small spaces.*

stone on the ground. It is a frequent inhabitant of the roofs of houses and common in most cities. Roosts in houses are frequently shared with other species, particularly the Chocolate Wattled Bat and the Little Freetail-bat. Most maternity colonies consist of about 10–15 females, possibly limited by the space available in natural roost sites. In buildings, where an unusually large space may be available, colonies may contain several hundred bats.

It emerges from the roost just after dark to forage for insects. In cold, wet conditions, it may hunt for only an hour but in late spring and summer it usually does not return to the roost until almost dawn. During winter it is torpid during the day and slow to become aroused. In summer it is likely to be active even in the roosting site and is capable of taking to the wing almost immediately if disturbed.

It emits echolocation calls of low intensity and usually forages quite close to, or on, the ground. The large ears suggest that it is able to detect prey by passive listening as well as by active echolocation. Its visual acuity has not been tested, but the large eyes seem likely to function in its prey-catching behaviour. Having relatively short, broad wings, it is able to fly slowly while searching for prey. Like other long-eared bats, it

*Nyctophilus geoffroyi*

has an undulating and fluttering flight. Because of its habit of flying close to the ground and its abundance in cities, it is a common victim of domestic Cats. It eats a wide variety of invertebrates, both flying and flightless. Unlike many other insectivorous bats it readily eats food from a tray and adapts well to captivity if adequate food and space is provided.

Its social behaviour varies with geographic location and may be influenced by the availability of suitable habitats. Roosts may be maintained continuously despite extreme disturbance, such as the complete replacement of an iron roof; or changed frequently, the decision probably being determined by the availability of alternative sites.

Breeding biology also varies with geographical location. Maternity colonies begin to form in spring and births occur in late spring or early summer. In south-eastern South Australia, maternity colonies are made up almost solely of pregnant females but Tasmanian colonies are known to contain both adult males and females.

Twins are common and the young are naked at birth. Although mothers are capable of flying freely while carrying their suckling young, the infants are usually left in the roosting site while the females are out foraging. Young bats can fly at about six weeks of age but do not venture far from the roost on their early flights. Later, while still less agile than the adults, they join the female parents in hunting but there is no evidence of parent–offspring relationships persisting after weaning.

The development of agriculture and the urbanisation of Australia appear to have had no adverse effect on this species and the opposite may well be the case. In most areas, the introduction of buildings appears to have increased the range of suitable habitats and farming has led to increased insect food. This situation should not, however, lead to complacency: bats depend upon an abundance of insects, and the use of chemical insecticides could reverse any beneficial effects of European settlement.

T.H. MADDOCK AND C.R. TIDEMANN

## *Nyctophilus geoffroyi*

**Size**
*Head and Body Length*
40–50 mm
*Forearm Length*
33–41 (37) mm
*Weight*
6–12 (8) g

**Identification**  Light grey-brown on dorsal surface; paler, tending to whitish on ventral surface. Ears very long (longer than in *N. gouldi*) and joined together above forehead; may be furled in live animal. The most useful diagnostic feature is the two lobes behind the noseleaf; in *N. geoffroyi* these are joined by a median elastic membrane, which is more extensive than in other Australian *Nyctophilus*.

**Recent Synonyms**  None.

**Other Common Names**  Geoffroy's Long-eared Bat.

**Status**  Very common over most of its range.

**Subspecies**  *Nyctophilus geoffroyi geoffroyi*, Western Australia.

*Nyctophilus geoffroyi pallescens*, northern South Australia.

*Nyctophilus geoffroyi pacificus*, eastern mainland, Tasmania.

**References**

Green, R.J. (1965).  Notes on the lesser long-eared bat *Nyctophilus geoffroyi* in northern Tasmania. *Rec. Queen Victoria Mus. Launceston* (n.s) 22: 1–4.

O'Neill, M.G. and R.J. Taylor (1986).  Observations on the flight patterns and foraging behaviour of Tasmanian bats. *Aust. Wildl. Res.* 13: 427–432.

Tidemann, C.R. and S.C. Flavel (1987). Factors affecting choice of diurnal roost site by tree-hole bats (Microchiroptera) in south-eastern Australia. *Aust. Wild. Res.* 14: 459–473.

# Gould's Long-eared Bat

## *Nyctophilus gouldi*

### Tomes, 1858

gule'-dee: 'Gould's night-lover'

Relatively broad wings allow this bat to use a slow and manoeuvrable flight to forage in dense undergrowth along watercourses in sclerophyll forests and woodlands. It is known to glean insects from low foliage and even to sit and wait before dropping on prey in the forest litter. This occasionally leads to capture by a domestic cat.

In eastern Australia it is found as far north as Atherton in Queensland and throughout the eucalypt forests and woodland of eastern New South Wales and south-eastern Victoria. It occupies a variety of roost-sits from rooftops to tree-holes and under peeling bark. There is some evidence, based on observations of captive and wild colonies, to suggest that males and females do not roost together. A typical colony probably accommodates 10–20 bats, but males might be solitary.

*Nyctophilus gouldi*

**Size**
*Head and Body Length*
55–65 (58) mm
*Tail Length*
45–55 (48) mm
*Forearm Length*
34–48 (42) mm
*Weight*
9–13 (11) g

**Identification** Fur can range from dark brown to dark grey on the back with typically a light grey on the undersurface. Noseleaf shape and the size of the post-nasal bump vary between *Nyctophilus* species and can confuse identification.
**Recent Synonyms** *Nyctophilus timoriensis gouldi*.

*Nyctophilus gouldi*

P. GERMAN

*Above: Gould's Long-eared Bat frequently produces twins which are carried by the mother.*

**Other Common Names** Greater Long-eared Bat formerly applied to this species and to *Nyctophilus timoriensis*.

**Status** Common.

**Subspecies** None.

**References**

Parnaby, H.E. (1986). Distribution and taxonomy of the Long-eared Bats, *Nyctophilus gouldi* Tomes, 1858 and *Nyctophilus bifax* Thomas, 1915 (Chiroptera: Vespertilionidae) in eastern Australia. *Proc. Linn. Soc. N.S.W.* 109: 153–174.

Phillips, W.R. and S.J. Inwards (1985). The annual activity and breeding cycles of Gould's Long-eared Bat, *Nyctophilus gouldi* (Microchiroptera: Vespertilionidae). *Aust. J. Zool.* 33: 111–126.

Tidemann, C.R. and S.C. Flavel (1987). Factors affecting choice of diurnal roost site by tree-hole bats (Microchiroptera) in south-east Australia. *Aust. Wildl. Res.* 14: 459–173.

During late summer or early autumn in southern Australia, reserves of body fat are laid down—to be expended during winter, when the bats hibernate. As in most tree-roosting bats at these latitudes, males produce sperm in summer and retain it in the epididymis until the following spring. Mating begins in autumn and—at least in captivity—continues until activity is reduced in winter. Females are able to store sperm during prolonged periods of torpor, enabling fertilisation to take place in spring.

In captivity, 54 per cent of births resulted in twins which were weaned at the age of six weeks. In wild populations, the first young have been seen to fly in January.

W. PHILLIPS

*Left: Gould's Long-eared Bat has a southern distribution in well-watered environments.*

# Lord Howe Island Bat

## *Nyctophilus howensis*

### McKean, 1973

how-en'-sis: '(Lord) Howe (Island) night-lover'

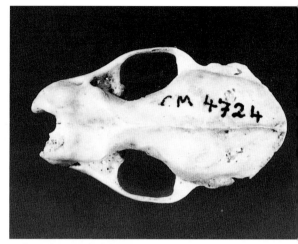

*The only record of the Lord Howe Island Bat is a single skull.*

This species, if alive, would be the largest of Australia's long-eared bats. It may be another example of the loss of a species made vulnerable because it inhabited a small island. It was described as a species after the discovery of a single skull found just inside a cave, perhaps not because it died there but having been eaten by an owl that regurgitated the skull in a pellet in its roost. Until recently, the skull was considered to be of sub-fossil age, but re-examination indicates that the animal was alive in the twentieth century.

There have been no other records of this species since the skull was collected. One hypothesis for its demise relates to the introduction of rats and owls to Lord Howe Island early this century. The Brown Rat, *Rattus norvegicus*, was inadvertently introduced from the ship *Makambo* which was grounded on Ned's Beach in June

1918. This partially arboreal rat has been blamed for the extinction of five bird species on Lord Howe Island and it may also have preyed upon bats roosting in tree hollows or in the dense foliage of trees such as Kentia palms.

Rat populations became so high on the island that control measures were required, since they were also destroying the commercially lucrative seeds of the Kentia palms. In a poorly

*Nyctophilus howensis*

### *Nyctophilus howensis*

**Size**
No information available, but head-and-body is probably around 80 mm and forearm around 50 mm. Skull length 23.2 mm.

**Identification** Unknown, probably similar to mainland *Nyctophilus* species.
**Recent Synonyms** None.
**Other Common Names** Lord Howe Long-eared Bat.
**Status** Probably extinct.
**Subspecies** None.
**Reference**
McKean, J.L. (1975). The bats of Lord Howe Island with a description of a new Nyctophiline bat. *Aust. Mammal.* 1: 329–332.

considered attempt at 'biological control', owls, including several Boobook Owls were introduced in the early 1920s, eight Barn Owls in 1923, ten Californian Barn Owls in 1927 and approximately 100 Tasmanian Masked Owls between 1922 and 1930.

It is possible that the Lord Howe Island Bat existed until the early 1920s when some of the population—perhaps the majority—was reduced by rat predation. The additional pressure from the owls (known to include long-eared bats in their diet elsewhere in Australia) may have reduced the population to extinction.

G.C. RICHARDS

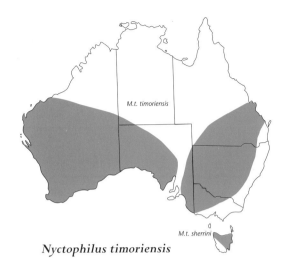

*Nyctophilus timoriensis*

# Greater Long-eared Bat

## *Nyctophilus timoriensis*

### (Geoffroy, 1806)

teem'-or-ee-en'-sis: 'Timor night-lover'

Although it was one of the first members of its genus to be named, the biology and taxonomy of the Greater Long-eared Bat remains poorly understood. The species was erected on the basis of specimens of uncertain origin obtained early in the nineteenth century and thought to be from Timor. It now appears that the Greater Long-eared Bat does not occur outside Australia, where it is widely distributed south of the Tropic of Capricorn, but uncommon and localised.

Recent studies indicate that, in addition to the Tasmanian subspecies, three geographically distinct forms exist across southern Australia, indicating that what is called *Nyctophilus timoriensis* may represent a species complex. Apart

from its occurrence in the tall forests of far south-western Western Australia and Tasmania, it is found in woodland and mallee across the arid and semi-arid regions. Its distribution overlaps extensively with that of the Lesser Long-eared Bat, from which it is immediately distinguished by its larger body size. In parts of its range it occurs together with Gould's Long-eared Bat. Large individuals of the latter species (from higher rainfall areas) approach the size of the Greater Long-eared Bat but individuals from

*The Greater long-eared Bat may represent a complex comprising as many as three species.*

P. GERMAN

## Nyctophilus timoriensis

**Size**

Considerable geographic variation

*Head and Body Length*

50–75 mm

*Tail Length*

35–50 mm

*Forearm Length*

39–49 mm

*Weight*

11–20 g

**Identification** Easily distinguished from *N. geoffroyi* by larger size and less developed Y-shaped groove on snout. Larger than *N. gouldi* from the same areas, outer distance across upper canines at gum-line more than 5.2 mm. Distinguished from *N. bifax daedalus* by darker fur colour and larger size.

**Recent Synonyms** *Nyctophilus major.*

**Other Common Names** None.

**Status** Uncommon, localised.

**Subspecies** *Nyctophilus timoriensis timoriensis*, mainland Australia.

*Nyctophilus timoriensis sherrini*, Tasmania.

(Recent studies recognise three geographically distinct forms on the Australian mainland.)

**Reference**

Goodwin, R.E. (1979). The bats of Timor: systematics and ecology. *Bull. Am. Mus. Nat. Hist.* 163: 73–122.

# Pygmy Long-eared Bat

## *Nyctophilus walkeri*

### Thomas, 1892

wawk'-er-ee: 'Walker's night-lover'

Since the type specimen of the Pygmy Long-eared Bat was taken at Adelaide River, Northern Territory, in 1891, only five additional specimens

## Nyctophilus walkeri

**Size**

*Head and Body Length*

38–45 (41) mm

*Tail Length*

30–36 (33) mm

*Forearm Length*

33–35 (34) mm

*Weight*

4–4.5 (4) g

**Identification** Fawn above; buff below. Basal half of body hairs dark brown. A few cream hairs around the anal region and throat. Skin of forearm, ears and mouth buff; wings dark brown; tail light fawn. Smallest Australian species of *Nyctophilus.*

**Recent Synonyms** None.

**Other Common Names** Territory Long-eared Bat, Little Northern Territory Bat.

**Status** Rare, limited. Possibly underestimated.

**Subspecies** None.

**Reference**

Churchill, S.K., L.S. Hall and P.M. Helman (1984). Observations on long-eared bats (Vespertilionidae: *Nyctophilus*) from Northern Australia. *Aust. Mammal.* 7: 17–28.

areas where both species exist are easily distinguished by their smaller size. It roosts in tree-hollows and under loose bark but its basic ecological and reproductive biology have not been studied.

The habitat of the Greater Long-eared Bat in arid and semi-arid Australia has undergone drastic modification or destruction and its survival status warrants close attention.

H. PARNABY

B.G. THOMSON

*Little is known of the biology of the Pygmy Long-eared Bat.*

**Nyctophilus walkeri**

have been collected, all subsequent to 1975. Four were from Western Australia and one from Kapalga, Northern Territory.

It appears to be closely associated with watercourses. Two males collected in Drysdale River National Park were flying over a rocky pool in a watercourse fringed by paperbarks and *Syzygium* and surrounded by pandanus, acacias and figs over spinifex tussocks. Two females collected at Mitchell Plateau were over a waterhole surrounded by pandanus and paperbarks.

Two females collected on 19 October 1976 had a large foetus in each uterine horn and their stomachs contained bugs, beetles and wasps.

D.J. KITCHENER

## Subfamily Vespertilioninae

This group of 'typical' vespertilionids cannot be defined on external features.

# Large-eared Pied Bat

## *Chalinolobus dwyeri*

### Ryan, 1966

kah'-lin-oh-lobe'-us dwie'-er-ee:
'Dwyer's bridle-lobe'

---

### *Chalinolobus dwyeri*

**Size**
*Head and Body Length*
47–53 mm
*Tail Length*
42–46 mm
*Forearm Length*
38–42 mm
*Weight*
7.5–9.5 g (pregnant females to 12 g). Males may be slightly smaller than females.

**Identification** Fur glossy black with fringe of white around body beneath wings and tail-membrane. Ears moderately large. Distinguished from *Chalinolobus picatus* by larger ears and longer forearm.
**Recent Synonyms** None. (Before 1966 mistakenly referred to as *C. picatus.*)
**Other Common Names** Large Pied Bat.
**Status** Rare, scattered; possibly underestimated.
**Subspecies** None.
**Reference**
Dwyer, P.D. (1966). Observations on *Chalinolobus dwyeri* (Chiroptera: Vespertilionidae) in Australia. *J. Mammal.* 47: 716–718.

---

In the early 1960s some alert residents of Copeton, New South Wales, noticed a group of jet-black, velvet-furred bats clustered on the roof of a disused mine tunnel. These proved to be representatives of a previously undescribed species of lobe-lipped bat. Now known as the Large-eared Pied Bat, it has since been recorded from scattered localities from near Rockhampton in central coastal Queensland to Bungonia in southern New South Wales. It is found in a variety of drier habitats, including the dry sclerophyll forests and woodlands to the east and west of the Great Dividing Range. Isolated records from subalpine woodland above 1500 metres and at the edge of rainforest and moist eucalypt forest, suggest it may tolerate a greater range of habitats than has so far been recorded.

Daytime roosts include caves, mine tunnels and the abandoned, bottle-shaped mud nests of Fairy Martins. Within the shallow sandstone caves in which it has been recorded to roost, it often selects positions close to the entrance to the cave in the 'twilight' zone where individuals huddle together. This contrasts with most other cave-dwelling bats with which it shares its range, these being normally encountered in the deeper and darker parts of cave systems.

The combination of a relatively short, broad wing and a low weight per unit area of wing is

*Chalinolobus dwyeri*

indicative of manoeuvrable flight: it probably forages for small flying insects below the forest canopy.

Knowledge of the biology of the Large Pied Bat is based on intermittent observations and banding records obtained over a few years at the Copeton mine tunnels. Here, up to 13 adult females and a few adult males assembled deep inside a tunnel in early spring. The females gave birth to one or two young (average 1.8 per female) during late November and early December and these were suckled until late January. The colony dispersed during autumn and, through this season and winter, individuals were difficult to locate. The few found were in deep torpor and it is likely that the species hibernates through the coolest months. During autumn and early winter, males had enlarged testes and both sexes had swollen glands on the muzzle, an indication that their secretions may be sexual attractants. It is uncertain whether mating occurs early in winter or in spring. Banding records show that females can give birth when 12 months old.

Regrettably, these mine tunnels were drowned with the opening of Copeton Dam, ending further investigation of that colony.

G.A. HOYE AND P.D. DWYER

*The Large-eared Pied Bat has velvety fur.*

C. ANDREW HENLEY/LARUS

*511*

G.B. BAKER

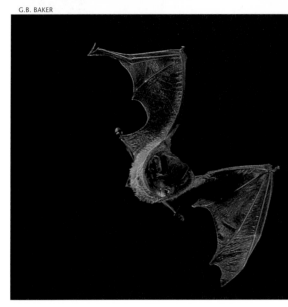

# Gould's Wattled Bat

## *Chalinolobus gouldii*

### (Gray, 1841)

gule'-dee-ee: 'Gould's bridle-lobe'

Widespread through Australia and occasionally captured on Norfolk Island, Gould's Wattled Bat is an inhabitant of open forest, mallee, dense forest, tall shrubland and urban areas. Gould noted that 'it frequents the outskirts of the brushes [dense forests] and the wooded borders of the great rivers'.

Recorded roosting sites include tree spouts, birds' nests, ceilings or basements of buildings, rolled-up canvas blinds and even the exhaust pipe of a tractor. Bats emerge from the roost about 20 minutes after sunset and forage below the tree canopy in woodland areas, seldom flying more than 20 metres above the ground but descending as low as 1 metre.

A study from 1974 to 1981 in a house at Ivanhoe, Victoria, revealed numerous aspects of its biology. A maternity colony of 20–30 bats, occupied the ceiling joists in the basement until rebuilding led to their dispersion.

Gould's Wattled Bat is active throughout the year, except in cooler climates where it becomes

*Gould's Wattled Bat flies with agility below the forest canopy.*

torpid in winter. In the Ivanhoe house, few bats were active during winter hibernation, which extended from May to late August or early September. Females outnumbered males by three to one and were larger and heavier than males. Gestation is approximately three months and twins are often produced. Bats born in the roost usually stayed there for less than a year. All long-term residents were females, up to five years old. Adult males were not seen in the colony for periods longer than a year.

The only detailed study of reproduction was made in Western Australia, where mating occurs in late August or early winter. Sperm is stored by the female and fertilisation is delayed until she ovulates at the end of winter or early spring. Gestation lasts about three months, the time of birth varying with latitude. At Ivanhoe, pregnant females were most common in September or October; lactating and dependent juveniles in November or December; and young able to fly in December or January. Twins were common and remained attached to the mother's thoracic teats during flight. Development was rapid, adult appearance being reached at six weeks. At Ivanhoe, the diet consisted mainly of moths and beetles but flies and orthopterans were included. Ectoparasites were common, including two forms of mites and a wingless dipteran.

Bats emit high-pitched chirps when flying; low chittering noises when roosting; and a range

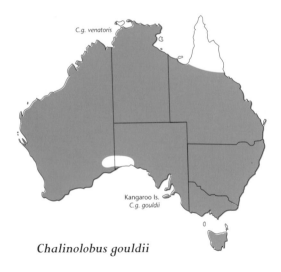

*Chalinolobus gouldii*

P. GERMAN

*The Chocolate Wattled Bat has a uniformly brown colouration.*

## *Chalinolobus gouldii*

**Size**

*Head and Body Length*
65–75 (70) mm
*Tail Length*
40–50 (45) mm
*Forearm Length*
40–48 (44) mm
*Weight*
10–18 (14) g

**Identification** Black above; black or brown below; upper surface often grading to medium brown posteriorly. Distinguished from other species of *Chalinolobus* by having outer margin of ear turned downwards from corner of mouth in flap of skin; tragus wide and posterior margin is rounded (not pointed as in *C. morio*).

**Recent Synonyms** None.

**Other Common Names** Gould's Lobe-lipped Bat.

**Status** Abundant.

**Subspecies** *Chalinolobus gouldii gouldii*, southern Australia.
*Chalinolobus gouldii venatoris*, northern Australia.
Central Australian populations are variable and probably represent a cline between the northern and southern forms.

**References**

Kitchener, D. (1975). Reproduction in female Gould's Wattled Bat, *Chalinolobus gouldii* (Gray) (Vespertilionidae) in Western Australia. *Aust J. Zool.* 23: 29–42.

Dixon, J.M. and L. Huxley (1989). Observations on a maternity colony of Gould's Wattled Bat *Chalinolobus gouldii* (Chiroptera: Vespertilionidae). *Mammalia* 53: 395–414.

# Chocolate Wattled Bat

## *Chalinolobus morio*

### (Gray, 1841)

mor'-ee-oh: 'Moros' bridle-lobe'
(Moros is Greek mythological 'son of night')

The vernacular name of this species refers to its colour and the small wattles at the corner of its mouth. It is common over a wide range of habitats in southern Australia from montane forests in the east to the treeless Nullarbor Plain. In high

of squeaks, chirps, clicks and a persistent low buzz when handled.

Gould's Wattled Bat is preyed on by owls and the feral Cat, as well as by the Pied Butcherbird and Pied Currawong.

J.M. Dixon

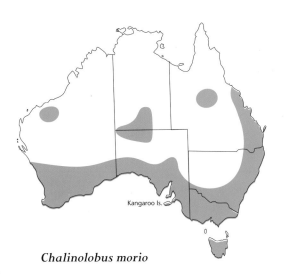

*Chalinolobus morio*

A. YOUNG

*The Chocolate Wattled Bat roosts in tree-hollows and sometimes in roofs.*

altitudes, where winter temperatures are low, it is the last of the resident bats to enter hibernation. It usually forms colonies in tree-hollows but it also inhabits roof cavities, often sharing these with the Little Freetail-bat. Colonies vary from about 20 to several hundred individuals. In some parts of the Nullarbor Plain and in the vicinity of Alice Springs, it forms quite large colonies in caves.

As in most vespertilionids, sperm is stored and fertilisation delayed. Unlike most other members of the family, which produce twins, the Chocolate Wattled Bat usually gives birth to a single young. In subtropical Queensland, births occur in mid-October; in the southern parts of the range, births do not take place until November or later.

There is an area of south-eastern Australia where the Chocolate Wattled Bat competes with at least seven other bat species. Lactation termi-

nates later than in these species, so the weaned young and their mothers have less time to forage for insects and to deposit a store of fat in preparation for hibernation. However, since the other bats hibernate earlier, the Chocolate Wattled Bat is able to forage, virtually alone, for the last insects of autumn. This life strategy is finely balanced, for the Chocolate Wattled Bat deposits so little fat that it is the first of the local bats to emerge from hibernation. At this time, insect numbers are low but, again, there is little competition for the resource.

C.R. TIDEMANN

### *Chalinolobus morio*

**Size**

*Head and Body Length*
50–61 (57) mm
*Tail Length*
45–50 (48) mm
*Forearm Length*
35–42 (38) mm
*Weight*
8–11 (9) g

**Identification** Uniform chocolate brown. No colour contrast between head and body as in (southern) *Chalinolobus gouldii*; wattles considerably smaller. Similar to *Vespadelus darlingtoni* but distinguished by small wattles. Forehead of living animal bulges over muzzle, giving domed appearance. Tragus has pointed tip.
**Recent Synonyms** None.
**Other Common Names** Chocolate Bat, Chocolate Lobe-lipped Bat.
**Status** Common.
**Subspecies** None.

**References**
Hall, L.S. (1970). A collection of the bat, *Chalinolobus morio* (Gray), from the Nullarbor Plain, Western Australia *Helictite* 8: 51–57.
Young, R.A. (1979). Observations on parturition, litter size and foetal development at birth in the Chocolate Wattled Bat, *Chalinolobus morio* (Vespertilionidae). *Vic. Nat.* 96: 90–91.
Young, R.A. (1980). Observations on the vulnerability of two species of wattled bats *Chalinolobus* to diurnal avian predators. *Vic. Nat.* 96: 258–262.

# Hoary Wattled Bat

## *Chalinolobus nigrogriseus*

### (Gould, 1856)

nig'-roh-griz-ay'-us: 'black-grey bridle-lobe'

In 1856, John Gould named this species from a small bat collected at Moreton Bay, Queensland, but for more than a century the name was virtually ignored and the species was regarded as the same as the earlier-named Little Pied Bat, *Chalinolobus picatus*. The distinction between the two species was not confirmed until 1966.

Confusion has also risen because of the erection in 1909 of a species, *C. rogersi*, for a bat from Wyndham, Western Australia, very similar to *C. nigrogriseus* but with more conspicuous white tipping to the long hairs of the body fur, giving it a 'hoary' or 'frosted' appearance. There has been some disagreement about the status of *C. rogersi* but it is now generally agreed

*Chalinolobus nigrogriseus*

**Size**
*Head and Body Length*
45–55 (47) mm
*Tail Length*
35–42 (40) mm
*Forearm Length*
32–38 mm
*Weight*
7.5–10 (8.8) g

**Identification** Blackish-grey above, hairs with white tips of variable length; sometimes paler below. Distinguished from *Chalinolobus gouldii* by having only a horizontal lobe between the mouth and ear and no vertical lobe at the corner of the mouth; from *C. picatus* by size and the absence of a white border between the body and undersurface of the wings; and from *C. dwyeri* by shorter ears and 'frosted' fur.
**Recent Synonyms** None. (Confused in the past with *Chalinolobus picatus* and the supposed *C. rogersi*.)

*The Hoary Wattle Bat has white-tipped hairs.*

J. LOCHMAN

that it is a smaller, western subspecies of the Hoary Wattled Bat.

The Hoary Wattled Bat has been collected at a number of sites in northern Australia and has usually been shot when flying over water or open spaces. Occurring in a wide range of habitats, from wet sclerophyll forest to open woodland and even over scrub on sand-dunes, it is a common species and one of the first to appear on the wing at dusk. Roosts have been reported in rock crevices but, in areas devoid of these, it is quite likely to roost in tree-hollows or similar sites.

The diet includes a wide range of insects and other invertebrates: flies and moths, which may be caught on the wing; mantids, earwigs, beetles, weevils, crickets and cicadas, which are poor fliers; and spiders and ants, which do not fly at all. Thus it seems that many prey are taken

*The Hoary Wattled Bat takes insects on wing and from the ground or other solid surfaces.* G.B. BAKER

**Other Common Names** Eastern Wattled Bat, Pied Bat, Blackish-grey Bat (*Chalinolobus nigrogriseus nigrogriseus*); Hoary Bat, Frosted Bat (*Chalinolobus nigrogriseus rogersi*).

**Status** Common, widespread.

**Subspecies** *Chalinolobus nigrogriseus nigrogriseus*, New Guinea through Cape York to the Clarence River, New South Wales. *Chalinolobus nigrogriseus rogersi*, Western Australia, Northern Territory and western Queensland.

**Extralimital Distribution** New Guinea.

**References**

Van Duesen, H.M. and K.F. Koopman (1971). Results of the Archbold Expeditions No. 95. The genus *Chalinolobus* (Chiroptera: Vespertilionidae). Taxonomic review of *Chalinolobus picatus*, *C. nigrogriseus* and *C. rogersi*. *Amer. Mus. Novit.* No. 2468: 1–30.

Van Deusen, H.M. (1969). The hoary wattled bat of North Queensland. *N. Qld. Nat.* 36: 5–6.

*Chalinolobus nigrogriseus*

from the ground or from the surfaces of trees or rocks. The ability of many bats to run over solid surfaces, and the food-gathering opportunities that this confers, is often underestimated.

Very little is known of reproduction in the Hoary Wattled Bat. The only indication of a breeding season is that a female collected in December and another in early January had well-developed mammary glands and may have been lactating.

F.R. ALLISON

Above: *The Little Pied Bat has distinctive black and white fur.*
Right: *The Little Pied Bat inhabits semi-arid regions, usually roosting in caves.*

G.B. BAKER

# Little Pied Bat

## *Chalinolobus picatus*

### (Gould, 1852)

peek-ah'-tus: 'pied (black and white) bridle-lobe'

This distinctive black and white bat lives only in the dry areas of southern Queensland, New South Wales and South Australia. So little is known about it that, until recently, it was thought to roost only in dry caves or mine shafts and the sparseness of such roosts in the range was offered in explanation of its apparent rarity. Records of subterranean colonies indicated that roosts never included more than ten or 15 individuals, but the discovery of a colony of nearly 40, tightly clustered behind the door of an abandoned house, shows that the species is much more adaptable than had been thought. This group sustained day temperatures in excess of 40°C, proving that the species is capable of dealing with heat and aridity, provided that water is obtainable within flight range. Females normally bear two young in the summer.

G.C. RICHARDS

## Chalinolobus picatus

**Size**

*Head and Body Length*
42–50 (47) mm
*Tail Length*
circa 35 mm
*Forearm Length*
31–36 (33) mm
*Weight*
4–8 (6) g

**Identification** Glossy black fur on back extends onto tail-membrane: dark grey below with white fur along flanks forming a V-shape in the pubic region. Outer margin on short, broad ear terminates in horizontal lobe at corner of mouth. Distinguished from *Chalinolobus dwyeri* by shorter ear.

**Recent Synonyms** *Scotophilus picatus.*

**Other Common Names** Pied Bat.

**Status** Uncommon in limited habitat.

**Subspecies** None.

**Reference**

Van Duesen, H.M. and K.F. Koopman (1971). Results of the Archbold Expeditions No. 95. The genus *Chalinolobus* (Chiroptera, Vespertilionidae). Taxonomic review of *Chalinolobus picatus, C. nigrogriseus* and *C. rogersi. Amer. Mus. Novit.* 2468: 1–30.

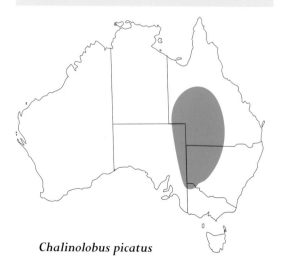

*Chalinolobus picatus*

# Western False Pipistrelle

## *Falsistrellus mackenziei*

### Kitchener, Caputi and Jones, 1986

fol'-see-strel'-us  mak-ken'-zee-ee:
'McKenzie's false-pipistrelle'

*Falsistrellus mackenziei*

Despite an extensive distribution through the forests and tall woodlands of south-western Australia, the Western False Pipistrelle has seldom been captured. The first specimens were taken in 1961 and, until 1986, it was regarded as conspecific with the very similar and closely related Eastern False Pipistrelle, *F. tasmaniensis.*

Its range extends northward nearly to Perth and eastward to the western margin of the wheatbelt. Most of the 30 or so sites at which it has been collected are in wet sclerophyll forests of Karri, *Eucalyptus diversicolor*, or in the high-rainfall zones of Jarrah forest, *E. marginata*. It has also been collected in Tuart forest, *E. gomphocephala* and in mixed Tuart–Jarrah woodlands on the adjacent coastal plain.

J. LOCHMAN

*The Western False Pipistrelle lives mainly in wet sclerophyll forests.*

Recent data suggest that it is locally common in Karri forests. A total of 24 individuals were captured in five mist-nets set near Windy Harbour during three nights in February 1992. They were caught 4–8 metres above ground, where their flight was fast and direct as they foraged in the cathedral-like spaces among the massive tree trunks, between the understorey and the Karri canopy. The echolocation calls are well-suited to this microhabitat, with a lowest frequency of 30 kilohertz, a duration of 12 milliseconds and an average call interval of 125 milliseconds. It is gregarious. Five bats released during the day aggregated around the canopy of nearby Karri trees; within ten minutes they had settled together on a dead branch and become inactive. A colony found in a hollow 15 metres up a Karri tree comprised about 30 bats. Five others have been taken from a hollow log in Jarrah forest. The available data suggest that sexes segregate for roosting and foraging, at least during spring and summer.

## Falsistrellus mackenziei

**Size**
*Head and Body Length*
55–67 (62) mm
*Tail Length*
40–53 (46) mm
*Forearm Length*
48–54 (51) mm
*Weight*
17–26 (21) g

**Identification** Dark brown above; light cinnamon below. Largest vespertilionid in Western Australia. Slightly larger, more rusty hued than *F. tasmaniensis* of south-eastern Australia.
**Recent Synonyms** *Pipistrellus tasmaniensis* (part), *Falsistrellus tasmaniensis* (part).
**Other Common Names** None.
**Status** Little known, but apparently widespread in forests of south-western Australia, including sites in conservation reserves such as D'Entrecasteaux National Park.
**Subspecies** None.
**Reference**
Kitchener, D.J., N. Caputi and B. Jones (1986). Revision of Australo-Papuan *Pipistrellus* and of *Falsistrellus* (Microchiroptera: Vespertilionidae). *Rec. West. Aust. Mus.* 12: 435–495.

The Western False Pipistrelle is widespread in forested areas of south-western Australia and may be more common than hitherto supposed. It occurs in several nature conservation reserves where it is secure. Elsewhere, its greatest threat is loss of roost sites from clearing and timber harvesting. Young are born in spring or early summer.

A.N. START AND N.L. MCKENZIE

# Eastern False Pipistrelle

## *Falsistrellus tasmaniensis*

### (Gould, 1858)

taz-mane'-ee-en'-sis:
'Tasmanian false-pipistrelle'

*The Eastern False Pipistrelle is a fast-flying predator on insects.*

Two observations have been made of roosts in stem holes of living eucalypts. In two study areas near Canberra an unusually high sexual bias was found among trapped animals. At one site the ratio was 35:1 in favour of males; at the other, the ratio was almost same, but in favour of females. At a third site in the same region the sex

Relatively little is known of the biology of this large vespertilionid, which seems to be in low numbers throughout its range, though more common at cool elevations.

In southern Australia, the Eastern False Pipistrelle apparently hibernates during winter months: activity is reduced following a period of increasing body weight in autumn, probably due to accumulation of body fat. Consistent with other hibernating vespertilionids, males produce sperm in late summer and store this in the epididymides over the colder period. Females are pregnant during late spring and early summer and lactating in mid-January.

*Falsistrellus tasmaniensis*

## *Falsistrellus tasmaniensis*

### Size

*Head and Body Length*
55–70 mm

*Tail Length*
40–50 mm

*Forearm Length*
48–54 (52) mm

*Weight*
14–26 (22) g

**Identification** Dark brown on the back with a dark grey undersurface. Relatively elongated ears with a 'keel' section deflected posteriorly. Two upper incisors on each side, the second very small; latter characteristic helps to distinguish it from *Scoteanax rueppelli* which has similar appearance.

**Recent Synonyms** *Pipistrellus tasmaniensis*.

**Other Common Names** Tasmanian Pipistrelle, False Pipistrelle.

**Status** Uncommon.

**Subspecies** None.

**References**

Adams, M., P.R. Baverstock, C.H.S. Watts and T. Reardon (1987). Electrophoretic resolution of species boundaries in Australian microchiroptera. 11. The *Pipistrellus* group. (Chiroptera: Vespertilionidae). *Aust. J. Biol. Sci., Zool.* 40: 163–170.

Phillips, W.R., C.R. Tidemann, S.J. Inwards and S. Winderlich (1985). The Tasmanian Pipistrelle: *Pipistrellus tasmaniensis* Gould 1858: annual activity and breeding cycles. *Macroderma* 1: 1.

ratio was almost 1:1. These puzzling statistics may indicate sexual segregation in roosting behaviour for at least part of the year.

Given the size and shape of its wings, the Eastern False Pipistrelle could be expected to be highly mobile, with a comparatively large foraging range. It is not very manoeuvrable and probably forages mostly above the forest canopy, in open woodland or over water.

W. PHILLIPS

B.G. THOMSON

*The Large-footed Myotis roosts in a variety of locations, seldom far from water.*

# Large-footed Myotis

## *Myotis adversus*

(Morsfield, 1824)

mie-oh'-tis ad-vers'-us:
'opposed mouse-ear' (meaning unknown)

Unusual social behaviour and foraging adaptions make the Large-footed Myotis one of Australia's most interesting vespertilionids. Colonies of 10–15 individuals, occasionally of several hundred, are found in caves, mines or tunnels, under bridges and buildings and even in dense foliage in the tropical part of its range. Within these breeding colonies it occurs in relatively small clusters,

## Myotis adversus

**Size**
*Head and Body Length*
52–56 (54) mm
*Tail Length*
36–40 (38) mm
*Forearm Length*
38–41 (40) mm
*Weight*
7–12 (10) g

**Identification** Normally grey-brown above; paler below; but old bats have ginger-coloured fur; and albinism, or partial albinism, is not uncommon. Feet exceptionally large (10–14 mm).
**Recent Synonyms** *Myotis macropus*, *Vespertilio macropus*.
**Other Common Names** Large-footed Mouse-eared Bat.
**Status** Comparatively rare over its limited range.
**Subspecies** Many subspecies extending over the tropical islands to the north and in the Pacific, Australian populations comprise:
*Myotis adversus macropus*, eastern Australia.
*Myotis adversus moluccarum*, northern Australia.
**References**
Dwyer, P.D. (1970). Foraging behaviour of the Australian large-footed *Myotis* (Chiroptera). *Mammalia* 34: 76–80.
Dwyer, P.D. (1970). Social organisation in the bat *Myotis adversus*. *Science* 168: 1006–1008.

*The Large-footed Myotis catches insects and small fishes by flying over the surface of the water and taking prey with its rake-like hindfeet.*

Wales have only one young each year, usually in November or December. They are torpid in winter, remaining in roosts that are separate from the maternity sites. In the middle latitudes, represented by south-eastern Queensland, two young are born each year; one in early October and the other in late January; male fertility follows a similar cycle. In northern Queensland, a typical tropical pattern of breeding is found and females may have three successive births in one year.

The young of most bats leave their mothers after weaning but a strong mother–young bond develops in the Large-footed Myotis and continues for about four weeks after lactation ceases. The young probably stay with their mothers until they have become adept in their unusual method of catching food.

Colonies of the Large-footed Myotis never occur far from bodies of water, ranging from rainforest streams to large lakes and reservoirs. Swooping over the water, they rake its surface with the sharp, recurved claws of their large feet to catch the aquatic insects which make up

where each male establishes a territory, excludes other males and forms a harem of females during the breeding periods. Males roost alone when not breeding, each defending its territory from others. Older males can be recognised by the scars on their ears resulting from aggressive encounters.

The reproductive cycle varies according to latitude. Females from Victoria and New South

Melville Is.
*M.a. moluccarum*

*M.a. macropus*

**Myotis adversus**

P. GERMAN

most of their diet. They may forage individually or hunt together over the water and in the air, where they are sometimes seen in downward spiralling flight as they search for flying insects. Their method of water-foraging is found in several other species of *Myotis* and in the Central American fish-eating bats of the genera *Pizonyx* and *Noctilio*, which are also characterised by large feet and long toes.

Although *Myotis* is more widely spread around the world than any other bat genus, usually with at least two of its 60-odd species in any large geographical region, only one species is known for certain in Australia. A single specimen from the Kimberley region may represent a second, very rare, species. The potential diversity of the genus in Australia appears to have been limited by the aridity of the continent and the relative lack of areas of permanent fresh water.

G.C. RICHARDS

# Cape York Pipistrelle

## *Pipistrellus adamsi*

### Kitchener, Caputi and Jones, 1986

pip'-ee-strel'-us ad'-am-zee:
'Adams' pipistrelle'

L. LUMSDEN

*The Cape York Pipistrelle is one of the smaller of the Ausralian bats.*

One of the smallest of Australian bats, the Cape York Pipistrelle varies in colour from dark brown to bright rufous. It is restricted to Cape York and the top of the Northern Territory, but is often plentiful where it occurs. On Cape York, it is a common bat along the creek and river systems flowing through savanna woodland on the central and eastern parts. In the rainforests of the tip and eastern portions of Cape York it invades gaps in the forest along tracks and rivers but does not occur in the rainforest proper. The few records from the Northern Territory are from coastal areas, including the Darwin region. It probably roosts in trees.

Pregnant females have been recorded from early September in the Northern Territory and

---

### *Pipistrellus adamsi*

**Size**
*Head and Body Length*
35–44 (40) mm
*Tail Length*
28–35 (32) mm
*Forearm Length*
29.8–32.5 (31.0) mm
*Weight*
3–6 (4.5) g

**Identification** Colour varies from dark brown, through grey-brown to rufous. Larger than *P. westralis* but could be confused with species of *Vespadelus*. Smaller than *V. troughtoni*; differs from *V. finlaysoni* and *V. caurinus* in shape of glans penis and baculum.

**Recent Synonyms** *Pipistrellus papuanus*, *Pipistrellus tenuis*.
**Other Common Names** Northern Pipistrelle.
**Status** Common in limited range.
**Subspecies** None.
**References**
Adams, M., P.R. Baverstock, C.H.S. Watts and T. Reardon (1987). Electrophoretic Resolution of Species Boundaries in Australian Microchiroptera. II. The *Pipistrellus* Group (Chiroptera: Vespertilionidae). *Aust. J. Biol. Sci.* 40: 163–170.
Kitchener, D.J., N. Caputi and B. Jones (1986). Revision of Australo-Papuan *Pipistrellus* and of *Falsistrellus* (Microchiroptera: Vespertilionidae). *Rec. West. Aust. Mus.* 12: 435–495.

from mid-September to mid-October on Cape York Peninsula, indicating that births of a single young occur in late October or early November. The presence of juveniles in the population at this time suggests that breeding may occur more than once a year.

G.A. HOYE

*Pipistrellus adamsi*

*The Northern Pipistrelle frequents mangrove swamps.*
B.G. THOMSON

# Northern Pipistrelle

## *Pipistrellus westralis*

### Koopman, 1984

west-rah'-lis: 'Western Australian pipistrelle'

During warm tropical nights, this small bat hunts in the dark passages where tidal creeks incise the mangrove forests of Australia's north coast. Fluttering, aerobatic flight allows it to follow the irregular contours of the forest's outer foliage as it forages in the airspace within 2 metres of the mangroves, catching flying insects such as small beetles and moths.

*Pipistrellus westralis*

The Northern Pipistrelle, which was first collected in 1895 by Knut Dahl at Roebuck Bay, near Broome, occurs along Australia's northern coastline from Cape Bossut in Western Australia to Karumba in western Queensland. Marine mangrove communities are its primary habitat, but it has also been captured in vegetation just behind mangroves, along narrow tracks through dense pindan thickets in the south-western Kimberley, and in riverine forests of paperbarks, *Pandanus* and *Barringtonia,* in the Northern Territory. It is not known from rainforest, a habitat favoured by its north-eastern Australian counterpart, *P. adamsi.*

Young are probably born throughout the year. Of eight adult females examined during the height of the dry season, one had a foetus in its left uterine horn and another had recently weaned a single young.

It is cryptic and probably more common than available records suggest. Secure populations occur in the vast, pristine mangrove communities adjacent to the Prince Regent Nature Reserve and the Keep River and Kakadu National Parks.

N.L. McKenzie

## *Pipistrellus westralis*

**Size**

*Head and Body Length*
34–42 (37) mm
*Tail Length*
29–37 (32) mm
*Forearm Length*
27.4–30.3 (28.9) mm
*Weight*
2.7–3.3 (3.1) g

**Identification** Fur of upper surfaces tipped light brown; fur on undersurfaces tipped buff. Basal fur on body greyish-black. Fur of face and neck mustard-brown. Skin of snout, ears, forearms and feet light brown. Wings dark brown. Smaller than *Pipistrellus adamsi* in all dimensions, particularly forearm length (29.8–32.2 mm in *P. adamsi*) and width across the upper canine teeth (3.4–3.8 mm in *P. westralis*; 3.9–4.3 mm in *P. adamsi*). Rear edge of tragus only slightly convex.
**Recent Synonyms** *Pipistrellus tenuis*, *Pipistrellus tenuis westralis*.
**Other Common Names** Western Pipistrelle, North-western Pipistrelle, Mangrove Pipistrelle.
**Status** Common in suitable habitat; habitat limited.
**Subspecies** None.
**References**

Kitchener, D.J., N. Caputi and B. Jones (1986). Revision of Australo–Papuan *Pipistrellus* and *Falsistrellus* (Microchiroptera: Vespertilionidae). *Rec. West. Aust. Mus.* 12: 435–495.

McKenzie, N.L. and J.K. Rolfe (1986). Structure of the bat guilds in the Kimberley mangroves, Australia. *J. Anim. Ecol.* 55: 401–420.

# Greater Broad-nosed Bat

## *Scoteanax rueppellii*

(Peters, 1866)

skoh'-tee-an'-ax  rue-pel'-ee-ee: 'Rüppell's darkness-chief'

G.A. HOYE

*The Greater Broad-nosed Bat feeds on flying insects and on other bats.*

Emerging just after sundown, the Greater Broad-nosed Bat usually flies slowly and directly at a height of 3–6 metres, deviating slightly from its path to catch beetles and other large, slow-flying insects. Recent evidence indicates that it also preys on other bats: in captivity and during capture, it has been observed to kill and eat bats of eight other species. It shares skull characteristics with bats from other countries that are known to include bats in their diet.

Its stronghold is in the gullies and river systems draining the Great Dividing Range, from north-eastern Victoria to the Atherton Tableland in tropical Queensland, but it extends to the coast over much of its range. Although inhabiting a variety of habitats from woodland through moist and dry eucalypt forest to rainforest, it does not occur at altitudes above 500 metres (except, perhaps, in the very north of its range where temperatures at such altitudes are milder). The open nature of eucalypt woodlands and forests suits its direct flight pattern, and the more cluttered environments of the wetter forests are overcome by utilising natural and human-made

### *Scoteanax rueppellii*

**Size**

*Head and Body Length*
80–95 (86) mm
*Tail Length*
40–55 (45) mm
*Forearm Length*
50–56 (54) mm
*Weight*
25–35 (30) g

**Identification** Dark reddish-brown to dark brown above; slightly paler below. Distinguished from other broad-nosed bats by greater size. Externally similar to *Pipistrellus tasmaniensis* but differs in having two, not four, upper incisors.

**Recent Synonyms** *Scoteinus ruppellii*, *Nycticeius ruppellii*.
**Other Common Names** Rüppell's Broad-nosed Bat.
**Status** Uncommon to rare.
**Subspecies** None.
**References**
McKean, J.L. (1966). Some new distributional records of broad-nosed bats (*Nycticeius* spp.). *Vic. Nat.* 83: 25–30.
Woodside, D.P. and A. Long (1984). Observation on the feeding habits of the Greater Broad-nosed Bat, *Nycticeius rueppellii* (Chiroptera: Vespertilionidae). *Aust. Mammal.* 7: 121–29.

openings in the forest. Creeks and small rivers are favoured corridors where it hawks backwards and forwards for its prey.

Little is known of its reproductive cycle but a single young is born in January, slightly later than in most of the other vespertilionids that share its range. Prior to birth, females congregate at maternity sites, located in suitable trees, where they appear to exclude males during the birth and raising of the single young. Usually roosting in tree-hollows, it has also been found in roof spaces of old buildings.

G.A. HOYE AND G.C. RICHARDS

*Scoteanax rueppellii*

# Inland Broad-nosed Bat

## *Scotorepens balstoni*

### (Thomas, 1906)

skoh'-toh-rep'-enz   bawl'-stun-ee: 'Balston's darkness-creeper'

This species inhabits drier, more inland environments where it is often present in lower numbers than the Little Broad-nosed Bat with which it occurs over much of its range. It roosts in tree-hollows and in roofs, often in a horizontal position. On a number of occasions, it has been found roosting in low numbers with colonies of several dozen Southern Freetail-bats.

**Scotorepens balstoni**
**Size**
*Head and Body Length*
42–60 mm
*Tail Length*
29–42 mm
*Forearm Length*
31–41 mm
*Weight*
7–14 g

**Identification** Light grey-brown above, pale brown below. Broad, square-shaped muzzle when viewed from above. Ears have short (4 mm) tragus. Larger than *S. greyi* and *S. sanborni*. Ears longer and fur often paler than in *S. orion*.
**Recent Synonyms** *Nycticeius balstoni*, *Scoteinus balstoni*, *S. b. caprenus*.
**Other Common Names** Western Broad-nosed Bat.
**Status** Common.
**Subspecies** None.

B.G. THOMSON

*The Inland Broad-nosed Bat inhabits dry inland environments.*

Only fragmentary information is available about the biology of the Inland Broad-nosed Bat. In Victoria, mating occurs around late April and early May. After about seven months of gestation, births of single young or twins usually occur in November. The young cling to the mother until about ten days old, when they weigh 4–5 grams and are too heavy to carry during flight. They are then left in the roost while the mother forages

### References

Kitchener, D.J. and N. Caputi (1985). Systematic revision of Australian *Scoteanax and Scotorepens*, with remarks on relationships to other Nycticeiini. *Rec. West. Aust. Mus.* 12: 85–146.

Ryan, M.R. (1966). Observations on the Broad-nosed Bat, *Scoteinus balstoni*, in Victoria. *J. Zool. Lond.* 148: 162–166.

at night. The eyes open and fur begins to grow during the third week of life. The young can fly and hunt independently during the second month of life.

This species was first described in 1906 from five specimens collected in south-western Australia; a geographical race from the Kimberley region was described 21 years later. Large individuals, previously thought to be a distinct species (*Nycticeius influatus*, the Hughenden Broad-nosed Bat), now seem to represent the large end of a size cline of *S. balstoni*, but regional variation in this species warrants further examination.

H. PARNABY

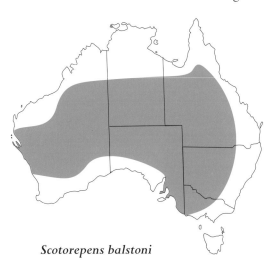

**Scotorepens balstoni**

# Little Broad-nosed Bat

## *Scotorepens greyii*

### (Gray, 1843)

gray'-ee-ee': 'Grey's darkness-creeper'

*The Little Broad-nosed Bat feeds on lying insects and other bats.* G.B. BAKER

The habits of the Little Broad-nosed Bat are similar to those of the closely related Western Broad-nosed Bat. It roosts mainly in tree hollows or disused buildings and has also been found beneath the metal caps on telegraph poles and in the hollow centres of fence posts. Colonies range in size from a pair to about 20 individuals.

Open woodlands and plains are the usual habitat; waterholes and creeks are favoured feeding areas. It begins to forage just on dusk and takes occasional drinks by skimming the surface of still waters.

Living in a hot, arid environment, this bat must suffer a high rate of water loss, particularly when roosting in relatively exposed situations such as on thin posts or underneath metal roofs.

Its survival depends upon access, probably every evening, to a body of water.

There is evidence to suggest that, in southern regions, mating occurs before the onset of winter and that births, usually of twins, occur in December. Whether, as in some other vespertilionids, mating occurs later in tropical latitudes, is not yet known. The taxonomy of the genus

*The Little Broad-nosed Bat inhabits dry inland environments.*

C. ANDREW HENLEY/LARUS

---

### *Scotorepens greyii*

**Size**
*Head and Body Length*
45–55 (51) mm
*Tail Length*
25–40 (30) mm
*Forearm Length*
28–34 (31) mm
*Weight*
8–12 g

**Identification** Variable chestnut to grey-brown above, paler below than in *S. balstoni*. Broad, square muzzle when viewed from above. Ears erect and broadly curved behind the tips. Calcaneum lobed at tip.
**Recent Synonyms** *Scoteinus greyii, Scotophilus greyii, Scotorepens balstoni caprenus, Nycticeius greyii.*
**Other Common Names** Grey's Bat.
**Status** Common.
**Subspecies** None.
**Reference**
Koopman, K.F. (1984). Taxonomic and distributional notes on tropical Australian bats. *Am. Mus. Novit.* 2778: 1–48.

G.A. HOYE

*The Eastern Broad-nosed Bat roosts in tree-hollows.*

may not yet be satisfactorily resolved and the close similarities between the Little, Western and Little Northern Broad-nosed Bats make further studies necessary.

G.C. RICHARDS

*Scotorepens greyii*

# Eastern Broad-nosed Bat

## *Scotorepens orion*

(Troughton, 1937)

oh-rie'-on: 'eastern darkness-creeper'

The Eastern Broad-nosed Bat generally occurs in low numbers over its range along the east coast from about Brisbane to Melbourne. It is restricted almost completely to areas east of the Great

*Scotorepens orion*

It roosts in tree-hollows, but the type specimen was taken from the roof space of a building in Sydney. One recorded roost was in a hollow in a Manna Gum, *Eucalyptus mannifera*, about 7 metres from the ground. No information is available on colony size.

The species has been little studied, but appears to have a reproductive cycle typical of other vespertilionids in the region (except *Miniopterus*), sperm storage occurring in both males and females and ovulation and fertilisation being delayed until spring. A single young is born in late spring or early summer.

A second species, similar to the Eastern Broad-nosed Bat may coexist, particularly in the southern half of the range. The two have similar forearm lengths, but the form suspected to be undescribed is much heavier and very robust.

Dividing Range, often in tall, wet forest or rainforest, but four individuals of this species were caught in low open forest near Googong Dam, about 25 kilometres south of Canberra.

C.R. TIDEMANN

---

## Scotorepens orion

### Size
*Head and Body Length*
43–54 (49) mm
*Tail Length*
29–38 (35) mm
*Forearm Length*
32.5–37.5 (35.5) mm
*Weight*
7–15 (11) g

**Identification** Dorsal fur warmish brown, not markedly bicoloured; ventral surface more drab. Body substantially more robust than in other *Scotorepens* species. Tragus with leading edge slightly concave and posterior edge more convex; apex slightly more acute than in congeners. Ratio of tibia length to forearm length usually less than 41 per cent. Most likely to be confused with *S. balstoni*, but distributions appear to be mutually exclusive.

**Recent Synonyms** *Nycticeius balstoni orion*.
**Other Common Names** None.
**Status** Relatively uncommon in most parts of its range.
**Subspecies** None.
**References**
Baverstock, P.R., M. Adams, T. Reardon and C.H.S. Watts (1987). Electrophoretic resolution of species boundaries in Australian Microchiroptera. III. The Nycticeiini—*Scotorepens* and *Scoteanax* (Chiroptera: Vespertilionidae). *Aust. J. Biol. Sci.* 40: 417–433.
Kitchener, D.J. and N. Caputi (1985). Systematic revision of Australian *Scoteanax* and *Scotorepens* (Chiroptera: Vespertilionidae), with remarks on relationships to other Nycticeiini. *Rec. West. Aust. Mus.* 12: 85–146.

G.B. BAKER

# Northern Broad-nosed Bat

## *Scotorepens sanborni*

(Troughton, 1937)

san'-born-ee: 'Sanborn's night-creeper'

As the sun sets, the Northern Broad-nosed Bat is frequently the first of the winged mammals to appear. Its favoured haunts, particularly in the drier parts of its range, are along watercourses but it also feeds along the edges of quiet coastal bays. It is frequently seen in summer as it skims just above still, fresh water, occasionally dipping its mouth below the surface to drink. For such a small bat, it appears to fly rather swiftly, changing direction as frequently as twice a second as it dives and twists in search of the mosquitos, midges and mayflies that constitute most of its diet. Normally an inhabitant of the understorey of coastal forests, it seldom flies more than 4–5 metres above the ground. In urban areas, it may feed around street lights.

While its usual home is a tree-hollow, it also roosts in houses or sheds of timber or fibro-cement construction, gaining access through cracks or under corrugated roofing iron and usually selecting a site where the building faces

*The Northern Broad-nosed Bat frequents water courses, where it catches small insects such as midges.*

northwards or eastwards. Spaces of 3–10 centimetres between walls and major roof supports are often selected as roosts and can be identified by the dung that accumulates beneath these. Dung from small colonies dries out but, if the colony is large, it stays moist and, with the addition of urine, becomes heavy and smelly: there are colourful stories from northern Queensland of ceilings laden with this foul mess descending upon unsuspecting households. The bats themselves are pugnacious and, when caught, usually bite quite fiercely (although their teeth cannot penetrate human skin) and they are reluctant to release a grip on clothing or soft parts of the hand.

*Scotorepens sanborni*

S. DONNELLAN

*Scotorepens sanborni*

**Size**
*Head and Body Length*
48–52 mm
*Tail Length*
31–34 mm
*Forearm Length*
31–35 mm
*Weight*
6–8 g (females usually larger than males)

**Identification** Brown with a slight reddish tinge above; slightly lighter below. Dark snout. Only two upper incisors.
**Recent Synonyms** *Scoteinus sanborni.*
**Other Common Names** Little Northern Broad-nosed Bat.
**Status** Common, limited.
**Subspecies** Uncertain.
**Extralimital Distribution** New Guinea.
**Reference**
Troughton, E. le G. (1937). Six new bats (Microchiroptera) from the Australian region. *Aust. Zool.* 8: 274–281.

*The Inland Forest Bat inhabits grassland and savanna and is able to roost in quite low trees.*

Following extensive clearing of coastal forests for sugar cane farming in northern Queensland, houses are becoming frequent roost sites. Such roosts can contain up to several hundred bats but most colonies in natural roosts, such as tree-hollows, consist of 10–20 bats.

Little is known of breeding but, since pregnant females have been captured in August and September and the high-pitched cries of young bats have been heard in November, it appears that most young are born towards the end of the year. As with many insectivorous bats, colonies are difficult to locate in winter.

Females are larger than males but size is variable, even among members of one colony. The distribution of the species overlaps with that of the Eastern Broad-nosed Bat in the coastal strip between Cairns and Ingham and separation of the two species on external characters is very difficult.

L.S. HALL

# Inland Forest Bat

## *Vespadelus baverstocki*

### Kitchener, Jones and Caputi, 1987.

vesp'-ah-day'-lus  bav'-er-stok'-ee:
'Baverstock's hidden-bat'

Common and widely distributed in central Australia, this bat inhabits grassland, savanna and shrubland. In adaptation to stunted vegetation, it roosts in tree-hollows that can be extremely small and in trees that may be only a few metres high. Abandoned buildings are sometimes used as roost sites.

Colonies range in size from a few individuals to as many as 50 or so. In early summer, females congregate to raise young in colonies from which males are excluded except (apparently) for mating visits. The single young is carried by its mother until its weight affects her flight and is then left in the roost.

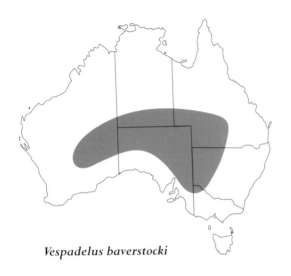

*Vespadelus baverstocki*

### *Vespadelus baverstocki*

**Size**
*Head and Body Length*
36–43 (40) mm
*Tail Length*
26–34 (30) mm
*Forearm Length*
26–32 (29) mm
*Weight*
3–6 (5) g

**Identification** Dorsal fur pale sandy brown, much paler ventrally where the hair shafts are bicoloured, being dark brown at the base with cream or very pale brown tips. Similar in appearance to other *Vespadelus*, but skin of ears and membranes is pale grey (not dark grey as in other species). The pale tragus is distinctive, as is the funnel-shaped glans of the penis. The femur of both the Inland Forest Bat and Finlayson's Cave Bat (*V. finlaysoni*) is furred halfway to the knee; but is furred two-thirds of the way in the Eastern Cave Bat (*V. troughtoni*); and only sparsely haired in the Little Forest Bat (*V. vulturnus*). These three species may be sympatric with the Inland Forest Bat at the boundaries of the arid zone.
**Recent Synonyms** None.
**Other Common Names** None.
**Subspecies** None.
**References**
Kitchener, D.J., B. Jones and N. Caputi (1987). Revision of Australian *Eptesicus* (Microchiroptera: Vespertilionidae). *Rec. West. Aust. Mus.* 13: 427–500.
Reardon, T.B. and S.C. Flavel (1987). *A guide to the bats of South Australia.* South Australian Museum, Adelaide.

The foraging behaviour of the Inland Forest Bat is appropriate to its open habitat. It flies rapidly, covering a broad search area, but is sufficiently manoeuvrable to pursue moths that may attempt to avoid capture.

G.C. RICHARDS

*535*

L. LUMSDEN

*The Western Cave Bat inhabits rocky country in the northern Kimberley and Northern Territory.*

# Western Cave Bat

## *Vespadelus caurinus*

### (Thomas, 1914)

kor-een'-us: 'north-western hidden-bat'

The Western Cave Bat occurs only in the far northern parts of the Northern Territory and Kimberley of Western Australia. It is reasonably common throughout its range and is frequently encountered in caves and disused mines. In this respect, it differs from other northern species of *Vespadelus*, which are mainly tree-dwelling. Other species, such as the Dusky Leafnosed-bat and Common Sheathtail-bat, often share its cave roosts.

*Vespadelus caurinus*

---

### *Vespadelus caurinus*

**Size**
*Head and Body Length*
30–40 mm
*Forearm Length*
26–35 mm
*Weight*
3–5 g

**Identification**  Small in size. Distinguished by darkly pigmented bare skin regions and cave-dwelling habit. Separated from other species of *Vespadelus* on cranial, dental and baculum characteristics.

**Recent Synonyms**  *Eptesicus pumilus caurinus*, *Eptesicus caurinus*.

**Other Common Names**  Northern Brown Bat, Little Brown Bat, Little Northern Cave Bat.

**Status**  Common.

**Subspecies**  None.

**References**

Kitchener, D.J., B. Jones and N. Caputi (1987). Revision of Australian *Eptesicus* (Microchiroptera: Vespertilionidae). *Rec. West. Aust. Mus.* 13: 427–500.

Thomson, B.G. (1989). *A Field Guide to Bats of the Northern Territory.* Conservation Commission of the N.T., Darwin.

At dusk, it emerges to forage for small flying insects in tropical woodland, grassland or patches of vine thicket. Within a roost, it is usually found in small groups or as solitary individuals. Maternity colonies may contain several hundred individuals. One young (rarely two) is born in October or November and is initially carried on the mother while she forages. As the young increase in size, they are left in small creches on the cave roof.

Maternity colonies are particularly sensitive to disturbance by human intrusion. Predators include various pythons, the Brown Tree-snake, goannas and owls. Ghost Bats also prey on the Little Northern Cave Bats, sometimes taking them in flight near roost entrances.

B. THOMSON

*The Large Forest Bat has longer fur than any other member of its genus.*

# Large Forest Bat

## *Vespadelus darlingtoni*

### (Allen, 1933)

darl'-ing-tun-ee: 'Darlington's hidden-bat'

The long fur and greater size that separate the Large Forest Bat from other members of its genus enable it to live in regions of cooler climate on the continent. Restricted to montane areas above 300 metres in the north of its range, it extends to lower altitudes towards the southern limits: in Tasmania, it occurs at sea-level. Extending into Queensland near the New South Wales border, it is common in the extensive rainforests that cover the McPherson Ranges. It utilises a wide variety of habitats from rainforests, through wet and dry eucalypt forests and subalpine woodland to alpine moors. At lower altitudes it is replaced by the Small Forest Bat.

Within the last seven million years it has colonised Lord Howe Island, where it occurs to sea-level. The moderating effect of the surrounding ocean on the climate of this small island has allowed the Large Forest Bat to utilise elevations which it does not occupy on the adjacent mainland.

Adapted to cool climates, it is able to forage for small insects during mild nights in winter when other species are in hibernation. It is thus able to feed with little competition at a time of year when food is scarce. Colonies of up to 60 bats have been found in hollows in old eucalypts

*Vespadelus darlingtoni*

S. JOLLY

*The little-known Yellow-lipped Bat roosts in caves.*

## *Vespadelus darlingtoni*

**Size**

*Head and Body Length*

38–50 (44) mm

*Tail Length*

31–38 (34) mm

*Forearm Length*

32–37 (34.5) mm

*Weight*

6–10 (8) g

**Identification** Largest *Vespadelus* species inhabiting forests of east coast of Australia. Distinguished from *V. pumilus* and *V. vulturnus* by longer forearm (more than 32.5 mm) and longer fur. Males separated from those of *V. regulus* on shape of glans penis. Females difficult to identify except by skull characteristics.

**Recent Synonyms** *Eptesicus pumilus*, *Scotophilus pumilus*, *Eptesicus sagittula*.

**Other Common Names** Large Forest Eptesicus, Large Forest Vespadelus.

**Status** Restricted to montane areas above 300 m in northern part of range but more widespread in southern end of distribution, extending to sealevel in Tasmania. Often locally common.

**Subspecies** None.

**References**

Kitchener, D.J., B. Jones and N. Caputi (1987). Revision of Australian *Eptesicus* (Microchiroptera: Vespertilionidae). *Rec. West. Aust. Mus.* 13: 427–500.

McKean, J.L., G.C. Richards and W.J. Price (1978). A taxonomic appraisal of *Eptesicus* (Chiroptera: Mammalia) in Australia. *Aust. J. Zool.* 26: 529–537.

in southern New South Wales and Tasmania.

Study of a colony on the central coast of New South Wales showed that single young are born from late November to early December and that juveniles are free-flying by late January or early February. Males are sexually active in late February and early March.

G.A. HOYE

# Yellow-lipped Bat

## *Vespadelus douglasorum*

### Kitchener, 1976

dug'-las-or'-um': 'Douglas' hidden-bat'

Because this bat is known only from the remote Kimberley region of Western Australia and has only recently been recognised as a species, little is known of its biology. It is usually collected over creeks or near pools from those north-western parts of the Kimberley receiving annual rainfall in excess of 800 millimetres and has not been recorded in areas receiving less than 600 millimetres.

The Yellow-lipped Bat is insectivorous and appears to be a strict cave-dweller, roosting in colonies of up to 50 individuals. Although it has been caught in the same mist-nets as the Little Cave Eptesicus, it is not known to share roosts with that species. That it forms maternity groups is indicated by the fact that seven or eight individuals trapped in a cave in March were females with enlarged teats; the other being a juvenile male. Like many other northern vespertilionids it appears to give birth to a single young within the period from the end of the dry season to the middle of the wet season.

D.J. KITCHENER

# Finlayson's Cave Bat

## *Vespadelus finlaysoni*

### Kitchener, Jones and Caputi, 1987

fin'-lay-sun-ee: 'Finlayson's hidden-bat'

Finlayson's Cave Bat was first described as a separate species in 1987 during a taxonomic revision of Australian species of *Eptesicus*. It was named after the South Australian biologist, Hedley Finalyson in honour of his work on mammals of arid Australia. Throughout its present range, it is easily distinguished by the combination of its small size and dark brown-rust red colour. However, present distribution records show that its range abuts that of other *Vespadelus* and *Pipistrellus* species from which it is less readily distinguished. Should further collecting show that its range does overlap with these species, characteristics such as humerus length and shape of teeth and penis will need to be examined for identification.

Of the four cave-dwelling *Vespadelus* species, Finlayson's Cave Bat has by far the broadest geographic range. Its habitat extends from the deserts of Central Australia (where annual rainfall can be less than 100 millimetres) to coastal eucalypt-dominated shrublands with rainfall

---

### *Vespadelus douglasorum*

**Size**

*Head and Body Length*
34–43 (39) mm
*Tail Length*
32–39 (36) mm
*Forearm Length*
34–38 (36) mm
*Weight*
circa 5 g

**Identification** Body greyish to olive buff; head, feet and forearms yellow; patch of light yellow fur on the undersurface, posterior to arms. Distinguished from *V. pumilus* by longer forearm, yellow colour of head, feet and forearms. Light colouration of face highlights glandular pads at sides of mouth, giving face a less pointed appearance that in *V. pumilus*.

**Recent Synonyms** *Eptesicus douglasi*.

**Other Common Names** Large Cave Eptesicus, Yellow-lipped Eptesicus.

**Status** Rare, scattered. Possibly underestimated.

**Subspecies** None.

**Reference**

Kitchener, D.J. (1976). *Eptesicus douglasi*, a new vespertilionid bat from Kimberley, Western Australia. *Rec. West. Aust. Mus.* 4: 295–301.

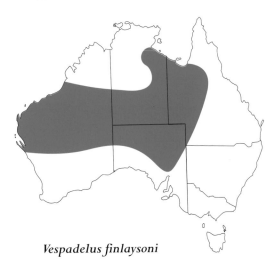

*Vespadelus douglasorum*

*Vespadelus finlaysoni*

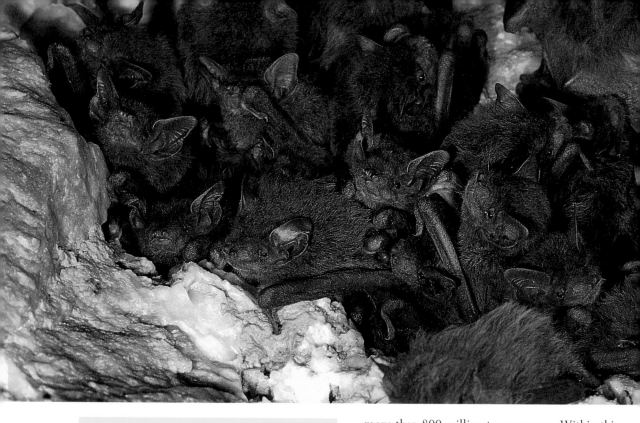

## Vespadelus finlaysoni

### Size

*Head and Body Length*
34.3–46.4 (40.1) mm
*Tail Length*
30.7–42.0 (35.2) mm
*Forearm Length*
29.8–36.7 (32.8) mm
*Weight*
3–7 g

**Identification** Back fur dark brown but
with rust-red hue. Belly fur lighter in contrast.
Long fur extends from forehead well onto ros-
trum. Skin of ears and wings dark. The only
other member of the genus in the range,
*V. baverstocki*, is readily distinguished by having
light brown fur.
**Recent Synonyms** *Pipistrellus finlaysoni,
Eptesicus finlaysoni, Eptesicus pumilus, Scotophilus
pumilus*.
**Other Common Names** Little Cave
Eptesicus, Little Brown Bat.
**Status** Abundant.
**Subspecies** None.

more than 800 millimetres per year. Within this
broad range, it is never far from rocky outcrops
or hilly terrain where caves, cracks and fissures
provide natural roost sites. Abandoned mine
shafts and adits are also utilised. A single colony
may exceed 500 individuals but much smaller
groups are usual. In caves, fissures and mines,
it occasionally roosts in the abandoned nests of
Fairy Martins, at the entrance or under rocky
overhangs. In the northern half of its range,
Finlayson's Cave Bat often cohabits with the much
larger Hill's and Common Sheathtail-bats. It has
also been recorded as sharing caves and mines
with the Ghost Bat, even though this is one of its
few natural predators.

### References

Maddock, T.H. and A.N. McLeod (1976).
Observations on the Little Brown Bat, *Eptesicus
pumilus caurinus* Thomas in the Tennant Creek
area of the Northern Territory. Part I:
Introduction and breeding biology. *S. Aust.
Nat.* 50: 4250.
Kitchener, D.J., B. Jones and N. Caputi (1987).
Revision of Australian *Eptesicus* (Micro-
chiroptera: Vespertilionidae). *Rec. W. Aust.
Mus.* 13: 427–500.

B.G. THOMSON

*Finlayson's Cave Bat roosts in caves, in colonies up to 500 or more individuals. It was not described until 1987.*

# Eastern Forest Bat

## *Vespadelus pumilus*

### (Gray, 1841)

poom'-il-us: 'dwarf hidden-bat'

It has a fluttering flight with fast and seemingly erratic changes of direction and is commonly observed at waterholes, foraging for insects and occasionally dipping into the water to drink. Its diet has not been determined but it is likely that small moths, mosquitos and flying ants are among the insects eaten.

There is some variation in the time of reproduction. The most studied population is at Tennant Creek in the Northern Territory. In this locality, young may be born throughout the year, with peaks in March and September or October, suggesting that females may breed twice a year. Although a single young is usual, nearly 20 per cent of births are twins. Further, south, where seasons are more distinct, there is only one breeding season, young being born in late spring to early summer.

Unlike most other cave-dwelling species, the Tennant Creek bats do not form maternity colonies. Mothers carry their young for a short time after birth. Young bats develop quickly and are capable of trial flights when three to four weeks old.

Despite its apparently harsh and varied habitat, Finlayson's Cave Bat is more fragile than the forest-dwelling species of its genus. Captured animals dehydrate quickly, are very easily stressed and therefore require careful handling. However, its survival appears not to be under any immediate threat.

T.B. REARDON

Although this species is frequently captured in bat-traps in the southern part of its range, little is known about its ecological requirements and general biology. This is partly due to its redefinition in 1987, before which it had been confused with *Eptesicus darlingtoni* (both species being combined as *E. sagittula*). It now becomes obvious that, prior to 1987, the name *pumilus* was applied to several other species.

### *Vespadelus pumilus*

**Size**

*Head and Body Length*
35–44 mm
*Tail Length*
27–34 mm
*Forearm Length*
30.1–33 (31.6) mm (females)
29.0–32 (30.5) mm (males)
*Weight*
3.5–6 (4.9) g (females)
3.5–4.5 (3.9) g (males)

**Identification** Similar to *V. darlingtoni* but smaller. Males distinguished from other *Vespadelus* (except *V. darlingtoni*) by short, angled penis. Usually distinguished from *V. darlingtoni* by smaller size: forearm length of *V. pumilus* adult males less than 32.5 mm, adult females less than 33 mm. Females sometimes difficult to distinguish from *V. regulus*, but head appears more domed; fur colour and skin pigmentation of face and wings usually darker. Distinguished from *V. vulturnus* by darker fur, skin of face and membranes and by lack of a distinctly swollen tip of penis.

**Recent Synonyms** *Eptesicus sagittula* (part), *Eptesicus pumilus pumilus.*

**Other Common Names** None.

**Status** Common.

**Subspecies** None.

**Reference**

Kitchener, D.J., B. Jones and N. Caputi (1987). Revision of Australian *Eptesicus* (Microchiroptera: Vespertilionidae). *Rec. West. Aust. Mus.* 13: 427–500.

*Vespadelus pumilus*

In New South Wales the Eastern Forest Bat occurs from sea-level to the top of the Great Dividing Range, generally in moister forest types. It is not known to inhabit caves and presumably roosts in tree hollows. Twinning appears to be common and, in New South Wales, births occur around December. In the northern part of its range, it is known from the Atherton Tableland, Kirrima near Cardwell and Kroombit Tops, but it probably occurs in isolated populations over a more extensive distribution in Queensland.

H. PARNABY

*Little is known of the biology of the Eastern Forest Bat.*

J. LOCHMAN

# Southern Forest Bat

## *Vespadelus regulus*

### (Thomas, 1906)

reg'-yue-lus: 'King (River) hidden-bat'

Both the common and scientific names of this bat relate to the King River, near Albany, Western Australia, where the species was first collected before the turn of the century. (The specimen chosen as the type of the species was later shown to consist of the skull of a *Vespadelus* and the skin of a Chocolate Wattled Bat.) The Southern Forest Bat is one of nine species in the genus, which is endemic to Australia.

It occurs in two disjunct areas of the mainland, separated by the Nullarbor Plain: in southwestern Australia, and in south-eastern Australia south of the New South Wales–Queensland border and also Tasmania. In far south-western Western Australia, the only species with which it is likely to be confused is the Chocolate Wattled Bat. In the eastern part of its range, it may coexist with the Little Forest Bat, the Large Forest Bat, Eastern Cave Bat and the Little Cave Bat.

*Vespadelus regulus*

## *Vespadelus regulus*

**Size**

*Head and Body Length*
36–46 (42) mm
*Tail Length*
28–39 (32) mm
*Forearm Length*
28–35 (32) mm
*Weight*
4–7 (5.5) g

**Identification** Differs from other members of the genus in having flatter skull, with braincase not markedly above level of muzzle. In some areas, has dorsal fur similar to that of *Chalinolobus morio*. Firm identification depends on penis morphology (males) and/or electrophoresis (both sexes).

**Recent Synonyms** *Pipistrellus regulus*, *Eptesicus regulus*, *Eptesicus pumilus* (part).

**Other Common Names** King River Little Bat, Little Bat.

**Status** Common.

**Subspecies** None.

**References**

Kitchener, D.J. and S.A. Halse (1978). Reproduction in female *Eptesicus regulus* (Thomas) (Vespertilionidae), in south-western Australia. *Aust. J. Zool.* 26: 257–267.

Kitchener, D.J., B. Jones and N. Caputi (1987). Revision of Australian *Eptesicus* (Microchiroptera: Vespertilionidae). *Rec. West. Aust. Mus.* 13: 427–500.

Tidemann, C.R. (1993). Reproduction in the bats, *V. vulturnus*, *V. regulus* and *V. darlingtoni* (Microchiroptera: Vespertilionidae) in coastal south-eastern Australia. *Aust. J. Zool.* 41: 21–35.

It occurs in a wide range of habitats: wet and dry sclerophyll forest, shrubland, low shrub woodland, mixed temperate woodland and mallee. It is common in most areas where it

P. GERMAN

P. GERMAN

Above: *The Southern Forest Bat is a fast-flying predator, taking insects on the wing.*
Left: *The Southern Forest Bat inhabits environments ranging from wet sclerophyll to shrubland.*

Long-eared Bat. Males tend to roost separately from females except during the mating period.

The Southern Forest Bat is an aerial insectivore, feeding on a wide variety of insects, most of which are taken between the tree canopy and the undergrowth. Like many small vespertilionids, it tends to favour moths as prey. No captures are made on the ground or from foliage.

The reproductive cycle involves the insemination of females in autumn, but ovulation and fertilisation do not occur until spring. Sperm is stored over winter in both male and female reproductive tracts and matings may continue during this season. A copulatory plug forms in the vagina after mating. Gestation takes about three months, after which a single young is born in early summer. It is weaned about six weeks later.

C.R. Tidemann

occurs. Colonies of up to 100 individuals are formed in tree-hollows and, in the Adelaide region, it is a common inhabitant of the roofs of houses, often coexisting with the Chocolate Wattled Bat, the Little Mastiff Bat and the Lesser

# Eastern Cave Bat

## *Vespadelus troughtoni*

### (Kitchener, Jones and Kaputi, 1987)

traw'-tun-ee: 'Troughton's hidden-bat'

G.A. HOYE

This species remains one of the least known members of its genus in eastern Australia; little is understood of its biology.

It is a cave-dweller, known from drier forest and tropical woodlands from the coast and Dividing Range to the semi-arid zone. It has been found roosting in small groups in sandstone

*The Eastern Cave Bat can be separated from similar species by the structure of the penis.*

overhangs and in mine tunnels, occasionally in buildings. In all situations, the roost sites are frequently in reasonably well-lit areas. Although it is widely distributed, relatively few records of it exist, particularly in the southern part of its range where it appears to be localised.

---

### *Vespadelus troughtoni*

**Size**

*Head and Body Length*
37–43 mm

*Tail Length*
31–38 mm

*Forearm Length*
33–37 mm

*Weight*
4–7 g

**Identification** Similar to *V. regulus* but larger. Generally larger than *V. finlaysoni* which has a proportionally longer tibia. Definite identification by penis morphology.

**Recent Synonyms** *Eptesicus troughtoni*.

**Other Common Names** Troughton's Vespadelus.

**Status** Uncommon.

**Subspecies** None.

**Reference**

Kitchener, D.J., B. Jones and N. Caputi (1987). Revision of Australian *Eptesicus* (Microchiroptera: Vespertilionidae). *Rec. West. Aust. Mus.* 13: 427–500.

H. PARNABY

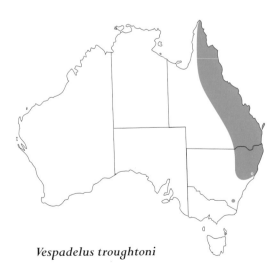

*Vespadelus troughtoni*

# Little Forest Bat

## *Vespadelus vulturnus*

### (Thomas, 1914)

vul-turn'-us: 'vulture-like hidden-bat'

The Little Forest Bat is extremely common below about 1000 metres across most of the Great Dividing Range and associated escarpments south of the Queensland–New South Wales border; in Tasmania; on the Fleurieu Peninsula in South Australia; and slightly north of Adelaide. It occurs in temperate mixed woodland and both wet and dry sclerophyll forest. Over most of its distribution it is sympatric with the Southern and Large Forest Bats. In the north-eastern part of its range, it may also occur sympatrically with the Eastern Forest Bat and Eastern Cave Bat. It is replaced in drier inland areas by the Inland Forest Bat.

It usually inhabits tree hollows but also roosts in the roofs of houses. Colonies are typically small and segregated by sex. Males often roost singly. Maternity colonies generally number less than 20 individuals and may occupy the

**Vespadelus vulturnus**

*Vespadelus vulturnus*

**Size**
*Head and Body Length*
34–48 (40) mm
*Tail Length*
27–35 (31) mm
*Forearm Length*
24–33 (28.5) mm
*Weight*
3.5–6 (5) g

**Identification**  Single individuals, especially females, difficult to identify. In general, smaller and paler than *V. pumilus*. More distinctly swollen tip of glans penis and paler fur than *V. regulus*. More truncated glans penis than in *V. baverstocki*. Smaller than *V. troughtoni*.
**Recent Synonyms**  *Eptesicus vulturnus, Pipistrellus vulturnus, Eptesicus pumilus* (part).
**Other Common Names**  Small Forest Eptesicus, Little Bat, Little Brown Bat, Vulturine Little Bat.
**Status**  Common.
**Subspecies**  None.

same site on a year-round basis. Activity is very temperature-dependent, particularly at higher latitudes: animals may not emerge from roosts for several weeks at a time during cold spells. Males arouse from winter torpor more frequently than females.

The Little Forest Bat feeds exclusively on flying insects, mostly below the tree canopy: it does not take insects from the ground or from foliage. It has an agile, fluttery flight, often involving complicated twists and turns and prey is usually eaten on the wing. Larger prey may be caught in the wing or tail membrane and taken to a feeding roost to be consumed. It is difficult to maintain in captivity for lengthy periods, partly because it is very slow at learning to feed in any manner other than on the wing.

D. WHITFORD

*The Little Forest Bat usually roosts in tree hollows but also utilises the roofs of houses.*

Mating begins in autumn and sperm is stored in the female reproductive tract for fertilisation in spring. Stored sperm is sometimes lost over the intervening period but 'top-up' matings during winter ensure that this is replaced. A single young, born in early summer, is usually left in the roost while its mother forages; rarely it is carried by the mother. Lactation continues for six to seven weeks, after which the young forages with its mother (mother and young are often caught together in surveys at this time).

C.R. TIDEMANN

### References

Kitchener, D.J., B. Jones and N. Caputi (1987). Revision of Australian *Eptesicus* (Microchiroptera: Vespertilionidae). *Rec. West. Aust. Mus.* 13: 427–500.

Tidemann, C.R. 1993. Reproduction in the bats *V. vulturnus, V. regulus* and *V. darlingtoni* (Microchiroptera: Vespertilionidae) in coastal south-eastern Australia. *Aust. J. Zool.* 41: 21–35.

Volleth, M. and C.R. Tidemann (1991). The origin of the Australian vespertilionine bats, as indicated by chromosomal studies. *Z. Saugtierkunde* 56: 321–330.

# ORDER RODENTIA

## *Rodents*

*The Cape York Melomys feeds on a variety of soft plant material, including fruits of rainforest trees.*

The largest living rodent, the Capybara of Central America, weighs about 50 kilograms but most rodents are intermediate in size between the mice and rats that are familiar pests of human habitations. Although diminutive, they are extremely successful animals: with about 2000 species so far described, they constitute more than half of the extant mammals.

Rodents are characterised by having only one pair of upper and one pair of lower incisors. These teeth are long, recurved and grow continuously throughout life. Since the front surface consists of tough enamel and the back is made up of softer dentine, an incisor is continuously sharpened to a chisel-edge as it is worn away by use. The combination of rapid wear and continuous growth of their incisors enables rodents to be persistent gnawers of woody plant material. In captivity, they may even gnaw through a metal cage.

Canine teeth are never present and premolars are lacking in most species. Thus there is a long gap (diastema) between the incisors and the grinding molar teeth—characteristic of mammals that feed on fibrous plant material. Also in adaptation to such a diet, rodents have a large caecum, in which cellulose is broken down by micro-bial fermentation and a long intestine to facilitate further digestion. It is clear from their anatomy and behaviour that rodents evolved as herbivores, but specialisation for gnawing has not prohibited many species from an omnivorous or even predominantly insectivorous way of life. As a group and as species, rodents are adaptable animals.

Small rodents constitute a major source of food for carnivorous mammals and birds. As is usually the case in animals subject to heavy predation, they tend to reach sexual maturity at an early age, to have a short gestation and to produce frequent, large litters. It seems, however, that the size of most rodent populations is limited more by the availability of food and shelter than by predation. A high reproductive rate seems to be a means of exploiting a habitat in which the food supply fluctuates markedly. Rodents adapted to stable environments tend to have greater individual longevity and a lower rate of reproduction.

The order Rodentia, which includes such diverse animals as agoutis, porcupines, cavies, squirrels, flying squirrels and beavers, is classified into four suborders and 24 families. However, all the Australian species fall into one family, the Muridae. The vast majority—the so-called 'old endemics'—are further restricted to a single sub-family, the Hydromyinae, found only in the Australasian region. Seven species of rats—comprising the 'new endemics', the ancestors of which came to Australia much later, are members of the subfamily Murinae. The House Mouse and the two species of rat that were introduced by Europeans also belong to the Murinae.

Elsewhere in this book, we have not been concerned to consider classification of animals below the level of subfamilies, but the native Australian rodents are so closely related that little sense can be made of them without taking their classification further. Present evidence suggests that the Hydromyinae—characterised, among other fea-tures by the presence of only four teats in the females—fall into three designated tribes: the Hydromyini, Uromyini and Conilurini and probably a fourth (not yet des-ignated) to include the little understood prehensile-tailed rats.

Rodents have aptly been termed 'the late invaders' among the Australian mammal fauna. A small number of mouse-like species crossed from eastern Indonesia into the Australia–New Guinea region no more than 15 million years ago and their subsequent evolutionary radiation has led to less variety than is found in other continents—where rodents have been resident for at least four times this period. Nevertheless, rodents comprised almost a quarter of the mammal species living on the continent at the time of European settlement. They include species of terrestrial, arboreal and amphibious habits, which, between them, occupy virtually every habitat on the continent.

In common with marsupials, many native rodents have suffered reduction in distri-bution and numbers since European settlement. Seven species—one tree-rat, one stick-nest rat, one mouse and four hopping-mice—appear to have become extinct and a further dozen or so are reasonably assumed to be at risk.

# FAMILY MURIDAE
## *Murids*

*The Western Mouse is locally common, but its habitat is at risk.*

With more than 1100 species in some 267 recent genera, this is the largest and most widespread family of mammals, possibly having originated in South-East Asia. The group is difficult to define but most members lack canines and premolars and have three pairs of upper and lower molars. Differentiation from other rodent families is largely on the basis of complex details of the molar cusps.

Some 16 subfamilies are recognised, of which two, the Hydromyinae and Murinae, are represented in Australia.

## Subfamily Hydromyinae

This group cannot be defined on external characters and it is difficult to do so in terms of cranial or dental criteria. Much of its justification is based on chromosomal similarities and serological criteria. However, in the Australian context, hydromyines have only four teats (six in *Pogonomys*), whereas members of the Murinae have eight to twelve teats.

Within the Australian members of the subfamily, three tribes are recognised: the Conilurini, Hydromyini and Uromyini. These are defined by a combination of subtle cranial, dental and chemical characters.

## TRIBE CONILURINI

With a predominantly Meganesian distribution (eight genera and 40 species in Australia) the Conilurini is the largest and most diverse group of the 'old endemic' Hydromyinae. Unlike the Hydromyini and Uromyini, it includes a number of species adapted to arid conditions.

By far the largest tribe of the Hydromyinae, this group appears to have had an evolutionary radiation in the drier parts of the continent. Of the eight genera, the arboreal *Mesembriomys* and *Uromys*, plus the rock-rats, *Zyzomys*, form a closely related group. The essentially arid-adapted *Pseudomys* and wet-adapted *Mastacomys* and, possibly, the stick-nest rats of the genus *Leporillus*, are another assemblage. The arid-adapted native mice of the genera *Pseudomys* and the hopping-mice of the genus *Notomys* appear to represent separate lines of evolution from a common ancestor.

The genus *Pseudomys* has some 17 species, usually referred to simply as native mice or, in literal translation, 'false-mice'. They are anatomically unspecialised and lack any outstanding features by which they may be readily recognised and it is probably because they are 'generalised' rodents that they have been able to spread over the whole continent. Related to *Pseudomys*, but distinguished by their notably shorter tails, are the two species of *Leggadina*. The Broad-toothed Rat, a single species in the genus *Mastocomys*, is a plump animal with broad incisors and very broad molars that are associated with its diet of grass and sedges. Two species in the genus *Leporillus* are rather large rodents that construct piles of sticks and branches within which a family resides. The nests are added to by successive generations.

Anatomically the most specialised of the conilurine rodents, the hopping-mice of the genus *Notomys* comprise nine species. All have very long hindlegs, on which they hop, and slender, tufted tails that are longer than the head and body—usually considerably so. They are social animals that inhabit communal burrows, mostly in arid regions.

# White-footed Tree-rat

## *Conilurus albipes*

(Lichenstein, 1829)

kon'-il-yue'-rus al'-bee-pez:
'white-footed rabbit-tail'

This little-known and probably extinct species was discovered in the early 1800s in 'New South Wales' (which then extended from Cape York to Van Diemen's Land). The original specimens were taken to England by the ornithologist George Caley between 1800 and 1810 and it seems that they were acquired by the Linnaean Society of London in 1818. Neither their precise localities nor their dates of collection are known. A specimen reached the Berlin Museum by 1824 and a few others are known from museums in Europe.

In the Museum of Victoria, Melbourne, there are two specimens, labelled 'Coopers Creek', but the accuracy of this Central Australian location is doubtful. Several fossil locations are known from

*White-footed Tree-rat. Illustration in John Gould's* The Mammals of Australia.

areas in southern Australia, excluding Tasmania.

Sir George Grey, Governor of South Australia from 1840 to 1845, sent a specimen to John Gould in London, noting that: 'This animal lives among the trees. The specimen I send to you, a female, had three young ones attached to its teats when it was caught. While life remained in the mother they remained attached to her teats by their mouths. On pulling the teats of the dead mother, they seized hold of my glove with the

*Conilurus albipes*

## *Conilurus albipes*

**Size**

*Head and Body Length*
230–260 mm
*Tail Length*
220–240 mm
*Weight*
circa 200 g

**Identification** Grey-brown above; white below. Long, soft fur. Distinguished from all its large relatives by bicoloured, long brushy tail, dark brown above and white below for its whole length.

**Recent Synonyms** None.

**Other Common Names** White-footed Rabbit-rat.

**Status** Presumed extinct. Not recorded as common since European settlement.

**Subspecies** None.

**Reference**

Gould, J. (1863). *The Mammals of Australia, Vol. 3*. The author, London. Reissued (1976) as *Placental Mammals of Australia*, with modern commentaries by J.M. Dixon, Macmillan, Melbourne.

mouth and held on so strongly that it was difficult to disengage them'.

The notes of John Gould, written over a century ago, provide the best information available on the species. He observed that 'it is dispersed over all parts of New South Wales, Port Phillip and South Australia, but is nowhere very abundant'; that there was some colour variation between animals from different regions; and that it was strictly nocturnal: '... it sleeps during the day in the hollow limbs of prostrate trees, or such hollow branches of the large Eucalypti as are near the ground in which situations it may be found curled up in a warm nest of dried leaves. More than once have I, after detecting the animal in its retreat, sawn off the hollow limb and secured it without injury'.

J.M. DIXON

# Brush-tailed Tree-rat

## *Conilurus penicillatus*

### (Gould, 1842)

pen'-iss-il-ah'-tus: 'brush (-tailed) rabbit-tail'

This attractive rodent lives in monsoonal northern Australia, adjacent islands and southern New Guinea. Its stronghold appears to be the Top End of the Northern Territory, where it is known from many localities. In the drier Kimberley region it has been found only in localities quite near the coast.

It lives in mixed eucalypt open forest or woodland and *Casuarina* and *Pandanus* stands, seeming to prefer habitats with a grassy understorey and sparse shrub layer. It is an active climber but also spends a considerable time on the ground. During the day, it rests in hollows high up and at the base of trees; in logs on the ground; and in *Pandanus* fronds. It is often active at dusk. Very agile on the ground, it bounds rapidly with tail held high and flicking from side to side. When disturbed or threatened it utters growling noises similar to the Black-footed

*Conilurus penicillatus*

*The Brush-tailed Tree-rat is an active climber but spends much time on the forest floor.*

### *Conilurus penicillatus*

**Size**
*Head and Body Length*
135–227 (165) mm
*Tail Length*
102–235 (188) mm
*Weight*
116–242 (163) g (males)
102–202 (144) g (females)

**Identification** Grizzled grey to golden brown on back, with rufous patch on neck. White to cream on belly, often with grey chest-patch. Tail blackish-brown with black or white brush towards tip. Distinguished by rabbit-like appearance, small to medium size and long, finely brushed tail. Animals from Melville and Bathurst Islands have small ears and hindfeet and are more dark-bellied than in other regions.

**Recent Synonyms** None.

**Other Common Names** Brush-tailed Rabbit-rat, Rabbit-eared Tree-rat.

**Status** Sparsely distributed but can be locally abundant.

**Subspecies** *Conilurus penicillatus penicillatus*, mainland Australia, islands in Gulf of Carpentaria. *Conilurus penicillatus melibius*, Melville and Bathurst Islands. *Conilurus penicillatus randi*, New Guinea.

**Extralimital Distribution** Southern New Guinea.

**References**
Kemper, C.M. and L.H. Schmitt (1992). Morphological variation between populations of the brush-tailed tree-rat *Conilurus penicillatus* in Northern Australia and New Guinea. *Aust. J. Zool.* 40: 437–452.
Friend, G.R., C.M. Kemper and A. Kerle (1992). Rats of the tree tops. *Landscope* 8: 10–15.

Tree-rat, White-tailed Rat and Sugar Glider. Soft, high-pitched vocalisations have been heard among captive individuals. Its diet is largely vegetable matter, particularly grasses and seeds, but termites are also eaten.

The long breeding season (March to October) is probably timed so as to take maximum advantage of the nutritious and abundant green grasses and seeds during the late-wet and mid-dry seasons. There is a post-partum oestrus and several litters are born in one season. The gestation period averages 36 days in females that are not lactating, but up to 12 days longer when a litter is being suckled. One to four young are born per litter, usually three. These are large at birth and develop quickly, being well-furred by about seven days and with open eyes at about 12 days. Although the young can be weaned as early as three weeks, litters in the laboratory have suckled until four to seven weeks old. This and the fact that males nest with the mother and her young, suggests that social bonding and tolerance are features of its biology. Near-maximum weight is reached by about three months and females can be mature when only six weeks old but, on average, maturity is reached at 11 weeks. Some laboratory animals have survived for five years.

The Brush-tailed Tree-rat is patchily distributed throughout its range and appears to have quite specific habitat requirements. Although it is not under immediate threat, local extinctions could occur if its habitat is altered or destroyed.

C.M. KEMPER

# Forrest's Mouse

## *Leggadina forresti*

### (Thomas, 1906)

leg'-ah-deen'-ah fo'-rest-ee:
'Forrest's *Leggada*-like (animal)'
(*Leggada* is a genus of Indian rodents.)

Forrest's Mouse, an inhabitant of some of the harshest environments of inland Australia, is a small, solidly built rodent distinguishable from all other inland mice by having a tail that is markedly shorter than its head and body. Its blunt nose, small rounded ears and thick pelt of close-cropped coarse hair also aid in identification, although individuals vary in appearance across the extensive range of the species. In fact, the taxonomy of the genus is in considerable confusion, since recent results of electrophoresis indicate that its range may be far more extensive across northern and north-western Australia than earlier thought, being replaced there by *L. lakedownensis*. A thorough taxonomic review of the genus, combining morphological and molecular data, must be undertaken before the northern limits of the range of *L. forresti* can be defined precisely.

Forrest's Mouse is restricted to the arid inland, usually inhabiting tussock grasslands and low chenopod shrublands on plains with loam, clay or stony soils. It is occasionally found in mulga woodlands, spinifex grasslands, rocky hills and sand-dune environments. Individuals have been dug from shallow burrows about 15 centimetres deep, containing nests of grass and one or two blind tunnels. These are inhabited by solitary individuals or by females with young. The preferred diet has not been established but it is known to eat seeds, arthropods and green vegetation, probably in that order of preference. Its

*Forrest's Mouse inhabits the arid interior.*

---

### *Leggadina forresti*

**Size**
*Head and Body Length*
80–100 (90) mm
*Tail Length*
50–70 (60) mm
*Weight*
15–25 (20) g

**Identification** Grey-brown above; white below. Tail length 60–70 per cent of head and body length. Small ears.

**Recent Synonyms** *Leggadina messoria*, *Leggadina waitei*, *Gyomys berneyi*.

**Other Common Names** Melrose Desert Mouse, Waite's Mouse, Berney's Queensland Mouse, Short-tailed Mouse, Southern Short-tailed Mouse, Forrest's Territory Mouse.

**Status** Sparse.

**Subspecies** None.

**References**

Philpott, C.M. and D.R. Smyth (1967). A contribution to our knowledge of some mammals from inland Australia. *Trans. Roy. Soc. S. Aust.* 91: 115–129.

Read, D.G. (1984). Diet and habitat preference of *Leggadina forresti* (Rodentia: Muridae) in western New South Wales. *Aust. Mammal.* 7: 215–17.

C. ANDREW HENLEY/LARUS

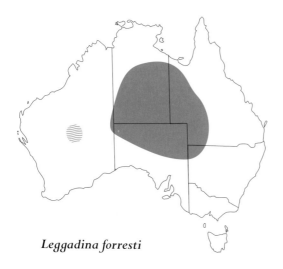

*Leggadina forresti*

nocturnal behaviour and burrowing habits enable it to avoid high temperatures and, like other well-adapted desert rodents, it is probably able to extract sufficient moisture from its food to survive without drinking.

Reproduction may be more dependent upon rainfall than season, for females have been found breeding after summer and winter rains. Studies in the laboratory show that the oestrous cycle is seven days and, although the gestation period is unknown, it is probably between 30 and 40 days. The usual litter size is three to four, and young can attain independence when about four weeks old.

Little is known concerning populations of Forrest's Mouse and there is no direct evidence that it reaches the high levels characteristic of other species of inland rodents that occur as plagues from time to time. On the other hand, indirect evidence from the occurrence of Forrest's Mouse in the diet of the Barn Owl suggests that populations increase to some degree following periods of extended rainfall and increased vegetation growth. The apparent sparseness of most populations makes it difficult to determine whether or not there have been dramatic changes in its distribution or status since European settlement, but the range has not changed markedly to our knowledge and the species is probably secure.

J.W.R. REID AND S.R. MORTON

# Lakeland Downs Mouse

## *Leggadina lakedownensis*

### Watts, 1976

lake'-down-en'-sis: 'Lakeland Downs Station *Leggada*-like (animal)'

This secretive and apparently rare species, known only from a few specimens taken near Cooktown, Queensland, has been seen in its natural habitat by only two naturalists. The first specimen was found in September 1969 on 'William Island' (the local name for an area cut off during periodical flooding

---

**Leggadina lakedownensis**

**Size**
*Head and Body Length*
60–75 mm
*Tail Length*
40–45 mm
*Weight*
15–20 g

**Identification** Distinguished from *Leggadina forresti* by incisive foramina which are wider, rather than narrower posteriorly and by upper incisors which point forwards.
**Recent Synonyms** *Leggadina lakedownensis* has been confused in the past with *L. forresti*.
**Other Common Names** None.
**Status** Rare, scattered; populations apparently fluctuate dramatically.
**Subspecies** None.
**Reference**
Watts, C.H.S. (1976). *Leggadina lakedownensis* a new species of murid rodent from north Queensland. *Trans. Roy. Soc. S. Aust.* 100: 105–108.

HANS & JUDY BESTE

of the Bizant and North Kennedy Rivers) when the mice were so common that a specimen was taken by hand. In 1973, two more specimens were collected in the same way from a horde of mice running about on a road through seeding sorghum fields at Lakeland Downs.

'William Island' is an area of open grassland with pockets of savanna woodland on alluvial clay and sandy soils. Palms, mounds of the magnetic termite and spear grass are conspicuous landscape

Note: western populations may be an undescribed species of *Leggadina*

**Leggadina lakedownensis**

*Little is known of the biology of the Lakeland Downs Mouse which, like Forrest's Mouse, has a noticeably short tail.*

features. Lakeland Downs is an isolated area of basalt and derived red and brown soils which support a natural vegetation of woodland and kangaroo grass and spear grass.

Virtually nothing is known about the biology of the Lakeland Downs Mouse but, since it thrives in captivity on a diet of small seeds, it seems reasonable to assume that native and introduced grass seeds are eaten in the wild. Like most Australian rodents, it is nocturnal. It breeds readily in captivity, producing two to four young. The gestation period is about 30 days.

The only specimens of the Lakeland Downs Mouse collected were taken when the species occurred in virtual plague proportions on well-seeded, watered grasslands and on sorghum fields. Its populations rise and fall dramatically, probably in response to climatic fluctuations and the availability of seeds.

J. COVACEVICH

557

# Lesser Stick-nest Rat

## *Leporillus apicalis*

(Gould, 1853)

lep'-or-il'-us  ah'-pee-kah'-lis:
'pointed little-hare'

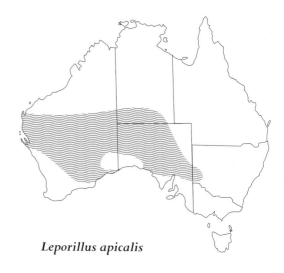

*Leporillus apicalis*

The Lesser Stick-nest Rat is now almost certainly extinct and in its passing, almost nothing was recorded of its natural history. Gerard Krefft, virtually the only person to write of its habits, found it in the years 1856–57 'in great numbers upon Sir Thomas Mitchell's old track on both sides of the River Murray' and noted that: 'This pretty little animal is nocturnal and gregarious in its habits. I have frequently taken eight to ten out of

a hollow tree and tamed them so that they kept about the camp, mounting the supper table at teatime for their share of sugar and damper. I believe this is the same social animal of which Bourke and Wills complained, naming a certain locality 'Rat Point' after it'. Krefft had the unique distinction among white men of having eaten the animal and noted that 'the flesh is white and of excellent flavour'.

At this time, although the Greater Stick-nest Rat had already gone from the Murray River area, Krefft noted that stick-nests that he assumed had been built by the larger species were 'either uninhabited or occupied by *Hapalotis apicalis* [Lesser Stick-nest Rat], a species always at war with the larger and apparently stronger, but not so numerous *Hapalotis conditor* [Greater Stick-nest Rat]'. This account led to some uncertainty about whether the smaller species did in fact build stick-nests but the question can be resolved by considering its distribution. The Lesser Stick-nest Rat appears to have had a wide range through central Australia south to about 26° latitude, the south-eastern portion of its former range extending much further south and overlapping with the northern edge of the Greater Stick-nest Rat's distribution along the Murray River, where Krefft made his observations.

The discovery of numerous old stick-nests in caves and overhangs throughout Central Australia, particularly in the Gibson Desert, confirms

---

### *Leporillus apicalis*

**Size**
*Head and Body Length*
170–200 (190) mm
*Tail Length*
220–240 (230) mm
*Weight*
up to 150 g

**Identification** Smaller and more lightly built than *L. conditor* and distinguished by a pencil of white hairs on the distal quarter of the tail.
**Recent Synonyms** None.
**Other Common Names** White-tipped Hapalotis, White-tipped House-building Rat, White-tipped Stick-nest Rat, White-tailed Stick-nest Rat.
Indigenous names: Tchujalpi (Pitjanjatjarra group), Turulpa (Arunta group), Tweealpi (Pintupi group), Tillikin (Yarra Yarra group).
**Status** Rare, possibly extinct.
**Subspecies** None.

*Lesser Stick-nest Rat. Illustration in Gould's* The Mammals of Australia, *wrongly indicated as aboreal.*

that this species also was a builder. One large nest found on De Rose Hill Station in South Australia measured about 3 metres by 2 metres and was 1 metre high. Faecal pellets collected from this and other nearby nests consisted almost entirely of vegetable matter, in particular the fleshy-leaved perennial shrub *Sclerolaena eriacantha*. Like its larger relative, the Lesser Stick-nest Rat appears to have been a herbivore and to have been unable to compete with the sheep, cattle and rabbits introduced to this country by European settlers.

There is a faint glimmer of hope that this attractive little animal may still survive. In 1970 an experienced bushman deposited some equipment in a cave west of the Canning Stock Route in Western Australia. He covered his gear with a tarpaulin and noticed an old stick-nest in the back of the cave. Returning several weeks later,

he found a large, attractive rat living happily under his tarpaulin, caught the animal to examine it more closely and then let it go. He may have caught a Lesser Stick-nest Rat. More recently there have been tantalising reports from the same general area of stick-nests in caves with pieces of fresh green vegetation woven into their structure, but the last specimen of which we can be sure was collected near Mount Crombie in 1933.

A.C. ROBINSON

**References**

Krefft, G. (1866). On the vertebrated animals of the Lower Murray and Darling, their habits, economy and geographical distribution. *Trans. Phil. Soc. NSW.* 1862–1865: 1–33.

Watts, C.H.S. and B.M. Eves (1976). Notes on the nests and diet of the white-tailed stick-nest rat *Leporillus apicalis* in northern South Australia. *S. Aust. Nat.* 51: 9–12.

W.G. BREED

# Greater Stick-nest Rat

## *Leporillus conditor*

(Sturt, 1848)

kon'-dit-or: 'builder little-hare'

Above: *Abandoned nest of the Greater Stick-nest Rat.*
Right: *Once widespread on the mainland, the Greater Stick-nest Rat survives only on Franklin Island.*

Sturt, Mitchell and other early Australian explorers were impressed by the communal nests, up to 1 metre high and 1.5 metres in diameter, built by this diligent, but placid, rodent. An appealing

*Leporillus conditor*

**Size**
*Head and Body Length*
170–260 (220) mm
*Tail Length*
145–180 (170) mm
*Weight*
180–450 (350) g

**Identification** Yellowish-brown to grey above; creamy white below. Fur fluffy. Hindfeet with distinctive white markings on upper surface. Ears long; eyes large; snout rather blunt. Tail evenly furred, dark brown above and light brown below. Animals rest in hunched posture reminiscent of a rabbit.
**Recent Synonyms** *Leporillus jonesi.*
**Other Common Names** House-building Rat, Large Stick-nest Rat, Stick-nest Rat, Franklin Island Stick-nest Rat, Franklin Island House-building Rat.
Indigenous names: Wopilikara (Wongkanguru group), Kulunda (Dieri gorup), Karnanyuru (Pangkala group), Kohla (Yarra Yarra group), Wiranja, Balgoolya, Walkla, Tjarrluda (Nullarbor group).
**Status** Rare, limited.
**Subspecies** None.

feature is that it seldom attempts to bite when handled but few people now have the opportunity to take it in their hands.

The nest is usually constructed around a bush which eventually becomes part of the structure. Branches are dragged to the site in the rat's jaws, very large branches being gnawed to manageable size. A nest of grass or other soft vegetation is constructed in the centre of the accumulating pile and tunnels lead from this to the perimeter. Captive animals continue to add sticks to their nest over long periods of time and it seems that this is also the case in the wild, where pieces of green vegetation are occasionally seen woven into the fabric of old dead sticks. The construction is added to or modified by succeeding generations and very large nests may contain communities of 10–20 animals, secure against eagles and the Dingo.

It is exclusively herbivorous, eating the leaves and fruits of such succulent plants as pigface and nitre bush, the digestion of which is aided by a sacculated stomach and very large caecum. Breeding can occur in the wild in any month of the year but there seems to be a peak in the autumn and winter. Oestrus cycle length is 14 days and after a gestation period of about 30 days, one to three well-developed young are born. Attaching themselves firmly to the mother's teats, they are dragged around beneath her until becoming weaned and independent at the age of one month.

Although once widely distributed, it had become rare by the middle of the nineteenth century, being found only where sheep and cattle had

not penetrated. It is now extinct on the Australian mainland and exists only on Franklin Island where a population of 1000–1500 survives, sharing this last stronghold with an island race of the Short-nosed Bandicoot, a colony of seals and a number of bird species, including the Short-tailed Shearwater, the burrows of which honeycomb the island and are used also by the rats and the bandi-coots. Here it is preyed upon extensively by Barn Owls and possibly the Black Tiger-snake but the population appears to be stable.

Following a detailed ecological study of the rats on the Franklin Islands, the South Australian National Parks and Wildlife Service removed some animals in 1985 to begin a captive breeding program. In 1990, animals from this program were released onto Reevesby Island in the Sir Joseph Banks Group Conservation Park in South Australia and Salutation Island in the Small Islands, Shark Bay Nature Reserve in Western Australia. In 1993 and 1994, further captive-bred animals were released onto St Peter Island in the Nuyts Archipelago Conservation Park in South Australia. To date (1994) these three populations seem to be establishing successfully.

A.C. ROBINSON

### References

Aslin, H.J. (1972). Nest-building of *Leporillus conditor* in captivity. *S. Aust. Nat.* 47: 43–46.

Robinson, A.C. (1975). The stick-nest rat, *Leporillus conditor* on Franklin Island, Nuyts Archipelago, South Australia. *Aust. Mammal.* 1: 319–327.

Pedler, L. and P.B. Copley (1994). *New introduction of Stick-nest Rats to Reevesby Island, South Australia WWF Project No. 175*. Biological Conservation Branch, Department of Environment and Land Management, Adelaide.

Copley, P.B. (1988). *The Stick-nest Rats of Australia, a final report to World Wildlife Fund (Australia)*. National Parks and Wildlife Service, Department of Environment and Planning, South Australia.

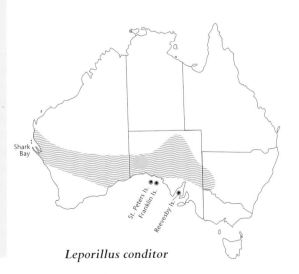

*Leporillus conditor*

H. & J. BESTE

# Broad-toothed Rat

## *Mastacomys fuscus*

Thomas, 1882

mas'-tah-koh-mis fus'-kus:
'dusky chewing-mouse'

The Broad-toothed Rat, named in reference to the size of its molars, has a wide, rounded head, long, fluffy fur and a relatively short tail. It is an uncommon animal, found in alpine and subalpine heath-lands and open eucalypt woodlands of the Snowy Mountains and Victorian Alps; in clearings with dense undergrowth in the wet sclerophyll forests of the Dandenong and Otway Ranges; and in wet sedgelands and subalpine heathlands in Tasmania. These habitats, which range from sea-level to 2200 metres, are characterised by high rainfall, a cool summer, cool to cold winter and a moderate to dense ground cover of grasses, sedges and shrubs. Areas with these conditions are now limited but fossils show that the Broad-toothed Rat had a more extensive distribution in the Pleistocene when cooler climates were more widespread.

Populations appear to be restricted to patches of optimum habitat. In the Snowy Mountains, for

*The Broad-toothed Rat has notably large jaw muscles which give it a broad face.*

example, it is often found close to streams and steep banks where there is an abundant growth of grass and rope-rush and where dense cover is provided by many species of alpine shrubs. Some apparently suitable areas are unoccupied.

Runways are constructed under the vegetation and large well-insulated nests of shredded grass are built under logs and dense undergrowth. In subalpine regions, the runways are comparatively cool even in summer. In winter, the vegetation holds the blanket of snow off the ground, forming a space in which the temperature remains between $0°C$ and $-2°C$, even though blizzards may be raging above. The animals move freely among these spaces throughout the winter and seem not to hibernate.

The diet consists mainly of grasses and, to a lesser extent, seeds and the leaves of shrubs. An individual eats 50–70 per cent of its body weight of fresh vegetable matter each day and produces many large, green, fibrous faecal pellets. The Broad-toothed Rat drinks freely, but also obtains substantial amounts of water in its food.

Breeding is seasonal: pregnancies have been recorded from October to February in Tasmania and births occur from December to March in the Snowy Mountains. The gestation period is about five weeks, mean litter size is 1.9 (range 1–4) and the young are weaned at about five weeks. Although subadult weight is attained at about 12 weeks, full adult weight is not achieved until the following summer. Adult females probably have only two litters each season and juveniles do not breed in the summer of their birth.

The home range, activity patterns and nesting behaviour of resident individuals varies according to season. During the breeding season, the average home range of females in Kosciusko National Park is about 1600 square metres; the home ranges of several females may overlap, especially in optimal habitats close to streams. In autumn, after the breeding season and when fewer resources are required, the home range decreases to about 1000 square metres. Males have larger home ranges which overlap those of several females and these ranges also decline from about 2700 square metres

in summer to about 1000 square metres in autumn. Activity in both sexes is mainly nocturnal in summer, although up to 40 per cent of activity may occur during the daytime. In autumn, there is a decline in both diurnal and nocturnal activity.

In summer and autumn, individuals nest alone although each adult female nests with her young until they are weaned. In winter, however, groups of up to five individuals, including males and females, congregate in communal nests; feeding is confined to a small area close to the nest and

## *Mastacomys fuscus*

**Size**
*Head and Body Length*
142–175 (161) mm
*Tail Length*
100–130 (116) mm
*Weight*
97–145 (122) g

**Identification** Light brown to dark brown above; paler below; feet dark. Fur fine, soft and dense. Tail short, dark with small bristles. Head broad, rounded, with well-developed cheeks. Incisors and molars broad.

**Recent Synonyms** *Mastacomys wombeyensis* (fossil).

**Other Common Names** None.

**Status** Common to sparse, limited.

**Subspecies** *Mastacomys fuscus fuscus*, Tasmania. *Mastacomys fuscus mordicus* (including supposed *M. f. brazenori* and *M. wombeyensis*), Victoria, New South Wales.

**References**

Bubela, T.M., D.C.D. Happold and L.S. Broome (1991). Home range and activity of the Broad-toothed Rat, *Mastacomys fuscus*, in subalpine heathland. *Wildl. Res.* 18: 39–48.

Calaby, J.H. and D.J. Wimbush (1964). Observations on the broad-toothed rat *Mastacomys fuscus* Thomas. *CSIRO Wildl. Res.* 9: 123–33.

Happold, D.C.D. (1989). Small mammals of the Australian Alps (in) R. Good (ed.) *The scientific significance of the Australian Alps.* Australian Academy of Science, Canberra. pp. 221–239.

Carron, P.L., D.C.D. Happold and T.M. Bubela (1990). Diet of two sympatric subalpine rodents, *Mastacomys fuscus* and *Rattus fuscipes. Aust. Wildl. Res.* 17: 479–489.

**Mastacomys fuscus**

M.f. mordicus

M.f. fuscus

# Black-footed Tree-rat

## *Mesembriomys gouldii*

### (Gray, 1843)

mez-em'-bree-oh-mis gule'-dee-ee:
'Gould's southern-mouse'

activity is minimal and usually occurs only in the afternoon and evening.

Nomadic individuals, which include males looking for mates, adult females searching for suitable nest sites and dispersing subadults, are capable of crossing alpine grasslands from one patch of heathland to another. As a result, Broad-toothed Rats can recolonise optimal patches when these become available after the death or emigration of previous residents.

In captivity, a pair will live and nest together peacefully until the female gives birth, but subsequently the male is attacked and driven away. In the wild, the location of many male home ranges in suboptimal habitats, the frequency of minor injuries to males and the large number of nomadic males, suggest that in spring and summer the females determine the social structure and spatial arrangement of the population. This structure appears to be resource-based and related to reproductive efficiency. In contrast, huddling together in communal nests in winter, when the quality and availability of food is reduced and the climate is very cold, is probably related to survival.

The scarcity of the species is puzzling. Contributing factors may include the sparseness of suitable habitat, excessive predation by the Fox and feral Cat, low reproductive rate, or competition from other native rats that occur in large numbers in most of its habitats.

D.C.D. HAPPOLD

One of Australia's largest rodents, the Black-footed Tree-rat is a spectacular creature, perhaps the Australian equivalent of a squirrel. It inhabits tropical woodlands and open forests in coastal areas of the Kimberley in Western Australia; the Top End of the Northern Territory; the east and west coastal areas of Cape York Peninsula between the Iron Range and Townsville inland to the Lynd Junction; and the lower Archer River on the west coast of Cape York Peninsula. There are records from the lower McArthur River area in the Gulf of Carpentaria. Early distribution records suggest that populations have contracted from more inland areas of the Northern Territory around Daly Waters. Patches of tall *Eucalyptus miniata* and *E. tetrodonta* open forest on deep loamy soils, supporting a moderately dense understorey of shrubs and small trees—for example, *Gardenia*, *Terminalia*, *Petalostigma* and *Pandanus* species—are its preferred

Melville Is.
M.g. melvillensis

M.g. rattoides

M.g. gouldii

**Mesembriomys gouldii**

*The Black-footed Tree-rat is an agile climber.*

habitat. Such areas are often associated with perennial soaks where moisture levels and food resources are relatively stable throughout the year.

The Black-footed Tree-rat is nocturnal, sheltering during the day in hollows in standing or fallen trees, or even under the roofs of buildings. A subadult male radio-tracked on the Mitchell Plateau in the Kimberley used nest-sites in four different trees. Most of these consisted of hollows about 10 centimetres in diameter and about 6–8 metres above ground in *Eucalyptus porrecta* and *E. miniata* trees. Its main nest was in the fronds of a *Livistona eastonii* palm where it leaned against a *E. tetrodonta*. Large distances (more than

## *Mesembriomys gouldii*

### Size

*Head and Body Length*
251–308 (290) mm (males)
251–290 (279) mm (females)

*Tail Length*
334–412 (371) mm (males)
320–392 (358) mm (females)

*Weight*
650–830 (728) g (males)
580–882 (705) g (females)

**Identification** Grizzled medium grey and black above, overall colour darker on Melville Island. White (Arnhem Land), medium grey (Melville Island), or pale grey (northern Queensland) below. Hindfeet usually mottled black and white (Arnhem Land), all black (Melville Island), or mostly black (northern Queensland). Fur long, shaggy; long ears; long black hairy tail with white tip. Distinguished from the following with which it sometimes occurs:
*Mesembriomys macrurus*, somewhat smaller, rich buffy-brown above, white feet and largely white hairy tail;
*Hydromys chrysogaster*, dark brown dorsally, short dense glossy fur, very small ears and partly webbed hindfeet;
*Uromys caudimaculatus*, naked tail, mottled black and white and an inhabitant of rainforest.

500 metres) were covered each night between refuge trees and feeding areas on the scree slope at the edge of the plateau.

Little is known of its diet. Analyses of stomach contents and droppings indicate that fleshy and hard fruits and large seeds are a major food item, supplemented by grass and invertebrates such as termites and even molluscs. The large fruits of *Pandanus* are particularly favoured. When the fruit is ripe in the middle of the dry season (July) the Black-footed Tree-rat may be observed in a tree, squatting on or near the fruits and gnawing these; as the fruits break up and fall, they are eaten on the ground. Occasional individuals have been seen in small flowering *Grevillea* trees in July, eating the flowers, which are rich in nectar. Although many of these food plants are patchily distributed, most have prolonged fruiting periods and provide a relatively stable food resource. This may largely explain its patchy distribution: shortage of refuge hollows is unlikely to be a factor limiting populations in northern tropical forests.

Data collected over three years from a population in Kakadu National Park showed that breeding occurs throughout the year, peaking in the late-dry season (August and September) and declining somewhat during the mid-wet season (January–March). Females are capable of con-

W.G. BREED

# Golden-backed Tree-rat

## *Mesembriomys macrurus*

### (Peters, 1876)

make-rue'-rus: 'long-tailed southern-mouse'

**Recent Synonyms** *Mesembriomys hirsutus.*
**Other Common Names** Long-haired Rabbit Rat, Shaggy Rabbit-rat.
**Status** Fairly common in Arnhem Land, including Cobourg Peninsula and on Melville Island. Apparently rare elsewhere.
**Subspecies** *Mesembriomys gouldii gouldii*, mainland Northern Territory.
*Mesembriomys gouldii melvillensis*, Melville and Bathurst Islands.
*Mesembriomys gouldii rattoides*, northern Queensland.
**References**
Crichton, E.G. (1969). Reproduction in the pseudomyine rodent *Mesembriomys gouldii* (Gray) (Muridae). *Aust. J. Zool.* 17: 785–797.
Finlayson, H.H. (1961). A re-examination of *Mesembriomys hirsutus* Gould, 1842 (Muridae). *Trans. Roy. Soc. S. Aust.* 84: 149–162.
Friend, G.R. (1987). Population ecology of *Mesembriomys gouldii* (Rodentia: Muridae) in the wet-dry tropics of the Northern Territory. *Aust. Wildl. Res.* 14: 293–303.

ceiving about every nine months and influxes of young animals follow breeding pulses that occur at intervals of roughly two to three months. The oestrous cycle (averaging 26 days) and gestation period (43–44 days) are long for such a murid, but the rapid development and early weaning (about four weeks) of the young (usually one to three per litter) allow populations to increase relatively rapidly. Newborn young are precocious (averaging 35 grams in weight), sparsely haired and with erupted upper incisors. Growth is rapid: the eyes open at 11 days and a weight of about 400 grams is attained by 40 days. Young cling to the mother's teats and are dragged along by, or run after, her while still attached. Adult status is reached at about 80 days, females then weighing about 580 grams and males about 650 grams. Females outnumber males throughout most of the year and available data suggest that most animals lead a relatively solitary existence.

G.R. FRIEND AND J.H. CALABY

The type specimen of this tropical rodent was collected in 1875 on the shore of Mermaid Strait near Roebourne, Western Australia. This locality, the wettest part of the relatively arid Pilbara coastline, has a mean annual rainfall of 320 millimetres. All other specimens are from areas in the Kimberley and Top End of the Northern Territory that have a mean annual rainfall of more than 600 millimetres.

Knut Dahl, who acquired specimens near Broome, Western Australia, in 1895, wrote: 'According to the evidence of the natives ... it frequents the hollow trees of the *Eucalyptus* shrubs [eucalypt–acacia woodlands on red sandy plains]'. Although this south-western Kimberley population now seems to be extinct, the Golden-backed Tree-rat still occurs in a variety of near-coastal north-western Kimberley habitats:

*The Golden-backed Tree-rat inhabits a wide variety of environments across northern Australia.*

rainforest patches on volcanic, lateritic, sandstone and floodplain surfaces; eucalypt-dominated woodlands over tussock or hummock grasslands on volcanic hill country, lateritic uplands (with *Livistona* palms), blacksoil plains (with *Pandanus* trees), rugged sandstone screes; and coastal beaches adjacent to the above communities, in one instance adjacent to a mangrove swamp.

The last specimen from the Northern Territory was collected among *Pandanus* along a watercourse in sandstone country at Deaf Adder Creek in 1969. Subsequent attempts to recapture the species at this site have failed.

It is nocturnal, but individuals have been seen foraging along the tide-line of an open beach soon after sunrise. It is arboreal, but spends a lot of time on the ground and can quickly move long distances, often through dense grass, scampering along with the tail held high. Radio-tracking of two pairs during the dry season on Mitchell Plateau, Western Australia, revealed that the home range may be up to 600 metres in length. A home range is occupied by one pair of adults and, probably, some immatures. The colony studied at Mitchell Plateau occupied the edge of rainforest

### Mesembriomys macrurus

**Size**
*Head and Body Length*
188–245 (220) mm
*Tail Length*
291–360 (315) mm
*Weight*
207–330 (267) g

**Identification** Grey above and on sides; creamy white below. Basal third of tail grey, remainder white, terminal third with white brush. Distinctive mid-dorsal stripe of orange-brown fur distinguishes it from all other tree-rats.
**Recent Synonyms** None.
**Other Common Names** Golden-backed Rabbit-rat, Western Rabbit-rat.
**Status** North-western Kimberley: sparse in suitable habitat, habitat abundant; possibly underestimated. South-western Kimberley and Pilbara: presumed extinct. Northern Territory: only three records, most recent in 1968.
**Subspecies** None.

patches, feeding in trees in both the rainforest and the adjacent eucalypt woodland.

Flowers, fruits and termites were the most important dry-season foods of the Mitchell Plateau population, but some grasses, leaves, ants and beetles were eaten occasionally. Most of its preferred flowers and fruits are from species associated with rainforest edges, termites being more common in the adjacent eucalypt woodlands. Near Broome, in 1895, it was a resident pest in the roofs of buildings, raiding storerooms for rice and flour.

In the wild, pregnant females and immatures have been found in August and October. In October 1976, a female and a partly furred juvenile were taken from a nest in the foliage of a *Pandanus* tree. The nest was loosely woven from strips of *Pandanus* leaf. In captivity, it breeds throughout the year and it probably does so in the

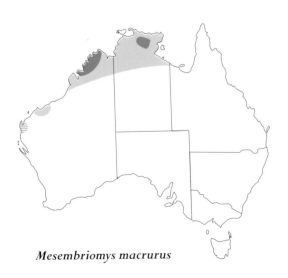

*Mesembriomys macrurus*

**References**

Dahl, K. (1897). Biological Notes on North Australian Mammalia. *Zoologist* Ser. 4, 1: 189–216.

Kitchener, D.J., L.E. Keller, A. Chapman, N.L. McKenzie, A.N. Start and K.F. Keneally (1981). Observations on mammals of the Mitchell Plateau Area (in *Biological Survey of the Mitchell Plateau and Admiralty Gulf, Kimberley, Western Australia*. Western Australian Museum, Perth.

Morton, C.M. (1991). *Diets of three species of Tree-rat,* Mesembriomys gouldii *(Gray),* M. macrurus *(Peters) and* Conilurus penicillatus *(Gould) from Mitchell Plateau, Western Australia.* Honours Thesis, University of Canberra.

wild. Females have four abdominal teats and usually bear two young, but litters of one to three have been recorded. The gestation period is about 47 days and there is post-partum mating. The young are weaned at about six or seven weeks and are fully grown at four months. One female first bred in captivity when she was ten months old.

The species has declined in the drier parts of its range, all records since 1903 having come from remote high rainfall areas dominated by country sufficiently rugged to prevent intensive pastoral usage. Until 1970, only nine localities had been recorded and less than 25 specimens were available; the Golden-backed Tree-rat is now known to be reasonably common in coastal areas of the remote north-western Kimberley.

Secure populations persist in the vast and inaccessible Prince Regent Nature Reserve of Western Australia and on uninhabited coastal islands of the Bonaparte and Buccaneer Archipelagos (Carlia, Unwins, Wollaston and Hidden). Other populations near Mitchell Plateau are in a proposed national park.

N.L. MCKENZIE AND J.A. KERLE

# Spinifex Hopping-mouse

## *Notomys alexis*

Thomas, 1922

noh'-toh-mis ah-lex'-is:
'Alexandria Downs southern mouse'

The Spinifex Hopping-mouse occurs throughout much of the arid zone of central and western Australia and is known as the Dargawarra to the Aboriginal people of Central Australia. Finlayson reported, in 1940, that it was chiefly found on loamy mulga flats, but that 'in January 1933 a few miles east of Mount Conner a considerable area of *Triodia* was crossed which had made luxuriant growth after a local rain … around the base of nearly every clump was strewn a mat of severed stalks from which the seed had been removed and the sand was reticulated with *Notomys* tracks'. Observations over the last 20 years have shown that the Spinifex Hopping-mouse is, in fact, characteristic of the spinifex-covered sandflats and stabilised sandhills

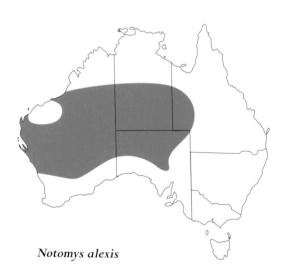

*Notomys alexis*

K. ATKINSON

*The Spinifex Hopping-mouse lives mainly close to spinifex hummocks in the arid interior.*

of this area and much of central Australia. Its numbers fluctuate greatly, low population density for many years being often followed after heavy rains with population explosions, as in south-western Northern Territory in 1974–75 and 1988–89. It is probably restricted largely to sandy areas during dry times, but extends into other habitats as populations increase after good rains.

Individuals avoid the heat of the desert day by sheltering in deep, humid burrows which may be over a metre beneath the surface of the soil. A typical burrow system consists of a large, horizontal nest chamber, lined with small twigs,

---

### *Notomys alexis*

**Size**
*Head and Body Length*
95–112 (102) mm
*Tail Length*
131–150 (137) mm
*Weight*
27–45 (35) g

**Identification** Uniform light brown above; grey-white below. Both sexes with small throat-pouch, weakly ridged behind and with central area of bare skin. Externally very similar to *Notomys aquilo*.
**Recent Synonyms** None.
**Other Common Names** Brown Hopping Mouse, Northern Hopping Mouse. Indigenous name: Dargawarra.

**Status** Common, widespread.

**Subspecies** *Notomys alexis alexis*, Western Australia, Northern Territory.

*Notomys alexis reginae*, Queensland.

*Notomys alexis everardensis*, north-western South Australia.

**References**

Breed, W.G. (1990). Copulatory behaviour and coagulum formation in the female reproductive tract of the Australian hopping mouse, *Notomys alexis*. *J. Reprod. Fertil.* 88: 17–24.

Breed, W.G. and M. Adams (1992). Breeding systems of spinifex hopping mice (*Notomys alexis*) and plains rats (*Pseudomys australis*): a test for multiple paternity within the laboratory. *Aust. J. Zool.* 40: 13–20.

Breed, W.G. (1992). Reproduction of the Spinifex hopping mouse (*Notomys alexis*) in the natural environment. *Aust. J. Zool.* 40: 57–71.

leaves and other plant material, from which several vertical shafts ascend to the surface. At its entrance, a shaft is circular and has an opening characterised by a lack of loose sand around the aperture.

Animals emerge at dusk and, when travelling at slow speed over the surface of the sand, move on all fours. They are most commonly seen bounding across open ground on their long hindfeet with the body projected forward and the long, brush-tipped tail trailing behind.

The diet appears to be quite variable. Finlayson states that 'at permanent camps in the Everard Hills where vegetables were grown near the soaks it became for a time in 1932 a nuisance owing to its depredations on the young shoots of cucumbers and beets' and that 'according to the blacks the large round woody seeds of the quondong (*Eucarya acuminata*) ... are also eaten ... the seed case is neatly drilled on one side only with a small hole and the con-

tents extracted'. Clearly, the Spinifex Hopping-mouse is omnivorous, eating a variety of seeds, roots, shoots and invertebrates in response to their relative abundance and availability.

Reproduction can take place at any time of year but appears to be most frequent in spring. The usual litter is three to four but up to six young can be reared on the four teats. Young are left in the nest while a female is foraging and both males and females retrieve young if they wander from the nest. Pregnancy is usually 32 days but this may be significantly prolonged since suckling extends the period during which unattached embryos remain in the uterus. Sexual maturity is usually attained at about 60 days.

Hopping-mice are unusual (perhaps unique among rodents) in having minute testes, about one-tenth of the average size of those of other mammals of equivalent body mass. Relatively few, highly variable sperm are produced and all accessory sex glands other than the ventral prostates are extremely small, so that no hard vaginal plug is formed after mating (as in most other murids). The penis is also unusual in having a thin shaft with large spines and the vagina has a narrow lumen and thick muscular coat. These morphological features relate to locking during copulation, during which time violent struggling between the pair may take place, the female sometimes biting the male. A mating pair may eventually turn back-to-back before disengaging.

The social organisation of the Spinifex Hopping-mouse in the natural environment is not known. Early observations indicated that animals might travel in groups, but recent findings indicate solitary foraging; group-foraging may be a response to unusual abundance of food. The reproductive anatomy of the male is indicative of monogamy but, in captivity, several individuals of both sexes cohabit and multiple insemination of a female can occur.

W.G. BREED

# Short-tailed Hopping-mouse

## *Notomys amplus*

### Brazenor, 1936

am'-plus: 'large southern-mouse'

This little-known hopping-mouse has a relatively short tail and ears that are almost as long as the head. Only two specimens have been collected, both by P.M. Byrne, postmaster at Charlotte Waters, Northern Territory. These were sent to Professor W.B. Spencer at the University of Melbourne in June 1896. One specimen (the type of the species) is a skin and skull, the other is a whole animal preserved in alcohol. Both are females; no breeding information is available.

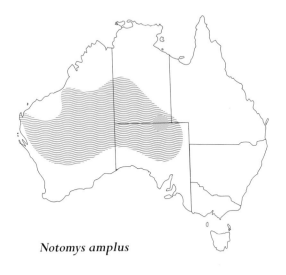

*Notomys amplus*

*Notomys amplus*

**Size**
*Head and Body Length*
143 mm
*Tail Length*
153 mm
*Weight*
*circa* 100 g

**Identification** Mid-brown above; grading to yellow-white below. Tip of tail white. Distinguished from other members of the genus by large size, relatively short tail with white tip. Gular area well marked.
**Recent Synonyms** None.
**Other Common Names** Brazenor's Hopping-mouse.
**Status** Extinct.
**Subspecies** None.
**Reference**
Brazenor, C.W. (1936). Two new rats from Central Australia. *Mem. Natn. Mus. Vic.* 9: 5–8.

The country around Charlotte Waters is mainly open gibber plain, with desert grasses and low shrubs. The sand ridge country of the Simpson Desert extends close to Charlotte Waters and the Short-tailed Hopping-mouse could have lived in either habitat.

Remains of the species have been found in fairly recent owl pellet deposits in the northern Flinders Ranges, South Australia, extending its range into regions that are mainly open plains.

J.M. DIXON

# Northern Hopping-mouse

## *Notomys aquilo*

### Thomas, 1921

ah-kwil'-oh: 'eagle southern-mouse'
(significance unknown)

The only hopping-mouse to inhabit northern Australia, this species was very poorly known until recently. Its distribution, as presently understood, is limited to three coastal or near-coastal localities around the Gulf of Carpentaria.

The type specimen was collected before 1867, somewhere on Cape York. There seems to be no reason to doubt the accuracy of the record but the species has never been recorded from Cape York since. It is widespread in coastal Arnhem Land and plentiful on Groote Eylandt, a large, partly sandy island in the western part of the Gulf of Carpentaria.

Donald Thompson secured specimens from the south-western corner of Groote Eylandt in 1942 but his extensive collections and field notes

*Notomys aquilo*

**Size**
*Head and Body Length*
91–112 mm
*Tail Length*
158–173 mm
*Weight*
35–44 g

**Identification** A rat-sized hopping-mouse; sandy brown in colour; with a relatively long tail and hindfeet. Both sexes have a small throat-pouch.
**Recent Synonyms** *Notomys carpentarius* Johnson, 1959.
**Other Common Names** None.
**Status** Vulnerable.
**Subspecies** None.
**References**
Dickson, J.M. and L. Huxley (1985). *Donald Thompson's Mammals and Fishes of Northern Australia.* Thomas Nelson, Melbourne.
Johnson, D.H. (1964). Mammals of the Arnhem Land Expedition (in) R.L. Specht (ed.) *Records of the American-Australian Scientific Expedition to Arnhem Land, 4, Zoology.* Melbourne University Press, Melbourne. pp. 427–515.

*Notomys aquilo*

remained unpublished until 1985. Thompson had a captive colony for around 12 months in northern Australia and maintained it for a further three years at Eltham, Victoria. He found that it tamed easily and ate a wide variety of seeds and green vegetation, but refused virtually all insect food. It is easily maintained on a seed-mix designed for small parrots, with some greens such as grass, beans or parsley. It breeds readily in captivity but mortality of the young is high, often through cannibalism.

It is extremely social and Thompson found that any older animal would groom and care for

P. GERMAN

*The Northern Hopping-mouse is restricted to the northern part of the continent.*

young. A twittering sound is made just before emerging in the evening (occasionally at other times) and a high-pitched squeal when alarmed. Both Thompson and I have excavated burrows containing three individuals; Thompson found burrows containing only females and their young. The burrow arrangement is distinctive. There is a spoil mound, usually 30–40 centimetres in diameter, quite free of any signs of a burrow. The entrance, up to 2 metres from the centre of the spoil heap, is a vertical shaft, up to 5 centimetres in diameter, with no sign of spoil around it. Between this shaft and the spoil heap there may be other, smaller pop-holes or depressions in the sand, leading to a shaft that terminates just below the surface. Active burrows are easily detectable by the trackways present around the pop-holes. The vertical shafts are dug from below, the displaced sand being used to back-fill the initial part of the burrow leading from the large spoil heap. The nest is rudimentary, usually no more than a few scraps of vegetation in an enlarged part of the burrow, but elaborate nests are made in captivity. Johnson records very extensive burrow systems, nearly 3 metres long and more than 1 metre deep, at Umbakumba in the north-east of Groote Eylandt, but the burrows that I examined in the south-western part of the island were much smaller, usually 1–2 metres long and 30 centimetres deep. When I first located one such burrow, two individuals burst from the sand, emerging from shafts dug to just below the surface and scattered into vine thicket. This burrow was made in a coastal foredune developed over base-rock and backed by dense vine thicket that extended to within 5 metres of the high-tide mark. It seemed to be typical of those in the region.

The Northern Hopping-mouse is extremely reluctant to enter traps: the best indication of its presence are the tracks formed by the entire length of the feet, impressing the substrate as the animal hops. The distance between footprints varies, but is usually 20–60 centimetres. During a survey of Groote Eylandt in May 1991, tracks were found to be widespread wherever sandy substrate occurred, but places where vine thicket backed coastal dunes seemed to be particularly favoured.

Gestation is approximately seven weeks. Females have four teats and one to five young are reared at one time. Mortality of the young appears to be high. Thompson records eight births for his captive colony. Five occurred between mid-March and mid-May (the late-wet season on Groote Eylandt), others in mid-February, late June and early September. Johnson collected two females near Umbakumba in June, pregnant with four and five foetuses respectively. The young are born virtually hairless and helpless but do not cling persistently to the mother's teats as do many Australian murids: the female carries them in her mouth when moving them. The eyes open 21–22 days after birth.

T.F. FLANNERY

# Fawn Hopping-mouse

## *Notomys cervinus*

### (Gould, 1853)

ser-veen'-us: 'tawny southern-mouse'

D. WHITFORD

*The Fawn Hopping-mouse manages to dig a burrow in the rock-hard substrate of gibber plains.*

An inhabitant of the vast gibber plains of the Lake Eyre basin and the channel country of south-western Queensland, the Fawn Hopping-mouse lives during the day in burrows up to 1 metre deep. The form of these burrows is unknown since it is extremely difficult for investigators to dig more than a few centimetres into the hard gibber soil, but the surface evidence suggests a burrow system similar to, but less extensive than, those of other hopping-mice. Presumably these are constructed when the clay soil is moist after heavy rain.

It appears to live in small family groups of two to four individuals, foraging outwards from the burrow at night for hundreds of metres. These small groups are widely, but thinly, spread over the habitat. In areas where the expanse of gibber is broken up by sandhills, the Fawn Hopping-mouse seldom ventures onto the sand, which is the domain of the Dusky Hopping-mouse.

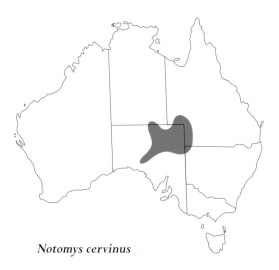

*Notomys cervinus*

On the chest of the male is a distinct, raised, flat, naked gland a few millimetres in diameter. In sexually active males the surface of the gland is often covered with a pale yellow, flaky secretion, the purpose of which is a mystery. Animals have never been observed to use the gland, although it must in some way be associated with scent-marking. Females lack a glandular area except for a small patch of enlarged sebaceous glands on the chest which may develop during pregnancy and lactation.

The Fawn Hopping-mouse feeds primarily on seeds but also eats quite a high proportion of green plant material and insects when these are readily available. As befits an occupant of the driest area of Australia, it can live without drinking fresh water, obtaining water only from green plants and the bodies of insects. It will accept water twice as salty as seawater, managed by reducing the amount of water lost in urine and faeces to a minimum, excreting excess salt,

C. ANDREW HENLEY/LARUS

*Right: The Dusky Hopping-mouse is one of the rarest of the Australian rodents.*

manufacturing water within the body from the breakdown of carbohydrates in the food and by avoiding high temperatures by remaining in a cool, moist burrow during the day.

The small amount of information on reproduction in the wild suggests that the Fawn Hopping-mouse is opportunistic, breeding whenever conditions are good. In captivity, breeding occurs throughout the year. Litters of one to five are born after a gestation period of 38–43 days. They are well-furred at birth, their eyes open at the age of about three weeks and they can be weaned when about four weeks old.

C.H.S. WATTS

### *Notomys cervinus*

**Size**
*Head and Body Length*
95–120 (110) mm
*Tail Length*
105–160 (150) mm
*Weight*
30–50 (35) g

**Identification** Female lacks well-marked glandular area on throat or chest; male has discrete raised glandular area of naked skin between forelegs.
**Recent Synonyms** *Podanomalus aistoni*, *Notomys aistoni*.
**Other Common Names** Oorarrie.
**Status** Common.
**Subspecies** None.
**Reference**
Finlayson, H.M. (1939). On mammals from the Lake Eyre Basin, Part IV. The Monodelphia. *Trans. Roy. Soc. S. Aust.* 63: 354–364 (as *Notomys* aistoni).

# Dusky Hopping-mouse
## *Notomys fuscus*
### (Jones, 1925)

fus'-kus: 'dusky southern-mouse'

With its large ears, prominent dark eyes and rich fur, the Dusky Hopping-mouse is perhaps the most beautiful of Australian rodents, but unfortunately one of the rarest and least known. It is typically a warm, light orange above and pure white below but other colour varieties exist, ranging from light fawn to a russet orange. Similar colours occur in the Fawn Hopping-mouse and the two species have often been confused.

Both live in the same area of inland Australia, stretching from Ooldea in South Australia past Lake Eyre to south-western Queensland, the

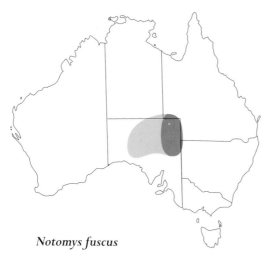

*Notomys fuscus*

Fawn Hopping-mouse living on the gibber plains while the Dusky Hopping-mouse occupies the large sand-dunes. On the flat top of a dune the Dusky Hopping-mouse digs a burrow with a single drive about 10 centimetres in diameter, more than a metre below the surface and up to 5 metres long, connected to the surface by up to six vertical entrance shafts about 2–3 centimetres in diameter. The shafts are dug from below and may end up anywhere on the surface. A nest, consisting of a pad of finely chewed vegetation, is centrally placed in a small alcove off the main shaft and well-marked pads connect different burrows along the top of the dune. The width of the entrance shafts is critical, because the animals climb them by hopping—bracing their backs against the sides with their forefeet between hops. Progress when coming down a shaft is a headlong dive, with the back braced against the wall of the shaft and the forelegs used as a brake. If the entering animal finds that the burrow is blocked with loose sand it will often squeak several times. This usually attracts other occupants and together, from opposite sides, they rapidly clear the blockage.

Living in groups of about five individuals in one or two adjacent burrow systems, the Dusky Hopping-mouse spends all its life on the dunes. During the day, it rests in a burrow, insulated from the hot surface. At night, it forages on the top and sides of a dune, seldom venturing onto the surrounding gibber a few metres away. Even in apparently favourable localities, population

## Notomys fuscus

**Size**
*Head and Body Length*
80–115 (105) mm
*Tail Length*
115–155 (140) mm
*Weight*
30–50 (35) g

**Identification** Orange to grey above; white below. Both sexes have well-developed throat-pouch with distinct fleshy margins covered with stiff white hairs.
**Recent Synonyms** *Ascopharynx fuscus*, *Notomys fuscus eyrius*, *Notomys filmeri*.
**Other Common Names** Birdsville Hopping-mouse, Wood Jones' Hopping-mouse, Willkintie.
**Status** Rare, limited.
**Subspecies** None.
**Reference**
Watts, C.H.S. and H.J. Aslin (1981). *The Rodents of Australia*. Angus and Robertson, Sydney.

density is low, with individuals congregated in a loose colony in a small area of a sandhill, adjacent sandhills being seemingly devoid of mice. This may not always have been so, since this is the species referred to by the explorer, Sturt, when he wrote of Aboriginal hunters returning to camp with 'bags full of jerboas which they had captured on the hills. They could not indeed have had less than 150 to 200 of these beautiful little animals so numerous are they on the sandhills'.

Like other hopping-mice, it does not need to drink, but obtains all its water and food requirements from a diet of seeds, green plants (when available) and occasional insects.

Little is known of its breeding biology in the wild. In captivity, it breeds throughout the year. The gestation period is 38–41 days and litters of one to five are reared. Eyes open at 20 days and weaning is at about 30 days.

C.H.S. WATTS

# Long-tailed Hopping Mouse

## *Notomys longicaudatus*

### (Gould, 1844)

lon'-jee-kaw-dah'-tus:
'long-tailed southern-mouse'

*Long-tailed Hopping-mouse. Illustration in Gould's*
The Mammals of Australia.

Soon after the Horn Expedition of 1894 to Central Australia, specimens were obtained from Burt Plain, Northern Territory and, in 1901–2, W.B. Spencer and F.J. Gillen obtained them from Barrow Creek, Northern Territory. The Spencer material is held in the Museum of Victoria, Melbourne.

These specimens are the only authenticated records of live material. A fragment of a small

The first specimens of the Long-tailed Hopping-mouse were collected in the vicinity of the Moore River, Western Australia, in about 1843 by John Gilbert, collector for John Gould, who was in the area at the time. The location of the specimens, immediately north of the Victoria Plains, was near where New Norcia stands today. Two more specimens, from the same area have been traced to the Leiden Museum; another collected further south, from Toodyay, is in the Liverpool Museum. During his explorations in the interior, Captain Charles Sturt collected the Long-tailed Hopping-mouse in the Coonbaralba Range near Broken Hill, New South Wales.

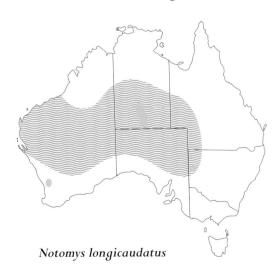

*Notomys longicaudatus*

### *Notomys longicaudatus*

**Size**

*Head and Body Length*
102–164 mm
*Tail Length*
132–204 mm
*Weight*
*circa* 100 g

**Identification** Tawny brown above; grading to grey-white below. Tail brown above, white below, terminal brush black. Males have large neck-gland. Distinguished from other hopping-mice by large size and extremely long tail.
**Recent Synonyms** None.
**Other Common Names** None.
**Status** Unknown, possibly extinct.
**Subspecies** None.
**References**
Gould, J. (1863). *The Mammals of Australia Vol. 3*. The author, London. Reissued (1976) as Placental Mammals of Australia, with modern commentaries by J.M. Dixon, Macmillan, Melbourne.
Brazenor, C.W. (1934). A revision of the Australian jerboa-mice. *Mem. Nat. Mus. Vic.* 8: 74–89.

rodent skull found in recent owl pellets from a cave 96 kilometres south-west of the Granites, Northern Territory, in 1977, may be of this species. The Long-tailed Hopping-mouse was recorded by Baynes in 1982 in archaeological material at Wilgie Mia, an Aboriginal ochre mine in the upper Murchison region. Recent fossils have been located also in the Flinders Ranges, South Australia. In his *Mammals of Australia*, Gould quotes the comments of Gilbert: 'This species differs considerably in its habits from the *Djyr-dow-in* [Mitchell's Hopping Mouse], for while that animal burrows in sandy districts, the favourite haunt of the present species is a stiff and clayey soil'. Gilbert noted that, in common with the latter species, it was extremely fond of raisins.

Once widespread throughout arid and semi-arid country where the vegetation included acacia and eucalypt woodlands, hummock grassland and low shrubland, the Long-tailed Hopping-mouse is now unknown and possibly extinct. Little is known of its biology but Brazenor described an oval gland with raised margins in the throat region of males.

J.M. DIXON

# Big-eared Hopping-mouse
## *Notomys macrotis*
### Thomas, 1921

mak-roh'-tis: 'long-eared southern-mouse'

This species is known from two damaged specimens held in the Natural History Museum, London: one from the Moore River, Western Australia; the other from an unknown locality. The former, designated as the type specimen, was purchased from John Gould by the Museum and its label bears the information 'J. Gilbert, 19th July, 1843'. This is probably the collection date,

*Notomys macrotis*

## Notomys macrotis

**Size**

*Head and Body Length*
117–118 mm

*Tail Length*
129–136 mm

**Identification** Faded grey-brown above; dull white below. No neck glands in either sex. A shallow groove extends along the upper surface of the upper incisor. Similar to *Notomys cervinus* from which it differs in having a longer hindfoot and more robust skull.

**Recent Synonyms** None.

**Other Common Names** None.

**Status** Presumed extinct; possibly rare before European settlement.

**Subspecies** None.

**References**

Mahoney, J.A. (1975). *Notomys macrotis* Thomas (1921), a poorly known Australian Hopping mouse (Rodentia: Muridae). *Aust. Mammal.* 1: 367–374.

Thomas, O. (1921). Notes on the species of *Notomys*, the Australian jerboa-rats. *Ann. Mag. Nat. Hist.* Ser. 9, 8: 536–541.

as Gilbert was collecting for Gould in Western Australia during that year. The species was first described by Thomas in 1921, who gave the locality as 'Interior of Western Australia, on Moore's River'. Most of the land in the vicinity of the Moore River has long been cleared for agriculture, but it was originally coastal heathland, woodland and open forest.

The second specimen was also purchased from John Gould by the British Museum. Its date of collection is likely to be earlier than 1844.

The Big-eared Hopping-mouse and Fawn Hopping-mouse appear to be closely related but, in the absence of adequate biological data, they should be treated as separate species.

J.M. DIXON

# Mitchell's Hopping-mouse

## *Notomys mitchelli*

### (Ogilby, 1838)

mitch'-el-ee: 'Mitchell's southern-mouse'

Marginally the largest extant species of *Notomys*, Mitchell's Hopping-mouse is found throughout the mallee areas of southern Australia from the Western Australian coast to the Big Desert area of Victoria. Previously it was distributed further eastwards along the scrubs of the Murray River. It has the distinction of being the first of the Australian hopping-mice to be described by a European, having been collected by the explorer Mitchell somewhere near the junction of the Murray and Murrumbidgee Rivers in southern New South Wales in 1836. Scientists at first considered it to be a jerboa (family Dipodidae) and in view of its hopping gait and long, tufted tail, in which it closely resembles these African rodents, this was understandable. Further study soon showed the two groups to be quite distinct and the resemblance between them is now recognised to be one of the classic examples of convergent evolution.

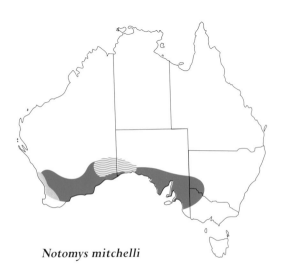

*Notomys mitchelli*

*The arid-adapted Mitchell's Hopping-mouse is the largest living member of its genus.*

D. WHITFORD

Like all Australian desert mice, Mitchell's Hopping-mouse is nocturnal, sheltering during the day in deep burrows. However, it is occasionally reported abroad in the daytime by farmers, particularly during scrub-clearing. It seems either that the clearing process scares individuals from their burrows or that surface nests are occasionally utilised, perhaps by dispersing individuals.

Its diet of seed and a few insects and green plants is similar to that of other hopping-mice.

---

### *Notomys mitchelli*

**Size**

*Head and Body Length*
100–125 (115) mm
*Tail Length*
140–155 (150) mm
*Weight*
40–60 (52) g

**Identification** Fawn to dark grey-brown above; often grizzled, greyish-white below. Tail brown or grey above, lighter below, with tuft of dark hairs at tip. Wide tract of shiny white hairs extends from throat to chest in both sexes.

**Recent Synonyms** None.

**Other Common Names** None.

**Status** Common.

**Subspecies** None.

**References**

Brazenor, C.W. (1957). Mitchell's Hopping-mouse. *Proc. Roy. Zool. Soc. NSW.* 1956–57: 19–22.

Watts, C.H.S. and H.J. Aslin (1981). *The Rodents of Australia*. Angus and Robertson, Sydney.

---

In times of drought, roots are eaten as a source of both food and water. Laboratory studies show that it is less able than other hopping-mice to exist without free water, which may be why it is restricted mainly to mallee scrub on the fringe of the desert.

Little is known of its social structure. When not breeding, animals live in mixed-sex groups of up to four individuals per burrow system, each group utilising several burrows which may be up to 150 metres apart. In the laboratory, it lives for up to five years but the only record of longevity in the field is in respect of one individual which lived for at least two years.

The litter size is one to five (usually three or four). Gestation is 38–40 days and young are weaned at around 35 days. There seems to be no definite breeding season.

C.H.S. WATTS

# Darling Downs Hopping-mouse

## *Notomys mordax*

Thomas, 1922

mor'-dax: 'biting southern-mouse'

The British Museum (Natural History) contains the skull of a hopping-mouse that lacks an associated skin or even a locality label, but the museum records strongly suggest that it was collected on the Darling Downs of south-eastern Queensland either by John Gilbert or Charles Coxen in the 1840s.

The skull is similar to that of Mitchell's Hopping-mouse, but is sufficiently different to suggest that it represents a distinct species. Apart

### *Notomys mordax*

**Size**
*Skull*
33 mm long, 17 mm wide
**Identification** Differs from *N. mitchelli* in having broader incisor teeth, larger molars and a supplementary cusp on the front of the first upper molar.
**Recent Synonyms** None.
**Other Common Names** None.
**Status** Extinct.
**Subspecies** None.
**Reference**
Mahoney, J.A. (1977). Skull characters and relationships of *Notomys mordax* (Rodentia: Muridae), a poorly known Queensland hopping-mouse. *Aust. J. Zool.* 25: 749–754.

from this specimen, no hopping-mice have ever been collected from the Darling Downs, so the chance of taking further specimens seems remote. It is possible that owls, which are very efficient collectors of small mammals, may have already undertaken the task and that a cache of Darling Downs Hopping-mice skulls awaits discovery in a cave or rock crevice near the type-locality. Without such material, no final decision on specific status of the specimen can be made.

C.H.S. WATTS

*Notomys mordax*

# Great Hopping-mouse

## *Notomys* sp.

This large hopping-mouse was never recorded as a living animal. Although it has been known from relatively recent and complete remains for many years, it has not yet been scientifically described. Its bones have been found in regurgitated pellets under owl roosts at Aroona Dam and Chamber's Gorge, northern Flinders Ranges, South Australia. Some pellets include the bones of the introduced House Mouse (*Mus musculus*), indicating that it survived into historic times. Indeed, Tunbridge (1991) suggests that it became extinct in the Flinders Ranges between 1850 and 1900. Its absence from nineteenth century mammal collections may be due to chance, other *Notomys* species, such as the

*Notomys* sp.

### *Notomys* sp.

**Size**
Unknown, but based on skull size, probably similar to, or larger than, the largest described *Notomys*.

**Identification** External appearance unknown. Skull heavy and robust with small bullae. As in *N. amplus* and *N. aquilo*, the suture between the frontal bones is straight (crenulate in other *Notomys* species).

**Recent Synonyms** None.

**Other Common Names** None.

**Status** Extinct.

**Subspecies** None.

**References**

Tunbridge, D. (1991). *The story of the Flinders Ranges Mammals*. Kangaroo Press, Kenthurst, NSW.

Watts, C.H.S. and Aslin, H.J. (1981). *The Rodents of Australia*. Angus and Robertson, Sydney.

Darling Downs, Short-tailed and Big-eared Hopping Mice being known from one or two specimens or even a single skull.

It occurs in deposits together with the Long-tailed and Short-tailed Hopping mouse but its remains are rarer. The habitat around Chamber's Gorge is mostly tussock grassland growing on clayey soils: this suggests that, like the Long-tailed and Short-tailed Hopping-mouse, it preferred areas with clayey soil to the sandy country occupied by many living species.

T.F. FLANNERY

BABS & BERT WELLS

# Ash-grey Mouse

## *Pseudomys albocinereus*

### (Gould, 1845)

sude'-oh-mis al'-boh-sin-er-ay'-us:
'white-grey false-mouse'

The Ash-grey Mouse inhabits low heath and shrubland vegetation on sandy soils of various depths in south-western Western Australia. On Bernier Island, it lives in inland heaths, whereas the rare Shark Bay Mouse, *P. fieldi*, occupies the coastal sandy fringes. There is little evidence to suggest that the Ash-grey Mouse has declined significantly in recent times. Some populations have been lost on the Swan coastal plain around Perth because of clearing for residential developments. However, in most coastal parts of its

*The Ash-grey Mouse inhabits low heath and shrubland.*

---

### *Pseudomys albocinereus*

**Size**

*Head and Body Length*
63–95 (79) mm (males)
63–85 (74) mm (females)

*Tail Length*
95–105 (100) mm (males)
85–97 (91) mm (females)

*Weight*
30–40 (35) g (males)
14–29 (26) g (females)

**Identification** Silver-grey above, tinged with distinct fawn; white below. Tail pale with some pigmentation near base. Paws pink. Not separable on external characters from *P. apodemoides*.
**Recent Synonyms** *Gyomys albocinereus*, *Pseudomys squalorum*.
**Other Common Names** Ashy-grey Mouse.

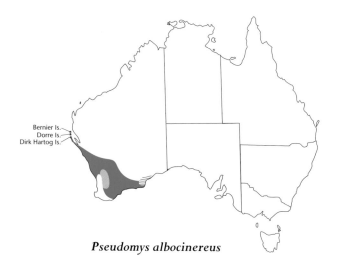

Bernier Is.
Dorre Is.
Dirk Hartog Is.

*Pseudomys albocinereus*

**Status** Sparse but widespread.
**Subspecies**
*Pseudomys albocinereus albocinereus*, mainland.
*Pseudomys albocinereus squalorum*, Bernier Island,
Western Australia.

**References**
Thomas, O. (1907). List of further collections
    of mammals from Western Australia, includ-
    ing a series from Bernier Island obtained for
    Mr W.E. Balston: with field notes by the
    collector, Mr G.C. Shortridge. *Proc. Zool. Soc.
    Lond.* 1906: 763–777.
Morris, K.D. and S.D. Bradshaw (1981). Water
    and sodium turnover in coastal and inland
    populations of the Ash-grey Mouse *Pseudomys
    albocinereus* (Gould) in Western Australia.
    *Aust. J. Zool.* 29: 519–532.
Ride, W.D.L. and C.H. Tyndale-Biscoe (1959).
    Mammals (in) A.J. Fraser (ed.) *The Results of
    an Expedition to Bernier and Dorre Islands, Shark
    Bay, Western Australia in July 1959.* WA
    Fisheries Dept. Fauna Bulletin No. 2.

mainland distribution, suitable habitat is relatively abundant in both National Parks and
Nature Reserves. In the wheatbelt, it is restricted to smaller reserves of uncleared vegetation
surrounded by cleared agricultural land. The
more eastern extensions of its distribution (beyond the 300 millimetre isohyet) occur in mixed
heaths and mallee shrublands on sandy soils with
variable gravel content, much of which will be
cleared in the future if the wheatbelt extends
eastward.

It is nocturnal and sociable, resting during
the day in a burrow that may be 60 centimetres
deep and extend horizontally for 3–4 metres with
a system of interconnecting tunnels and nesting
chambers. Some burrows on Bernier Island are
occupied by breeding pairs and their partly grown
young. In the laboratory, during the day, adults
sleep huddled together in shelter boxes.

Two forms of the Ash-grey Mouse have
been recognised. In 1907 Thomas described the
animals from Bernier Island as *P. albocinereus
squalorum* in recognition of their smaller size
compared to mainland animals, *P. a. albocinereus*.
In the eastern areas of its distribution, the Ash-
grey Mouse grows to 60 grams, has darker fur
and pigmented tail and is often confused with
the rare Western Mouse *P. occidentalis*.

Throughout its range, the Ash-grey Mouse
survives for four to seven months each year with

little free water. It is predominantly herbivorous
during winter but includes arthropods in the
diet in summer when the water content of the
vegetation is insufficient for its needs.

At least in the western part of the range,
breeding occurs in spring. Females produce two
to six young after a gestation of 37–38 days.
Young grow throughout the summer and following winter, attaining maximum body weight in
August; reproductive activity then commences.
In the eastern parts of the range, breeding is
more opportunistic. Populations appear to remain relatively constant with densities of two to
three animals per hectare at Cockleshell Gully,
near Jurien, Western Australia. It appears that
the entire population is replaced annually.

Predators of the Ash-grey Mouse include the
poisonous Gwardar and Dugite snakes, Stimson's
Python, Barn Owl, Cat and Fox.

K.D. MORRIS

I.R. MᶜCANN

# Silky Mouse

## *Pseudomys apodemoides*

### Finlayson, 1932

ah'-poh-dem-oy'-dayz: '*Apodemus*-like false-mouse'
(*Apodemus* is a genus of European fieldmice)

The presence of the Silky Mouse can readily be recognised by the large spoil-heaps it creates during the construction of its extensive burrows. The nest chamber, up to 3 metres below the surface, has lateral tunnels radiating from it to vertical shafts which lead to pop-holes at the surface. Pop-holes can be located by carefully searching the area between 1 and 4 metres from the spoil-heap. Burrows are often placed at the base of a shrub of the Desert Banksia, the nectar of which is a major source of food during winter and the leaves and roots of which interact to create local areas of moist soil.

Although banksia nectar is important during winter and the Silky Mouse may eat cockroaches,

*The Silky Mouse eats seeds, supplemented by nectar in winter and cockroaches in summer.*

which irrupt from time to time in this habitat, it feeds primarily on seeds throughout the year. It is patchily common throughout the dry mallee-heathlands of north-western Victoria and eastern South Australia wherever the low shrub communities contain a sufficiently diverse array of plants to produce seeds all through the year—conditions that are normally produced two or three

### *Pseudomys apodemoides*

**Size**

*Head and Body Length*
65–80 (75) mm
*Tail Length*
90–110 (105) mm
*Weight*
16–22 (20) g

**Identification** Silver-grey above, flecked with light brown; white below. Tail white-haired on pink skin, often with 10–15 grey-brown bands. Eyes bulging; ears much larger than in House Mouse.

**Recent Synonyms** *Gyomys apodemoides*, *Pseudomys glaucus*. (Has recently been confused with *P. albocinereus* from Western Australia.)

**Other Common Names** Silky-grey Mouse, Silky-grey Southern Mouse, Finlayson's Mouse.

**Status** Common, limited.

**Subspecies** None.

**References**

Cockburn, A. (1981). Diet and habitat preference of the silky desert mouse, *Pseudomys apodemoides* (Rodentia). *Aust. Wildl. Res.* 8: 475–497.

Cockburn, A. (1981). Population processes of the silky desert mouse, *Pseudomys apodemoides* (Rodentia), in mature heathlands. *Aust. Wildl. Res.* 8: 499–514.

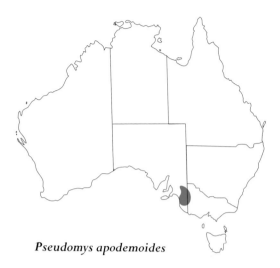

*Pseudomys apodemoides*

# Plains Rat

## *Pseudomys australis*

### Gray, 1832

ost-rah'-lis: 'southern false-mouse'

Populations of the Plains Rat were once known from the western edge of the Nullarbor Plain to the inland slopes of the Great Dividing Range and from central Queensland south to the mouth of the Murray River, but the species is now restricted to the barren gibber plains of the Lake Eyre basin.

It constructs shallow, complex burrow systems dug into the rock-hard gibber or the softer

years after a fire. Invasion and saturation of such a habitat are rapid and, provided that conditions remain favourable, successive litters of two to five young (commonly four) may be produced throughout the year. As the vegetation matures and plant productivity declines, breeding becomes seasonal but, for reasons that are not fully understood, some populations breed in winter while others do so in late spring and summer.

Adult animals share burrows and there does not appear to be a fixed social system. Several litters at various stages of development may be found when a burrow is excavated. In some areas the young stay with the adult group until just before the breeding season, when they leave the nest and try to locate an unoccupied site for breeding. Throughout the year, however, there is some turnover of individuals particularly in marginal habitat. The flexible social and breeding system appears to have developed in response to the uncertainty of an environment where widespread fires or drought may reduce survival of offspring to very low levels. Despite a precarious balance, the Silky Mouse seems to be successful in its environment and under no conceivable threat unless humans encroach further into the mallee-heathlands, the soil of which is too poor to offer much encouragement to farmers.

A. COCKBURN

---

### *Pseudomys australis*

**Size**

*Head and Body Length*
100–140 (135) mm
*Tail Length*
80–120 (110) mm
*Weight*
50–80 (65) g

**Identification**  Grey to grey-brown above; white or cream below. Tail shorter than or equal to head and body, becoming lighter towards tip. Ears relatively large.

**Recent Synonyms**  *Pseudomys auritus*, *Pseudomys minnie*.

**Other Common Names**  Plains Mouse, Eastern Mouse, Eastern Rat.

**Status**  Rare.

**Subspecies**  None.

**Reference**

Smith, J.R., C.H.S. Watts and E.G. Crichton (1971). Reproduction in the Australian desert rodents *Notomys alexis* and *Pseudomys australis* (Muridae). *Aust. Mammal.* 1: 1–17.

*The Plains Rat inhabits gibber plains in the Lake Eyre basin.*

D. WHITFORD

soil built up around the bases of the stunted bushes. Burrow systems are grouped about 10 metres apart to form colonies, interconnected by surface runways, which often spread over vast areas. One stretched south from Charlotte Waters for over 40 kilometres and covered a minimum area of 50 square kilometres. When animals are breeding, a burrow system seldom contains more than one adult male and two or three females; when not breeding, groups as large as 20 may form.

Although the density of animals in a colony is not great, the total number must be considerable, yet a most striking aspect of such an aggregation is its impermanence: a large colony can disappear completely in as little as three months, even when environmental conditions appear to be favourable. It is known that predators, mainly the Dingo, Fox and owls, congregate near such colonies and must exert an immense toll on these small rodents, but is unlikely that predation alone could cause such rapid changes in numbers and distribution. A combination of predators, weather conditions and characteristics of the animals themselves, such as increased stress from fighting, are thought to be involved, but it is not known how these interact.

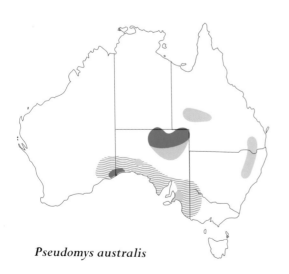

***Pseudomys australis***

With its soft, silvery fur, the Plains Rat is among the loveliest of Australian rodents, but in captivity the sheen is easily lost and the fur can become greasy and unattractive. Few studies of the diet have been made, but the little that is known suggests it is predominantly a seed-eater, supplementing this food with vegetable matter and insects.

As befits an inhabitant of the driest regions of Australia, it is well-adapted to desert life, living in cool burrows during the heat of the day and foraging abroad only at night. Water is obtained from the metabolism of starches in the food and, aided by the production of a highly concentrated urine and the absence of sweat glands, the Plains Rat is able to survive without drinking.

Under laboratory conditions, young are born throughout the year. The situation in the wild is not known but it is thought that breeding only occurs following heavy rain. Litter size is one to seven, usually three to four. Young are born after a gestation of 30–31 days and are ready to leave the nest some 28 days after birth.

C.H.S. WATTS

# Bolam's Mouse

## *Pseudomys bolami*

## Troughton, 1932

boh'-lah-mee: 'Bolam's false-mouse'

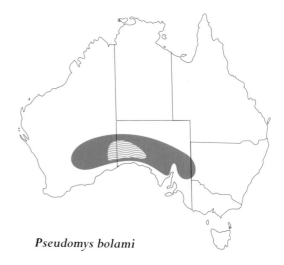

*Pseudomys bolami*

This small nocturnal rodent, which lives in the semi-arid areas of southern Western Australia and South Australia, was long confused with the

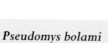

**Size**

*Head and Body Length*
57–77 mm
*Tail Length*
78–96 mm
*Weight*
10–21 g

**Identification**  Size of small House Mouse with large eyes, ears and feet. Approximately 70–100 hairs on each scale of tail. Very similar to *P. hermannsburgensis*.
**Recent Synonyms**  Sandy Inland Mouse.
*Pseudomys hermannsburgensis bolami*.
**Other Common Names**  None.
**Status**  Uncommon.
**Subspecies**  None.
**Reference**
Kitchener, D.J., M. Adams and P. Baverstock (1984). Redescription of *Pseudomys bolami* Troughton, 1932 (Rodentia: Muridae). *Aust. Mammal.* 7: 149–159.

Sandy Inland Mouse which shares its habitat, particularly in the Goldfields area of Western Australia. Very similar to the Sandy Inland Mouse, it has larger ears, longer hindfeet, generally longer tail and more heavily furred tail (about 80 hairs on each tail scale, 40 in the Sandy Inland Mouse). Despite close physical resemblance, these species differ by 20–30 per cent in those of their proteins that have been examined. At first glance, both species can also easily be confused with the introduced House Mouse but this species has smaller ears and eyes and a notch in the upper incisors.

Little is known of the habits of Bolam's Mouse other than that it is nocturnal and usually found in loamy to clay soils in sparse mallee or acacia woodland with scattered dwarf shrubs. Its diet is likely to be varied, including seeds, plant material and insects. In South Australia, young animals have been recorded only in the second half of the year, indicating that most breeding occurs in spring.

C.H.S. Watts

M. ARMSTRONG

The Kakadu Pebble-mound Mouse is one of three species that construct piles of stones on and around their nesting burrows.

# Kakadu Pebble-mound Mouse

## *Pseudomys calabyi*

### Kitchener and Humphreys, 1987

kal'-ah-bee-ee: 'Calaby's false-mouse'

The Kakadu Pebble-mound Mouse is one of several small mammals discovered recently in the Top End of the Northern Territory. It was first collected in 1973 but there were no additional records until 1988. In 1987 the original six specimens were described as a distinctive subspecies of the Kimberley Mouse, *Pseudomys laborifex*. The collection of more than 20 specimens over the period of 1988–90 has provided

### *Pseudomys calabyi*

**Size**
*Head and Body Length*
69–95 (84.4) mm (males)
68–82 (78.8) mm (females
*Tail Length*
72–94 (78.4) mm (males)
64–85 (74.0) mm (females)
*Weight*
15–31 (20.6) g (males)
12–23 (17.0) g (females)

**Identification** Greyish above, tinged pale orange on flanks; white below. Differs from *P. delicatulus* in larger size and a tail slightly shorter than head-body length.
**Recent Synonyms** *Pseudomys laborifex calabyi*.
**Other Common Names** Calaby's Mouse.
**Status** Uncertain, reasonably abundant in a very limited known range, all of which is now within National Park.
**Subspecies** None.

**References**

Kitchener, D.J. and W.F. Humphreys (1987). Description of a new subspecies of *Pseudomys* (Rodentia: Muridae) from Northern Territory. *Rec. West. Aust. Mus.* 13: 285–295.

Woinarski, J.C.Z. (1992). Habitat relationships for two poorly known mammal species *Pseudomys calabyi* and *Sminthopsis* sp. from the wet-dry tropics of the Northern Territory. *Aust. Mammal.* 15: 47–54.

sufficient material to confirm its specific status.

It occurs mostly on gravelly slopes carrying an open woodland with tall grass understorey. The substrate of gravel and small stones may be important in allowing these granivorous rodents access to fallen seeds during the long dry season. All records have been from Stage III of Kakadu National Park, in the foothills of the upper South Alligator River. Concern about the status of the Kakadu Pebble-mound Mouse ensured that this fairly nondescript rodent was a prominent participant in the lengthy deliberations about a large proposed mine at Coronation Hill, in this area.

Little else is known of its ecology and behaviour in the wild, but observations on captive individuals held at the Territory Wildlife Park have demonstrated that the species is a pebble-mound builder, surrounding and blocking its tunnel

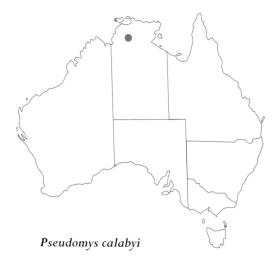

*Pseudomys calabyi*

entrance with large collections of small stones and pebbles. Captive individuals showed remarkable fecundity, an initial pair producing seven litters (total 22 young) over a nine-month period.

The Kakadu Pebble-mound Mouse is superficially quite similar to the more abundant and widespread Delicate Mouse and its limited known distribution may possibly reflect some past errors in field identifications of small *Pseudomys* species in this region.

J.C.Z. WOINARSKI, R.W. BRAITHWAITE
AND M. MAGUIRE

# Western Pebble-mound Mouse

## *Pseudomys chapmani*

### Kitchener, 1980

chap'-man-ee: 'Chapman's false-mouse'

Few Australian mammals make lasting structures, but piles of small stones mark the past or present residences of pebble-mound mice. Mounds of the Western Pebble-mound Mouse cover areas of 0.5 to 9.0 square metres. Sixteen hundred stones from 21 mounds in the Hamersley Range ranged in weight from 0.05 to about 10 grams (94 per cent under 5 grams) with a mode of 1.0–1.5 grams.

'Active' mounds are characterised by volcano-like craters where mice access entrances to a burrow system under the mounds. Nest chambers are built at the ends of short side passages and lined with leaves and other plant material. There may be more than one nest under a mound.

Biologists have noted pebble-mounds for many years. Until recently, these were assumed to be the work of the Sandy Inland Mouse, although

BABS & BERT WELLS

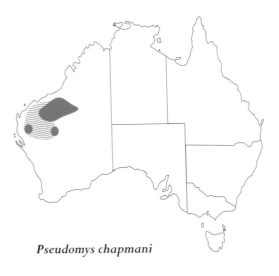

M. LOCHMAN

Above: *Western Pebble-mound Mouse constructing mound.*

Left: *Mound of Western Pebble-mound Mouse.*

Mounds are most common on spurs and lower slopes of rocky hills. The ground is often covered by a stone mulch (including many pebbles of the preferred size) and vegetated by hummocks of Hard Spinifex, *Triodia basedowii*. There are usually a few Snappy Gums or Bloodwoods over a sparse scatter of shrubs, typically cassias, acacias and *Ptilotus*; sharply incised drainage lines may be marked by ribbons of dense *Acacia*-dominated scrub.

there are no mounds over most of the range of that species. In the late 1970s, specimens of a *Pseudomys* taken in trap-lines near 'active' pebble-mounds in the Pilbara proved to be a new species. Observation of captive animals showed that they carry pebbles up to half their own weight in their mouths and arrange these in piles, shuffling them into position with their forelimbs.

Extant populations are widespread but appear to be patchily distributed in the more extensive mountainous ranges from the west of the Pilbara, eastward to the Rudall River. The persistence of abandoned mounds in the adjacent Gascoyne and Murchison regions as well as on small, isolated and coastal ranges in the Pilbara, indicate a considerable recent decline in distribution.

*Pseudomys chapmani*

591

## *Pseudomys chapmani*

**Size**

*Head and Body Length*
52–67 (60) mm
*Tail Length*
73–79 (75) mm
*Weight*
10–15 (12) g

**Identification** Head, back and sides buff brown; paws cartridge buff; white below (including throat and edge of mouth). Similar to *P. hermannsburgensis* but shorter tail, ears and feet. Readily distinguished by pattern of foot-pads.

**Recent Synonyms** None, but confused in the past with *P. hermannsburgensis*.

**Other Common Names** Pebble-mound Mouse.

**Status** Uncertain; scattered; contracting range.

**Subspecies** None.

**References**

Kitchener, D.J. (1980). A new species of *Pseudomys* (Rodentia: Muridae) from Western Australia. *Rec. West. Aust. Mus.* 8: 405–14.

Dunlop, N.J. and I.R. Pound (1980). Observations on the Pebble-mound Mouse *Pseudomys chapmani* Kitchener 1980. *Rec. West. Aust. Mus.* 9: 1–5.

Cooper, N.K. (1993). Identification of *Pseudomys chapmani, P. hermannsburgensis, P. delicatulus* and *Mus musculus* using foot-pad patterns. *Rec. West. Aust. Mus.* 19: 69–73.

Little is known of the reproductive biology of the Western Pebble-mound Mouse. A captive female produced a litter of four young in late May. Following good rains, a female suckling a 4 gram infant was pit-trapped in September 1991 and a lone juvenile (6.5 grams) as well as one pregnant and two lactating adults (one with four young weighing 5.25–5.5 grams) were caught in May 1992.

A.N. START AND D.J. KITCHENER

# Delicate Mouse
## *Pseudomys delicatulus*
## (Gould, 1842)

del'-ik-at'-yue-lus: 'delicate false-mouse'

In the mid-nineteenth century, John Gould described the Delicate Mouse as the smallest and most beautiful species of *Mus* from Australia: it reminded him of the English Harvest Mouse. It is now known that the Sandy Inland Mouse and Pebble-mound Mouse from the Western Desert are similarly small. The Delicate Mouse is easily killed while being handled, possibly giving rise to its name.

This tropical species has an almost perfect Torresian distribution, occurring irregularly across northern Australia and down the east coast to southern Queensland. There is no evidence of any range reduction since European settlement. It survives in areas that have been heavily grazed and trampled by cattle, but seems to prefer more friable soils suitable for burrowing. The burrows are simple, usually with a single entrance and no deeper than 40 centimetres. Sometimes termite mounds are used as above-ground burrow sites.

Although it has a varied diet, the Delicate Mouse eats a large amount of annual grass, making it one of the most granivorous of Australian rodents. Although seed production varies enormously, it is difficult to imagine that this alone explains the dramatic fluctuations in its abundance. In large samples from Kakadu, at the same site, capture rates have varied from one in 8000 trap-occasions to one in 50. During a two-year period in the Northern Territory, a population bred continuously and with two dramatic, but brief, irruptions in November and September in successive years. In another three-year period of regular sampling, the species was exceedingly uncommon and was not seen to breed. In North Queensland, it is most common each year in June and July.

M. ARMSTRONG

*The Delicate Mouse feeds mostly on the seeds of native grasses.*

The oestrous cycle is about six days and the gestation period of 28–31 days is followed by a post-partum oestrus. A litter usually comprises three young, sometimes four. Neonates weigh about 1 gram but development is very rapid. Weaning occurs at about 30 days and sexual maturity is attained at a body weight of about 6 grams. A pair maintained in captivity on canary seed produced six litters in six months.

Populations appear to increase rapidly after episodic events such as unusual rainfall or intense fire. It is tolerant of open areas with little vegetation and this may allow the species to steal a march on the larger terrestrial rodents that subsequently exclude them as the ground vegetation returns. The Delicate Mouse is often associated with the permanently sparse vegetation on coastal sand-dunes, a habitat generally devoid of other species of small mammals. Its occurrence in more open ground vegetation may account for its common presence in owl pellets. Its ecology appears to resemble that of the exotic House Mouse in heathland in southern Australia. Perhaps this is why the House Mouse, although present in towns in northern Australia, is not known to have become established in bushland in the tropical north.

R.W. BRAITHWAITE AND J. COVACEVICH

## *Pseudomys delicatulus*

**Size**
*Head and Body Length*
55–75 mm
*Tail Length*
55–80 mm
*Weight*
6–15 g

**Identification** Fur yellow-brown to grey-brown above; with white or cream fur below. Tail slender and slightly longer than head-body length. Nose and feet pink. Overall size small.

**Recent Synonyms** *Leggadina delicatulus, Leggadina mimula, Leggadina patria, Gyomys pumilus.*

**Other Common Names** Indigenous names: Mo-lyne-be (Port Essington), Bobyek (Mayali, Kakadu area), Djekung (East Arnhem), Kalla (Cape York), Wurrendinda (Groote Island), Kirrubeejben (Lordil, Cape York).

**Status** Sparsely scattered in abundant suitable habitat; marked fluctuations.

**Subspecies** *Pseudomys delicatulus delicatulus,* mainland and most offshore islands.
*Pseudomys delicatulus mimulus,* Groote Eylandt, Sir Edward Pellew Group.

**Reference**
Braithwaite, R.W. and P. Brady (1993). The Delicate Mouse, *Pseudomys delicatulus:* a continuous breeder waiting for the good times. *Aust. Mammal.* 16: 94–98.

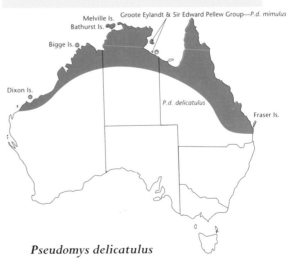

Melville Is.
Bathurst Is.
Bigge Is.
Dixon Is.
Groote Eylandt & Sir Edward Pellew Group—*P.d. mimulus*
*P.d. delicatulus*
Fraser Is.

*Pseudomys delicatulus*

# Desert Mouse

## *Pseudomys desertor*

## Troughton, 1932

dez-ert'-or: 'desert false-mouse'

The Desert Mouse is typical of many rodents of the Australian arid zone. During dry conditions, which may occur for prolonged periods, small populations survive in pockets of a preferred habit. After suitable rains, these populations expand and more habitats are occupied. Extended and extensive drought conditions of the 1950s and 1960s, together with a lack of systematic surveys in the arid zone, probably explain why this rodent was once considered to be rare.

---

### *Pseudomys desertor*

**Size**

*Head and Body Length*
70–105 (83) mm
*Tail Length*
67 -103 (83) mm
*Weight*
15–35 (25) g

**Identification** Bright chestnut brown above with long dark guard hairs giving it a spiny appearance; belly light grey-brown. Tail slightly bi-coloured, with scaly appearance and equal to or shorter than head-body length. Large eyes with conspicuous pale orange eye-ring. May be confused with *P. nanus* in the northern Tanami Desert, where sympatric.
**Recent Synonyms** *Gyomys desertor*.
**Other Common Names** Brown Desert Mouse.
**Status** Secure and widespread in arid areas.
**Subspecies** None.

---

**References**

Happold, M. (1976). Social behaviour of the Conilurine rodents (Muridae) of Australia. *Z. Tierpsychol.* 40: 113–182.

Reid, J.R.W., J.A. Kerle and S.R. Morton (1992). The distribution and abundance of vertebrate fauna of Uluru (Ayers Rock–Mount Olga) National Park. *Kowari* No. 4. Australian National Parks and Wildlife Service, Canberra.

Its range covers the vast arid Australian interior and parts of the semi-arid tropics in the northern Tanami, Northern Territory and the Bungle Bungles, Western Australia. It was collected near the Murray River by the Blandowski expedition of 1856–57. It has also been recorded from Bernier Island, Western Australia. Outside this modern distribution, subfossil remains have been found at Cape Range, Jurien Bay, on the Nullarbor Plain and the south-western Kimberley of Western Australia.

Its abundance varies greatly throughout its range: it is most frequently trapped in areas where spinifex has matured to form large hummocks. In the Tanami Desert, Northern Territory, it is common in dense spinifex on sand plain and less common on sand-dunes, stony hills and rises, or in riparian woodland. In the northern Simpson Desert, Northern Territory, it is generally uncommon except in some localities. These sites are associated with dry watercourses or salt lakes and have a dense grass or shrub cover including cane grass on adjacent dunes. In the rocky ranges to the east and west of Alice Springs it is found on the spinifex-covered limestone and conglomerate scree slopes. One dense population on Anna Creek Station, South Australia, lives in thick sedge growth surrounding a bore drain.

A survey of Uluru National Park, originally showed the Desert Mouse to be uncommon and very restricted in distribution but, after two

D. MATTHEWS

*Populations of the Desert Mouse vary greatly in response to rainfall.*

years and extensive heavy rains, it became widespread and common. Its most important habitat was dense Kangaroo Grass at the base of Uluru but it was not re-trapped there after this habitat had been burnt by wildfire. By the end of the survey the Desert Mouse was found to be widely distributed through the mature spinifex communities.

The reproductive potential of the Desert Mouse (as measured by litter-size, gestation, weaning period and age to sexual maturity) is greater than that of the sympatric and widely distributed Sandy Inland Mouse. The oestrus cycle is seven to nine days, mean litter size three (range one to four) and the gestation period 27–28 days. There is a post-partum oestrus. Reproductive maturity is attained at about ten weeks. In captivity, breeding can occur throughout the year. In the wild, breeding may be more sporadic and dependent upon suitable conditions. Births have been recorded in March, June and August to December. The high reproductive potential after good rains was illustrated by a 24-fold increase in trapping

success recorded for two sites near Uluru over a period of less than two years.

In the wild, some individuals have been dug from shallow burrows; in captivity, burrows are actively dug. When the vegetation is dense, such as around bores and in Kangaroo Grass, nests have been found in the middle of the thickest tussocks. In captivity, individuals display solitary behaviour patterns but this does not appear to be the case in the wild, particularly at high densities.

The Desert Mouse appears to be herbivorous, feeding on the succulent shoots, rhizomes, seeds and flowers of a sedge at Anna Creek and on grasses and seeds around Yuendumu. In the laboratory, it is able to live without water and even to gain weight on a diet of dry seeds.

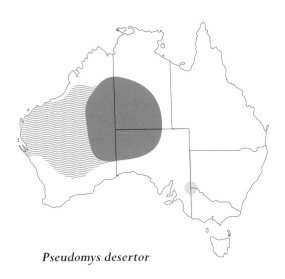

**Pseudomys desertor**

J.A. KERLE

# Shark Bay Mouse

## *Pseudomys fieldi*

### (Waite, 1896)

feel'-dee: 'Field's false-mouse'

For much of this century the Shark Bay Mouse was referred to as *Pseudomys praeconis* and thought to be restricted to Shark Bay, Western Australia. However, its subfossil remains have been found in cave deposits all along the west coast (south of Shark Bay), the upper Gascoyne, northern Goldfields, Gibson Desert and, more recently, at Uluru in the Northern Territory. It is now apparent that because *P. praeconis* and the extinct Alice Springs Mouse, *P. fieldi*, were continuous in distribution and showed no significant morphological differences, they should be regarded as conspecific. It has also been suggested that Gould's Mouse, *P. gouldii,* may be part of the same species group but this matter is not addressed here. Because *P. fieldi* was named in 1896, before *P. praeconis*, the scientific name of the Shark Bay Mouse is *P. fieldi*.

Since European settlement, the Shark Bay Mouse has been recorded only twice on the mainland: once in 1858 on the Peron Peninsula, Shark

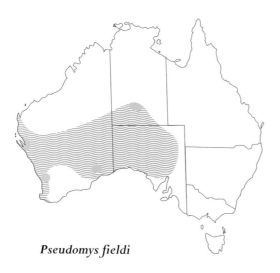

*Pseudomys fieldi*

## *Pseudomys fieldi*

**Size**
*Head and Body Length*
85–115 (100) mm
*Tail Length*
115–125 (120) mm
*Weight*
30–61 (45) g

**Identification**  Grizzled dark brown above, grading into buff on the sides; pure white below. Robust with a particularly long and shaggy coat. Tail fully furred, grey above, distinctly separate from white undersurface.
**Recent Synonyms**  *Thetomys praeconis*, *Pseudomys praeconis*.
**Other Common Names**  Shaggy Mouse, Shaggy-haired Mouse, Alice Springs Mouse.
**Status**  Rare, limited.
**Subspecies**  None.
**References**

Baynes, A (1990). The mammals of Shark Bay, Western Australia (in) P.F. Berry, S.D. Bradshaw and B.R. Wilson (eds) *Research in Shark Bay: Report of the France–Australe Bicentenary Expedition Committee*. Western Australian Museum, Perth.

Robinson, A.C., J.F. Robinson, C.H.S. Watts and P.R. Baverstock (1976). The Shark Bay Mouse *Pseudomys praeconis* and other mammals on Bernier Island, Western Australia. *West. Aust. Nat.* 13: 149–155.

Bay and again at Alice Springs in 1895 (but then called the Alice Springs Mouse). In 1989, a search for the species was undertaken on the mainland in the Shark Bay area, but none was found. It is now one of Australia's most restricted native mammals, occurring naturally only on Bernier Island in Shark Bay, one of Australia's most important nature reserves, which also supports remnant populations of several marsupials that are now extinct on the mainland. In June 1993, it was translocated to Doole Island, in Exmouth Gulf.

G.B. BAKER

*Once widely distributed, the Shark Bay Mouse is now restricted to some offshore islands.*

On Bernier Island, the Shark Bay Mouse prefers to inhabit the coastal fringes of the island, where it makes tunnels and runways in heaps of seagrass piled at the top of the beach during winter storms. Short, shallow burrows have been found underneath shrubs on sand-dunes. It also occurs at lower densities on the inland open steppe comprising *Triodia* and *Acacia* species. (The Ash-grey Mouse, *P. albocinereus,* also occurs here.) The Shark Bay Mouse eats flowers (probably of the Coastal Daisy Bush, *Olearia*) with some leaves and stems of an unknown fleshy plant, fungi, insects and spiders.

Breeding occurs during winter and spring on Bernier Island. The gestation period is 18–30 days and litter size is three to four. The young cling to their mothers' teats and are dragged around behind her when she moves. Their eyes open at 15 days and by 30 days they are weaned and independent.

The reasons for the disappearance of this once widespread species are unknown. Its decline may have started prior to European settlement and relate to subtle climatic change. The early introduction of the Cat to much of its former range and the effects of the pastoral industry may also have contributed. The lack of more complex, deep burrow systems may have made the Shark Bay Mouse more vulnerable than other native rodents to predation by the Cat and damage through trampling of nest sites by domestic stock. The Western Australian Department of Conservation and Land Management is now implementing a recovery plan for the Shark Bay Mouse, including its reintroduction to parts of the former range, together with appropriate predator-control programs for exotic predators.

K.D. MORRIS AND A.C. ROBINSON

# Smoky Mouse

## *Pseudomys fumeus*

### Brazenor, 1934

fue-may'-us: 'smoky false-mouse'

The Smoky Mouse now occurs mainly in the Grampians, the highlands north-east of Melbourne and the coastal woodlands of eastern Gippsland. Individuals have been trapped in the Brindabella Range, Australian Capital Territory and in south-eastern New South Wales, but the extent of these colonies is unknown. Subfossil deposits suggest that the range of this beautiful rodent has declined since arrival of Europeans in Australia.

On the summit of Mount William in the Grampians, the Smoky Mouse may attain high densities. Here, during summer, it forages for the seeds of the shrubby legumes common in the area, berries of epacrids and Bogong Moths. In winter, when few seeds are pro-

---

### *Pseudomys fumeus*

**Size**
*Head and Body Length*
85–100 (90) mm
*Tail Length*
110–145 (140) mm
*Weight*
45–90 (70) g

**Identification** Pale grey to blue-grey to black above; belly grey to white; feet pink with white fur. Tail similar to top of body, with white
lateral stripes.
**Recent Synonyms** None.
**Other Common Names** None.
**Status** Rare; limited; at risk.

---

**Subspecies** No formal subspecies have been described but the following types will ultimately warrant distinction on size, cranial and pelage characters:
*Pseudomys fumeus* (western form), west of Melbourne, now only known from the Grampians ranges; large, more darkly coloured.
*Pseudomys fumeus* (eastern form), east of Melbourne; small, pale grey.
**References**
Cockburn, A. (1981). Population regulation and dispersal of the smoky mouse, *Pseudomys fumeus*. 1. Dietary determinants of microhabitat preference. *Aust. J. Ecol.* 6: 231–254.
Cockburn, A. (1981). Population regulation and dispersal of the smoky mouse, *Pseudomys fumeus*. 2. Spring decline, breeding success and habitat heterogeneity. *Aust. J. Ecol.* 6: 255–266.

duced, it switches to underground truffle-like fungi that are common around the roots of certain shrubs and grasses. All these foods are rich in nitrogen.

Reliance on three very distinct food resources creates problems for animals that must have access to high-quality food resources throughout the year. As the soil on Mount William dries out in late spring, the fruiting bodies of the fungi disappear and, in this subalpine environment, there are few areas where plants flower between winter and the mid-summer flush of productivity. Only in these restricted habitats, where Bogong Moths are attracted to the spring blossoms and new seeds are set, can the Smoky Mouse cope with the nutritional crisis from September to November: those animals whose home-ranges lie outside the sites of early flowering do not survive. The rapid decline in population involves more males than females, apparently because females are more selective in their choice of habitats.

G.B. BAKER

Once this population rearrangement is complete, breeding commences and females produce one to two litters, each of three to four young. Females often live to breed in more than one season and older animals breed slightly earlier than younger ones.

The pattern of life of the Smoky Mouse on Mount William is probably representative of the species throughout its range, since surveys in eastern Victoria and south-eastern New South Wales indicate that its preferred habitat is ridge-top sclerophyll forest with a diverse understorey of heath, dominated by legumes. Since this vegetation complex is generated by fire, it seems that the Smoky Mouse, like many of its close relatives, is dependent upon post-fire succession for its continued survival.

A. COCKBURN

*The rare Smoky Mouse feeds mainly on seeds in the summer and underground fungi in winter.*

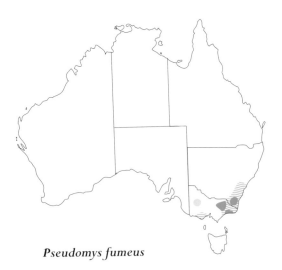

***Pseudomys fumeus***

# Gould's Mouse

### *Pseudomys gouldii*

### (Waterhouse, 1839)

gule'-dee-ee: 'Gould's false-mouse'

The first specimen of this species was collected in New South Wales, probably from the Upper Hunter River district or the Liverpool Plains. John Gould, who claimed to have seen specimens from the Liverpool Plains, South Australia and the neighbourhood of the Moore River, Western Australia, noted that it evinced a preference for the plains and sandhills of the interior. He also referred to two specimens sent to G.R. Waterhouse by the collector Frederick

Strange, found between the Coorong and Lake Albert, South Australia.

The animals were reputed by Gould to 'make their burrows under bushes'. Gilbert, Gould's collector, noted that in Western Australia it was

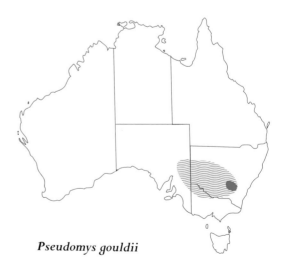

*Pseudomys gouldii*

*Gould's Mouse. Illustration in John Gould's*
The Mammals of Australia.

found on loose soil on the sides of grassy hills.
Extensive burrows were constructed about
15 centimetres below the surface and, usually,
families of four to eight individuals occupied the
same nest of soft, dried grass. According to
Gould, the long, slender, white hindfeet are
characteristic of the species.

Subfossil remains have been found at
Chambers Gorge in the Flinders Ranges, South
Australia, and Lake Victoria, New South Wales.
Much of this material is from owl pellets.

J.M. DIXON

### *Pseudomys gouldii*

**Size**
*Head and Body Length*
100–120 mm
*Tail Length*
90–100 mm
*Weight*
*circa* 50 g

**Identification**  Yellow-brown above; grey to
buffy white below. Soft fur. Similar to *P.
australis*, from which it is distinguished by the
presence of an accessory cusp on the first upper
molar.
**Recent Synonyms**  *Thetomys gouldii*.
**Other Common Names**  Gould's Native
Mouse, Gould's Eastern Mouse.
Indigenous name: Kurn-dyne (Moore River).
**Status**  Rare, probably extinct.
**Subspecies**  None.
**Reference**
Waterhouse, G.R. (1839). *Voyage of HMS.*
    Beagle, *Zoology*. Vol. 1: 67.

# Eastern Chestnut Mouse

## *Pseudomys gracilicaudatus*

### (Gould, 1845)

gras'-il-ee-kaw-dah'-tus:
'slender-tailed false-mouse'

Relatively recent discovery of the Eastern
Chestnut Mouse in a few isolated localities along
the northern New South Wales coast considerably extended its southern range. On the basis of
electrophoretic studies, supposed records from
the Northern Territory are now attributed to
the closely related Western Chestnut Mouse.

The Eastern Chestnut Mouse has been
recorded in northern Queensland from open
woodland with a grassy understorey. In New
South Wales it is more often found in heathland
and is most common in dense wet heath and
swampy areas, usually shared with the Swamp
Rat. Distribution is patchy, existing populations
being isolated by substantial distances, often
encompassing apparently suitable habitat from
which the species is absent. It has usually been
reported at low densities but, at one site, the

*Pseudomys gracilicaudatus*

## Pseudomys gracilicaudatus

### Size

*Head and Body Length*
115–145 mm (males)
105–138 mm (females)
*Tail Length*
82–120 mm (males)
80–114 mm (females)
*Weight*
55–118 g (males)
45–81 g (females)

**Identification** Grizzled chestnut above, particularly on the flanks; greyish below. A pale ring around the eye is less prominent than in *P. nanus*. The smaller *P. nanus* from the Northern Territory differs markedly from the larger *P. gracilicaudatus* from New South Wales and also exhibits lighter coat colour, but biochemical tests provide the clearest differences. Long, grey-white hairs on the dorsal surface of the feet extend beyond the claws. Ventrally light-coloured feet distinguish this species from the leaden grey feet of *Rattus lutreolus*, which often occurs in similar habitat.

**Recent Synonyms** None.

**Other Common Names** None.

**Status** Common, limited.

**Subspecies** Two subspecies have been described but their status is uncertain:
*Pseudomys gracilicaudatus gracilicaudatus*, south-eastern Queensland and New South Wales.
*Pseudomys gracilicaudatus ultra*, north-eastern Queensland.

*The Eastern Chestnut Mouse reaches maximum population density in heathland that is regenerating after fire.*

and the Swamp Rat is numerically and behaviourally dominant. The Eastern Chestnut Mouse exhibits shifts in habitat use and diet in response to experimental removal of the Swamp Rat.

Although some individuals have been observed to move up to 250 metres, most show strong attachment to sites and have home ranges less than 0.5 hectares. It is largely nocturnal, but crepuscular and very limited daytime activity has been observed. Runway systems are used and maintained among the dense sedge cover present in wet heathlands. The nest may be constructed of grass above ground or be part of a burrow complex.

The Eastern Chestnut Mouse includes a wide range of items in its diet and could be termed either a granivore–omnivore or a generalist granivore. The seeds favoured by many other species of *Pseudomys* are an important component of its diet, contributing 45 per cent in summer (when seeds are most numerous) but reduced in importance to 20 per cent in autumn. There is a high proportion of plant material in the diet: stems are a major component, varying from 25 per cent in winter to 40 per cent in autumn: leaf material is a minor component contributing only

### References

Mahoney, J.A. and H. Posamentier (1975). The occurrence of the native rodent *Pseudomys gracilicaudatus* (Gould 1845) Rodentia: Muridae in New South Wales. *Aust. Mammal.* 1: 333–346.

Higgs, P. and Fox, B.J. (1993). Interspecific competition: a mechanism for rodent succession after fire in wet heathland. *Aust. J. Ecol.* 18: 193–201.

Luo, J., B.J. Fox and E. Jefferys (1994). Diet of the eastern chestnut mouse (*Pseudomys gracilicaudatus*): I. Composition, diversity and individual variation. *Wildl. Res.* 21: 401–417.

density increased sixfold in the first 18 months after an intense wildfire, reaching a peak of 6 per hectare. Its optimal habitat is provided by young regenerating heath vegetation and the fire history of areas may influence whether or not it is present. Experimental studies have shown that it is competitively displaced by the Swamp Rat as the heath vegetation ages until, in mature heath, the Eastern Chestnut Mouse occurs at low densities

D. WHITFORD

4–7 per cent. Fungi comprise 25 per cent of the diet in winter and around 20 per cent in other seasons, but contributes only 2 per cent to the diet in summer. Insects comprise up to 20 per cent of the diet in summer and autumn (when they are most abundant) but only 3–4 per cent in spring and winter.

The oestrous cycle is seven to eight days and gestation takes about 27 days. The litter size is one to five (commonly three). There is a post-partum oestrus and up to three litters may be produced by a female during the breeding season which, in New South Wales, normally extends from September to March. Males have descended testes from September to January. In good seasons, mating extends from mid-August to the end of March, lactation continuing until the end of April. The young are precocial at birth, being partially furred and with lower incisors already erupted. They are fully furred by six days, open their eyes in 11 days and are weaned during their fourth week. Adults are sexually dimorphic and this is clearly evident in the growing young after two months. Adult size is reached by six months.

In the field, some females have been observed to breed in two successive years and several males have lived through a second summer. Many animals show a drop in weight over the winter and a further increase in weight during their second summer: second-year animals are markedly heavier than those in the first year. No individual has been observed to breed in the season of its birth.

Short gestation, precocial young, rapid post-natal development and early weaning accelerate the reproductive rate which is appropriate for an opportunist or fire-specialist species that reaches maximum abundance in early stages of vegetation succession following fire. The rapid early development probably favours survival of juveniles.

B.J. FOX

# Sandy Inland Mouse

## *Pseudomys hermannsburgensis*

### (Waite, 1896)

her'-mans-berg-en'-sis:
'Hermannsburg Mission false-mouse'

B. MILLER

*Populations of the Sandy Inland Mouse vary considerably in response to rainfall.*

The Sandy Inland Mouse resembles several other closely related species (from which it has only recently been separated) as well as the introduced House Mouse, *Mus musculus*, from which it differs in lack of musty smell, larger ears and eyes, less pointed head and absence of a notch on the upper incisors. It has a wide distribution over much of central and western Australia, often occurring sympatrically with the House Mouse and the Spinifex Hopping-Mouse, *Notomys alexis*. It is found on sandhill country associated with spinifex and on loamy soil with mulga scrub

The day is spent in a burrow up to half a metre below the surface of the soil, sometimes shared with the Central Knob-tailed Gecko, *Nephrurus levis*. After sunset, it emerges to feed on seeds, shoots, roots, small tubers and invertebrates.

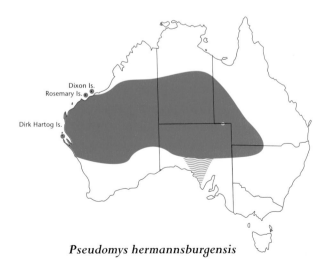

Dixon Is.
Rosemary Is.
Dirk Hartog Is.

*Pseudomys hermannsburgensis*

### *Pseudomys hermannsburgensis*

**Size**
*Head and Body Length*
65–85 (75) mm
*Tail Length*
70–90 (80) mm
*Weight*
9–14.5 (12) g

**Identification**  Sandy brown to grey-brown. Similar to *Mus musculus* but distinguished by larger ears, eyes and tail and by smooth inner surface of upper incisors (notched in *Mus*). Lacks 'mousy' odour of House Mouse.

**Recent Synonyms**  *Leggadina hermannsburgensis, Leggadina brazenori*.

**Other Common Names**  Hermannsburg Mouse.

**Status**  Common, widespread.

**Subspecies**  None.

**References**

Breed, W.G. (1986). Comparative morphology and evolution of the male reproductive tract in the Australian hydromyine rodents (Muridae). *J. Zool. Lond.* (A) 209: 607–629.

Breed, W.G. (1990). Comparative studies on the timing of reproduction and foetal number in six species of Australian conilurine rodents (Muridae: Hydromyinae). *J. Zool. Lond.* 221: 1–10.

D.WATTS

*The Long-tailed Mouse lives in the cool, wet forests of Tasmania.*

# Long-tailed Mouse

## *Pseudomys higginsi*

### (Trouessart, 1897)

hig'-in-zee: 'Higgins' false-mouse'

Populations fluctuate greatly and animals can become extremely common after a period of high rainfall. Reproductively active individuals have been obtained at all times of year and breeding has occasionally been recorded during very dry periods. Nevertheless, non-pregnant adult females are frequently caught.

The gestation period is 30–34 days and the usual litter size is three to four but up to seven young can be produced, although only four teats are present. Little is known of its social behaviour but small groups of reproductively active animals have been found together and up to 22 reproductively inactive individuals have been recorded from a single burrow system. It is frequently caught in trap-lines together with the House Mouse. It seems to be holding its own and, at present, is still widely distributed throughout much of the arid zone.

W.G. BREED

This endemic Tasmanian rodent lives mainly in Southern Beech rainforests where annual rainfall may exceed 2000 millimetres. In this wet, deeply shaded habitat it makes tunnels and runways beneath the deep moss-covered accumulation of forest debris. It also is found a little beyond the rainforest in dense, wet sclerophyll forests and deep fern gullies. While the temperature in this region may rise as high as 30°C in summer, it regularly falls to −10°C in winter, when a blanket of snow may confine the animals below the surface for days or weeks at a time. The Swamp Rat and three species of dasyurid marsupials share this habitat, but the interaction between them and the Long-tailed Mouse is unknown.

The Long-tailed Mouse is most active at night but may make daylight forays, particularly during the winter. An underground nest of plant material, shredded into fibres by the incisors, is usually occupied by a permanent pair and their juvenile offspring, if any. Such groups are strongly social and territorial.

Studies on captive animals indicate that the Long-tailed Mouse is omnivorous, eating a wide variety of plant material as well as insects and spiders which it holds in its forepaws while nibbling the soft parts. If provided with succulent food, it does not need to drink.

Breeding occurs from November to April and one or two litters, each of three to four young, are reared annually. The young are born clothed in natal fur and with a head and body length about one-third of that of the mother. The incisors, which erupt before birth, are at first directed inwards and are used by the young to attach themselves firmly to the mother's inguinal teats when she is disturbed. The mother may then 'explode' from the nest, dragging her litter behind like an antechinus. Normally, however, the young are left in the nest while the parents are foraging, the entrance being temporarily plugged.

The young are completely dependent upon the mother until about 25 days old. Weaning then commences, their incisors soon wear to the

### *Pseudomys higginsi*

**Size**

*Head and Body Length*
115–150 (133) mm
*Tail Length*
145–200 (162) mm
*Weight*
50–90 (67) g

**Identification** Soft fur; dark grey above; paler below. Readily distinguished by bi-coloured tail, grey above, white below, almost 50 per cent longer than the head and body and often carried in a gentle curve, clear of the ground.
**Recent Synonyms** None.
**Other Common Names** Higgins' Mouse, Tasmanian Mouse, Tasmanian Pseudo-rat, Long-tailed Rat.
**Status** Common, limited.
**Subspecies** None.
**References**
Green, R.H. (1968). The murids and small dasyurids in Tasmania, Parts 3 and 4. *Rec. Queen Vict. Mus.* No. 32: 1–11.
Green, R.H. (1993). *The fauna of Tasmania: Mammals.* Potoroo Publishing, Launceston.

typical rodent chisel-like condition, and molar teeth erupt to equip the animals for the imminent change to an adult diet. At the age of about three months, the juvenile fur is moulted and they leave the nest as small adults. Although captive individuals have survived for up to four years, it is unlikely that those in a natural state live longer than about 18 months.

Populations of the Long-tailed Mouse are relatively dense, possibly as high as ten families per hectare in ideal habitats. Its survival appears to be ensured so long as Tasmanian rainforests remain intact.

R.H. GREEN

*Pseudomys higginsi*

M.W. GILLAM

*A recently described species, the Central Pebble-mound Mouse is one of the three that nest in mounds of pebbles.*

# Central Pebble-mound Mouse

## *Pseudomys johnsoni*

Kitchener, 1985

jon'-sun-ee: 'Johnson's false-mouse'

The discovery of a new mammal species is an exciting and relatively uncommon event in recent times. The Central Pebble-mound Mouse was first described as a distinct species in 1985 from specimens collected in 1983. A perception that these specimens 'looked different' from the similar Sandy Inland Mouse and were found in an unusual habitat, led to its identification. Little is known of its biology.

The first specimens were collected from the Davenport and Murchison Ranges south-east of Tennant Creek in central Northern Territory. Subsequently it has been found only on Mittiebah Station in the Barkly region of the Northern Territory (as skull fragments in owl pellets). It is

### *Pseudomys johnsoni*

**Size**
*Head and Body Length*
60–74 (69) mm
*Tail Length*
75–95 (84) mm
*Weight*
9–17 (12) g

**Identification** Grey-brown on back and sides with black guard hairs; head clay or fuscous; white below, including sides of mouth and feet. Difficult to distinguish from *P. hermannsburgensis* but has slightly longer foot and tail, shorter ears and fewer hairs per scale row on the tail.

**Recent Synonyms** None (may have previously been confused with *P. hermannsburgensis*).

**Other Common Names** None.

**Status** Unknown but apparently uncommon in the two known localities.

**Subspecies** None.

**Reference**

Kitchener, D.J. (1985). Description of a new species of *Pseudomys* (Rodentia: Muridae) from Northern Territory. *Rec. West. Mus.* 12: 207–221.

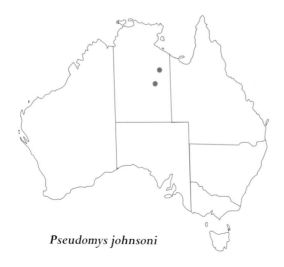

*Pseudomys johnsoni*

# Kimberley Mouse

## *Pseudomys laborifex*

### Kitchener and Humphreys, 1986

lab-or'-ee-fex:
'work-making false-mouse'

This rodent is widely distributed in the Kimberley, Western Australia, but nowhere is it abundant. It is found in a wide range of habitats: laterite plateaus dominated by open low woodlands of

---

**Pseudomys laborifex**

**Size**
*Head and Body Length*
58–76 (65) mm
*Tail Length*
63–87 (76) mm
*Weight*
7–20 (12) g

**Identification** Restricted to the Kimberley. Differs from *P. hermannsburgensis* in having skull with a shorter tympanic bulla (0.19–0.21 mm against 0.21–0.24 mm), first upper molar with anterior loph more elongate and labial cusps more reduced. Differs from *P. delicatulus* in being heavier (8.3–17.0 g against 6.0–10.0 g) and longer skull (22.2–23.5 mm against 20.0–21.0 mm).
**Recent Synonyms** None.
**Other Common Names** None.
**Status** Widespread, uncommon.
**Subspecies** None.
**Reference**
Kitchener, D.J. and W.F. Humphreys (1986). Description of a new species of *Pseudomys* (Rodentia: Muridae) from the Kimberley Region, Western Australia. *Rec. West. Aust. Mus.* 12: 419–434.

---

likely that specimens identified as the Sandy Inland Mouse in 1905 from Alroy Station by John Stalker also refer to the Central Pebble-mound Mouse. He describes piles of stones that mice had excavated from their burrows.

In the Davenport and Murchison Ranges, the Central Pebble-mound Mouse is sympatric with the Sandy Inland Mouse on stony ridges, rises and gravelly plains. The habitat is very harsh, vegetation being mature spinifex, *Triodia longiceps,* and grevilleas, especially *G. wickhamii.* In 1983 and early 1984, the species appeared to be quite abundant, but in 1985, after two years of drought, the population had declined dramatically. Since then the population has not been surveyed.

Although some pebble-mounds have been seen in this arid range country, they are seldom found. Observations of the behaviour of *P. johnsoni* in captivity revealed that it excavates stones from its burrows and dumps them in a rubble heap near one entrance, which is later closed. Other vertical pop-holes with no accumulation of stones are opened and used in preference: this behaviour may explain the absence of obvious mounds. Pebbles carried in the mouth include some equal to the body weight of the mouse. Stones are manipulated with the forefeet and, sometimes, the hindfeet. The burrow system is quite complex, and one instance is known of a simple platform nest of spinifex being built.

J.A. KERLE

C. KEMPER/ W.A. MUSEUM

*Described in 1986, the Kimberley Mouse is widespread but sparse.*

*Eucalyptus latifolia* or *E. tetradonta*; valley woodlands with a ground cover of spinifex and cane grass dominated by *Melaleuca* species and *Cochlospermum fraseri* on heavy soils; and *Eucalyptus* species, *Erythrophleum chlorostachys*, *Acacia* species, *Owenia vernicosa*, and *Ficus* species on sandy soils.

Females appear to give birth towards the end of the dry season through to the early wet season. Two females with, respectively, two and three near-term foetuses (crown to rump length more than 20 millimetres) were collected in late August. Another collected in late November had recently given birth; it was lactating and had four uterine implantation scars. Males collected in January, February and September were actively producing sperm.

D. J. KITCHENER

*Pseudomys laborifex*

# Western Chestnut Mouse

## *Pseudomys nanus*

## (Gould, 1858)

nah'-nus: 'dwarf false-mouse'

In 1843, John Gilbert collected a number of specimens of this chestnut-coloured mouse in the vicinity of Moore River and New Norcia, Western Australia, but although fossil and sub-fossil remains have been discovered in caves in the general area, no further living specimens have been found there and it seems now to be restricted to northern Australia from Barrow Island to the islands of the Sir Edward Pellew Group in the Gulf of Carpentaria. It is found in a range of habitats developed on sandy volcanic and lateritic soils and, wherever it has been found, there is always a dense ground cover of tussock grasses, often beneath the low eucalypt woodland characteristic of much of tropical Australia. Although it may occur along the edges of watercourses it appears to be less dependent on this moister habitat than some of the other native rodents with which it shares this habitat. Even at the peak of the dry season the Western Chestnut Mouse can be found on grass-covered lateritic plains in the Kimberleys, well away from the nearest creek.

Despite its abundance there is little information on its natural history. It appears to eat a variety of grasses and it is known to build grass nests, but whether or not it constructs a burrow remains to be determined. In the Kimberleys, the Western Chestnut Mouse probably gives birth to young throughout the year, except for the very dry period from September to November, and it is capable of responding to favourable conditions by a very rapid breeding rate. The oestrous cycle is five to six days and gestation occupies 22–24 days. The young are

K. ATKINSON

*The Western Chestnut Mouse builds a nest of grass;
whether it makes a burrow is not known.*

well furred at seven days, open their eyes at 12
days and can be weaned at 21 days. Normally
three young are born in a litter but as many as
five have been recorded. This extremely efficient
breeding cycle is an excellent adaptation to life in
the grassy woodlands of northern Australia where
frequent summer fires, flooding and rapid vegetation growth following the onset of the wet season provide patches of highly productive habitat
which can be readily exploited by a small fast-
breeding, grass-eating rodent.

A.C. ROBINSON

*Pseudomys nanus*

---

### *Pseudomys nanus*

**Size**
*Head and Body Length*
80–140 (100) mm
*Tail Length*
70–120 (100) mm
*Weight*
25–50 (34) g

**Identification** Light orange-fawn above, mixed
with long darker brown hairs giving the animal a
grizzled appearance; white below, grading into
light orange-brown on the sides of the body.
Pronounced lighter coloured eye-ring. Tail blackish above and white below. Very short ears.

**Recent Synonyms** *Thetomys nanus*, *Thetomys
ferculinus* (specimens from the Northern
Territory have been confused with *Pseudomys gracilicaudatus*).

**Other Common Names** Little Mouse,
Barrow Island Mouse.

**Status** Abundant.

**Subspecies** *Pseudomys nanus nanus*, western and
north-western mainland.

*Pseudomys nanus ferculinus*, Barrow Island.

**Reference**

Taylor, J.M. and B.E. Horner (1972).
Observations on the reproductive biology of
*Pseudomys* (Rodentia: Muridae). *J. Mammal.*
53: 318–328 (as *P. gracilicaudatus*).

610

# New Holland Mouse

## *Pseudomys novaehollandiae*

(Waterhouse, 1843)

noh'-vee-hol-and'-ee-ee:
'New Holland false-mouse'

Originally described in the mid-nineteenth century, this small rodent lived unnoticed for over 100 years on the edge of Sydney before it was rediscovered in Ku-ring-gai Chase National Park in 1967. Since then it has been found in many dry heath and open forest localities in coastal New South Wales, Victoria and north-eastern Tasmania, including Flinders Island. Its patchy distribution within this range is probably the result of marked preferences for soft substrates (usually sand), a heath-type layer of leguminous perennials less than 1 metre high and sparse ground cover and litter. These characteristics are associated with a habitat in the early and middle stages of regeneration after such disturbances as fire or sandmining. The New Holland Mouse begins to recolonise burned areas after about one year of regeneration; mined areas after four to five years.

The diet changes, seasonally and with locality. The New Holland Mouse eats seeds (preferred in

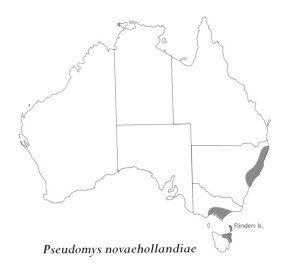

*Pseudomys novaehollandiae*

---

### *Pseudomys novaehollandiae*

**Size**
*Head and Body Length*
65–88 (78) mm
*Tail Length*
81–107 (87) mm
*Weight*
12–20 (16) g (New South Wales)
21–25 (23) g (Victoria)
20–26 g (Tasmania)

**Identification**  Grey-brown above; greyish-white below. Readily distinguished from *Mus musculus*, the only other small rodent found in the same areas, by tail, which is dusky brown above, white below and 10–15 per cent longer than the head and body and lack of notched upper incisors. Ear smaller in Tasmania (12 mm) than on the mainland (16 mm).

**Recent Synonyms**  *Gyomys novaehollandiae*.

**Other Common Names**  None.

**Status**  Common, limited (New South Wales, Tasmania). Endangered (Victoria).

**Subspecies**  None.

**References**

Kemper, C.M. (1990). Small mammals and habitat disturbance of open forest of coastal New South Wales. I. Population parameters. *Aust. Wildl. Res.* 17: 195–206.

Norton, A.W. (1987). The ecology of small mammals in north-eastern Tasmania. II. *Pseudomys novaehollandiae* and the introduced *Mus musculus. Aust. Wildl. Res.* 14: 435–441.

Wilson, B.A. (1991). The ecology of *Pseudomys novaehollandiae* (Waterhouse, 1843) in the eastern Otway Ranges, Victoria. *Wildl. Res.* 18: 233–247.

---

spring and summer), insects and other invertebrates (preferred in winter), leaves, flowers and fungi. Since the introduced House Mouse is about the same size and can have a similar diet, there is a possibility of competitive interaction between the two species: this may depend on the situation.

*The New Holland Mouse feeds mainly on seeds in the summer and invertebrates in the winter.*

It is nocturnal and constructs burrows in sandy soils for daytime refuge. Some appear to be temporary shelters while other deep tunnels, up to 5 metres long and with a nest chamber, are probably permanent residences. It is not known how many animals occupy a burrow but it seems that small family groups may share a dwelling. In the wild, the home ranges of breeding females may overlap (sometimes greatly) but those of mature males do not.

In New South Wales, the breeding season normally lasts five months (late winter to early summer) but breeding can be extended or delayed by several months if weather conditions are unusual. Since the life span is one-and-one-half to two years, there are two classes of females: those in their first year that produce only one litter; and those in their second year that may produce three to four litters in the normal five month breeding season. The period of gestation is about 32 days and the litter size two to six, most young being born in spring and early summer. Sexual maturity can be attained at the age of seven weeks but the proportion of males and females becoming mature during the season in which they are born may depend on the density of a population. Animals that mature in the season of birth grow more quickly than those that remain immature until the following season. Animals in their first year seldom have a head and body length more than 75 millimetres; those in their second year are usually larger.

In Victoria and Tasmania, breeding is later and more protracted, occurring from late spring to early summer, so some of the data applicable to New South Wales animals may not hold for these populations.

The size of a population is highest in autumn, lowest in spring. Under favourable conditions, the New Holland Mouse may attain a density of 17 animals per hectare. Although it appears to benefit from habitat disturbance, the future of a population becomes threatened if the habitat is grossly altered or if no refuge patches remain from which colonisation can occur. Too frequent control-burning, urbanisation and permanent clearing of land are detrimental. In Victoria, it is known from only a few fragmented populations and in recent years has become extinct in some small reserves where the successional age of the vegetation has not been appropriate.

C. KEMPER

HANS & JUDY BESTE

*The rounded snout (roman nose) of the Western Mouse is characteristic.*

# Western Mouse

## *Pseudomys occidentalis*

### Tate, 1951

ox'-ee-dent-ah'-lis: 'western false-mouse'

First described from a specimen collected at Tambellup, Western Australia, in 1930, the Western Mouse was known by only five other specimens until the Western Australian Museum began extensive mammal surveys in the wheatbelt in 1971. Since then a number of other specimens have been collected from 11 semi-isolated conservation reserves within the southern wheatbelt and south coast, the smallest of which is about 200 hectares.

Most of the sites of capture have long-unburnt vegetation (30–50 years) on sandy clay loam or sandy loam, frequently with a matrix of gravel. On three reserves, populations are located in dense vegetation surrounding granite rocks on similar soil types. The vegetation is extremely variable, including sparse low shrubland, tall dense shrubland, sparse to dense shrub mallee and mid-dense woodland. Most capture sites contain patches of extremely dense vegetation, 0.5–2.5 metres high. Dominant plants of the upper vegetation layers include species of *Eucalyptus*, *Isopogon*, *Acacia*, *Casuarina* and *Melaleuca* and it is

### *Pseudomys occidentalis*

**Size**

*Head and Body Length*
90–110 (97) mm
*Tail Length*
120–140 (129) mm
*Weight*
33–53 (34) g

**Identification** Mixture of dark grey and yellowish-buff fur with black guard hairs above; greyish-white below; distinctive dark hairs on dorsal surface of tail. Paws white. May be confused with *P. albocinereus* (which is larger and darker in the eastern part of its range).

**Recent Synonyms** None.

**Other Common Names** None.

**Status** Common, limited. At risk.

**Subspecies** None.

**References**

Kitchener, D.J. and A. Chapman (1977). Biological Survey of the Western Australian Wheatbelt. Part 3: Vertebrate fauna of Bendering and Western Bendering Nature Reserves. *Rec. West. Aust. Mus.* Suppl. No. 5.

Kitchener, D.J., A. Chapman, J. Dell, R.E. Johnstone, B.G. Muir and L.A. Smith (1976). Biological Survey of the Western Australian Wheatbelt, Preface to the Series and Part 1: Tarin Rock and North Tarin Rock Reserves. *Rec. West. Aust. Mus.* Suppl. No. 2.

McKenzie, N.L., A.A. Burbidge and N.G. Marchant (1973). *Results of a Biological Survey of a Proposed Wildlife Sanctuary at Dragon Rocks near Hyden, Western Australia*. Dept. Fish. Fauna. Report No. 12, Perth.

thought that the Quandong, *Santalum acuminatum* and various sedge species are also important habitat requirements within the northern part of the range.

The Western Mouse is largely nocturnal, living during the day in a burrow approximately 20–40 centimetres deep. Burrows consist of a single vertical entrance shaft (similar to that of Mitchell's Hopping Mouse), connected to a loop 2–3 metres in diameter and a nesting chamber

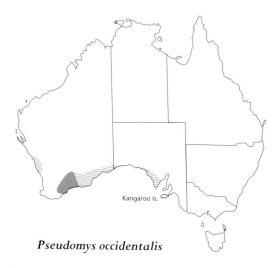

**Pseudomys occidentalis**

# Hastings River Mouse

## *Pseudomys oralis*

Thomas, 1921

oh-rah'-lis: 'notable-mouthed false-mouse'
(meaning unknown)

located opposite the entrance. Nesting material consists largely of loose, fibrous plant material. It is communal: in the laboratory, animals sleep during the day huddled together in shelter boxes. Ten animals were collected from one burrow system in the wild. Individuals have been recorded to travel 600 metres overnight from a trap-site to a burrow.

The diet consists largely of fibrous plant material, flowers of *Acacia* and *Hibbertia* and various invertebrates including beetle and moth larvae. It also eats the seeds of *Hakea* and *Dryandra* and the flowers of *Comespermum*, *Pultenea* and *Hibbertia*. It eats immature fruit directly off the bushes, easily climbing small bushes and shrubs to acquire these foods. Large accumulations of Quandong nuts found under dense bushes and vegetation are chewed to extract the kernels which are rich in proteins and oils.

Pregnant females are found in early to mid-spring and young are born from mid- to late spring, entering the population in early summer. A body weight of 32–37 grams is attained by May or June. Juveniles are able to breed in July or August in the year of birth.

Trapping data suggest that the population density varies between seasons and from year to year but is generally one to seven individuals per hectare. Populations seem to be restricted to selected areas and vegetation types within conservation reserves.

L. WHISSON AND D.J. KITCHENER

In 1921, the Hastings River Mouse was first described (as a subspecies of the Plains Mouse) from a specimen in the British Museum, purchased from a dealer (who gave the locality simply as 'Australia') and another in the Liverpool Museum which was recorded as having coming from the Hastings River, New South Wales. These two specimens were the only known representatives of the species until 1969, when a third animal

---

**Pseudomys oralis**

**Size**
*Head and Body Length*
130–170 mm
*Tail Length*
110–150 mm
*Weight*
90–100 g

**Identification** Brownish-grey above; buff to greyish-white below. Distinguishable from rodents of similar size occurring in the same region by large, protuberant eyes; strongly rounded snout (Roman nose); distinctly white feet; and tail which is dark above, distinctly separate from the white, furred undersurface.
**Recent Synonyms** *Pseudomys australis oralis*.
**Other Common Names** None.
**Status** Rare, limited.
**Subspecies** None.

R. & A. WILLIAMS

*The Hasting's River Mouse is rare and elusive.*

was trapped beside a creek in the western foothills of the Great Dividing Range near Warwick, Queensland, some 350 kilometres north of the Hastings River. Further trapping there and at nearby locations led to the capture of several more examples of this little-known species. During 1981 and 1982, several specimens were trapped from the upper Hastings River and the area in which they were found has been incorporated into a new national park. Other specimens have recently been collected near Dorrigo, New South Wales, some 160 kilometres north of the Hastings River site. The species is clearly less rare than was once thought, but it remains extremely local and elusive: attempts to recapture it from known sites are often unsuccessful.

All recent collections have been made near low creek banks in tall, open eucalypt forest with dense ground cover of sedges, grasses and/or ferns. Little is known of its habits in the wild but captive animals dig into soil, climb readily on sticks and stones and have been maintained successfully on a diet consisting mostly of grain.

Breeding appears to be opportunistic: pregnant or lactating females have been recorded in all months from June to February. Two or three young comprise a litter.

It is of interest that skeletal remains of the Hastings River Mouse (almost certainly derived from accumulated owl pellets) are common in geologically recent cave deposits along the Great Dividing Range in Victoria and New South Wales, indicating that the species once had a much wider distribution. Identifiable bones have also been found in relatively fresh owl pellets found near Maleny, Queensland, about 100 kilometres north of Brisbane.

T.H. KIRKPATRICK

*Pseudomys oralis*

### References

Thomas O. (1921). On three new Australian rats. *Ann. Mag. Nat. Hist.* Ser. 9, 8: 618–622.

Tate, G.H.H. (1951). The rodents of Australia and New Guinea. Results of Archbold Expeditions No. 65. *Bull. Amer. Mus. Nat. Hist.* 97: 189–430.

Read, D. (1989). Elusive mouse lives up to its reputation. *Aust. Geographic* 15: 21.

D. WHITFORD

*The Pillaga Mouse appears to be very rare.*

land unit and the alluvials of the Gwabagar land unit and occurs in both the Pilliga Nature Reserve and the Pilliga State Forest. The species is very sparsely distributed: four nights of trapping from three sites in November 1979 (representing 1000 trap-nights) yielded only three animals; and an extensive, increased trapping

# Pilliga Mouse

## *Pseudomys pilligaensis*

### Fox and Biscoe, 1980

pil'-ig-ah-en'-sis: 'Pillaga Scrub false-mouse'

First captured in 1975, the Pilliga Mouse was described in 1980 from three museum specimens and three living animals which together comprised all that were known at the time. Subsequently, there has been some suspicion that the Pilliga Mouse was conspecific with the New Holland Mouse, but the morphology of their sperms is markedly different and captive breeding has demonstrated that hybrids between the two species are infertile.

The Pilliga Mouse has been found only in an isolated area of low-nutrient, deep sand which has long been recognised for its distinctive vegetation. Referred to as the Pilliga Scrub, this consists predominantly of Cypress Pine forests with some associated eucalypts and a sparse understorey of heath species, but also includes woodland and true heath. Within this restricted area of less than 8000 square kilometres, it is limited to dissected sandstone plateaus of the Cubbo

---

### *Pseudomys pilligaensis*

**Size**
*Head and Body Length*
63–79 mm
*Tail Length*
67–79 mm
*Weigh*
10–12 g

**Identification** Mouse-grey above, grading to russet on sides; white below. Tail darker above than below with small, black terminal tuft. Distinguished by length of hindfoot (18–19 mm) from *P. delicatulus* (15–18 mm) and *P. novaehollandiae* (20–22) mm.

**Recent Synonyms** None.

**Other Common Names** None.

**Status** Rare, limited, vulnerable.

**Subspecies** None.

**References**

Fox, B.J. and D.A. Briscoe (1980). *Pseudomys pilligaensis*, a new species of murid rodent from the Pilliga Scrub, northern New South Wales. *Aust. Mammal.* 3: 109–126.

Lim, T.L. (1992). *Conservation and biology of the rare Pilliga Mouse*, Pseudomys pilligaensis, *Fox and Briscoe 1980 (Muridae: Conilurini)*. Report to Zoological Parks Board of New South Wales.

*Pseudomys pilligaensis*

effort two years later failed to find any animals at all. A similar trapping effort of 1050 trap-nights in September 1988 produced 11 animals with a further seven animals captured in seven trapping sessions spread over the period to August 1990. A total of 2540 trap-nights in four trapping sessions from November 1990 to November 1991 produced no animals at all.

Understandably, little is known of the biology of the Pilliga Mouse. It is terrestrial and appears to live in burrows, which may explain its restriction to a sandy substrate. In many respects it is similar to the New Holland Mouse which has a coastal distribution limited to sandy areas. It is nocturnal and inhabits areas with sparse ground cover. In captivity, animals live in family groups.

The breeding season extends at least from October to mid-February and the gestation period is between 24 and 31 days. Young are born with the incisors erupted and eight laboratory-born litters had a modal size of three young.

B.J. Fox

# Heath Rat

## *Pseudomys shortridgei*

### (Thomas, 1907)

short'-rid-jee: 'Shortridge's false-mouse'
Heath Rat

Mature Australian heathlands are unproductive environments but the fires that frequently raze them release nutrients stored in the vegetation, leading to a surge of new plant growth and temporary high productivity. The Heath Rat (and some other members of the genus *Pseudomys*) are able to exploit this ephemeral habitat and much of what is known of their natural history reflects this ecological role. The Heath Rat now occurs almost exclusively in recently burnt, species-rich, dry heathlands in south-western Victoria, but a colony has recently been located in the Fitzgerald River National Park in Western Australia. It colonises patches of suitable heathland when the initial flush of growth provides adequate cover and disappears when important food species and productivity decline.

*The rare Heath Rat is found in recently burnt heathland.*

J. SEEBECK

*Pseudomys shortridgei*

Breeding takes place in late spring and summer when animals are eating flowers, seeds and berries. One or two litters of three young are produced in the season. As food resources dwindle at the end of summer, animals switch to the stems and leaves of grasses, sedges and lilies, a more difficult diet for these small animals to process. Growth and breeding cease and juveniles survive only if they have attained adult weight. Following the onset of autumn rains, the Heath Rat turns to eating the subterranean fruiting bodies of truffle-like fungi which persist until flowers and seeds become available again.

Because any patch of habitat is available for only a short period, the behaviour of the Heath Rat must be directed towards two ends: dispersing and finding newly burnt areas; and saturating these areas once they are located, the relative emphasis on these demands changing through time. Soon after colonisation, when conditions are good locally, offspring survive very well, but adults disappear during, or after, the first breeding season. As the vegetation matures and conditions deteriorate, adults cling tenaciously to their home sites and the young born each year must leave in search of unoccupied area. By living for up to four years and expelling the young, adults increase the chance that at least some of their offspring will locate a new patch of suitable habitat, the availability of which is unpredictable.

Because the Heath Rat is dependent on recently burnt areas, populations disappear as local

## Pseudomys shortridgei

**Size**
*Head and Body Length*
90–120 (110) mm
*Tail Length*
80–110 (95) mm
*Weight*
55–90 (70) g

**Identification** Grey-brown above, flecked with buff and black; paler below. Feet brown, paler than body. Face blunt, with bulging eyes. Ears covered with fine hairs; tail hairy and non-annulated (distinguishing it from species of *Rattus*).
**Recent Synonyms** None.
**Other Common Names** Blunt-faced Rat, Blunt-faced Mouse, Shortridge's Native mouse, Heath Mouse.
**Status** Rare, limited.
**Subspecies** None.
**References**
Cockburn, A. (1978). The distribution of *Pseudomys shortridgei* (Muridae: Rodentia) and its relevance to that of other heathland *Pseudomys*. *Aust. Wildl. Res.* 5: 213–219.
Cockburn, A.R., W. Braithwaite and A.K. Lee (1981). The response of the heath rat, *Pseudomys shortridgei* to pyric succession: a temporally dynamic life-history strategy. *J. Anim. Ecol.* 50: 649–666.

patches of heathland mature and, unless suitable habitat is available nearby, more widespread extinction ensues. The survival of the species requires a mosaic of areas of differing maturity but the advent of Europeans has reduced the size of patches of native bush and imposed fire controls which increase the chance that any large area will have a uniform fire history. The Heath Rat has consequently disappeared from much of its former range and is restricted to localities where human impact is light.

A. COCKBURN

# Basalt Plains Mouse

## *Pseudomys* sp.

The Basalt Plains Mouse was never recorded as a living animal. Its remains are known only from owl-roost deposits and Aboriginal middens on the basalt plains of western Victoria, particularly in lava tubes associated with dormant volcanoes such as Mount Napier, Mount Hamilton and Mount Eccles. At the latter of these localities it is the single most common species in the owl-pellet deposits, 580 individuals being represented.

Watts and Aslin (1981) suggest that it became extinct prior to European settlement, but Wakefield, who excavated its remains, believed that it survived into recent times, this opinion being supported by remains that are associated with those of the Plains Rat, Eastern Chestnut Mouse, New Holland Mouse and Pale Field Rat. Along with much of the mammal fauna from the deposits, none of these species were collected in the area in the nineteenth century. Indeed no concerted effort had been made by scientists to inventory the fauna of the region until recent years. All of these native rodents probably vanished during the nineteenth century, as a result of the widespread and rapid changes brought about by sheep.

*Pseudomys* sp.

### *Pseudomys* sp.

**Size**
Unknown but, based on skull size, probably the largest *Pseudomys* species.

**Identification**  External appearance unknown, but skull very distinctive, larger than that of other *Pseudomys*, with broad molar rows, narrow interorbital region and broad anterior palatal foramina that terminate posteriorly in line with root of the first molar.

**Recent Synonyms**  None.

**Other Common Names**  None.

**Status**  Extinct.

**Subspecies**  None.

**References**

Anon. (1971). The Journal of Granville Williams Chetwynd Stapylton (in) Douglas, H.M. and L. O'Brien (eds) *The Natural History of Western Victoria*. Aust. Inst. Agric. Sci., Horsham, Victoria. pp. 85–116.

Wakefield, N.A. (1971). Mammals of Western Victoria (in) Douglas, H.M. and L. O'Brien (eds) *The Natural History of Western Victoria*. Aust. Inst. Agric. Sci., Horsham, Victoria. pp. 35–51.

Watts, C.H.S. and H.J. Aslin (1981). *The Rodents of Australia*. Angus and Robertson, Sydney. pp 175-176.

Stapylton, a member of Mitchell's 1836 expedition to the basalt plains area, described the region as: 'fine grassy plains…the country is covered in *Xanthonia* [presumably *Danthonia* the generic name for Wallaby Grass], but wet…the broken nature of the ground…covered with small pools [in July], which must always retain considerable moisture'. He also comments that: 'The settler will have no timber to clear away'. This country was first settled in 1834 and by 1844 the entire region had been taken up and was carrying large flocks of sheep. Destruction of the natural tussock grassland was extremely rapid. Virtually no natural grassland survives in the region today.

T.F. FLANNERY

G.A. HOYE

# Common Rock-rat

## *Zyzomys argurus*

### (Thomas, 1889)

ziz'-oh-mis  arg-yue'-rus: 'silver-tailed
zyzomys' (meaning unknown)

Rock-rats are a little-known group of rodents, endemic to northern Australia. The tail is the most distinctive feature, being very thick, especially towards the base (leading to the earlier name, thick-tailed rats). Compared with other rodents, they lose the skin of their tails very easily and the tail then withers away, leaving a stump

*Zyzomys argurus*

*The Common Rock-rat lives among rocky outcrops, usually of sandstone.*

that does not regrow. Shortened or missing tails are common in the wild.

The Common Rock-rat has the widest distribution of the five species that make up the genus, being found in suitable habitats across northern Australia from the Dampier Archipelago in West Australia to Ayr on the Queensland coast. The western and eastern populations appear to be isolated but populations are continuous through the Kimberley, Top End and hinterland of the Gulf of Carpentaria. An isolated population has been found on the Stokes Range in central Queensland. When present it is usually abundant.

It occupies a wider range of habitats than the other rock-rats but is always associated with rocky outcrops, particularly sandstone formations, but is occasionally found in igneous, lateritic or limestone scree slopes. Rocks provide the cover needed for nesting and it does not burrow into the shallow, hard soils associated with rocky ground. The vegetation of these habitats varies from riverine communities through open forest to woodland with an understorey of spinifex or tropical grasses. The Common Rock-rat may occupy monsoon forest patches and dry vine thickets where the larger rock-rat species may also occur.

Breeding is intensive throughout the year, with a slight suppression in pregnancies during the height of the wet season (January). The gestation period is 35 days and litters comprise one to four young. These are not carried around attached to the mother's teats, as in some related rodents, but are left in the nest (perhaps to allow the mother greater mobility on the rocky terrain). Growth is rapid and both sexes become mature at five to six months, at which time the testes of the males descend into the scrotum. Some individuals may live for more than two years, but few survive beyond one breeding season.

The Common Rock-rat eats plant stems, leaves, seeds, small amounts of insect material and fungi. Small hard-shelled seeds such as the

J.A. KERLE

*The Arnhem Land Rock-rat is one of two species that were described as recently as 1989.*

## *Zyzomys argurus*

**Size**

*Head and Body Length*
85–122 (108) mm
*Tail Length*
up to 125 mm (damaged in many specimens)
*Weight*
26–55 (36) g

**Identification** Varies from clay brown to hair brown above; with white below. Distinguished from all other *Zyzomys* by slender form, smaller size, distinctly lesser weight and lightly haired tail.

**Recent Synonyms** *Zyzomys indutus*.

**Other Common Names** White-tailed Rat, White-tailed Rock-rat.

**Status** Common in suitable habitat.

**Subspecies** None.

**References**

Begg, R.J. (1981). The small mammals of Little Nourlangie Rock, Northern Territory. IV. Ecology of *Zyzomys woodwardi*, the Large Rock-rat and *Z. argurus*, the Common Rock-rat (Rodentia: Muridae). *Aust. Wildl. Res.* 8: 307–320.

Calaby, J.H. and J.M. Taylor (1983). Breeding in wild populations of the Australian Rock-rats, *Zyzomys argurus* and *Z. woodwardi*. *J. Mammal.* 64: 610–616.

Billy Goat Plum, *Terminalia ferdinandiana*, show a characteristic gnawing pattern when opened by this rodent.

The wet season (December to March) is a very stressful period. It must contend with possible flooding of the nest and a reduction in the source of food (since few plants seed at this time). It is very hard to trap for the two months around the end of the wet season, either because individuals move around less or because of a change in social behaviour prior to mating, which may involve formation of male–female pair bonds.

M.R. FLEMING

# Arnhem Land Rock-rat

## *Zyzomys maini*

### Kitchener, 1989

may'-nee: 'Main's zyzomys'

It was once thought that a single form of large rock-rat occurred across northern Australia but a recent taxonomic study has separated three species; *Z. maini* is restricted to the Arnhem Land plateau of the Northern Territory. The limited knowledge of this species comes from studies along the western edge of the Arnhem Land Plateau where it can be locally abundant and animals are readily trapped. There is a single record of two specimens from the Cadell River in central Arnhem Land, but recent faunal surveys in eastern Arnhem Land have failed to find it again.

The Arnhem Land Plateau is a residual area of hard Kambolgi sandstone, deeply dissected by gorges and gullies that have cut down along fault lines. The rock scree slopes of the cliff lines and the boulder beds of the gorges provide the primary habitat for this rock-rat—large tumbled

## *Zyzomys maini*

**Size**

*Head and Body Length*
115–170 (146) mm
*Tail Length*
up to 150 mm (often damaged)
*Weight*
70–186 (94) g

**Identification** Grey-brown above; white below. Distinguished from *Zyzomys argurus* by larger size and well-haired tail. Not sympatric with the other large rock-rats.

**Recent Synonyms** *Laomys woodwardi*, *Zyzomys woodwardi* (part).

**Other Common Names** Woodward's Thick-tailed Rat, Woodward's Rock-rat, Western Thick-tailed Rat, Large Rock-rat.

**Status** Sparse but can be locally abundant.

**Subspecies** None.

**References**

Begg, R.J. (1981). The small mammals of Little Nourlangie Rock, Northern Territory. IV. Ecology of *Zyzomys woodwardi*, the Large Rock-rat and *Z. argurus,* the Common Rock-rat (Rodentia: Muridae). *Aust. Wild. Res.* 8: 307–321.

Calaby, J.H. and J.M. Taylor (1983). Breeding in wild populations of the Australian Rock-rats, *Zyzomys argurus* and *Z. woodwardi*. *J. Mammal.* 64: 610–616.

Kitchener, D.J. (1989). Taxonomic appraisal of *Zyzomys* (Rodentia: Muridae) with descriptions of two new species from the Northern Territory, Australia. *Rec. West. Aust. Mus.* 64: 331–373.

*Zyzomys maini*

boulders covered with deciduous vine thicket, with an overstorey of monsoon rainforest trees. Such a habitat, with little understorey and a thick leaf litter layer, often occurs in gullies or on scree slopes where seepage or protection from fire promote the retention of a rainforest community. Sandstone areas with scattered eucalypts are also occupied by the Arnhem Land Rock-rat. Where it occurs in the same general area as the Common Rock-rat, the Arnhem Land Rock-rat appears to occupy the wetter, more heavily vegetated areas and to exclude the smaller species.

The large seeds of certain rainforest trees and the much smaller seeds of perennial grasses provide most of the diet. The hard-shelled seeds of rainforest trees, such as Mango Bark, *Canarium australianum*, Emu Apple, *Owenia vernicosa*, Geebung, *Persoonia falcata* and Green Plum, *Buchanania obovata*, are collected and carried to crevices in the rocks where the thick seedcoats are gnawed through, out of danger from predators. Accumulations of these chewed seeds on ledges indicate the presence of this rock-rat.

The pattern of reproduction and life history is very similar to that of the Common Rock-rat. It breeds throughout the year, producing litters of two to three. Young become mature at five to six months and may live for up to two years, although few survive a second breeding season.

The species seems to be secure, since all its major populations are located within Kakadu National Park, but its strong preference for relict monsoon rainforest patches makes the Arnhem Land Rock-rat vulnerable to those threats (such as fire and tourists) which degrade the rainforest. Already one other species of rodent, the Golden-backed Tree-rat, appears to have been lost from this habitat in the Northern Territory.

M.R. FLEMING

# Carpentarian Rock-rat

## *Zyzomys palatalis*

### Kitchener, 1989

pal'-ah-tah'-lis: 'notable-palate zyzomys'

*Zyzomys palatalis*

The Carpentarian Rock-rat is known from only three gorges in rugged sandstone ranges near the border between the Northern Territory and Queensland. As with *Zyzomys woodwardi* from the Kimberley and *Z. maini* from the Arnhem Land area, it is associated mainly with thickets of monsoon rainforest which fringe some escarpments and scree slopes in this region. Woody plants in these thickets provide the large seeds and fruit

that dominate the diet. Much of this habitat is being degraded as a consequence of recent changes in fire regimes and the spread of the Pig and cattle.

The species was described as recently as 1989, based on the three specimens then known (of which only one was adult), collected at Echo Gorge in Wollogorang Station in 1986 and 1987. Subsequently it was found at a further two localities, about 30 kilometres from the original site.

*The Carpentarian Rock-rat was described in 1989 and little is yet known of its biology. The individual illustrated here is a juvenile.*

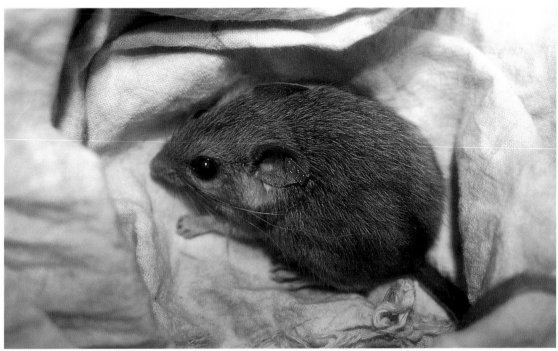

C.R. TRAINOR

## Central Rock-rat

### *Zyzomys pedunculatus*

Waite, 1896

ped-unk'-yue-lah'-tus:
'swollen (-tailed) zyzomys'

The Central Rock-rat is either one of Australia's rarest rodents or its most recently extinct mammal. It was first recorded by the Horn Scientific Expedition to Central Australia in 1894. Of the 33 specimens obtained to date, only five were collected in the twentieth century.

The last specimen was collected in 1960 by a stockman in the Western MacDonnell Ranges near Mount Leibig, about 300 kilometres west of Alice Springs, while it was attempting to break into the camp food supplies. It has been collected on only two other occasions in this century: in 1900 from the Alice Springs region and in 1952 from near The Granites in the Tanami Desert. All other specimens, except one from Illamurta, are from 'Central Australia' or 'Alice Springs'. Hedley Finlayson noted in 1961 that 'acceptable records' had come from Hugh Creek (1935), Napperby Hills (1950) and the Davenport Range (1953) but he had not seen these specimens.

The Central Rock-rat is very similar in appearance to other species of *Zyzomys*, displaying the characteristic 'roman nose', harsh fur and incrassated tail. The tail is quite densely furred, concealing the tail scales, and has a noticeable tuft at its tip.

No first-hand information is available on its biology, but it probably resembles other species of *Zyzomys*, found in the more tropical areas of northern Australia. Broad support for this assumption is provided by the records of collection sites. These vary greatly in geology and form, from granitic boulder fields and quartzite scree slopes to eroded sandstone cliff lines. As a consequence of this varied geology, the vegetation is

---

### *Zyzomys palatalis*

**Size**

*Head and Body Length*
126–197 (161) mm
*Tail Length*
95–150 (117) mm
*Weight*
111–136 (123) g

**Identification** A large rock-rat, very similar to *Z. maini* and *Z. woodwardi*, but tail shorter than head-body length. Anterior palatal foramen is broader than in the other large *Zyzomus* species edged at the external margins by sharp, low palatal ridge.

**Recent Synonyms** None.

**Other Common Names** None.

**Status** Endangered. Known distribution extremely limited and entirely in pastoral land; preferred habitat is being severely degraded by fire and introduced herbivores.

**Subspecies** None.

**References**

Kitchener, D.J. (1989). Taxonomic appraisal of *Zyzomys* (Rodentia: Muridae) with descriptions of two new species from the Northern Territory, Australia. *Rec. West. Aust. Mus.* 14: 331–373.

Menkhorst, K.A. and J.C.Z. Woinarski (1994). Further records of the Carpentarian Rock-rat *Zyzomys palatalis* from the Gulf region of the Northern Territory. *Northern Territory Naturalist*. 30: 40–41.

The smaller Common Rock-rat was also abundant at these locations.

Very little is known about the ecology of this species. As with all other rock-rats, the tail is usually greatly fattened and is very susceptible to breakage. The holotype, a female, was pregnant when collected in June and the two juveniles recorded were both collected in September, suggesting that breeding occurs towards the middle of the dry season.

J.C.Z. WOINARSKI, S. CHURCHILL AND
C. TRAINOR

## *Zyzomys pedunculatus*

**Size**

*Head and Body Length*
108–140 mm

*Tail Length*
110–140 mm

**Identification** Yellow-brown above, cream to white below. Distinguished from rodents of other genera by characteristic thickened tail. Similar externally to other *Zyzomys* species but well separated by known distribution.

**Recent Synonyms** *Laomys pedunculatus.*

**Other Common Names.** None.

**Status** Endangered, presumed extinct.

**Subspecies** None.

**Reference**

Kitchener, D.J. (1989). Taxonomic appraisal of *Zyzomys* (Rodentia, Muridae) with descriptions of two new species from the Northern Territory, Australia. *Rec. West. Aust. Mus.* 14: 331–373.

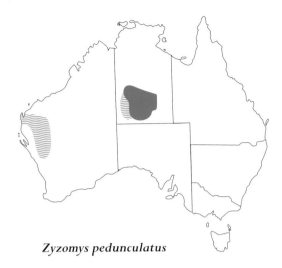

*Zyzomys pedunculatus*

also diverse, varying from sparse hummock grasses (*Triodia pungens* and *T. clelandii*), to shrublands of varied species such as *Grevillea wickhamii*, *Eremophila latrobei*, *Solanum* spp. and *Dodonaea* spp.). The local Quandong, *Santalum accuminatum*, may also have played an important role as a food resource for this rodent. Old, chewed Quandong seeds can still be found in rock crevices and on ledges in the central ranges, many of which may have been used by the Central Rock-rat. The Quandong has been greatly reduced in density due to grazing pressure from introduced herbivores.

The outlook for the Central Rock-rat appears very poor. An extensive survey during 1990 and some shorter surveys in the 1970s and early 1980s failed to find any evidence of its survival.

The decline of the Central Rock-rat cannot be readily attributed to any one factor. It appears to have been uncommon when first recorded by Europeans, so habitat alteration, caused largely by introduced herbivores, probably expedited its demise. Whatever the cause of its decline, it may now be extinct, confirming fears indicated by Finlayson in the mid-1930s: 'Neither specimens nor recognisable accounts of this interesting rat could be obtained'.

D. WURST

# Kimberley Rock-rat

## *Zyzomys woodwardi*

### (Thomas, 1909)

wood'-ward-ee: 'Woodward's zyzomys'

Restricted entirely to the North Kimberley, this large rock-rat is most abundant in the high rainfall country found along the district's north-west coastline. The type specimens were collected from a 'rough stony gorge' near Wyndham in 1908. These are still the only records eastward of Cambridge Gulf and mark the eastern extremity of its range.

Even in the north-western Kimberley, its distribution is patchy. It prefers the most rugged areas of rock scree, with massive boulders. On the mainland and adjacent islands, it favours rainforest patches on sandstone, volcanic or laterite scree, but it is also found in open sandstone country where a scattered cover of low trees—such as *Owenia*, *Terminalia*, *Xanthostemon*, *Planchonia* and *Buchanania*—over wattle shrubs and hummock grass, mantle the precipitous slopes and boulder-strewn gullies, or cling to ledges and crevices along gorges or behind mangroves.

Buccaneer & Bonaparte Archipelagos

*Zyzomys woodwardi*

---

*Zyzomys woodwardi*

**Size**
*Head and Body Length*
121–166 mm (143 mm)
*Tail Length*
111–130 mm (122 mm)
*Weight*
80–186 g (136 g)

**Identification** Cinnamon brown above; fur below neutral grey tipped white to give white belly. Tail moderately furred and thickened basally, but often missing. Differs from other *Zyzomys* (except *Z. palatalis*) in averaging larger in all measurements and tail distinctly shorter than head-body length. Differs from *Z. palatalis* in shorter, narrower anterior palatine foramina, without palatal ridges.

**Recent Synonyms** *Laomys woodwardi*.

**Other Common Names** Large Rock-rat, Woodward's Thick-tailed Rat.

**Status** Common but patchily distributed.

**Subspecies** None.

**References**

Kitchener, D.J. (1989). Taxonomic appraisal of *Zyzomys* (Rodential, Muridae) with descriptions of two new species from Northern Territory, Australia. *Rec. West. Aust. Mus.* 14: 331–373.

Kemper, C.M., D.J. Kitchener, W.F. Humphreys and R.A. How (1987). Small mammals of the Mitchell Plateau region, Kimberley, Western Australia. *Aust. Wildl. Res.* 14: 397–414.

Friend, G.R., K.D. Morris and N.L. McKenzie (1991). The mammal fauna of Kimberley rainforests (in) N.L. McKenzie, R.B. Johnston and P.G. Kendrick (eds) *Kimberley Rainforests of Australia*. Surrey Beatty and Sons, Sydney. pp. 393–412.

Local populations can be moderately dense and movements by individuals of up to 80 metres in two nights have been recorded. When handled, its tail is very fragile and patches of its crisp fur easily rub off to expose the skin. Throughout its range, it is frequently trapped in association with the Common Rock-rat and the Northern Quoll. In the Prince Regent Nature Reserve it is found in the same environments as the Golden Bandicoot, *Isoodon auratus*.

Breeding occurs throughout the year, but a study on Mitchell Plateau found that few females were pregnant late in the dry season. The Mitchell Plateau population fluctuated in numbers with a marked decline at the end of wet season (April). A scarcity of seeds during this period may cause the decline.

*The Kimberley Rock-rat feeds on the seeds of grasses and of rainforest trees.*

The diet is thought to resemble that of the Arnhem Land Rock-rat, comprising a mixture of hard-shelled seeds from rainforest trees and smaller seeds of perennial grasses. The stomach contents from a small sample contained only fibrous vegetable material. *Terminalia* seeds show a characteristic hole when chewed open by this species.

Rainforest patches are contracting on the Mitchell Plateau and on the inland periphery of its range. Cattle have opened up these patches to savanna grasses that carry the annual, dry-season bushfires into the rainforest. Stock removal would improve the Kimberley Rock-rat's chance of persisting. Secure populations occur in the Prince Regent Nature Reserve and on island groups along the length of the north-western coastline of the Kimberley.

M.R. FLEMING AND N.L. McKENZIE

J. LOCHMAN

## TRIBE HYDROMYINI

Much better represented in New Guinea, this group includes two Australian genera, *Hydromys* and *Xeromy*. They have two molars in each tooth row (three in other native rodents). Both have water-repellent fur and hunt other animals in and around water.

# Water-rat

## *Hydromys chrysogaster*

### Geoffroy, 1804

hide'-roh-mis kris'-oh-gas-ter:
'golden-bellied water-rat'

Unrelated to the Water Vole of Europe, which is also called a water-rat, this species is restricted to Australia, New Guinea and adjacent islands. Among its adaptations for aquatic life are broad, partially webbed hindfeet and waterproof fur. The dense, soft, lustrous pelage has led, particularly in the past, to its commercial exploitation as a fur-bearer.

The Water-rat usually lives in the vicinity of permanent bodies of fresh or brackish water and even on some marine beaches. It is an occasional vagrant to temporary waters. Nests are made at the end of tunnels in banks or occasionally in logs.

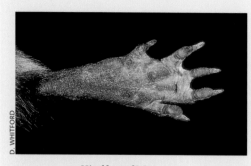

*Hindfoot of Water-rat.*

Slightly clumsy on land, it can climb hollow trees in search of prey but takes most of its food from the water, searching among vegetation along the shoreline and around submerged roots and sunken logs. It eats little plant material, being essentially an opportunistic predator on large aquatic insects, fishes, crustaceans and mussels. Frogs, lizards, small mammals, fresh carrion and water birds may also be taken. In winter, it spends less time in water and eats larger prey, often vertebrates. Prey is often carried to a regularly used feeding site.

The Water-rat is unusual among Australian rodents in not being entirely nocturnal; most activity takes place around sunset but animals may forage in full daylight. Individuals appear to be territorial and, where populations are dense, there is considerable fighting, particularly among males, many of which have damaged tails. In south-eastern Queensland, home ranges are 2–10 hectares, larger in estuaries than in fresh water.

Breeding can occur throughout the year but most litters are born between spring and late summer. Social factors and rainfall influence the time of breeding and the age at which individuals breed. Under favourable conditions a female may become sexually mature at the age of four months and breed in the season of its birth. Commonly, however, breeding does not begin until animals are about eight months old. Gestation occupies about 34 days and up to five litters (usually one to two) may be produced annually, each usually with three to four young. The female has four teats and suckles the young for about four weeks, after which they remain with her for up to another four weeks, gradually attaining independence. At the age of two to three months, the fur becomes

J. LOCHMAN

*Apart from the Platypus, the Water-rat is the only amphibious Australian mammal.*

waterproof and it is thereafter moulted twice a year, being coarser and less dense in summer than in winter. Growth continues throughout life, males becoming larger than females.

Young animals are preyed upon by snakes and large fishes; adults and young by birds of prey and cats. Populations are dense in some irrigated areas, particularly in drainage swamps, but appear to be sparse along rivers. Swamp reduction and flood mitigation practices have removed much habitat but the overall range of the Water-rat does not seem to have changed much since European settlement. It is regarded by some as a pest of inland fisheries and its burrows are said to cause damage to irrigation.

P.D. OLSEN

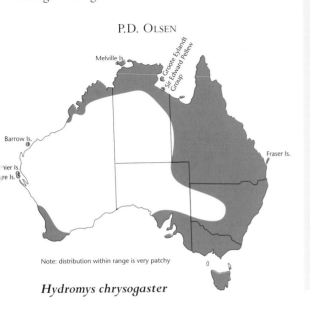

Note: distribution within range is very patchy

*Hydromys chrysogaster*

## *Hydromys chrysogaster*

**Size**
**South-eastern Australia**
*Head and Body Length*
231–345 (310) mm (males)
245–370 (290) mm (females)
*Tail Length*
227–320 (275) mm (males)
242–325 (272) mm (females)
*Weight*
400–1275 (755) g (males)
340–992 (606) g (females)

**Identification**  Colour variable with locality; almost black to slate grey above; white to orange below. Tail thick, well covered by dark hair, usually with white tip. Head and body somewhat compressed; body tapering gradually to tail. Small ears, eyes and nostrils set high on head. Hindfeet broad and partially webbed.

**Recent Synonyms**  *Hydromys lawnensis, Hydromys moae, Hydromys grootensis.*

**Other Common Names**  Beaver Rat.

**Status**  Sparse to common, limited.

**Subspecies**  Numerous subspecific forms, many originally regarded as separate species, have been described, but most are probably invalid.

**Extralimital Distribution**  New Guinea and some adjacent islands.

**References**

Fanning, F.D. and T.J. Dawson (1980). Body temperature variability in the Australian Water Rat, *Hydromys chrysogaster*, in air and water. *Aust. J. Zool.* 28: 229–238.

Harris, W.F. (1978). *An ecological study of the Water Rat* (Hydromys chrysogaster *Geoffroy) in South-east Queensland.* MSc. thesis. University of Queensland, Brisbane.

Olsen, P.D. (1982). Reproductive biology and development of the Water Rat, *Hydromys chrysogaster*, in captivity. *Aust. Wildl. Res.* 9: 39–53.

C. ANDREW HENLEY/LARUS

# False Water-rat

## *Xeromys myoides*

### Thomas, 1889

ksee'-roh-mis mie-oy-dayz:
'mouse-like dry-mouse'

Few living native rodents are as poorly known as this species. For over a hundred years since its description in 1889, it managed to elude collectors and there are no more than 14 specimens in museums around the world. These are from a variety of well-watered habitats including mangrove forests, freshwater lagoons, sedged lakes close to foredunes, swamps and crocodile stomachs, around the northern Australian coastline from the Western Australia–Northern Territory border to

Melville Is.

Fraser Is.

*Xeromys myoides*

*False Water-rat digging in mangrove swamp.*

the Coomera River, 50 kilometres south-east of Brisbane. In traps and in captivity it was known to relish fish, crabs and dog biscuits but its diet and habits in the wild remained a mystery.

In 1976, zoologists studying crocodiles on Melville Island, Northern Territory, described an extraordinary structure in which a female False Water-rat and her two young were sheltering. The nest, a 60 centimetre high mound of friable black soil, stood like a leaning termite mound against the base of a mangrove tree. Inside the mound was a labyrinth of narrow tunnels and, just under the upper surface, was a dry chamber where the rodents nestled among mangrove leaves.

In 1991, ten active nests were found in sedgelands of the Myora mangrove community on North Stradbroke Island. These varied from complex, domed, sedge-covered mounds built within the reed zone, to simple burrow systems incorporated into the existing supralittoral peat bank at the sedge-wallum interface. Over the

---

*Xeromys myoides*

**Size**
*Head and Body Length*
**Stradbroke Island population**
87–119 (100) mm (male)
82–120 (102) mm (female)
*Tail Length*
70–91 (81) mm (male)
71–87 (80) mm (female)
*Weight*
32–54 (41) g (male and female)

**Identification** Fur short and very silky; dark slate-grey to brown above; pure white below. Adults usually white-spotted dorsally (Stradbroke Island population). Ears rounded and short; eyes very small. Only two molars in each of the upper and lower rows. Strong musty odour.
**Recent Synonyms** None.
**Other Common Names** None
**Status** Rare, vulnerable.
**Subspecies** None.

*False Water-rat using its snout to compact mud at entrance to its retreat.*

course of 18 months over 100 individuals were recorded from the island's western mangroves. Complex mound-nests contained up to eight individuals represented by a mix of all age categories, juvenile, subadults and adults. Usually one adult male and more than one adult female were resident at the same mound.

Many prey items were recorded, including four species of crab, a mud lobster, marine shellfish and an undescribed genus of marine polyclads (mucous-covered flatworms universally known for their toxicity and previously unknown to be eaten by any animal). Radio-tracking of individuals at Myora demonstrated a dependence on mangroves for foraging areas; an avoidance of the drier wallum; home ranges, among reeds and mangroves, of around 0.6 hectares; and a frenetic disposition that could involve an animal travelling distances of more than 2 kilometres a night while scouring its territory. It is nocturnal and, as the tide recedes, it moves from its nest to forage among mangrove roots, hollow trunks and logs. It follows established pathways, feeding, sheltering and resting in favoured, identifiable hollows, usually in the bases of Grey Mangroves. These sites are often characterised by the presence of discarded fragments of crab and lobster shell and a pungent scent typical of the species. Individuals usually return to their nests before tides inundate

### References

Magnussen, W.D., G.J.W. Webb and J.A. Taylor (1976). Two new locality records, a new habitat and a nest description of *Xeromys myoides* Thomas (Rodentia, Muridae). *Aust. Wild. Res.* 3: 153–157.

Van Dyck, S. (1992). Parting the reeds on Myora's *Xeromys* kibbutz. *Wildlife Australia* 29 (4): 8–10.

Van Dyck, S.M. and Durbidge, E. (1992). A nesting community of False Water Rats *Xeromys myoides* on the Myora sedgelands, North Stradbroke Island. *Mem. Qld. Mus.* 32: 374.

the reed-zone but may occasionally take shelter up a suitable hollow mangrove spout to wait out the tide. No truly arboreal activity has been seen and, although it is an adept swimmer and does not hesitate to cross deep pools and inundated patches of low-lying ground, it seems not to be truly aquatic: it could best be described as a terrestrial 'puddler'. Births are thought to occur in any month of the year and up to four young have been found in a nest. Little more is known about its reproduction.

Swamp and mangrove 'reclamation', the alteration of adjacent water tables and offshore pollution are the biggest perceived threats to the species.

S. VAN DYCK

## TRIBE UROMYINI

The tree-rats of this group are largely arboreal in tropical forests. They are referred to as 'mosaic-tailed rats' in reference to the arrangement of scales on the tail, fitting together like the tiles of a mosaic rather than overlapping, as in most murids. The closely related genera, *Uromys* and *Melomys* are members of the tribe. The prehensile-tailed rats, of the genus *Pogonomys,* are also included here, but with reservations.

# Grassland Melomys

## *Melomys burtoni*

## (Ramsay, 1887)

mel'-oh-mis ber'-tun-ee:
'Burton's Melanesian-mouse'

Although one of the most common rodents of the tropical coastal region, this species was not described from coastal Queensland until 1916 and, from the Northern Territory, in 1945. The earliest specimens came from the Derby area in 1887, but it has been commonly found in Western Australia only since the 1970s. It occurs on many offshore islands.

It lives in a wide range of habitats but is aptly named the Grassland Melomys in Queensland, where it prefers the tall grasslands of the coast, sedgelands, open forest, woodland and grassy patches within rainforests. It has flourished in canefields and become a serious economic pest. In Western Australia and the Northern Territory, it has a clear preference for monsoon forests and vine thickets, swamps and riparian woodlands (with pandanus and paperbark trees) and mangroves. In north-eastern New South Wales, it has been found in heathland bordering eucalypt woodland.

In addition to sugar cane, the Grassland Melomys eats plant stems, seed, fruits and

---

### *Melomys burtoni*

**Size**
*Head and Body Length*
90–160 mm (male)
95–145 mm (female)
*Tail Length*
90–175 mm (male)
100–170 mm (female)
*Weight*
26–124 g (male)
26–97 g (female)

**Identification**  Very variable in colour, from dark grey to grey-brown and reddish brown above; paler below, white grey or cream, sometimes with pale orange on the sides; feet dark grey or buff above. In Northern Territory and Western Australia, it is the only described *Melomys* species, readily distinguished from other small rodents by mosaic scale pattern on the almost naked tail; in Queensland, smaller than other *Melomys* species.

**Recent Synonyms**  *Melomys cervinipes* (in the Northern Territory), *Melomys australis, Melomys callopes, Melomys littoralis, Melomys melicus, Melomys mixtus, Melomys murinus.*

---

insects, the proportions of which vary seasonally. One sample from the Northern Territory indicated a dry season diet of plant stems, a very varied wet season diet and a late-wet season preference for seeds. Individuals from the Archer River on

*The Grassland Melomys is one of the commonest rodents in coastal Queensland.*

Cape York ate large quantities of winged insects, especially grasshoppers.

The Grassland Melomys is an agile climber, having relatively broad hindfeet and a partially prehensile tail. It is often observed in trees or on grass stems. Nests are mostly built above ground and used by an individual or a female with young. The nest is generally spherical, 20–30 centimetres in diameter and often with two entrances. In grasslands and canefields it is built up to a metre above ground using grass and dead leaves. Elsewhere, nests made of shredded pandanus leaves or bark have been found in shrubs, pandanus and hollow trees; in a hollow log; and under sheets of bark on the rainforest floor. Short burrows have occasionally been found.

In the Kimberley and on Cobourg Peninsula, the Grassland Melomys breeds throughout the year with an increased incidence during the wet season. Breeding appears to be more seasonal in Queensland, occurring mostly in autumn and winter. The breeding pattern allows for a regular input of a few young into the population, rather than a sudden increase in numbers. Litters of up to five have been recorded, two or three being

most common. Young develop rapidly and are weaned when three weeks old.

Significant variation in body size and colour between populations of the Grassland Melomys led to its description as eight different species and several subspecies. Populations with the largest animals are found in the Kimberley monsoon forests and mangroves and in the vine thickets on Cobourg Peninsula, where they weigh about 80 grams. Individuals from Cape York are half this size and have a proportionally longer tail. In the riparian woodlands of the Kakadu area, Northern Territory, the average weight is

*Melomys burtoni*

633

# Cape York Melomys

## *Melomys capensis*

### Tate, 1951

kape-en'-sis: 'Cape (York) Melanesian-mouse'

This small, semi-arboreal rodent is found only on Cape York Peninsula and some offshore islands. It occurs in rainforest and adjacent habitats, such as monsoon forest, microphyll closed forest and eucalypt woodland. Its preferred habitat is rainforest with a dense understorey of saplings and vines. The Cape York Melomys is easily recognised by its long, almost naked tail, although it can be mistaken for its close relative, the Grassland Melomys, which is smaller and found mainly in grassland.

By day the Cape York Melomys nests in tree hollows lined with dried leaves. At night it forages on tree branches and the forest floor, feeding on a variety of leaves, shoots, fruits and seeds. Its semi-prehensile tail helps it climb with agility on small branches and vines.

Populations in the rainforest are relatively stable, but numbers increase slightly in the late dry to early wet season (November–January) when fruits are abundant. Breeding occurs

**Other Common Names** Banana Rat, Cape York Scale-tailed Rat, Groote Eylandt Melomys, Hayman Island Melomys, Khaki Rat, Little Cape York Melomys, Little Melomys, Long-tailed Melomys, Lonnberg's Scale-tailed Rat, Small Khaki Rat, Small Mosaic-tailed Rat, Tree Rat, Western Melomys, Burton's Melomys.

**Status** Abundant in sugar cane areas of north-eastern Queensland and in monsoon forest, riparian habitat and mangroves of Northern Territory and Western Australia. Rare in northern New South Wales.

**Subspecies** None, but some forms previously described as species may prove to be sub-species.

**References**

Baverstock, P.R., C.H.S. Watts, M. Adams and M. Gelder (1980). Chromosomal and electrophoretic studies of Australian *Melomys* (Rodentia: Muridae). *Aust. J. Zool.* 28: 553–574.

Kemper, C.M., D.J. Kitchener, W.F. Humphreys, R.A. How, A.J. Bradley and L.H. Schmitt (1987). The demography and physiology of *Melomys* spp. (Rodentia: Muridae) in the Mitchell Plateau area, Kimberley, Western Australia. *J. Zool. Lond.* 212: 553–562.

Begg, R.B. Walsh, F. Woerle and S. King (1983). Ecology of *Melomys burtoni*, the Grassland Melomys (Rodentia: Muridae) at Coburg Peninsula, N.T. *Aust. Wild. Res.* 10: 259–267.

42 grams, compared with 53 grams for the adjacent monsoon forests. The significant difference in size between animals occupying different habitats suggests some ecological separation but the taxonomic significance of this is unknown. In Western Australia, there is some doubt about the identity of this species and it is often referred to as '*Melomys* sp. cf. *burtoni*'.

J.A. KERLE

*Melomys capensis*

C. ANDREW HENLEY/LARUS

*The Cape York Melomys climbs with the aid of its long tail.*

## Melomys capensis

**Size**

*Head and Body Length*
120–162 (144) mm (males)
119–152 (135) mm (females)
*Tail Length*
129–172 (148) mm (males)
133–171 (149) mm (females)
*Weight*
45–116 (75) g (males)
45–96 (65) g (females)

**Identification** Light brown to orange-brown above; white to cream below. Tail brown to black, almost naked and usually longer than head and body. Feet fawn. Fur soft. Larger than *M. burtoni*. Indistinguishable externally from *M. cervinipes*.

**Recent Synonyms** *Melomys cervinipes capensis.*

**Other Common Names** None.

**Status** Common in suitable habitat; habitat limited.

**Subspecies** None.

**Reference**

Watts C.H. and H.J. Aslin (1981). *The Rodents of Australia*. Angus and Robertson, Sydney.

throughout the year and does not exhibit any strong seasonal trend. Litters are usually of two, although the female has four teats. Females give birth to their first litter as early as 80 days of age and continue to breed into their second year, producing a number of litters per year. Males are probably promiscuous, as they traverse home ranges of a number of females and have huge testes (indicative of sperm competition).

At birth, the young weigh 5–8 grams and their eyes and ears are closed. When disturbed, the mother flees the nest with her young clinging to her teats. The young hiss when lost, assisting their mother to locate and retrieve them. Young develop rapidly, emerging from the nest when they reach 20 grams, at about 13 days of age. Juveniles have distinctive grey fur.

The Cape York Melomys often enters houses located in its natural habitat, keeping people awake at night as it noisily scampers around the place. It has been known to shred books into nesting material and to damage cars by gnawing electric cables. It is not particularly wary of humans and can easily be followed at night using a torch-light. It can be captured by hand or by traps baited with fruit. An unorthodox method of getting rid of the animal is by keeping a White-tailed Rat: the melomys disappears as soon as the rat moves in.

L.K-P. LEUNG

# Fawn-footed Melomys

## *Melomys cervinipes*

### (Gould, 1852)

ser-vin'-ee-pez: 'tawny-foot
Melanesian-mouse'

The Fawn-footed Melomys is found in the closed
forests of eastern Australia from just south of the
border between Queensland and New South
Wales to Cape York. In the northern part of its
range, it seldom ventures out from the forest
into the adjacent grasslands which are inhabited
by its close relative, the Grassland Melomys. In
the south its habitat requirements are less strict
and it is found in areas of wet sclerophyll forest
and coastal mangrove forest as well as closed
forest. In all these habitats, it is restricted to
areas with a ground cover of leaf litter and logs.

It is very variable in colour and has been described a number of times under different names.
The commonest form is dull reddish-brown with
a long, brown tail but some individuals can be
striking bright russet or orange-brown with jet
black tails.

*Melomys cervinipes*

---

### *Melomys cervinipes*

**Size**

*Head and Body Length*
120–200 mm (males)
95–150 mm (females)
*Tail Length*
usually longer than body
*Weight*
45–110 g (males)
45–110 g (females)

**Identification**

Dull reddish-brown to orange-brown above;
buff white to grey below. Feet fawn; tail brown
to black. Fur soft. Usually larger than *Melomys
burtoni*, from which it is distinguished by having
four (rather than five) roots to the upper first
two molars. Indistinguishable externally from
*M. capensis*.

**Recent Synonyms**

*Uromys banfieldi*, *Melomys limicauda*.

**Other Common Names**

Large Khaki Rat, Fawn-footed Scale-tailed·Rat.

**Status**

Common in Queensland, rare in New South
Wales. Local extinction likely where rainforest
destroyed.

**Subspecies**

None currently recognised.

**References**

Baverstock, P.R., C.H.S. Watts, M. Adams and
   M. Gelder (1980). Chromosomal and elec-
   trophoretic studies of Australian *Melomys*
   (Rodentia: Muridae). *Aust. J. Zool.* 28:
   553–574.
Wood, D.H. (1971). The ecology of *Rattus
   fuscipes* and *Melomys cervinipes* (Rodentia:
   Muridae) in a south-east Queensland
   Rainforest. *Aust. J. Zool.* 19: 371–392.

The Fawn-footed Melomys is an excellent
climber, moving easily along the smallest
branches of trees to browse on leaves, shoots
and fruits. If vines entangle the trunk of a tree it

*The Fawn-footed Melomys makes a nest in a tree.*

is readily climbed, but a smooth-barked tree is inaccessible. Nests made from the leaves of pandanus, banana, or grass, are built in the trees. Limited data on the movements of the Fawn-footed Melomys suggest that stable breeding pairs may be formed and that individual animals range over an area about 50 metres in diameter. In the laboratory, unfamiliar individuals may fight fiercely if placed together, suggesting that territories may be maintained in the wild.

Reproduction occurs throughout the year, but the main breeding season is from early spring to early winter. The maximum litter is four with an average of a little less than two. The young are born after a gestation period of about 38 days and grow rapidly; at ten days of age they are fully furred and with open eyes; at about 20 days they are weaned. The young, like those of many other Australian rodents, can escape from predators by clinging to the teats of the mother as she flees the nest.

The Fawn-footed Melomys can be a minor pest of sugar cane, but only in canefields that adjoin forested areas.

T.D. REDHEAD

# Bramble Cay Melomys

## *Melomys rubicola*

### Thomas, 1924

rue'-bik'-oh-lah: 'Bramble (Cay)-dwelling Melanesian-mouse'

The most isolated and among the most vulnerable of Australian mammals, the stocky Bramble Cay Melomys is known only from Bramble Cay, a small

*Melomys rubicola*

## Giant White-tailed Rat

### *Uromys caudimaculatus*

### (Krefft, 1867)

yue'-roh-mis  kaw'-dee-mak-yue-lah'-tus:
'spotted-tailed tailed-mouse'

The Giant White-tailed Rat, one of Australia's largest rodents, is restricted to northern Queensland from just south of Townsville to the tip of Cape York Peninsula. It is found in rainforest and closed sclerophyll forest at all altitudes, but also utilises wetter open forests and woodlands surrounding the closed forests. In lowland areas, it can be found in more open habitats such as melaleuca forests and swamps and mangroves. It can survive in forest fragments as small as 7.5 hectares but does not readily move between fragments if separated by agricultural land. It forages outside the forest into disturbed habitat and remnant or regenerating forest. The species also occurs in New Guinea, where it is abundant in both lowland and mid-montane areas.

There is a striking degree of genetic variation in the Australian population, where two chromosome races have been identified. The first ranges from Cooktown to Townsville and is physically separated from the northern form, found in the Iron and McIlwraith Ranges. Adelaide Zoo has recorded interbreeding between captive individuals of the two races.

It climbs well, enabling it to forage on nuts and fruits while these are still on a tree. The tail is not prehensile but can curl over a surface, its rasp-like scales contributing to the grip. It climbs tree trunks by holding on by the forefeet and propelling itself up by powerful thrusts of the strong hindlegs. It also forages extensively on the ground and has a diet that includes fruits, nuts, fungi, insects, small reptiles, amphibians,

### *Melomys rubicola*

**Size**

*Head and Body Length*
140–160 mm
*Tail Length*
145–180 mm
*Weight*
circa 100 g

**Identification** Externally very similar to *Melomys capensis* and *Melomys cervinipes* but with much rougher tail.
**Recent Synonyms** None.
**Other Common Names** None.
**Status** Vulnerable.
**Subspecies** None.
**Reference**
Limpus C.J. and C.H.S. Watts (1983). *Melomys rubicola*, an endangered murid rodent endemic to the Greater Barrier Reef of Queensland. *Aust. Mammal.* 6: 77–79.

vegetated coral cay at the extreme northern end of the Great Barrier Reef, 50 kilometres from New Guinea. Here it lives among dense grass in company with numerous seabirds and turtles, foraging at night among the grass and out onto the beach. Its diet is unknown but is likely to be plant material: there is no evidence of its eating the eggs of birds or turtles. Although living closer to New Guinea than mainland Australia, it is a close relative of the Cape York Melomys, from which it differs in some blood proteins and in having a tail that is considerably rougher (due to individual scales standing out from the tail). Nothing is known of its breeding biology.

Its island home, 340 metres long and 150 metres wide, is decreasing in size due to the action of the sea (a literal case of habitat loss). Because the island cannot accommodate more than several hundred individuals, the long-term survival of the species is unlikely.

C.H.S. WATTS

D. WHITFORD

*Despite its size, the Giant White-tailed Rat is an agile climber.*

### Uromys caudimaculatus

**Size**
*Head and Body Length*
305–382 (352) mm (male)
275–363 (310) mm (female)
*Tail Length*
340–362 (352) mm (male)
323–357 (342) mm (female)
*Weight*
500–890 (667) mm (male)
500–800 (625) mm (female)

**Identification** Grey-brown above; cream to white below; paws pale. Long, naked tail, at least one-third of distal part white and remainder grey-black. Distinguished from species of *Mesembriomys* and *Hydromys* by naked tail (shaggy in *Mesembriomys*, furred in *Hydromys*) and white, unwebbed feet (feet black in *Mesembriomys gouldii*, hindfoot webbed in *Hydromys*).
**Recent Synonyms** *Uromys sherrini. Uromys exilis.*
**Other Common Names** White-tailed Rat, Giant Rat, Giant Naked-tailed Rat, Giant Mosaic-tailed Rat, Cape-York Uromys, Atherton Uromys, Hinchinbrook Island Uromys.
**Status** Common.
**Subspecies** *Uromys caudimaculatus, caudimaculatus*, north-eastern Australia.
*Uromys caudimaculatus aruensis*, Aru Islands.
*Uromys cauimaculatus multiplicatus*, northern New Guinea.
**Extralimital Distribution** New Guinea and the Aru Islands.

crustaceans and bird eggs. Strong jaws enable it to break into hard-shelled rainforest fruits that are impervious to other rodents (except the Masked White-tailed Rat, *U. hadrourus*).

Recent studies on the Atherton Tableland have shown that its consumption of hard-shelled, palatable tree seeds such as Yellow Walnut, *Beilschmiedia bancroftii*, Cream Silky Oak, *Athertonia diversifolia* and Hairy Walnut, *Endiandra insignis*, is close to 100 per cent, leading to a very low recruitment rate of these trees compared with those with unpalatable seeds. Seeds may be eaten on the spot; removed to a favoured buttress or fallen log where a midden of chewed seed cases may be found; or buried in the earth up to 60 metres from the parent tree. Buried (cached) seeds are carefully covered with one or two leaves that are sometimes weighed down with a stick. The cached seeds are usually removed the

following night but it is not known whether the rat that eats the seed is the one that cached it.

The Giant White-tailed Rat is a generalist feeder with pronounced seasonal variation in its diet. Rotting logs and branches are torn open in pursuit of large passalid beetles, a habit shared with the Striped Possum, *Dactylopsila trivirgata*,

**References**

Baverstock, P.R. (1976). Heterochromatin variation in the Australian rodent *Uromys caudimaculatus*. *Chromosoma* 57: 397–403.

Harrison, J.L. (1962). Mammals of Innisfail 1. Species and distribution. *Aust. J. Zool.* 10: 45–83.

Wellesley-Whitehouse, H.L. (1981). *A study on the comparative ecologies of two rainforest murid rodents,* Uromys caudimaculatus *and* Rattus fuscipes coriacius, *in North Queensland.* Honours Thesis, Department of Zoology, James Cook University, Townsville.

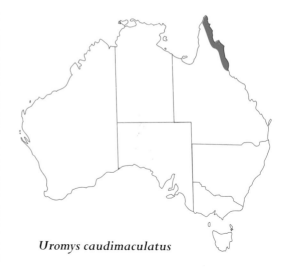

***Uromys caudimaculatus***

and Long-nosed Bandicoot, *Perameles nasuta*. Bark is a favoured food item at certain times of the year, signs of feeding being particularly noticeable on trees with buttresses. It also hunts mud crabs and smaller crabs in daylight in mangroves.

During the day, it takes refuge in a burrow or tree hollow; claw marks on tree trunks often draw attention to the latter sites, which are often within 10 metres of the ground and generally have a narrow, slit-shaped entrance. Burrows are normally located under logs, fallen trees or other debris and in the undercut banks of streams and dry gullies.

Breeding begins in September or October and peaks at the height of the wet season in December to January. The gestation period is 36 days, after which a litter of two to three (rarely four) young is born. These stay with the female for up to three months, by which time they weigh 250–300 grams. By March, freshly independent juveniles of around 300 grams are common. A study on the Atherton Tableland resulted in the capture of 300 adult females over two years but, although many of these were lactating, only one was carrying young on the teats. In this single instance the young were extremely small and certainly born in the cage during the previous night. It seems that the carrying of young is a rare event, occurring only as a result of disturbance. One instance in 1962 involved a female with two attached young which left a burrow after being frightened by a Brown Snake; one of the young subsequently fell to the ground.

It is difficult to determine the home range of the Giant White-tailed Rat, which has to move widely to find food in what is generally an erratically fruiting rainforest environment. The home ranges of marked animals overlap with those of their neighbours in complex patterns but overnight movements of 500 metres have been observed in some individuals. Studies on the Atherton Tableland have shown that individuals forage intensively over a minimum area of 4 hectares and extensively beyond that in response to adverse environmental conditions. Despite this mobility there is a high level of site fidelity, indicating a strong degree of territoriality in what is a relatively long-lived species. Recapture data from studies on the Mount Windsor Tableland show that the Giant White-tailed Rat often reaches an age of at least four years in the wild.

Known predators are the Lesser Sooty Owl, *Tyto multipunctata*, which take individuals up to 300 grams in weight; the Spotted-tailed Quoll, *Dasyurus maculatus*; and the Rufous Owl, *Ninox rufa* (seen roosting with a partly-eaten large adult in its claws). It is likely that the Dingo and feral Cat prey on this species.

L.A. MOORE

# Masked White-tailed Rat

## *Uromys hadrourus*

### (Winter, 1984)

had'-roh-yue'-rus: 'strong-tailed tailed-mouse'

The Masked White-tailed Rat is the smallest member of the genus and was originally placed in the genus *Melomys*. However, it is clearly a *Uromys* on the basis of the thickened tail, skull shape and dental characteristics.

It was first discovered in 1973, on the summit of Thornton Peak. For many years it was known from only seven specimens, all from Thornton Peak and the McDowall Range, a westerly spur of the peak, at elevations of 550–1220 metres. The Thornton Peak massif is isolated from other upland areas by the Bloomfield and Daintree River valleys and the species was thought to be restricted to this area. However, in August 1990, the skull of one was found in the pellet of a Lesser Sooty Owl on the Mount Carbine Tableland, south of the Daintree River; a month later, a second specimen from the Tableland was found dead on the Mount Lewis Road. In July 1991, another major extension of its known geographical range occurred when an individual entered a trap set for Musky Rat-kangaroos in the Lamins Hill area of the Atherton Tableland.

A study of small mammals in the Lamins Hill area had also begun to capture the species for the first time after 12 months of intensive trapping (*circa* 7000 trap-nights): the first capture coincided with a continuation of the previous year's drought conditions. Captures continued throughout 1991 but ceased shortly after the first effective rainfall of the wet season, suggesting that dry conditions may have caused the species to be more trappable than usual. It was never recorded

### *Uromys hadrourus*

**Size**
**Thornton Massif and Carbine Tableland**
*Head and Body Length*
171–180 mm
*Tail*
184–196 mm
**Atherton Uplands**
*Head and Body Length*
186 mm
*Weight*
202–220 (209) g (males)
170–204 (179) g (females)

**Identification** Fawn above (older individuals reddish-brown nape and shoulders); paler below with pure white on throat and chest. Adults and older subadults with distinctive dark eye ring. White-tipped (circa 35 mm), robust, scaly tail slightly longer than the head and body. Distinguished from *Melomys* by larger size and thicker, white-tipped tail. Easily confused with juvenile *Uromys caudimaculatus* but distinguished by smaller head size (skull length less than 43 mm), shorter foot (less than 39 mm) and lack of dark mottling on tail tip.
**Recent Synonyms** *Melomys hadrourus*.
**Other Common Names** Thornton Peak Rat, Thornton Peak Melomys.

in the course of more than 15 000 trap-nights in remnant fragments of rainforest of the Atherton Tableland. This compares with 20 individuals in continuous rainforest at the Lamins Hill sites over a similar number of trap-nights, indicating that the Masked White-tailed Rat is normally trap-shy and that it is unlikely to survive in rainforest fragments. Difficulty in trapping the species may account for the lack of previous records from the Atherton Uplands despite intensive trapping in the area since the early 1900s.

These latest records indicate that the species probably occurs throughout the Atherton Uplands, between Kuranda and the Herbert River gorge, west of Ingham. The Atherton

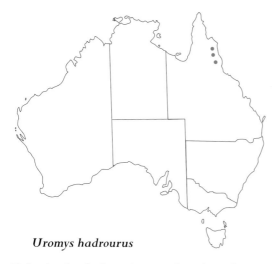

**Uromys hadrourus**

Upland individuals are heavier than those from the more northerly population, either because of better ecological conditions or because of genetic divergence in isolated populations.

Little is known of its habits in the wild. All captures have been in traps set on, or near, the ground and there is a single observation of an undisturbed individual walking past a group of Chowchillas, foraging on the forest floor in mid-afternoon. These observations suggest that it is terrestrial but difficulties in trapping the animal on the ground and a lack of traps set in trees leave open the possibility that it is also scansorial. The stomach of one individual from Thornton Peak contained the creamy coloured endosperm of a nut. In feeding trials on captive individuals from the Lamins Hill area, it readily attacked and extracted the kernel of hard-shelled rainforest fruits such as Creamy Silky Oak, *Athertonia diversifolia*, Yellow Walnut, *Beilschmiedia bancroftii* and Hairy Walnut, *Endiandra insignis*. The diet also includes insects, particularly larger beetles: individuals would rush out of their roost boxes to attack passalid beetles as soon as the beetle squeaked and would often take these insects from the hand. This suggests that they were familiar with the beetles and that they form a normal part of the diet. In the wild, the Masked White-tailed Rat would need to break into rotting branches and fallen logs to reach passalid beetles, a behaviour that is known from the closely related but larger Giant White-tailed Rat, *Uromys caudimaculatus*.

*The Masked White-tailed Rat was earlier known as the Thornton Peak Melomys.*

Adult males have been captured in breeding condition in September and December. The weight of juveniles and subadults indicate that the breeding season is centred on late spring and summer (early wet season), as in the other rain-forest murids with which they coexist. Two subadult males were caught at different study sites in June 1991, weighing 140 grams and 145 grams respectively: they were recaptured 88 days later, by which time they had each increased in weight by only 13 grams. Bearing in mind that this was in a time of drought, it nevertheless suggests that growth is slow and that longevity is greater than in similarly sized species of *Rattus* and similar to that of the White-tailed Rat.

In captivity, it is relatively vocal, females especially so. A raspy honking call, quieter than, but similar to, that of the White-tailed Rat, is the most common call, usually given when a roost box was disturbed. Males, although heard to give similar calls, rarely vocalised but were much more aggressive than females and would attack

**Status** Rare, limited.
**Subspecies** None.
**References**

Nix, H.A. and Switzer, M.A. (1991). Rainforest animals: Atlas of vertebrates endemic to Australia's Wet Tropics. *Kowari* 1. Australian National Parks and Wildlife Service, Canberra.

Winter, J.W. (1984). The Thornton Peak Melomys, *Melomys hadrourus* (Rodentia: Muridae): a new rainforest species from north-eastern Queensland, Australia. *Mem. Qld. Mus.* 21: 519–539.

Groves, C.P. and T.F. Flannery (1994). A revision of the genus *Uromys* Peters, 1867 (Muridae: Mammalia) with descriptions of two new species. *Rec. Aust. Mus.* 46: 145–169.

A. DENNIS

# Prehensile-tailed Rat

## *Pogonomys mollipilosus*

### Peters and Doria, 1881

poh-gon'-oh-mis mol'-ee-pil-oh'-sus:
'soft-haired tailed-mouse'

This beautiful, soft-furred rodent was first described from New Guinea in 1881 but was not recorded from Australia until 1974 when a domestic Cat brought a specimen into the tourist lodge at Lake Barrine on the Atherton Tableland, Queensland. Most specimens are from pellets of the Lesser Sooty and Boobook Owls; several have been brought in by Cats. Others have been shot, found near death on the ground following heavy, continuous rain, or obtained from felled trees.

| *Pogonomys mollipilosus* |
| --- |
| **Size** |
| *Head and Body Length* |
| 130–165 (144) mm (male) |
| 120–147 (135) mm (female) |
| *Tail Length* |
| 160–208 (178) mm (male) |
| 164–191 (181) mm (female) |
| *Weight* |
| 52–83 (69) g (male) |
| 42–68 (57) g (female) |

**Identification** Grey above; pure white below; narrow black ring around eye. Fur soft. Tail nearly 1.5 times length of head and body, tip smooth on upper surface. Distinguished from superficially similar *Cercartetus caudatus* by typical rodent incisors and hindfeet, smaller ears and lack of thickened base of tail; from *Melomys cervinipes* by white belly, dark eye-ring and longer tail.

handlers at the slightest provocation. When animals that had been involved in feeding trials were released back into the forest after three months in captivity, all were subsequently recaptured three months later at the exact sites at which they had originally been trapped.

A known predator is the Lesser Sooty Owl. Potential predators are the Rufous Owl and pythons. No clear-felling of rainforest has occurred in the range of the Thornton massif and Carbine Tableland populations and the species seems therefore to be secure in these areas. It appears to have disappeared from the fragmented rainforest of the Atherton Tablelands but it probably continues to exist in the large tracts of continuous rainforest in the area.

J.W. WINTER AND L.A. MOORE

HANS & JUDY BESTE

*The Prehensile-tailed Rat climbs with the aid of its tail.*

Surprisingly, not one has been found in the thousands of traps that have been set since mammal collecting for museums commenced in the late 1800s in its North Queensland range.

The known Australian range is restricted to the rainforests of Cape York Peninsula and the Wet Tropics region between Cooktown and Townsville. On Cape York Peninsula, it is known from only two records, both from Gordon Creek, near Iron Range, collected in semi-deciduous monsoonal forest on the edge of an alluvial flood plain at an elevation of about 40 metres. In the Wet Tropics, it is now known to occur from sea-level to the tops of the ranges, between Shiptons Flat near Helenvale in the north to Zillie Falls near Millaa Millaa in the south. Its southern distribution probably extends to at least the Herbert River gorge near Ingham and possibly as far south as the southern end of the Seaview Range near Townsville. All localities are from rainforest and its immediate fringes.

Strictly nocturnal, the Prehensile-tailed Rat is superficially similar to the Fawn-footed Melomys and the Long-tailed Pygmy-possum (with which it shares the rainforest), but its behaviour when caught in a spotlight at night is diagnostic. It often becomes confused in the light, running down understorey plants to near the ground then back up again, dashing back and forth along overhead branches, the very white underparts showing clearly. It rarely comes to the ground and, if it does, it immediately climbs back up a tree. Species of *Melomys* nearly always run towards a tree trunk and then down to the ground to escape, whereas the Prehensile-tailed Rat crosses from one understorey shrub to the next, apparently following known pathways, relying on its weight to depress those branches that enable it to cross gaps. Where pandanus is present, a disturbed animal hides among its leaves; otherwise it keeps moving away through the understorey. The tail is used for support by curling it around twigs, corkscrew fashion, for two or three turns: an individual can support itself completely by its tail. *Melomys* species and the Prehensile-tailed Rat are not good jumpers, unlike the Long-tailed Pygmy-possum, which, in low vegetation, may leap to the ground to escape.

The Prehensile-tailed Rat eats leaves and fruits. The stomachs of six specimens examined contained finely chewed leaves with no sign of any insect fragments or other animal matter. Fruits known to be eaten are those of the Red-leaved Fig (*Ficus congesta*), *Rhodomyrtus trineura*, *Amoora ferruginea*, *Pandanus monticola*, bananas and Wild Tobacco bush (*Solanum mauritianum*). Leaves observed to be eaten include those of the introduced Yellow Passionfruit vine, *Passiflora edulis flavicarpa*. The small fruits of the Red-leaved Fig, Tobacco Bush and *Rhodomyrtus* are often, but not always, nipped off and carried in the mouth to be eaten elsewhere (held in the forepaws). The larger fruits of pandanus and banana are eaten *in situ*. The stylar spines and underlying hard skin of a

**Pogonomys mollipilosus**

**Recent Synonyms** Included by Flannery (1995) within *P. macrourus* but specific identity of the Australian population in doubt; may represent an undescribed species.
**Other Common Names** Soft-haired Tree-mouse.
**Status** Common but rarely seen, restricted.
**References**

Dwyer, P.D. (1975). Observations on the breeding biology of some New Guinea murid rodents. *Aust. Wildl. Res.* 2: 33–45.

Flannery, T.F. (1995). *Mammals of New Guinea*, 2nd edn. Reed Books, Sydney.

pandanus fruit are bitten off and discarded to gain access to the seeds. Similarly, the green skin of a banana is discarded to gain access to the soft tissue. Leaves of the Passionfruit vine are bitten off at the stalk and the leaves held in the forepaws whilst eaten: the ground beneath a vine can be littered with partly eaten leaves. Although usually seen singly, up to three individuals at a time have been seen feeding on the leaves of a Passionfruit.

An extremely arboreal forager at night, the Prehensile-tailed Rat rests in burrows during the day. Three burrows have been found in Australia. One, with freshly dug soil extending from its entrance, was excavated and shown to consist of a tunnel 90 centimetres long, gradually sloping downwards from the entrance to a round nest chamber about 30 centimetres below the forest floor. It was about 15 centimetres in diameter and packed with fresh green leaves. A tunnel sloping gradually upwards and about 45 centimetres long led away from the nest chamber to an escape hole concealed by the forest floor litter. Both tunnels entered the upper part of the nest chamber. An adult male burst from the escape hole as the leaves of the nest chamber were being removed. Leaves were detected in a second burrow, when probed with a twig and a Prehensile-tailed Rat shot out of the concealed bolt-hole. The third burrow had its entrance

under the trunk of a small understorey tree. Animals were seen just above the entrance at 9:50 p.m. and when red light was used to watch the burrow entrance, a few nights later, one animal emerged at 9:45 p.m. In New Guinea, three to five animals usually share a burrow but as many as 15 have been taken from a single one.

Females have six teats and usually rear two to three young. In New Guinea, breeding occurs from October to January but little is known of the pattern of reproduction in Australia. The female taken from Gordon Creek in September had two embryos; two adult females from Lake Barrine and Kuranda, obtained respectively in July and February, showed no signs of breeding.

HANS & JUDY BESTE

*This Prehensile-tailed Rat is about to leap from one branch to another.*

The few males, obtained between June and September, had scrotal testes and were presumably capable of breeding.

The Prehensile-tailed Rat is undoubtedly more common than the few records would suggest: it is a consistent and frequent item in the diet of the Lesser Sooty Owl, as indicated by the presence of its skulls in the owl's pellets.

J.W. WINTER AND D. WHITFORD

## Subfamily Murinae

Currently the most successful of the rodents, this group comprises more than 400 species, spread over most of the world except the Americas. Of some 112 genera, only the ubiquitous *Rattus* is native to Australia. The founding stock apparently arrived relatively recently and there has been only minor speciation, mostly within rather well-watered environments: these are known as the 'new endemic' rodents. In historical times, Europeans have introduced two species of *Rattus*, both still associated with human habitations; and the House Mouse, which has become adapted to almost every Australian environment.

D. WATTS

*The introduced House Mouse inhabits dwellings but is also common in most Australian natural environments.*

# House Mouse

## *Mus musculus*

### Linnaeus, 1758

mus mus'-kue-lus: 'little-mouse mouse'

Because of its ability to exploit newly available habitats, the House Mouse has been referred to as a 'weed' species. Among mammals, it is second only to humans in the extent of its global distribution. The taxonomy of the *Mus musculus* complex of species is still unresolved but the Australian population is referred to *M. m. domesticus* or (according to some authorities) a separate species, *M. domesticus*.

In Europe and generally elsewhere (including northern Australia), it lives up to its common name, being found mostly in houses, barns and food stores. In these parts of the world, it may also occur in cultivated fields, but only in low density, being excluded by other small mammalian herbivores. Over most of Australia, however, it is virtually the only rodent occurring in cultivated fields, being able to exploit this situation to such an extent that populations periodically reach their reproductive potential and irrupt into 'plagues'. Approximately 20 per cent of Australia (particularly in the eastern grain-belt) is prone to such plagues and, on average, one occurs at least somewhere in the grain-belt every four years.

In Australia, the breeding season is usually October to April, but in good periods, it may extend throughout the year. Sexual maturity is reached at six to seven weeks and litter size is four to eight. Pregnancy lasts about 19 days and there is a post-partum oestrus. This implies an immense reproductive potential.

Factors that lead to a mouse plague vary from region to region and are often influenced by rainfall and farming practices—for example, irrigation enables production of both winter and summer crops. In the semi-arid mallee wheatlands, high autumn rainfall followed by a series of

Melville Is.
Bathurst Is.
Groote Eylandt Is.
Mornington Is.
Fraser Is.
Kangaroo Is.
Note: also on many other offshore islands
King Is.
Flinders Is.

**Mus musculus**

rainfall events over the next 12 months appears to be responsible for a plague after about 18 to 20 months, whereas in the higher rainfall regions of northern New South Wales and southern Queensland (where cropping is more intensive) low rainfall in late spring and early summer may be the necessary precursor of a plague some four months later. Other models suggest that plagues need to be preceded by a year-long drought, or have highlighted the coincidence of good burrowing conditions (of importance in some clay soils but not those that are predominantly sandy loams) and good food supply. Other key factors that have been proposed as limiting mouse numbers are social organisation, predators (mainly raptors) and disease.

Between population outbreaks, it can become rare over much of its former range. Some refugial habitats and 'donor' habitats (where they commence breeding earlier than elsewhere) have been identified in some regions. An understanding of the dynamics of its spatial use of the landscape holds the key to our ability to predict plagues and to develop management strategies.

In Australia, the House Mouse is not only of interest in cultivated regions. In areas of natural vegetation, it often becomes abundant and ubiquitous about 18 months after a fire, at a time when native species are in low density. Mouse densities can remain high for three to four years,

### Mus musculus

**Size**

*Head and Body Length*

60–100 mm

*Tail Length*

75–95 mm

*Weight*

8–25 g

**Identification** Yellowish-brown to blackish above; white to grey to pale yellow below. Distinguished from species of *Pseudomys* by presence of notch on inner surface of upper incisor and five pairs of teats in the female.

**Recent Synonyms** *Mus domesticus*.

**Other Common Names** None.

**Status** Abundant.

**Subspecies** Many described, recently consolidated into four or five subspecies. In Australia, genetic data indicate only the presence of *M. m. domesticus*.

**References**

Singleton, G.R. (1989). Population dynamics of an outbreak of house mice (*Mus domesticus*) in the mallee wheatlands of Australia—hypothesis of plague formation. *J. Zool. Lond.* 219: 495–515.

Singleton, G.R. and R.D. Redhead (1991). Structure and biology of mouse populations that plague irregularly: an evolutionary perspective. *Biol. J. Linn. Soc.* 41: 285–300.

decreasing as the population density of small native mammals increases: this phenomenon has been reported in arid, semi-arid and temperate regions. Also of interest are the very occasional simultaneous irruptions of the House Mouse with native rodents such as the Long-haired Rat, Spinifex Hopping-mouse, Alice Springs Mouse, Plains Rat and Sandy Inland Mouse.

G.R. Singleton

HANS & JUDY BESTE

# Dusky Rat

## *Rattus colletti*

### (Thomas, 1904)

rat'-us  kol'-et-ee: 'Collett's rat'

*The Dusky Rat lives on the alluvial flood plains of northern Australia*

The Dusky Rat, of the monsoonal subcoastal plains of the Northern Territory, lives on the broad, flat and treeless alluvial floodplains of meandering tidal rivers bordered by mangroves and low, wide levees. Its pattern of life is determined and regulated by the extreme seasonal variability of this environment. In the wet season, heavy rainfall and run-off from the adjacent wooded lowlands flood the plains for three to six months or even longer in the lowest areas.

At such times, the Dusky Rat must retreat to higher ground either on levee banks or the shallow flooded margins of the plains where it takes refuge in clumps of grasses and fallen sedge— often in high densities—and forages at night in the shallow water. Some may dig burrows where the soil is soft and not waterlogged; others invade the adjacent higher ground, finding shelter in tree roots and logs and foraging on the woodland floor. At this time they feed mainly on the stembases of grasses and the corms of sedges.

At the onset of the dry season (April–May) the water retreats and the plains begin to dry out. The rats abandon the higher ground and spread out over the grass and sedge of the now rapidly drying plain, dwelling by day in the abundant cracks forming in the shrinking clay soil. Underground sedge corms provide most of the Dusky Rat's food and water from this time until the end of the dry season.

Reproduction starts during the transition from the wet to the dry season. Litters are born in nests built in the humid cracks, and young emerge about three weeks after birth, mainly in June, although the duration of breeding depends upon the severity of the dry season. As aridity increases, the soil cracks widen, the dry vegetation

*Rattus colletti*

## *Rattus colletti*

**Size**

*Head and Body Length*
65–211 (132) mm

*Tail Length*
78–149 mm

*Weight*
22–213 (61) g

**Identification**

Grizzled dark brown or black above; darker than in *Rattus sordidus* and *R. villosissimus*. Dorsal hair harsh; ventral fur grey or yellowish. Three pairs of pectoral and three pairs of inguinal teats. Females often significantly smaller than males.

**Recent Synonyms**

*Rattus gestri colletti*, *Rattus sordidus colletti*.

**Other Common Names**

Collett's Rat, Territory Dusky Rat, Mulbu (Gunwinjgu language).

**Status**

Abundant; sometimes rare and scattered; reproductively irruptive.

**Subspecies**

None.

**References**

Redhead, T.D. (1979). On the demography of *Rattus sordidus colletti* in monsoonal Australia. *Aust. J. Ecol.* 4: 115–136.

Williams, C.K. (1987). Water physiology and nutrition in fluctuating populations of *Rattus colletti* in monsoonal Northern Territory, Australia. *Aust. Wildl. Res.* 14: 443–458.

compacts, breeding tends to cease and animals retreat to dry, shallow depressions in the plains. Growth slows down, ectoparasites increase, nocturnal activity above ground diminishes, older animals die and the population decreases—the severity of these responses depending upon the duration of the arid period. As the wet season approaches and the rainfall increases, animals become more active above ground and return towards the edges of the plains as the dry vegetation disintegrates, the soil swells to obliterate the cracks and the vegetation begins to grow again.

Under favourable conditions, a female Dusky Rat is capable of commencing breeding when one to two months old. Females have 12 teats, produce up to nine or more young and have the capacity to reproduce throughout the year. They are thus among the most fecund of the Australian species of *Rattus*, but the degree to which this potential can be utilised is determined largely by rainfall.

Mild flooding in the wet season and rainfall extending into the dry season promote earlier and extended breeding, rapid increase in the population and high densities of animals on the plains. This leads to intense social interactions and much daytime activity above ground. High mortality usually follows in the late dry season or following year if the seasons remain relatively mild. When flooding is severe or extended, few adults survive to recolonise the plains and there is little reproduction. If a long wet season is followed by extended dry seasons or droughts, the species may become extremely rare over most of its range for several years. Such population fluctuations are not cyclic but follow irregular variations in the annual rainfall pattern in the monsoonal climate.

C.K. WILLIAMS

# Pacific Rat

## *Rattus exulans*

(Peale, 1848)

ex'-yue-lanz: 'exiled rat'

The Pacific Rat is widespread in South-East Asia and the Pacific Islands, always in close proximity to humans. It is one of a small suite of rodents that have seemingly benefited from human alteration of the environment; the Asian Black Rat, *Rattus tanezumi,* and House Mouse, *Mus musculus,* are others. It is thought to have originated in mainland South-East Asia and to have been spread by humans throughout the South Pacific, reaching New Zealand around 1000 years ago in company with the Maoris.

In Australia, it is only known from a few specimens, collected in the nineteenth century from Adele and Mer Islands (almost certainly having reached there with human help), but we do not know whether it continues to exist on these islands. It is surprising that a rodent so widespread in South-East Asian and Pacific islands has not become established on the Australian mainland. Perhaps it reflects a lack of cultivation and close settlement by humans in northern Australia.

The Pacific Rat lives around and in huts, in cultivated fields and in regrowth forest. It is an agile and energetic climber but spends much of its time foraging on the ground for seeds, fruit, vegetable matter and insects. It is known to take young birds and eggs. It builds a neat spherical nest, usually in or under logs, in rotting tree stumps or in the discarded husks of coconuts.

Breeding occurs throughout the year, usually with a litter of three to four young. These are weaned when about three weeks and are sexually mature at ten weeks.

C.H.S. WATTS

*Rattus exulans*

**Size**

**Non-Australian specimens**

*Head and Body Length*
80–140 (120) mm
*Tail Length*
125–135 (130) mm
*Weight*
30–100 (40) g

**Identification**  Small rat. Tail longer than head and body; light greyish brown above; sharply defined from white underside; spiny fur. Upper incisor teeth not notched (as in *Mus*).

**Recent Synonyms**  None in Australia.

**Other Common Names**  Polynesian Rat, Kiore (New Zealand).

**Status**  Unknown; very limited Australian distribution.

**Subspecies**  None.

**Extralimital Distribution**  South-East Asia, the Pacific Islands.

**References**

Taylor, J.M. and B.E. Horner (1973). Results of the Archbold Expedition No. 98. Systematics of native Australian *Rattus* (Rodentia: Muridae). *Bulletin of the American Museum of Natural History* 150: Article 1.

Watts, C.H.S. and H.J. Aslin (1981). *The Rodents of Australia.* Angus and Robertson, Sydney.

*Rattus exulans*

K. ATKINSON

*Fungi are a major component of the winter diet of the Bush Rat in the winter and after bushfires.*

# Bush Rat

## *Rattus fuscipes*

### (Waterhouse, 1839)

fus'-kee-pez: 'dusky-footed rat'

---

### *Rattus fuscipes*

**Size**
*Head and Body Length*
111–214 mm
*Tail Length*
105–195 mm
*Weight*
40–225 g

Note: Females considerably smaller than males. On Glennie Island, off Victoria, males have a mean weight of 250 g and females 182 g. *R. f. coracius* is similar in size and weight to *R. f. assimilis* but *R. f. fuscipes* and *R. f. greyii* are 10–20 per cent shorter and weight about 30–40 per cent less.

**Identification** Fur dense and soft. Grey-brown to reddish-brown above; light grey to light brown below. Tail brown, grey or black; slightly shorter than the head and body. Conspicuous rounded ears pink, grey or brown. Feet white, pink, grey or brown, hindfeet often darker than forefeet; forefeet grasping. Distinguished from introduced *Rattus rattus* which has a relatively longer tail.

Note: Variable in weight and colour. *R. f. coracius* commonly has a white blaze between the ears and ventral fur can be bright white with rust-brown marking.

---

Widespread and common along the forested coast and ranges, with densities sometimes reaching ten individuals per hectare, the nocturnal, cover-seeking Bush Rat is seldom seen in the wild. Nevertheless, since it can readily be captured, it is familiar to those who study small Australian mammals. It is found in subalpine woodland (where it remains active under the snow in winter), coastal scrub, eucalypt forest and rainforest, selecting sites where there is dense ground cover. Unlike the Swamp Rat, which has a preference for grasses, sedges and reeds and moves about during the day, the Bush Rat prefers undergrowth of shrubs and ferns. Where there is a sharp

transition between these two types of ground cover, there is also a distinct boundary between populations of the two species. The densest understorey in a forest usually occurs in gullies and these are where the Bush Rat is sensitive to the density of ground storey cover: as this decreases, so does its density. Habitat disturbance through fire and logging is quickly reflected in reductions in local populations. Since it rarely coexists with humans, the persistence of the Bush Rat as a common species is dependent upon maintenance of its natural habitat.

Of the four subspecies, most is known of *Rattus fuscipes assimilis* from south-eastern Australia. It is capable of reproduction throughout the year but tends not to breed in winter, particularly in the southern part of its range. The

**Recent Synonyms** *Rattus greyii, R. murrayi, R. mondraineus, R. glauerti, R. assimilis coracius.*

**Other Common Names** Western Swamp Rat (*R. f. fuscipes*); Allied Rat, Southern Bush Rat (*R. f. assimilis*), particularly in earlier writings.

**Status** Common.

**Subspecies** *Rattus fuscipes fuscipes*, coastal south western Western Australia, including offshore islands.

*Rattus fuscipes greyi*, coastal from Eyre Peninsula to about Portland, Victoria and adjacent offshore islands.

*Rattus fuscipes assimilis*, coastal and subcoastal from western Victoria to Rockhampton, Queensland; southern high country and some offshore islands.

*Rattus fuscipes coracius*, coastal from Townsville to Cooktown, Queensland.

**References**

Lunney, D., B. Cullis and P. Eby (1987). Effects of logging and fire on small mammals in Mumbulla State Forest, near Bega, New South Wales. *Aust. Wildl. Res.* 14: 163–181.

Taylor, J.M. and J.H. Calaby (1988). *Rattus fuscipes*. Mammalian species. *American Society of Mammalogists.* 298: 1–8.

Warneke, R.M. (1971). Field study of the bush rat (*Rattus fuscipes*). *Wildl. Control Vic.* No. 14.

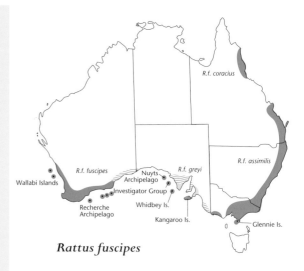

*Rattus fuscipes*

ter, Bush Rats in the high country become lighter in weight, increase their body fat and put on more hair.

The movement patterns of the Bush Rat and the Swamp Rat are similar. In both species, it seems that young rapidly disperse from the maternal territory and establish small individual home ranges that are necessary for survival through the winter. Females retain or expand their territories in spring while males either disperse or occupy vastly extended home ranges, overlapping those of other males and a number of

usual litter size is five and several litters may be produced in a good season. Young weigh about 5 grams at birth and become independent at the age of four to five weeks. Males and females are capable of breeding in the wild at the age of about four months and some individuals born in spring may breed before the onset of winter. Most, however, remain sexually inactive until the following spring.

In autumn, there is heavy mortality of individuals that have been sexually active during the previous six to eight months. With the death of almost all adult males and of most adult females, the overwintering population consists predominantly of young of the previous summer. These spend much time in burrows, making only short foraging excursions. With the approach of win-

*The Bush Rat is widely distributed in coastal forests.*

C. ANDREW HENLEY/LARUS

females. It is common for a male to cover a kilometre on a summer night.

The Bush Rat has been the fruitful subject of behavioural studies. Despite similar signals (odour, tactile, sound, visual) between the Bush Rat and the Brown Rat, *R. norvegicus*, the Bush Rat is never gregarious, unlike its introduced congener.

Of all the native rats, the Bush Rat is the most insectivorous, but it has such a varied diet that it can survive through a period of acute food shortage following a bushfire by eating the fungi that rapidly emerge. Although formerly regarded as an opportunistic omnivore, it is now seen to be selective in its food preference. One study found that fungi are a dominant component of the winter diet (along with fibrous stems and leaves of specific grasses and lilies) and that insects, fruits and seeds are important in the spring-summer diet. Dependence upon fungal mycorrhizae and leaf litter-dependent insects are reasons why it takes a year for a population of Bush Rats to become re-established in a forest or heath after fire. The higher density of populations in rainforest, compared with adjacent eucalypt forest, has been attributed to fire and logging disturbance of the eucalypt forest.

Fire removes much of the ground cover upon which the Bush Rat depends for shelter and food. Individuals usually survive a fire by remaining in their burrows, but females in burnt areas fail to breed in the following spring. A burnt area thus becomes available for recolonisation by juveniles dispersing from unburnt areas. After such recolonisation, populations have been observed to increase, in the course of several years, to four times the pre-fire level and, thereafter, within three to four years, to decline to the original size. Long-term research has shown that, although a catastrophic post-fire population crash can be attributed to fire, the subsequent population boom depends upon rainfall. Fire followed by continuing drought keeps the Bush Rat population at a low level. Otherwise, population fluctuations are largely related to rainfall.

D. LUNNEY

C. ANDREW HENLEY/LARUS

*The Cape York Rat is terrestrial, finding most of its food in forest litter.*

# Cape York Rat

## *Rattus leucopus*

### (Gray, 1867)

luke'-oh-poos: 'white-footed rat'

Found in Cape York and New Guinea, this rodent is one of a number of species that provide evidence of a past land connection across Torres Strait. Throughout its range it is restricted to rainforest and, in Cape York, its distribution is separated into two distinct areas by a wide dry

### Rattus leucopus

**Size**

*Head and Body Length*
155–214 (176) mm
*Tail Length*
142–209 (162) mm
*Weight*
95–207 (132) g

**Identification** Grey-brown to fawn above, paler below. Pointed face; elongate body; almost hairless scaly tail equal to or a little longer than the head and body. Southern subspecies distinguished from northern form of *R. fuscipes* by more pointed snout and the presence of a darkish eye-ring. Northern subspecies has white mottling on tail, white ventral fur and no eye-ring.

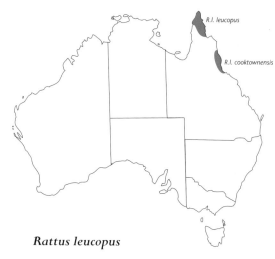

*R.l. leucopus*

*R.l. cooktownensis*

*Rattus leucopus*

corridor in the district around Laura. North of this barrier it occurs in most areas of rainforest from McIlwraith Range to Bamaga, at the tip of Cape York; and in a number of isolated patches of rainforest across the peninsula to the west coast. The southern population extends from Cooktown to the Paluma Range, just north of Townsville but is yet to be found in patches of semi-deciduous rainforest to the west of the true rainforest.

Isolation of the two populations has led to differences between them, recognisable in the chromosome complement and in external appearance. The northern subspecies is larger (*R. l. leucopus*, 115–207 grams; *R. l. cooktownensis*, 95–164 grams) and a high proportion possess brown tails with off-white mottling (leading to the common name, Mottle-tailed Rat). Most of the southern subspecies have a uniformly dark brown tail and a noticeable face pattern. The southern subspecies is very similar in appearance to the northern subspecies of the Bush Rat, *Rattus fuscipes coracius*, with which it shares much of its habitat, but is distinguished by a more pointed snout, the presence of a darkish eye-ring accentuated by prominent guard hairs and a flatter profile to the skull. The eye-ring is less distinct or

occasionally absent in subadults and in the southern reaches of the range, making separation more difficult. *R. l. cooktownensis* has a strong pungent smell which, once experienced, can help identify the animal.

The Cape York Rat is terrestrial, foraging at night in leaf litter and debris on the forest floor. It has a largely insectivorous diet which also includes fungi, fruits and nuts. In captivity it prefers insects to other food.

When released from capture in a trap, individuals have been seen to disappear into burrows in the forest floor, completely hidden under the leaf litter. Some burrows of *R. l. leucopus* appear to be communal, with several adults and juveniles often in residence. Radio-tracking has shown that some individuals utilise more than one burrow. Burrows usually have several entrances, consisting of a number of interconnecting chambers lined with dry leaves. Holes in rotting logs and in the bases of trees are also used.

A five-year study carried out in unlogged upland rainforest on the Mount Windsor Tableland showed that *R. l. cooktownensis* comprised up to 40 per cent of the *Rattus* population, the remainder being *R. f. coracius*. On the relatively disturbed Atherton Tableland, *R. f. coracius* is dominant in logged continuous rainforest. In forest fragments, *R. l. cooktownensis* becomes the dominant and, sometimes, the exclusive *Rattus* subspecies.

The northern subspecies is usually sparsely distributed but in areas such as the continuous rainforest on the summit of the McIlwraith Range, it may reach a high density. Further north, in Iron Range, it is common in areas close to creeks and rivers but attains its highest density (four to six times that in the continuous forest) in small patches of secondary rainforest. In New Guinea, the species is well represented in regenerating patches of rainforest.

Breeding occurs throughout the year but increases during the wet season (summer) and is subdued in the late dry season (winter). Females have six (rarely eight) teats and can produce up to three litters per year. In captivity, the litter size is two to five, which is low in comparison with other Australian *Rattus* species. The gestation period is 21–24 days but development is comparatively slow; the eyes become fully open at about 22 days of age, weaning is completed at about 25 days and sexual maturity is reached after three months.

L.A. MOORE AND L. LEUNG

**Recent Synonyms** None.
**Other Common Names** Mottle-tailed Rat, Mottle-tailed Cape York Rat, Spiny-furred Rat.
**Status** Common, limited.
**Subspecies** *Rattus leucopus leucopus*, northern Cape York Peninsula.
*Rattus leucopus cooktownensis*, south-eastern Cape York Peninsula.
**Extralimital Distribution** New Guinea.
**References**
Taylor, J.M. and B.E. Horner (1973). Results of the Archbold Expeditions No. 98. Systematics of native Australian *Rattus* (Rodentia, Muridae). *Bull. Amer. Mus. Nat. Hist.* 150: 1–130.
Winter, J.W., F.C. Bell, L.I. Pahl and R.G. Atherton (1984). *The specific habitats of selected north-eastern Australian rainforest mammals.* Report to World Wildlife Fund Australia. Sydney, NSW.

D.WATTS

*The Swamp Rat is primarily a creature of heathland and sedges, which are only irregularly swampy.*

# Swamp Rat
## *Rattus lutreolus*

(Gray, 1841)

lute'-ray-oh'-lus: 'otter-like rat'

Initially described in 1841 from specimens from the Hunter River, the Swamp Rat was soon collected over a wide area of south-eastern Australia. It was subsequently redescribed as at least three separate species and there was much confusion about its biology: Gould illustrated it as a semi-aquatic animal but mislabelled it as *Rattus (Mus) fuscipes*. It is now known that, although it can swim, it tires within minutes and readily grasps nearby vegetation. Swimming appears to be no more than a means of reaching refuge vegetation such as tall emergent shrubs growing above the floodwaters that irregularly cover its preferred habitats on river flats, sedge swamps and low-lying heaths. One study showed that, despite raging floodwaters across their habitat, residents were not displaced and, with the floodwaters still receding, individuals descended

## Rattus lutreolus

**Size**

*Head and Body Length*
122–197 mm
*Tail Length*
56–147 mm
*Weight*
56–156 g

Note: *R. l. velutinus* is about the same length as *R. l. lutreolus* but weighs about 15 per cent less.

**Identification** Dark grey or grey-brown above; underside paler, yellow-cream to brown. Fur on upper half of the body is golden-tipped. Ears small, almost concealed by body hair. Tail dark grey, scaly, sparsely haired and about two-thirds length of the head and body. Eyes less protuberent than in R. fuscipes.

**Recent Synonyms** *Rattus vellerosus, R. cambricus, R. imbril, R. velutinus, R. lacus, R. fuscipes.*

**Other Common Names** Eastern Swamp-rat, Tawny Rat, Tawny Long-haired Rat, Velvet furred Rat, Dusky-footed Rat, Water Rat.

**Status** Common.

**Subspecies** *Rattus lutreolus lutreolus* (including *R. l. imbril, R. l. cambricus*), coastal and subcoastal south-eastern Australia, Kangaroo Island. *Rattus lutreolus velutinus*, Tasmania and Bass Strait islands. *Rattus lutreolus lacus*, Atherton Tableland, Queensland.

**Rattus lutreolus**

decades has excluded the Swamp Rat except where the forest has a grass or sedge understorey, at which boundaries the Bush Rat may be caught if the trap is moved by as little as a metre, or reversed in direction.

In Tasmania, where the Bush Rat does not occur, the Swamp Rat occupies a much broader range of vegetation types, including alpine areas, wet sclerophyll forest, moist areas in dry sclerophyll forest and rainforest. It is most commonly trapped in heath and sedge areas; in one study of the distribution of Tasmanian mammals, it was found in 23 per cent of the grid cells, making it the most frequent and widespread of Tasmanian rodents.

The Swamp Rat is usually so cryptic that even researchers who have spent years studying the species rarely see it in its native habitats. Some studies have found it to move by day and night, others only by day. It makes tunnels through dense grass, sedge or heath vegetation by biting away and eating the obstructing plants. Dependence on such tunnels reflect its diurnal behaviour and dietary preference for sedge and grass. It usually constructs nests in burrows up to 1 metre deep but, where the ground is waterlogged, it makes nests above ground in tussock grass. So specific is the Swamp Rat to its tunnels (especially those leading to and from nests) that it is difficult to capture unless a trap is placed within the tunnel or runway. Dependence upon dense cover makes it vulnerable to clearing, grazing and

from the shrubs and were readily trapped on the sodden ground. When released from traps, they often swam through pools of water in a direct route to the nearest cover. Although such habitats are usually swampy or moist, they often dry out and, at such times, the Swamp Rat could be described as a rodent of heath, grass or sedge.

In favoured habitats, it may occur in densities as high as 15 individuals per hectare. It is regularly caught in field surveys, identified in scat or gut contents in studies of the diets of the Dog and Fox and encountered in educational trapping programs. Thorough studies of the species are rare: research on forest fauna in the last two fire: up to five years of regrowth after such disturbance is required before the habitat can again support a breeding population. In winter, prior to the breeding season, areas of regrowth after fire have been found to support many individuals in their own home ranges, but females fail to breed and disappear from the population unless part of their home range lies on unburnt land. An exception to this generalisation is provided by the case of a population which recolonised a burnt heath within two years, females breeding at less than the usual weight. This success is attributed to lack of competition from species of native *Pseudomys*.

The Swamp Rat feeds mainly on grasses and sedges, supplemented (except in winter) by fleshy fruits, seeds and arthropods; fungi are not significant in the diet.

Breeding usually extends from early spring to autumn but can occur throughout the year. After three weeks' gestation, three to five naked young, each weighing about 5 grams are born. These become independent when three to four weeks old, weighing 25–40 grams. In a good season, a female may produce several litters and three-month-old females from the first litter may, themselves, give birth in the same breeding season. Individuals born late in the season do not become reproductively active until the following spring.

A home range of about 0.2 hectares appears to be a prerequisite for both males and females to survive the winter. In spring, females defend a territory of about 0.5 hectares against other females, while males cease to be territorial and expand their home ranges up to 4 hectares. Active males range through adjacent, non-overlapping female territories. This great mobility indicates why large areas of habitat are required if populations are to continue to flourish. For a newly independent young female to succeed in breeding, it must establish a territory, the possibility of which depends upon the extent of spring and summer growth of grasses and sedges. In turn, this depends upon rainfall and, since this varies from year-to-year, so does the timing, rate of reproduction and population size. Individuals born in spring leave their home population in late summer and move more than 2 kilometres

### References

Braithwaite, R.W. (1980). The ecology of *Rattus lutreolus*, III. The rise and fall of a commensal population. *Aust. Wildl. Res.* 7: 199–215.

Norton, T.W. (1987). The ecology of small mammals in north-eastern Tasmania. I. *Rattus lutreolus velutinus*. *Aust. Wildl. Res.* 14: 415–433.

Taylor, J.M. and J.H. Calaby (1988). *Rattus lutreolus. Mammalian Species* 299: 1–7. The American Society of Mammalogists.

through burnt forest habitat to a new patch of suitable breeding habitat. Such dispersal ability is consistent with the elimination of patches of habitat that have been completely flooded or burnt out, providing the opportunities for re-colonisation.

Prior to European settlement, the Swamp Rat was hunted by coastal Aborigines, providing a significant food resource in winter, when fishes and molluscs were least abundant. This hunting pressure no longer exists, but most of its original habitat in fertile valleys and coastal heaths has now been farmed, grazed or cleared for housing. Although it is still common, its populations have become far smaller and much more fragmented since European colonisation. Its long-term survival is now dependent upon retention of undisturbed habitat in national parks and nature reserves. A study in south-eastern New South Wales showed a Swamp Rat population to be flourishing on a river flat that had been cleared, grazed and cropped, then left deserted for 40 years. After the farm had been abandoned, the area was incorporated into a Nature Reserve and, although exotic weeds persist, the native bush is returning. This recovery potential makes the Swamp Rat a good indicator of the quality of habitat rehabilitation at ground level—a characteristic that may be of importance as restoration becomes increasingly utilised in wildlife management.

D. LUNNEY

C. ANDREW HENLEY/LARUS

# Brown Rat

## *Rattus norvegicus*

(Berkenhout, 1769)

nor-vedge'-ik-us: 'Norwegian rat'

Dominant rodent of Europe, the Brown Rat arrived on the early fleets to Australia but failed to spread much beyond the major coastal cities and ports. Not being a climber, it is found mainly at ground level or below, in cellars or sewers: warehouses and wharfs around ports are a favoured locality but it is occasionally found around farm buildings and along creek banks in the wetter southern areas.

If relatively unpersecuted, the Brown Rat lives a colonial life in burrows. These can be deep and extensive and some have been known to remain in use in the same place for over 20 years. Within each such colony a ranking order exists between males, dominant individuals establishing territories around burrows that contain several breeding females. Young rats are forced to leave the colony of the dominant male and search elsewhere for a home.

*The introduced Brown Rat has not spread far beyond the coastal cities.*

### *Rattus norvegicus*

**Size**

*Head and Body Length*
180–255 (240) mm
*Tail Length*
150–215 (200) mm
*Weight*
200–400 (320) g

**Identification**  Grey-brown above; white or grey below. Solid build; small ears which only just reach eye when bent forward; tail stout, less than or just equal to head and body length. Female has six pairs of teats. Aggressive disposition.

**Recent Synonyms**  None in Australia.

**Other Common Names**  Norway Rat, Sewer Rat, Laboratory Rat, White Rat.

**Status**  Locally common.

**Subspecies**  None in Australia.

**Reference**

Calhoun, J.B. (1963). *The ecology and sociology of the Norway rat.* U.S. Dept. of Health, Education and Welfare, Bethesda.

It is thickset, with scruffy brown fur, small eyes, short ears and a short, thick tail which is often scabby in older individuals. A trapped or cornered Brown Rat will repeatedly launch itself towards its captor: in this behaviour it differs markedly from the Black Rat and Australian native rats which are timid and usually cower when trapped.

Around human habitation the Brown Rat eats most human foods, the consequent spoilage of which involves a considerable economic loss. Near the coast it is known also to eat shellfish, bird eggs, seeds and some insects. Given the opportunity, it is a successful predator on small mammals and birds.

In contrast to its role in the wild, the Brown Rat has proved useful to humans as a laboratory animal. Domesticated strains have lost the aggressiveness and unpleasant appearance of their wild ancestors and different strains and colour mutants are popular as pets.

The Brown Rat is a profile breeder. Females have 12 teats and produce litters of up to 18 young (usually seven to ten). The gestation period is 21–23 days and the young are weaned when about 20 days old. At birth they are blind and naked but by ten days they are well furred. The eyes are open when they are about 15 days old, at which time they begin to explore beyond the nest. They are weaned when 20 days old.

C.H.S. WATTS

*Rattus norvegicus*

# Black Rat

## *Rattus rattus*

(Linnaeus, 1758)

rat'-us: 'rat rat'

Probably introduced when the First Fleet disgorged its cargo in Sydney Cove, the Black Rat has spread through much of settled Australia and is now common over most of the agricultural land of southern Australia. Although reaching greatest numbers in areas near human habitation, it is not restricted to habitats altered by humans but is found in relatively unaltered country around the entire Australian coast. Probably because of its water requirements, it has not been able to penetrate into arid central Australia.

### *Rattus rattus*

**Size**
*Head and Body Length*
165–205 (190) mm
*Tail Length*
185–245 (230) mm
*Weight*
95–340 (280) g

**Identification** Body slender and elongated; tail longer than head and body. Ears large and thin, reaching past middle of eye when bent forward. Females usually with fives pairs of teats.
**Recent Synonyms** None.
**Other Common Names** Fruit Rat, Roof Rat, Tree Rat, Alexandrine Rat, Ship Rat, European Black Rat.
**Status** Abundant.
**Subspecies** None in Australia.
**Reference**
Yosida, T.H. (1980). *Cytogenetics of the Black Rat. Karyotype evolution and species differentiation.* University of Tokyo, Tokyo.

K. ATKINSON

*The introduced Black Rat is common on agricultural land in southern Australia. It is also a pest of stored food.*

The Black Rat has made a success of life in its new land and is well-known to virtually everyone, although often masquerading under such names as Roof Rat, Fruit Rat or Tree Rat. It is a sleek, long-tailed, large-eared, gentle animal, commonest around farms, watercourses, rubbish dumps, food warehouses and in older buildings in cities and ports. It is persecuted, with reason, as a carrier of such diseases as salmonellosis and leptospirosis, transmitted to humans through its urine and faeces. It is also the principal carrier of the plague bacillus which was responsible for much human death in medieval Europe. Since then, more hygienic living conditions have broken the link between bacteria, flea, rat and human, and plague has become a rare disease in most parts of the world.

Among human dwellings and farms, the Black Rat eats almost anything that is utilised by humans or domestic pets. In the forests of eastern

Australia, it is also omnivorous, at times living almost exclusively on underground fungi. Although proven guilty of destroying the eggs and young of birds in New Zealand, evidence of similar behaviour in Australia is less strong. In one study, young birds were found to have been eaten in only one month of the year; in another, there was no evidence that Black Rats living among a colony of ground-nesting birds ate more than the occasional bit of bird carrion.

Early naturalists felt that the Black Rat would displace native species where they came into contact but this has not happened; native species have held their own in areas that have not been too greatly degraded. The Black Rat seems to be exploiting niches that either are not occupied by any native species or have been newly created by human activities.

It climbs well, often living in house roofs or raiding fruit trees. It is a good swimmer and is often confused with the Water Rat. Its colour ranges from panther black to light brown and the name 'Black Rat' is a misnomer since black individuals are rare. Not surprisingly, it has accumulated a long list of common names, all undoubtedly referring to the same species. However, it is almost certain that there are two species of Black Rat in Australia, the European Black Rat (which has spread around Australia) and the Asian Black Rat, *R. tanzumi,* was discovered living alongside the European species in Brisbane in the 1970s. The two forms are externally similar, but differ in chromosomes and blood chemistry.

Female Black Rats have ten teats (rarely 11 or 12) and may have up to six litters of five to ten young in a year. The young are born blind and naked after a gestation of 21–22 days, but develop rapidly and may be weaned when 20 days old. They are reproductively mature when three to four months old and although the life span in the wild is seldom more than a year, they may live as long as three years in captivity.

C.H.S. WATTS

Kangaroo Is.

**Rattus rattus**

D. WATTS

*The Canefield Rat is primarily an inhabitant of tropical grasslands.*

# Canefield Rat

## *Rattus sordidus*

### (Gould, 1858)

sor'-did-us: 'dirty rat'

As its common name suggests, this rodent is common in canefields but, away from this source of introduced food, it inhabits tropical grassland, open forest and even grassy patches within clearings in dense rainforest. In this respect it resembles the Grassland Melomys. The two species are also similar in being serious pests of sugar cane in north-eastern Queensland, gnawing the hard outer rind of the cane and exposing the sap to bacteria that reduce the sugar content. The attack of the Canefield Rat is usually confined to ground level, where it may make a gash up to 50 millimetres long and 20 millimetres wide at the base of a cane, or even gnaw through it.

Burrows are dug where the soil is appropriately moist. Large discrete colonies may be formed, often at some distance from a canefield, which is not invaded until the young crop provides a complete canopy of leaves.

The Canefield Rat is omnivorous. Preferred foods are stems and seeds of grasses and some broad-leafed herbs, but insect remains are commonly found in the stomach.

It is capable of breeding throughout the year but females are rarely pregnant in late winter and early spring. With females capable of becoming

---

### *Rattus sordidus*

**Size**
*Head and Body Length*
135–210 (180) mm (males)
110–160 (150) mm (females)
*Tail Length*
Always shorter than head and body.
*Weight*
50–260 (190) g (males)
50–150 (125) g (females)

**Identification** Similar in appearance to *Rattus vilosissimus* and *R. colletti* but lighter in colour than the latter and lacking its yellowish colour on the flanks, especially in the throat region.
**Recent Synonyms** *Rattus conatus, Rattus gestri conatus, Rattus youngi.*

# Pale Field-rat

## *Rattus tunneyi*

## (Thomas, 1904)

tun'-ee-ee: 'Tunney's rat'

**Other Common Names** Field-rat, Field Ground Rat, Dusky Field-rat, Sordid Rat, Annam River Rat (a misspelling of Annan River), Sombre Downs Rat.
**Status** Abundant in the north; rare and locally extinct in the south.
**Subspecies** None.
**Reference**
Taylor, J.M. and B.E. Horner (1973). Results of the Archbold Expeditions No. 98. Systematics of native Australian *Rattus* (Rodentia: Muridae). *Bull. Amer. Mus. Nat. Hist.* 150: 1–130.

Yellow-brown in colour, with large eyes, the Pale Field-rat is an attractive native mammal. It is now distributed around the northern Australian coast but records from the nineteenth century indicate that it formerly occurred near Adelaide and

pregnant at nine to ten weeks of age, a gestation period of only three weeks and a mean litter size of about six, the Canefield Rat is one of Australia's most fecund rodents. The proportion of adult females that breed is strongly influenced by rainfall but whether breeding is a direct response to rain or to an increased food supply or to other related factors remains unknown.

The Canefield Rat is very similar to the Long-haired Rat, *R. colletti,* and hybridises with it in the laboratory, producing offspring of reduced fertility. The Canefield Rat, or a closely related species, also occurs in the savanna of southern New Guinea.

T.D. REDHEAD

### *Rattus tunneyi*

**Size**
*Head and Body Length*
118–198 mm
*Tail Length*
78–190 mm
*Weight*
42–206 g

**Identification** Yellow-brown to brown above; grey or cream below. Protuberant eyes. Tail shorter than head and body.
**Recent Synonyms** *Rattus culmorum apex.*
**Other Common Names** Tunney's Rat, Paler Field Rat.
Indigenous name: Chiiny chiiny (Wik Munghan).
**Status** Sparse, widespread, fluctuating; some pockets of abundance.
**Subspecies** *Rattus tunneyi tunneyi*, northern and north-western Australia.
*Rattus tunneyi culmorum*, eastern Australia.
**Reference**
Taylor, J.M. and B.E. Horner (1973). Results of the Archbold Expeditions No. 98. Systematics of native Australian *Rattus* (Rodentia: Muridae). *Bull. Amer. Mus. Nat. Hist.* 150: 1–130.

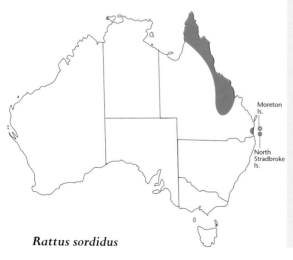

Moreton Is.

North Stradbroke Is.

*Rattus sordidus*

BABS & BERT WELLS

*Now limited to northern coastal Australia, the Pale Field-rat was once distributed over much of the mainland.*

Perth. Skeletal remains in owl pellets dating back a couple of hundred years indicate a former distribution throughout the western desert and it seems likely that its pre-European distribution covered almost all of continental Australia. Its distribution in northern Australia is now very patchy. In southern Kakadu National Park (Stage III), it was 67 times more common in the couple of hundred square kilometres of extremely heterogeneous habitat around the Marrawal-Arnhem Convergence (Coronation Hill area) than in the rest of Stage III.

Typically, it lives in tall grassland, usually associated with a small, often seasonal, watercourse. It appears to be strictly nocturnal, spending the day in a shallow burrow in loose, sandy soil: the need for such soil seems to limit its local distribution. It eats grass stems, seeds and roots but, despite the rich variety of available vegetation in the north, its diet is rather specialised. In Kakadu, it particularly favours the shoot bases of Alloteropsis grass, Sorghum seeds, Pandanus roots and some sedges. During the dry season it

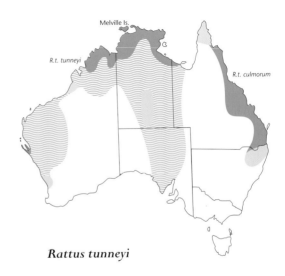

**Rattus tunneyi**

is known to feed on grass seeds lodged in small sandstone outcrops in the savanna. In some parts of Queensland, it is a pest of native Hoop Pine plantations, where it burrows down and chews the roots, killing half-grown trees.

Its current distribution is very patchy but, in the nineteenth century, Knut Dahl complained of substantial areas with such a density of burrows that horses found it difficult to keep their footing. Burrows often have accumulations of faeces at their entrance and a set of four or five entrances may be interconnected. Each burrow system is usually occupied by a single adult but may also contain a number of juveniles. Termite mounds are often used as nest sites, the horizontal entrances at ground level giving them the appearance of little castles. Presumably such situations provide shelter with less likelihood of flooding in riparian situations.

Breeding takes place in spring in north-eastern New South Wales and variously between January and August in the Northern Territory. There is a four to five day oestrous cycle, a gestation length of 21–22 days and a post-partum oestrus. Females have ten teats and produce 2–11 young (usually four). Young can be weaned at three weeks and attain sexual maturity as early as five weeks, creating a high reproductive potential. Densities are occasionally as high as 7.4 per hundred trap-nights in some Kakadu savannas and 50 per hundred trap-nights in north-eastern New South Wales.

The Pale Field-rat is one of several Australian rodents that exist over a vast area with substantial variation in abundance in response to great climatic variability. Its creekline refuges have proved vulnerable to destruction by introduced mammals, such as the Rabbit and domestic stock, with consequent loss of as much as 90 per cent of its former range. The northern coastal areas remain its last refuge, probably because this environment has been less modified than elsewhere and, perhaps, because climatic variation is less extreme.

R. W. Braithwaite and P. R. Baverstock

A.C. ROBINSON

*Populations of the Long-haired Rat vary enormously from being rare to eruption in plagues.*

# Long-haired Rat

## *Rattus villosissimus*

### (Waite, 1898)

vil'-oh-sis'-im-us: 'hairiest rat'

In the summer of late 1860, the expedition of Burke and Wills arrived at Cooper Creek, South Australia, where they found tranquillity after 23 days' march into the desert. At dusk the peace was turned to chaos by the invasion of the camp by thousands of rodents which ate almost every article that lay on the ground. Although some items could be protected by hanging them in trees, the expedition was forced to abandon camp and move downstream. The raiders were Long-haired Rats, first collected by Europeans only 13 years previously when Edmund Kennedy's expedition reached Cooper Creek some 250 kilometres upstream from the camp of Burke and Wills.

## *Rattus villosissimus*

### Size

*Head and Body Length*
150–220 (187) mm (males)
120–205 (167) mm (females)
*Tail Length*
125–180 (150) mm (males)
100–175 (141) mm (females)
*Weight*
65–280 (156) g (males)
54–200 (112) g (females)

**Identification** Very similar in appearance to *R. sordidus* and *R. colletti* but lighter grey in colour and without lateral tan or brown colouration which is often present in these species.

**Recent Synonyms** *Rattus sordidus villosissimus*.

**Other Common Names** Plague Rat.

**Status** Usually rare, living in small, widely scattered refuge populations; occasionally in plague proportions.

**Subspecies** None.

### References

Finlayson, H.H. (1939). On mammals of the Lake Eyre Basin. IV. The Monodelphia. *Trans. Roy. Soc. S. Aust.* 63: 88–118.

Carstairs, J.L. (1976). Population dynamics and movement of *Rattus villosissimus* (Waite) during the 1966-69 plague of Brunette Downs, Northern Territory. *Aust. Wildl. Res.* 3: 1–10.

Like the closely related Canefield Rat, the Long-haired Rat has an extremely high reproductive potential and, when favourable conditions lead to an abundance of grass seeds and green plants, this potential is realised. Populations rapidly increase and spread into areas where they are not found at other times but, as the country returns to its usual arid condition, most die and distribution again becomes restricted to small refuge areas where water and plant food is always available.

During rat plagues the Letter-winged Kite, Black Kite, Dingo, Fox, Cat and snakes reap the benefit of abundant prey and these predators increase in numbers. In the main, however, shortage of food or water, rather than increased predation, cannibalism or disease, is the major reason for the decline of each plague. It is at such times that the Long-haired Rat is most noticeable, starvation forcing it to forage for food in the daytime.

Like other small mammals of the desert, it is normally nocturnal and rests by day in a complex, shallow burrow containing a central nest of shredded grass. Despite its common name, the Long-haired Rat has body hair no longer than that of the introduced Black Rat.

C.H.S. WATTS

Since those early days, spasmodic hordes of the Long-haired Rat have inflicted misery on settlers in the inland regions of the Channel Country, Simpson Desert, the Barkly Tableland and north-western Queensland. Plagues occurred in 1860–61, 1869–70, 1887, 1916–18, 1930–32, 1940–42, 1948, 1950–52, 1956, 1966–69 and 1972, apparently following periods of exceptionally good rains or floods in a region where drought is the normal condition. Since 1972, plagues have continued to occur when conditions have been favourable.

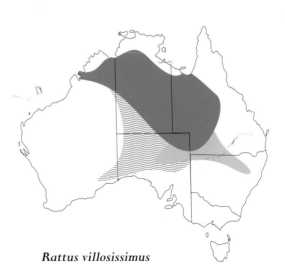

*Rattus villosissimus*

# ORDER SIRENIA

## *Dugong and Manatees*

**U**tterly unrelated to seals and whales, the sirenians are unique among the marine mammals in being herbivorous. As in whales, the tail-fin is horizontal. The two surviving members of the order have a tropical distribution.

# FAMILY DUGONGIDAE
## *Dugong*

This family, which has only one living species, is restricted to warm
waters of the Indian and south-western Pacific Oceans. The related
Manatees inhabit warmwaters of the western Atlantic Ocean. Manatees
have a platypus-like tail; the tail of the Dugong is like that of a whale.

# Dugong

## *Dugong dugon*

(Müller, 1776)

due'-gong due'-gon: 'dugong'

The Dugong is the only living herbivorous mammal that is completely marine. It inhabits shallow, warm (18°C or above) tropical and subtropical coastal waters of the Indian and western Pacific Oceans: in Australia, its normal range extends around the northern coast from Shark Bay, Western Australia, to Moreton Bay, Queensland. However, on the east coast, individual Dugongs are occasionally seen near Sydney and even further south along the New South Wales coast to as far south as Tathra.

*A young dugong accompanies its mother.*

T. PREEN

One record from Albany, Western Australia, indicates that individuals may stray south along the west coast. The Dugong spends most of its time in calm, sheltered, shallow, nutrient-rich waters where fine bottom sediments support meadows of seagrasses on which it feeds. It occurs in clear or muddy bays, broad channels, the lee sides of coastal islands and on, or close to, some coral reefs.

It feeds almost exclusively on seagrasses, marine algae being occasionally eaten if these are scarce. All species of seagrasses are browsed and there is a preference for tender new growth. In tall, luxuriant stands of seagrasses, only the leaves are grazed, but where these are sparse, the entire plants, including rhizomes and roots are consumed, this destruction resulting in distinctive elongated, trail-like scars through the aquatic vegetation. The Dugong is also known to feed on benthic invertebrates such as sea-pens, mussels and tunicates. Adaptations to this diet include a dense, heavy skeleton (which helps to keep it on the bottom while feeding); a downwardly deflected snout and ventrally positioned mouth; a broad, flexible, sensitive upper lip for

working seagrasses into the mouth; horny pads at the anterior ends of both upper and lower jaws for grasping and possible mastication of seagrasses; flattened peg-like cheek-teeth for further mastication; a very large sac-like stomach for storage of the fibrous food; and a very long intestine. Microbial digestion of plant fibre occurs in the relatively short caecum and the large intestine (which is up to 25 metres long). The mechanisms by which the Dugong copes with the high content of salt in its diet are not yet known.

When undisturbed, its daily movements and activities are largely determined by tide, weather and season. On rising tides, it moves into shores or shoals to feed; at low tide it rests or feeds in deeper water. In calm weather, individuals move from sheltered to exposed waters. Daily movements of up to 25 kilometres are common—longer daily movements are known, one Dugong being recorded to travel 200 kilometres in two days.

It is normally gregarious: in populations that have not been recently disturbed or decimated—as in Moreton Bay and eastern Cape York—herds of several hundred animals are found. Details of social behaviour are not well known, but Aboriginal hunters refer to 'whistler' Dugongs, especially large, strong old males that act as herdmasters and appear to use a whistling sound to keep their herds together. Underwater chirping sounds may also contribute to social behaviour, such as the maintenance of contact between a mother and her calf.

The ear apertures and eyes of a Dugong are small but it has an excellent sense of hearing and seems to have very good sight. Animals that have not been hunted or harassed show curiosity toward swimmers and divers and can readily be photographed underwater. The Dugong is active and alert, with an intelligence probably comparable to that of terrestrial herbivores such as deer.

The large, triangular, horizontally expanded tail-fluke propels the animal at a speed of up to 22 kilometres an hour by means of slow, powerful, up-and-down beats. The short, paddle-like forelimbs, which assist in locomotion, steering, feeding and maintenance of body position, are

## Dugong dugon

**Size**
**Townsville, Queensland**
*Head to tail*
109 (newborn)–315 (223) cm (male)
106 (newborn)–331 (219) cm (female)
*Weight*
20 (newborn)–420 kg (both sexes)

**Identification** Grey to bronze above; almost white below. Fusiform body with appearance of a rotund dolphin with horizontal dolphin-like tail flukes. Readily distinguished from dolphins by broad, rounded upper lip with bristles, paired nostrils, mobile paddle-like forelimbs and absence of dorsal fin.
**Recent Synonyms** *Dugong dugong*, *Dugong australis*.
**Other Common Names** Sea Cow.
**Status** Common, limited; vulnerable.
**Subspecies** None.

also employed in slow movement over the bottom while feeding or resting. During rapid swimming, they are held close to the sides of the body. The nostrils, near the front of the head, are opened above water for one or two seconds when an individual comes to the surface to breathe, then closed while it submerges. The average diving time is 76 seconds but an individual has been observed to remain submerged for about 8.5 minutes.

Reproductive maturity is reached when an individual attains a length of about 2.4 metres and a weight of about 250 kilograms. This does not occur before the age of nine to ten years and may be as late as 15–17 years. Males and females are similar in size and appearance, but the distance between the anal and genital openings is much greater in males than in females. Although the second upper incisors grow throughout life in both sexes, they seldom erupt in females; in males, they develop into short chisel-like tusks. A pair of teats and associated mammary glands, well

**References**

Marsh, H. (ed.) (1981). *The Dugong.* Proceedings of a Seminar/Workshop held at James Cook University, 8–13 May 1979. James Cook University, Townsville, Queensland.

Marsh, H. (1989). Dugongidae (in) D.W. Walton and B.J. Richardson (eds) *Fauna of Australia, Vol. 1B, Mammalia.* Aust. Gov. Publ. Serv., Canberra. pp. 1030–1038.

Nishiwaki, M. and H. Marsh (1989). Dugong, *Dugong dugon* (Müller, 1776) (in) S.H. Ridgeway and Sir R. Harrison (eds) *Handbook of Marine Mammals, Vol. 3.* The Sirenians and Baleen Whales. Academic Press, London. pp. 1–31.

developed in adult females, lie slightly ventral and posterior to each flipper.

In northern Queensland, males are sexually active from May to November. Gestation appears to be about 13–14 months, at the end of which a single young is born. Most births occur between September and April. A female about to give birth moves into an area of shallow water located on, or protected by, sandbars where it is safe from large sharks and other predators. The mother swims off with her young as the incoming tide or wave covers her, the calf riding just above her back except when suckling. A calf begins to feed on seagrasses within a few weeks of birth, while still suckling and stays with its mother for at least 18 months, possibly up to two years. Pregnancy rates are very low and the interval between the births of calves in Queensland populations is three to seven years. Individuals may reach an age of more than 70 years.

Dugongs are sometimes killed through strandings from storm surges during cyclones. Large sharks, the Estuarine Crocodile and the Killer Whale attack the Dugong, but its main threat is from humans. Australia has the largest Dugong populations left in the world but these are threatened by habitat loss, pollution, accidental capture in fish nets and possibly overhunting by poachers or legal hunting by Aborigines and Torres Strait Islanders using non-traditional methods. The vulnerability of Dugong habitats is exemplified by an occurrence in 1992 in Hervey Bay, Queensland. About 1000 square kilometres of seagrass meadows were smothered with silt after massive floods on the adjacent land. The disturbance may have been exacerbated by pollutants carried by the floodwaters and by cyclonic disturbances and wave action that excavated some seagrasses and covered others with resuspended silt. Whatever the contributing causes, this previously rich feeding ground was largely denied to Dugongs. Many died in Hervey Bay. Others moved into cooler waters of coastal New South Wales and died in an emaciated condition, apparently from starvation. More large marine sanctuaries or national parks, in which habitats are protected, hunting is prohibited and net-fishing is excluded, need to be established in important habitat areas to protect this declining species. A few sanctuaries—for example, within the Great Barrier Reef Marine Park—have been established, but more are needed throughout the Dugong's range.

G.E. HEINSOHN

*Dugong dugon*

# ORDER CARNIVORA

## Carnivorous Eutherian Mammals

One of the most successful of the extant mammalian groups, the Carnivora, has native terrestrial representatives—dogs, cats, bears, etc.—on every continent except Australia: the Dog, Cat and Fox have been introduced there by humans.

Marine carnivores, broadly referred to as seals, require access to land in order to breed but are less restricted in their distribution than terrestrial carnivores. Nevertheless, they are largely restricted to cold oceans and there has been extensive and separate evolution of seals in the northern and southern circumpolar regions. A few species come ashore on the Australian mainland and Tasmania to breed. Some more southern species are included here because they occur in the Australian sector of Antarctica.

There has been considerable debate about the relationship between the terrestrial and marine carnivores. In the first edition of this work, the seals and their relatives were placed in a separate order, the Pinnipedia, but the weight of evidence is in favour of their inclusion within the Carnivora.

Moreover, it has become increasingly difficult to regard seals as a homogeneous group with a common ancestor somewhere within the early terrestrial carnivores. It seems rather that, some 20 million years ago, seal-like mammals arose from two distinct groups of terrestrial carnivores. This being the case, we must regard the resemblence of the eared seals, family Otariidae, to the earless seals, family Phocidae, as the outcome of convergent adaptations to a marine way of life. Similar adaptations are seen in the clearly unrelated whales and dugong.

In addition to the marine Otariidae and Phocidae, the order Carnivora is represented in Australia by two other families of introduced mammals: the Canidae (dogs) and Felidae (cats).

# FAMILY OTARIIDAE
## Eared seals

C. ANDREW HENLEY/LARUS

*The Australian Fur-seal is one of the largest pinnipeds to come ashore on Australian coasts.*

The sea-lions and fur-seals that comprise this family are characterised by the possession of a small external ear (pinna) and the ability to employ all four limbs in terrestrial locomotion. The forelimbs (flippers) are large and are the main swimming organ. The neck is larger than in phocids and the dentition resembles that of terrestrial carnivores.

The otariids fall into two subfamilies. The Arctocephalinae comprises the fur-seals; the Otariinae the sea-lions.

## Subfamily Otariinae

Sea-lions resemble fur-seals in having a small external ear (pinna) but lack a dense underfur. Most species are restricted to the Pacific Ocean.

# Australian Sea-lion

## *Neophoca cinerea*

(Péron, 1816)

nee'-oh-foh'-kah  sin'-er-ay'-ah:
'ash-coloured new-seal'

J. LOCHMAN

The Australian Sea-lion is the only pinniped endemic to Australia, occurring on offshore islands from Houtman Abrolhos, Western Australia, to Kangaroo Island and on the mainland at Point La Batt, South Australia. However, fossil remains, bones from Aboriginal middens and the accounts of early explorers indicate that it formerly occurred further east to Bass Strait, in which region it was locally exterminated by sealers.

*An Australian Sea-lion pup is fed on very rich milk and is weaned at the age of about one year.*

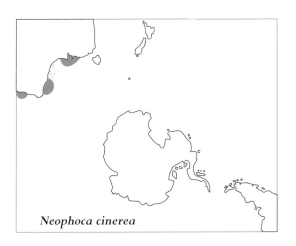

*Neophoca cinerea*

Relatively sedentary, it does not undertake definite migrations. It comes ashore on sandy beaches but uses rocky areas for breeding territories. Pupping sites are established in crevices and gullies. Little is known of the diet, but squid beaks have been found in the stomachs of dead animals.

*Male Australian Sea-lions battle to determine which has dominance over a group of females.*

## Neophoca cinerea

**Size**

*Head to Tail*
185–235 cm (males)
155–165 cm (females)
*Weight*
*circa* 300 kg (males)
*circa* 80 kg (females)

**Identification** Blunt snout; small, tightly rolled external ear. Males dark blackish or chocolate brown; crown of head and nape of neck white. Forequarters very large. Females silvery ash-grey above; yellow to cream below.
**Recent Synonyms** None.
**Other Common Names** White-naped Hair-seal.
**Status** Sparse.
**Subspecies** None.
**References**

Marlow, B.J. (1975). The comparative behaviour of the Australasian Sea-lions *Neophoca cinerea* and *Phocarctos hookeri* (Pinnipedia: Otariidae). *Mammalia* 39: 159–230.

Ling, J.K. and G.E. Walker (1978). An 18 month breeding cycle in the Australian Sea-lion? *Search* 9: 464–465.

The Australian Sea-lion has an extended and variable breeding season. On Dangerous Reef, South Australia, births have been recorded from mid-October to early January in some years but in others from August to early January. On Kangaroo Island, South Australia, the breeding season is even more extended and records from 1975 and 1978 show marked peaks at 18-month intervals. If indeed there is an 18-month breeding cycle, it is unique among the otariids, all other species having an annual cycle.

Both males and females are strongly territorial during the breeding season, often exhibiting marked aggression which may even be directed towards young pups and contribute significantly to their mortality. A strong bond is established between the female and her pup and this may last for some time after it has been weaned at the age of more than a year.

It is estimated that there are between 3000 and 5000 Australian Sea-lions and, in the absence of human predation, the population appears to be stable.

B.J. Marlow

## Subfamily Arctocephalinae

Fur-seals have a dense, insulating body fur beneath a sleek covering of long guard hairs. The group is essentially southern in distribution but two species extend into warm waters of the Pacific coast of Central America.

# New Zealand Fur-seal

## *Arctocephalus forsteri*

(Lesson, 1828)

ark'-toh-sef'-ah-lus forst'-er-ee: 'Forster's bear-head'

More widespread than its common name would suggest, this fur-seal breeds on rocky islands off South Australia, the southern coast of Western Australia and off the south-west coast of Tasmania. In New Zealand, it breeds along the southern and western coasts of the South Island, on rocky headlands of the south-east coast of the North Island, on Stewart and Chatham Islands and on the subantarctic islands. Recently it has

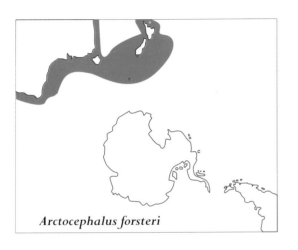

*Arctocephalus forsteri*

### *Arctocephalus forsteri*

**Size**
*Nose to Tail*
150–250 cm (male)
100–150 cm (female)
60–70 cm (pups at birth)
*Weight*
120–180 kg (male)
35–50 kg (female)
4–6 kg (pups at birth)

**Identification** External ears and long white vibrissae. Adult males: uniform dark grey to brown with pale muzzle fur and long, pointed snout. Massive neck with thick mane. Adult females: metallic grey to brown dorsal coat, paler ventrally with dark brown abdomen. Pups: dark brown to black natal coat with grizzled fur around head and neck.
**Recent Synonyms** *Arctocephalus doriferus* (Australian population only).
**Other Common Names** None.
**Status** Locally common in Australia, common in New Zealand and its subantarctic islands.
**Subspecies** None.

been reported to breed on Macquarie Island. Irregular rock platforms or large boulder-filled beaches are usually favoured sites for breeding colonies and haul-out areas. Vegetated areas and small grottos in rock crevices behind colonies are used as shelter, particularly by pups and juveniles during rough weather. Open flat areas, sand and cobblestone beaches, are avoided.

Females may give birth for the first time at four to five years of age, males become sexually

*The New Zealand Fur-seal has a thick coat of insulating fur.*

mature at a similar age but are not large enough or sufficiently socially mature to mate until about eight to nine years old. Most pups are born over a five-week period between late November and early January. On Kangaroo Island, the median date of birth is 21 December, some 10–12 days later than at colonies in the South Island of New Zealand. Adult males defend territories throughout the breeding season, establishing territories in October. Fighting to obtain or defend territories involves guttural and barking vocalisations and threatening postures which may culminate in chest-to-chest thrusting and attempts to bite the opponent on the face, neck or shoulder. Fights last only a few minutes but savage wounds are often inflicted by the lower canine teeth. Contests end when one combatant squeals submissively and backs away. The defence of territories requires constant vigilance and individuals have been observed to spend as long as 70 days ashore. Males do not feed during this time.

The New Zealand Fur-seal is polygynous, males defending territories usually containing five to eight females (but up to 16 have been reported in a single territory). Perhaps fewer than 10 per cent of males ever hold territories, each male for perhaps only one to three seasons. Throughout the non-breeding period, the spatial arrangement of adult males in breeding colonies is similar to that observed during the breeding season, but territorial behaviour is reduced.

### References

Crawley, M.C. and G.R. Wilson (1976). The natural history of the New Zealand fur seal (*Arctocephalus forsteri*). *Tuatara* 22: 1–29.

Crawley, M.C. (1990). New Zealand fur seal (in) C.M. King (ed.) *The Handbook of New Zealand Mammals*. Oxford University Press, Auckland. pp. 246–262.

Goldsworthy, S.D. and P.D. Shaughnessy (1994). Breeding Biology and Haul-out Pattern of the New Zealand fur seal, *Arctocephalus forsteri*, at Cape Gantheaume, South Australia. *Aust. Wildl. Res.* 21: 265–376.

Non-breeding age-classes are present throughout the year, but are absent from breeding colonies during the breeding season.

Pregnant females come ashore a day or two before giving birth to a single, dark-furred pup. As in most other otariid seals, lactating females are mated about one week after birth, usually by the nearest male. Embryonic diapause is thought to last approximately four months (until about April), followed by active gestation for about eight months. Females remain with their pups for about ten days, then depart the breeding colony to forage at sea, returning regularly to nurse the pup. Periods ashore usually last about two days, but foraging trips increase in duration throughout lactation in response to an increasing energy demand by the pup. Early in lactation, a female returns regularly (every three to five days) to nurse its pup. Growth rates during this period are highest and pups double their birth weight quickly (60–100 days). Such rapid growth places increasing energy demand on females, requiring them to forage for longer periods at sea. Late in lactation, females may forage from 6–16 days at a time.

As its rate of growth stabilises, a pup may stand a better chance of maintaining condition by independent feeding, rather than fasting for longer and longer periods, between less frequent nursing bouts. Mean weaning age is 285 days, but is variable (range 8–12 months). Most pups wean themselves and females continue to alternate between foraging trips and periods of shore attendance, but these foraging trips are of shorter duration than those of late-lactation females. Almost all pups have weaned prior to the next

K. GRIFFITHS

Above: *A juvenile New Zealand Fur-seal galloping into the sea.*
Right: *Like other eared seals, the New Zealand Fur-seal has mobile hindlimbs.*

G.W. JOHNSTONE

breeding season (late November), at which time the number of females ashore declines rapidly. During this period, females may be foraging intensively to build up their own energy reserves in preparation for the next breeding season.

Little is known about the diet of this species in Australia. On Kangaroo Island it preys principally on fishes and cephalopods, with birds (mainly the Little Penguin) making up a small proportion of the diet. The proportion of cephalopods in the diet is greatest during summer months; fishes predominate during winter. Recent work, utilising dive recorders, indicates that there is a marked seasonal variation in the foraging behaviour of lactating females on Kangaroo Island. Early in lactation, females dive in shallow-shelf waters at night near the sea-floor at 70–80 metres. Late in lactation they dive at more variable depths from 20–200 metres in waters near or just beyond the shelf-break, about 80–100 kilometres from shore. Females dive only at night, resting on the surface during the day. As prey species and their availability are likely to vary throughout the range, the diet and foraging behaviour of seals may be similarly variable.

In the nineteenth century, the New Zealand Fur-seal was indiscriminately harvested throughout its range by sealers. There is little information on the numbers of animals killed or the location of colonies prior to exploitation, but the range in Australia once extended into the Furneaux Group in eastern Bass Strait, where it was quite abundant. Some colonies in South Australia are increasing rapidly at 16 per cent per annum. The Australian population is estimated to number about 35 000 animals. The New Zealand population is estimated to number more than 50 000 and to be increasing at about 2–5 per cent per annum.

S.D. GOLDSWORTHY AND M.C. CRAWLEY

677

*The Antarctic Fur-seal was almost exterminated by hunting but is now recovering.*

# Antarctic Fur-seal

## *Arctocephalus gazella*

### (Peters, 1875)

gah-zel'-ah: '(HMS) Gazelle's bear-head'

Like other fur-seals, this species was indiscriminately hunted to near-extinction but its recovery from exploitation has been spectacular. Breeding colonies were first discovered in the late 1700s at South Georgia by Captain Cook and were commercially exterminated there by the 1820s; by the late 1800s it was thought to be extinct. It was rediscovered in 1915 at South Georgia and, in 1933, a small group of breeding animals was found on Bird Island, South Georgia. Between 1957 and 1972 the annual pup production increased from 5000 to more than 100 000 (17 per cent per year). About 95 per cent of the total population—about 1.5 million—breed at South

Georgia. There are smaller populations at Prince Edward Island, Iles Crozet, Iles Kerguelen and at three islands in Australian territory: Heard Island, the McDonald Islands and Macquarie Island. All breeding locations are north of 65° south and important ones are south of the Antarctic Convergence. Macquarie and Prince Edward Islands, Iles Crozet and Iles Kerguelen are north of the Antarctic Convergence.

Antarctic Fur-seals at Macquarie and Heard Islands were exterminated by 1820 and 1870, respectively. At Macquarie Island, approximately 200 000 fur-seals (species unknown) were killed. The islands have probably been recolonised by

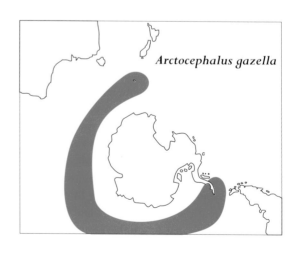

*Arctocephalus gazella*

C.A. HENLEY

*The Antarctic Fur-seal.*

animals dispersing from the major populations at South Georgia. Breeding Antarctic Fur-seals were first recorded on Heard Island in 1962–63 and, since then, annual pup production has increased rapidly and the breeding population now exceeds 1100. The status of fur-seals on the McDonald Islands is largely unknown, as they have rarely been visited. On Macquarie Island, the first pup since sealing ceased was recorded in 1955 and the population is recovering less rapidly. In addition to the Antarctic Fur-seal, the Subantarctic Fur-seal and smaller numbers of the New Zealand Fur-seal also breed on the island. When ashore, the Antarctic Fur-seal utilises a variety of habitats. On Heard Island, it uses flat, grassy meadows adjacent to beaches; on Macquarie Island, it uses open cobblestone beaches and steep tussock slopes.

### Arctocephalus gazella

**Size**

*Nose to Tail*
170–200 cm (male)
105–135 cm (female)
60–70 cm (pups at birth)

*Weight*
125–200 kg (male)
25–40 kg (female)
4–6 kg (pups at birth)

**Identification** Adult males: uniform silver-grey to brown coat; belly fur dark-brown. Well-developed mane, powerful chest and shoulders. Adult females: variable coloured fur, silver-grey to brown dorsally, paler cream to white fur ventrally, dark brown abdomen. Pups: ash-grey natal coat with grizzled fur around head and neck; muzzle and belly pale cream colour. External ears and long, white vibrissae.

**Recent Synonyms** None.

**Other Common Names** Kerguelen Fur Seal.

**Status** Locally common.

**Subspecies** None.

Adult males begin to come ashore and contest territories in early November. Males defend territories containing 5–15 females and fast during the breeding season until females have been mated. Females come ashore about a day prior to giving birth to a single pup, come into oestrus and are mated seven days later. Most pups are born over a three to four week period, peaking around 11 December. Adult females remain ashore with their pups for about ten days before leaving the colony to feed at sea. They then alternate between nursing bouts ashore and foraging trips at sea, until their pups wean at about four months. After the weaning period, adult females and pups abandon colonies; adult females do not come ashore until the next breeding season.

At Heard Island, lactating females make foraging trips of about five days, early in lactation, to about seven days towards the end of lactating. Overall, females spend 80 per cent of their time at sea and 20 per cent ashore. The attendance patterns of lactating females at Macquarie Island differ markedly, with females spending only about 65 per cent of their time at sea and foraging trips last only about three days. Differences in these attendance patterns are probably related to food availability.

Antarctic Fur-seals at Heard and Macquarie Islands feed predominantly on pelagic myctophid (lantern) fish and, to a lesser extent, on squid

**References**

Green, K., R. Williams, K.A. Handasyde, H.R. Burton and P.D. Shaughnessy (1990). Interspecific and intraspecific differences in the diets of fur seals, *Arctocephalus* species (Pinnipedia: Otariidae), at Macquarie Island. *Aust. Mammal.* 13: 193–200.

Shaughnessy, P.D., G.L. Shaughnessy and L. Fletcher (1988). *Recovery of the fur-seal population at Macquarie Island.* Pap. Proc. Royal Soc. Tas. 122: 177–187.

Shaughnessy, P.D. and S.D. Goldsworthy (1990). Population size and breeding season of the Antarctic fur-seal *Arctocephalus gazella* at Heard Island—1987/88. *Marine Mamm. Sci.* 6: 292–304.

(Heard Island only). This is in marked contrast to populations at South Georgia, which feed almost exclusively on Antarctic krill. Recent studies of foraging behaviour at Macquarie Island indicate that seals dive at night, usually to shallow depths (10–20 metres). This coincides with movements of myctophid fish, which make daily vertical migrations from deeper water (where they remain during the day), to near the surface at night.

Since no adult females are on islands during winter, it is assumed that they migrate to unknown locations. Males appear to remain near breeding colonies throughout winter. At Heard Island, large numbers of non-breeding males (up to 15 000) come ashore to moult after the breeding season. As this number is much greater than expected from the size of the breeding population (about 1100 animals), these immigrants may have travelled from large concentration at South Georgia (approximately 6600 kilometres from Heard Island), or possibly from a large, undiscovered population on the north-west coast of Iles Kerguelen. The species is occasionally sighted on the Antarctic mainland and in Antarctic pack-ice during winter. Despite an ability to travel great distances, there has been only one report from Australia, on Kangaroo Island.

S.D. GOLDSWORTHY AND P.D. SHAUGHNESSY

# Australian Fur-seal

## *Arctocephalus pusillus*

### (Schreber, 1775)

pue-sil'-us: 'little bear-head'

The Australian Fur-seal is the most abundant of the pinnipeds now resident in Australia. Fully mature males greatly exceed adult females in size and weight, having massive necks and shoulders and large canine teeth. Size, strength and teeth are employed in the intense competition for access to females during the breeding season.

The Australian Fur-seal formerly occurred on islands and reefs in south-eastern Australian waters, from central New South Wales to southern Tasmania. Breeding now occurs at only nine islands, all within Bass Strait, where it resorts to traditional sites: typically on rocky islands in exposed places close to the sea; on open slopes; shore platforms and reefs; on boulder stacks and pebble-cobblestone beaches; and in caves. Many individuals range widely and non-breeding groups of varying size are regularly found at many sites within the historical range. In recent years, it has been reported from northern New South Wales and Kangaroo Island.

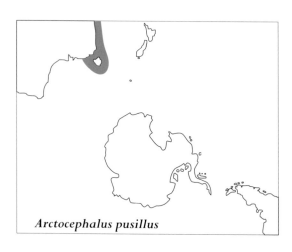

*Arctocephalus pusillus*

*Australian Fur-seals congregate ashore to breed over a period of six weeks.*

### *Arctocephalus pusillus*

**Size**

*Head to Tail*
201–227 (216) cm (males)
136–171 (157) cm (females)
*Weight*
218–360 (279) kg (males)
41–113 (78) kg (females)

**Identification** Males dark brown to brownish-grey when dry, with distinct mane of coarse hair. Females and immatures light brown to pale grey, shading to a contrasting fawn to cream throat. New-born pups black with variable silvering. External ears present.

Births and mating are closely synchronised within a period of six weeks in November to December. Females and immature animals are tolerant of a high degree of body contact and very dense aggregations (one animal per 1.5 square metres) can occur at pupping sites, which are dominated by the most powerful and experienced males. Each successful male maintains a sharply defined territory by means of vocal threats and ritualised aggressive posturing that involves a minimum of physical contact, although deliberate encroachment provokes a violent clash.

Territories range in size and shape according to individual prowess and the terrain, but average

**Recent Synonyms** *Gypsophoca tropicalis,*
*Gypsophoca dorifera, Gypsophoca tasmanica,*
*Arctocephalus doriferus, Arctocephalus tasmanicus.*

**Other Common Names** Large Fur-seal,
Tasmanian Fur-seal, South Australian Fur-seal,
Giant Fur-seal, Cape Fur-seal, South African
Giant Fur-seal.

**Status** Abundant, slowly increasing.

**Subspecies** *Arctocephalus pusillus pusillus,* South
Africa.

*Arctocephalus pusillus doriferus,* Australia.

**References**

Marlow, B.J. and J.E. King (1974). Sea-lions and
fur-seals of Australia and New Zealand—the
growth of knowledge. *Aust. Mammal.* 1:
117–135.

Warneke, R.M. and P.D. Shaughnessy (1985).
*Arctocephalus pusillus* the South African and
Australian Fur Seal: taxonomy, evolution, bio-
geography and life history (in) J.K. Ling and
M.M. Bryden (eds) *Studies of Sea Mammals in
South Latitudes.* South Australian Museum,
Adelaide. pp. 53–77.

about 60 square metres. Some are entirely
aquatic in deep tide pools or at sea adjacent to
the shoreline of the pupping areas. Harems are
not formed and females and immature animals
move freely between territories. Bachelor males
congregate where space is available at the
periphery of the breeding areas or on offshore
rocks.

Births usually begin early in November, reach
a peak around the first day of December and are
over by 20 December. One pup is born to each
female and is mothered solicitously, the maternal
bond being maintained by mutual recognition of
calls and odour. Oestrus occurs five to seven days
after birth and the female usually initiates mating
by approaching one of the territorial males. After
mating, a female departs to sea to forage, returning within several days to resume suckling her
pup. This pattern continues with increasing ab-
sences at sea for a period of six to eight months,
by which time the pup is able to accompany her
and begins to take solid food. Weaning normally
occurs before the onset of the next breeding
season.

Sexual maturity is reached at three to six
years in females and four to five years in males,
but the latter do not gain access to oestrous females until they achieve territorial status at 8–13
years. The breeding career of a male seldom exceeds three seasons (average 1.8, maximum six),
being terminated by defeat or (rarely) a mortal
wound. The White Pointer or Great White
Shark is a significant predator of fur-seals of all
ages and attacks by the Killer Whale have been
reported. Entanglement in marine debris is a
worsening problem and may become a significant cause of mortality. Longevity is 21 years for
females and 19 years for males.

When ashore, the Australian Fur-seal is sensitive to heat and avoids stress by moving into
shade, tide pools or back into the sea, but high
tides and storms force resting animals to higher
ground. On land it employs its four flippers in a
variety of gaits, from a slow shambling walk to a
fast gallop. It is a fair climber in rocky terrain.

A superb swimmer, it can dive to at least
200 metres in search of prey. Squids and school-
ing fishes are important foods, but a wide selection of sedentary bottom-dwellers, including
fishes, octopus and rock-lobsters are also taken.
Large prey are broken up by vigorous shaking,
and indigestible remains are later vomited.

Prior to European colonisation, the remote-
ness of most breeding colonies secured them
from Aboriginal hunters. By 1820, colonial and
foreign sealers had reduced the population to a
mere remnant. Even so, sealing persisted at a
low level in Victorian waters until protective
legislation was enacted in 1891, and limited
harvesting continued under regulation in the
Tasmanian region until at least 1923. By 1945,
the Bass Strait colonies had substantially recovered and recent estimates indicate a population
between 35 000 and 60 000.

R.M. WARNEKE

# Subantarctic Fur-seal

## *Arctocephalus tropicalis*

### (Gray, 1872)

trop'-ik-ah'-lis: 'tropical bear-head'
(based on erroneous locality)

As its name suggests, this species occurs on sub-antarctic islands: all breeding colonies are north of the Antarctic Convergence in the South Indian and South Atlantic Oceans. It prefers a rocky coastal habitat, including rock platforms and exposed boulder beaches. Like other fur-seals, it has recovered from earlier overexploitation. Its major concentration is at Gough Island in the South Atlantic Ocean, where the population is estimated to be 200 000. Other major populations are at Prince Edward Island (about 35 000), where it is increasing at about 15 per cent per annum and at Amsterdam Island (more than 35 000), where it is increasing at about 17 per cent per annum.

The only breeding colony in Australian territory is at Macquarie Island. Fur-seals were eliminated there within about ten years of its discovery in 1810, after some 200 000 had been harvested. The species of fur-seal on the island prior to sealing is unknown but, since sealers harvested there throughout the winter (by which time most Antarctic Fur-seals have left the islands), the original seals are more likely to have been Subantarctic than Antarctic Fur-seals. In

Left: *A breeding territory of Subantarctic Fur-seals showing an adult male, three adult females, two with pups.*
S. GOLDSWORTHY

Above: *A yellow chest and face, contrasting with dark dorsal fur is a characteristic of the male Subantarctic Fur-seal.* N. BROTHERS

recent years the Subantarctic Fur-seal has been breeding in small numbers on Macquarie Island, often in the same territories as the Antarctic and New Zealand Fur-seals. Although the species is recovering, the rate of increase is relatively slow (about 10 per cent per annum) compared with that of other populations. Interbreeding with

---

### *Arctocephalus tropicalis*

**Size**
*Head to Tail*
150–200 cm (male)
100–140 cm (female)
60–70 cm (pups at birth)
*Weight*
97–158 kg (male)
30–50 kg (female)
4–6 kg (pups at birth)

**Identification** Adult males: chocolate-brown to black dorsal fur with contrasting yellow chest and face. Crest of black fur on top of head becomes erect when excited. Well-developed mane, chest and shoulders. Adult females: dark grey or dark chocolate-brown dorsally. Chest and throat pale yellow with pale fur extending above the eyes. Belly fur rich chocolate brown. Pups: glossy-black natal coat with dark chocolate-brown belly. External ears and long white vibrissae.

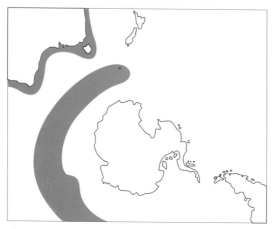

*Arctocephalus tropicalis*

**Recent Synonyms** *Arctocephalus gazella tropicalis*.

**Other Common Names** Amsterdam Fur Seal.

**Status** Locally common.

**Subspecies** None.

**References**

Bester, M.N. (1987). Subantarctic fur seal, *Arctocephalus tropicalis*, at Gough Island (Tristan da Cunha Group) (in) Croxall, J.P. and R.L. Gentry (eds) *Status, biology and ecology of fur-seals*. NOAA Technical Report NMFS 51: 57–60.

Goldsworthy, S.D. and P.D. Shaughnessy (1989). Subantarctic fur seals *Arctocephalus tropicalis* at Heard Island. *Polar Biology* 9: 337–339.

Shaughnessy, P.D. (1992). New mammals recognised for Australia—Antarctic and Subantarctic fur seals *Arctocephalus species. Aust. Mammal.* 15: 77–80.

other fur-seal species is a possible reason for this.

Pups are born in November and December with a peak around 10–15 December. Males compete for territories prior to the commencement of the breeding season, then fast and defend territories until all females are mated. Males hold territories, containing between 6–20 females. These come into oestrus and are mated about seven days after giving birth, then leave the colony a day or two later to feed. Like other eared seals, females alternate between periods of shore attendance (when they nurse their pups) and foraging trips at sea which vary in length depending on the availability of food. On Marion Island, females make foraging trips lasting about five days between bouts of shore attendance lasting about 2.5 days (spending about two-thirds of the lactation period at sea). This is in marked contrast to females on Macquarie Island, which make short nightly foraging trips lasting about eight hours, between longer foraging trips lasting about three days. Weaning age is estimated to be about 300 days. Male pups have typically higher growth rates and weaning weights than females.

On Macquarie Island, the Subantarctic Fur-seal feeds almost exclusively on myctophid (lantern) fish. It forages at night, usually at 10–20 metres, but occasionally dives to about 80 metres. Dives during twilight hours (dusk and dawn) tend to be deeper than those during darker hours. This behaviour is consistent with predation on myctophid fishes, which have a marked nocturnal vertical migration. Cephalopods are more important in the diet at Gough and Marion Islands than at Macquarie Island.

The Subantarctic Fur-seal is not known to be migratory, but individuals are known to make long movements. Two seals tagged as pups at Marion Island were sighted at Heard Island, almost 3000 kilometres distant. One pup was born at Heard Island in the 1987–88 summer, this being the first breeding record south of the Antarctic Convergence and possibly the first stage of colonisation of the island by this species. Adult males have been recorded occasionally at Juan Fernandez in the south-east Pacific Ocean and at South Georgia. Numerous juvenile animals have been stranded along the southern and eastern coasts of Australia in recent years. Presumably the recently established colony at Macquarie Island was founded by animals that had dispersed from one of the South Indian Ocean populations (Iles Amsterdam, Iles Crozet or Prince Edward Island) at least 5500 kilometres distant.

S.D. GOLDWORTHY AND P.D. SHAUGHNESSY

# FAMILY PHOCIDAE
## *'True' seals*

*Southern Elephant seals frequently wallow in muddy pools.* K. HANDASYDE

Phocids lack external ears. Their hindlimbs are directed backwards and do not contribute to terrestrial locomotion (which is caterpillar-like). Compared with otariids, the forelimbs are shorter. The group is represented in both hemispheres but with greater diversity in northern oceans.

# Southern Elephant Seal

## *Mirounga leonina*

### Gray, 1827

mi-roong'-ah  lay'-oh-neen'-ah:
'lion-like miouroung' (an Aboriginal name for
elephant-seals)

Reaching a length of more than 4 metres and a
weight of almost 4000 kilograms, the male
Southern Elephant Seal is the largest of the living
pinnipeds; females are much smaller, with a maximum weight of about 350 kilograms. It is a
predator on cephalopods and fishes.

Distribution is circumpolar, mainly in subantarctic latitudes, although individual records

extend from 16° south to 78° south. Lengthy
periods at sea are interspersed with regular
haul-outs on subantarctic islands and parts of the
Antarctic continent, from which individuals
appear to disperse randomly, rather than in a mass
migration.

Most adults feed in waters adjacent to the
Antarctic continent. While at sea, only brief periods (seldom more than two or three minutes) are
spent at the surface. Animals probably sleep beneath the surface. Very extensive dives, to 1200
metres or more and well in excess of an hour in
duration, have been recorded.

Breeding males come ashore in late August
and distribute themselves over the beaches of
such islands as South Georgia, Kerguelen and
Macquarie. They are followed by pregnant females which congregate in groups ('harems')
within the territory of an older male which expends much time and energy defending the territory and—with less success—the integrity of the
harem. The prominent proboscis of mature males

*The male Southern Elephant Seal is extremely bulky.*
K. HANDASYDE

is a resonator that assists in production of the loud, bubbling roar with which one male challenges another.

Elephant Seals come ashore to moult in the summer and autumn, the timing being different with sex and age. The moult involves shedding of hair attached to sheets of surface skin cells.

The single pup born to each female in September or October is suckled for about three weeks and must thereafter fend for itself. It fasts for about seven weeks while its aquatic skills are maturing as it plays and swims in freshwater pools and shallows of the beaches. It goes to sea when about ten weeks old.

Mating usually takes place in the harem just before weaning is completed, but a few females may mate in the sea, soon after leaving the harem. The ovum, fertilised about November, develops into a quiescent blastocyst which is not implanted until about the end of February.

Females become sexually mature at four to six years of age and live for 10–25 years. Males are sexually mature when about six years of age but do not attempt to mate until about ten years old and are unlikely to control a harem before the age of 14. They live for 20–25 years.

Subfossil bone deposits show that the Elephant Seal formerly colonised the north-western coast of Tasmania where it was eaten by Aborigines. A breeding colony on King Island in Bass Strait was hunted to extinction in the nineteenth century and heavy exploitation at Macquarie Island and other subantarctic islands well into the twentieth century led to a serious decline in total numbers. Apart from the birth of single pups on the coast of Tasmania in 1958 and 1975 and in South Australia in 1986, no breeding has taken place in Australia in this century. Only occasional stragglers reach our shores.

M.M. BRYDEN

## *Mirounga leonina*

**Size**
*Head to Tail*
350–420 cm (males)
200–260 cm (females)
*Weight*
2000–3800 kg (males)
250–350 kg (females)

Note: Males of 6.5 m and females of 3.5 m length are known.

**Identification** In water, uniformly dark grey; on land, dark brown above, paler below. Older males distinguished by size, extensive scarring of the neck and proboscis.

**Recent Synonyms** None.

**Other Common Names** None.

**Status** Abundant.

**Subspecies** *Mirounga leonina falclandicus*, Falkland Islands.
*Mirounga leonina macquariensis*, Macquarie Island, Chatham Island.
*Mirounga leonina crosetensis*, Crozet Island, Kerguelen Island, Heard Island.

**Reference**
Carrick, R. et al. (1962). Studies on the Southern Elephant Seal *Mirounga leonina* (L), I-V. *CSIRO Wildl. Res.* 7: 89–206.

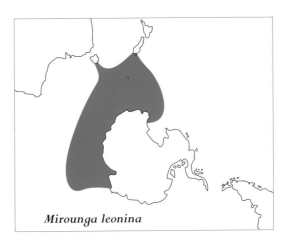

*Mirounga leonina*

R. PUDDICOMBE

# Leopard Seal

## *Hydrurga leptonyx*

### (de Blainville, 1820)

hide-rurg'-ah lept-on'-ix:
'slender-clawed water-worker'(?)

Of all the pinnipeds, only the Leopard Seal feeds extensively upon warm-blooded animals. A solitary animal, inhabiting the outer fringes of the Antarctic pack-ice, it ranges northwards in winter to many subantarctic islands and it is not uncommon for some animals to be stranded between August and October on the south-western coast of Western Australia and the

C. ANDREW HENLEY/LARUS

*The long snout of the Leopard Seal is an adaptation to its diet, which includes penguins.*

### *Hydrurga leptonyx*

**Size**
*Head to Tail*
up to 300 cm (males)
up to 360 cm (females)
*Weight*
up to 270 kg (males)
up to 450 kg (females)

**Identification** Long, slim body with disproportionately large head separated from body by marked constriction at neck. Dark grey above; lighter below; with light and dark grey spots on throat and sides. 'Reptilian' appearance of head, wide gape of jaws and characteristically three-pronged teeth make identification easy.
**Recent Synonyms** None.
**Other Common Names** Sea Leopard.
**Status** Common.
**Subspecies** None.
**Reference**
Hofman, R.J. (1979). Leopard Seal (in) *Mammals in the Seas*. FAO Fisheries Series 5: 2.

south-eastern Australian coast as far north as Coffs Harbour, New South Wales. Stranded animals are usually less than two years old and often in poor health but a number have survived for many years in Sydney's Taronga Zoo.

The Leopard Seal is the major predator of the Adelie Penguin and often patrols a stretch of coast where penguins enter and leave the water. The penguin is frequently seized from under water, the seal surfaces and with much head-tossing, the penguin flesh is bitten off and a relatively clean skeleton left behind. Sometimes, as the penguin is tossed about by the seal, the skin comes off the heavier body and may be found, inside out, at places where a Leopard Seal has been feeding. The Leopard Seal eats krill in large amounts and may also feed on carrion, fishes and cephalopods. It is known to attack young Crab-eater, Weddell and Elephant Seals.

It is believed to have an extended breeding season, the single young being born between September and January. It is about 1.5 metres long at birth with a soft, thick fur which is dark grey above with a darker central stripe and paler, with darker spots below. The exact

*The very elongate Leopard Seal inhabits Antarctic pack-ice but is frequently stranded on southern coasts of Australia.*

duration of lactation is not known but some observations suggest that it may be not much longer than 15 days. Mating is thought to occur between January and March.

J.E. KING

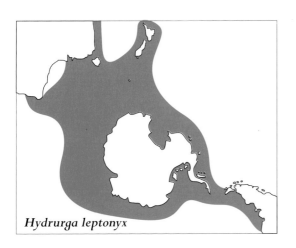

*Hydrurga leptonyx*

# Weddell Seal

## *Leptonychotes weddelli*

### (Lesson, 1826)

lep-on'-ik-oh'-tayz wed'-el-ee:
'Weddell's slender-claw'

Inhabiting the borders of the Antarctic continent and surrounding islands, the Weddell Seal is at home in regions of fast ice. It swims, feeds, mates and may even sleep below the ice but is dependent on holes for breathing and hauling-out. It is present, but not abundant, in pack-ice but the only specimen recorded from the Australian coast was caught in a fishing net in Encounter Bay, South Australia, in 1913.

The diet varies with locality but consists mainly of fish, cephalopods and small crustaceans. Although males and juvenile females feed throughout the year, breeding females appear to fast over most of the period during

## Leptonychotes weddelli

**Size**
*Head to Tail*
250–300 cm
*Weight*
350–450 kg

**Identification** Dark above; slightly paler below; liberally spotted. Head small, rather catlike.
**Recent Synonyms** None.
**Other Common Names** Weddell's Seal.
**Status** Common. Rare visitor to Australian coast.
**Subspecies** None.
**Reference**
Kooyman, G.L. (1981). *Leptonychotes weddelli* Lesson 1826 (in) S.H. Ridgway and R.J. Harrison (eds) *Handbook of Marine Mammals, Vol. 2*. Academic Press, London. pp. 275–296.

*The Weddell Seal is the most southerly of the Antarctic seals, often hunting below ice.*

day during the nursing period, towards the end of which the young spend more and more time in the water and gradually become independent. Mating takes place towards the end of the lactation period but has been observed only once, in December. A number of cut and bleeding males, injured in fights to establish breeding territories under the ice, are usually to be seen at this time. Development of the embryo is delayed for about two months and implantation does not occur until January or February, the young being born after a nine- to ten-month pregnancy.

The Weddell Seal is an excellent swimmer and the most accomplished diver of all the seals, being able to descend to 600 metres and to remain below the surface for as long as an hour without breathing. On the surface of the ice it is easy to approach and handle: when disturbed, it usually rolls onto its side with its belly towards the intruder and the

which they are lactating. Most feeding takes place at night, for the Weddell Seal is essentially nocturnal and, even at midsummer when the sun does not set, enters the water during the hours of minimum daylight. In winter, it spends almost all its time in the water.

Pregnant females leave the water and congregate near cracks in the fast ice between September and November. Each then gives birth to a single pup weighing about 28 kilograms and with a dense coat of silver-tan to grey hair that provides thermal insulation for three to six weeks while it is developing a thick layer of subcutaneous fat. The pup then moults to the adult pelage. Fed an extremely rich milk (up to 60 per cent fat), it grows rapidly and reaches a weight of about 115 kilograms at the age of six to seven weeks, when weaning occurs.

Mother and young usually begin to enter the water together when the pups are two to three weeks old and they swim together at least once a

*In common with other true seals, the Weddell Seal has hindlimbs that are directed backwards to form the equivalent of a tail fin.*

D.WATTS

G.W. JOHNSTONE

foreflipper raised. It moves on the ice with a sluggish, caterpillar-like motion.

The upper incisors and canines project more horizontally than in other seals and are used in a side-to-side sawing action to bite and abrade the ice to keep breathing holes open. Severe wear may expose the pulp cavity, leading to infection and abscesses.

R.A. TEDMAN

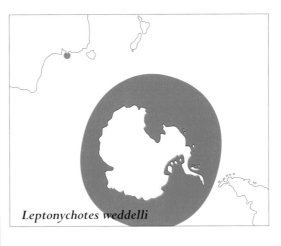

*Leptonychotes weddelli*

691

H. BURTON

*The Crab-eater Seal does not eat crabs. It filters krill through a sieve-like arrangement of its teeth.*

# Crab-eater Seal

## *Lobodon carcinophagus*

(Hombron and Jacquinot, 1842)

loh'-boh-don  kar'-sin-oh-fah'-gus:
'crab-eating lobed-tooth'

With an estimated population of 50 million, the Crab-eater Seal is by far the most abundant of the Antarctic pinnipeds. It is a gregarious animal found in large concentrations on drifting pack-ice, but individual stragglers are found occasionally in temperate waters of the Atlantic, Indian and Pacific Oceans. Over the past century, there have been seven records of the species on the eastern Australian coast, between Tasmania and Nambucca Heads, New South Wales, but many more visitors must have passed unnoticed.

The common name is misleading since this seal does not eat crabs but feeds almost exclusively on krill, small shrimp-like crustaceans which are abundant in Antarctic waters. The seal opens its mouth, taking in krill and water, then closes it, expelling the water through an elaborate sieve formed by the gaps between its multicusped cheek-teeth. Like the baleen whales, it is a filter-feeder and its vast numbers reflect its ability to harvest the ocean at a rather low level in the pyramid of food production.

Females do not reach sexual maturity until at least two years of age. Early in October, pregnant Crab-eater Seals come onto the ice, each followed by an adult male which is seeking to be her next mate. The female gives birth to a single pup, which is about 1.5 metres long and clothed in a soft, fluffy, greyish-brown coat of insulating fur. This is moulted when the pup is about two weeks old and the pup is weaned at the age of about four weeks. Following the departure of her pup the female permits the male, which she has previously kept at a distance, to approach and begin courtship, followed by copulation, after which

## *Lobodon carcinophagus*

**Size**

*Head to Tail*
up to 260 cm

*Weight*
up to 225 kg

**Identification** Streamlined, with moderately long snout. In winter, silvery brownish-grey with some spotting on shoulders and sides; in summer, almost uniform yellowish-white. Uniquely multicusped cheek-teeth.

**Recent Synonyms** None.

**Other Common Names** White Seal, Crabeating Seal.

**Status** Abundant.

**Subspecies** None.

**Reference**
Siniff, D.B., I. Stirling, J.L. Bengston and R.A. Reichle (1979). Social and reproductive behaviour of crab-eater seals (*Lobodon carcinophagus*) during the austral spring. *Canad. J. Zool.* 57: 2243–2255.

deterrent to Leopard Seals but the parallel scars borne by many individuals are evidence of encounters with this predator.

When approached, a Crab-eater Seal is nervous and, on being disturbed, lifts its head, opens its mouth and utters a hissing sound. It is more prone to run away than to attack and can reach the surprising speed of 24 kilometres an hour on snow-covered ice.

J.E. KING

# Ross Seal

## *Ommatophoca rossi*

### Gray, 1844

om'-ah-toh-foh'-kah ros'-ee:
'Ross's eyed-seal'

the breeding pair separates. Although the male takes no interest in the pup, being solely concerned to get to the female and repel other prospective mates, his presence probably helps to protect the young animal against predators such as the Leopard Seal. Juvenile seals often form large aggregations which also provide some

Normally an inhabitant of the circumpolar pack-ice, the Ross Seal is only very marginally a member of the Australian fauna. It has been recorded once, in January 1978, when a juvenile animal was stranded at Beachport, South Australia. It is one of the least known of the Antarctic seals but is become better understood since ice-breakers have been used to penetrate its domain. It was previously thought to be rare, but current estimates of a population in excess of 200 000 indicate that, although the species may not be abundant, it is not currently endangered. Of the few that have been measured, the largest male was 208 centimetres long and weighed 216 kilograms; the largest female was 236 centimetres long and weighed 204 kilograms.

When disturbed on the ice, the Ross Seal raises its head high, bulges its throat forwards, opens its mouth wide and emits an extraordinary series of bird-like trills, clicking, cooing and thumping noises as a threat display. When making

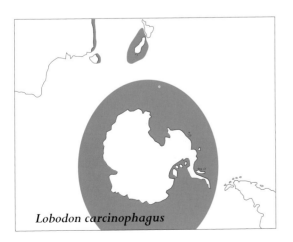

*Lobodon carcinophagus*

693

G.W. JOHNSTONE

*The Ross Seal is solitary and little is known of its biology.*

## Ommatophoca rossi

**Size**
*Head to Tail*
up to *circa* 250 cm (male)
up to *circa* 250 cm (female)
*Weight*
(unsexed) up to *circa* 250 kg

**Identification** Dark grey above; silver-grey below; spotted on sides. Throat and chest light with parallel streaks of grey from mouth to forelimbs. Body plump; head wide with very short, blunt snout and large, protruding eyes.
**Recent Synonyms** None.
**Other Common Names** Singing Seal, Big-eyed Seal.
**Status** Sparse, scattered.
**Subspecies** None.
**References**

King, J.E. (1969). *Some aspects of the anatomy of the Ross Seal,* (Ommatophoca rossi) *(Pinnipedia: Phocidae)*. Brit. Antarct. Surv. Sci. Rep. 63: 1–54.
King, J.E. (1983). *Seals of the World*. Oxford University Press.
Ray, G.C. (1981). Ross Seal *Ommatophoca rossi* Gray 1844 (in) S.H. Ridgway and R.J. Harrison (eds) *Handbook of Marine Mammals, Vol. 2*. Academic Press, London. pp. 237–260.

these sounds, its very large, soft palate bulges forward and meets the raised tongue so that the seal looks as if it has two tennis balls caught in its throat.

Large squids, which are probably swallowed whole, are the main food, but fishes, bottom-dwelling invertebrates and krill are also eaten. To catch its fast-moving prey the Ross Seal must be a rapid and agile swimmer and it is probably for this reason that its forelimbs are longer and more mobile than those of a typical phocid (not unlike those of a sea-lion). Its large eyes, in reference to which it has sometimes been referred to as the Big-eyed Seal, perceive movement of prey in the dim light under the ice: the sharply recurved, needle-like canines and incisors impale the squid,

which is then gripped and swallowed by the very powerful muscles of the throat. Given this method of feeding, it is not surprising that the teeth posterior to the canines are weakly developed.

Little is known of its breeding biology but mature animals probably associate in pairs, the female giving birth to light-coloured pups about 1 metre long in mid-November. Mating takes place a few weeks later. Adults moult annually in January and apparently do not feed at this time.

J.K. Ling

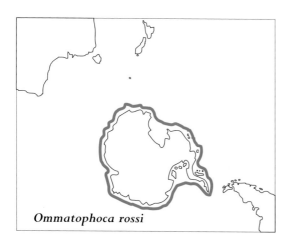

*Ommatophoca rossi*

# FAMILY CANIDAE
## *Dingo and Fox*

*Fox*   P. GERMAN

Canids are rather unspecialised carnivores with long muzzles, erect ears and strong claws (which cannot be retracted). They are opportunistic predators on a wide range of vertebrates and invertebrates but most species will also eat soft fruits. An adult male and female usually associate as a permanent pair and co-operate in rearing the young. Many canid species are social and may hunt in packs. The Fox was introduced by Europeans; the Dingo must have been brought to Australia by humans but the transporters (from the South-East Asia region) have yet to be identified with certainty.

L.K. CORBETT

*The Dingo is a descendant of the Asian Wolf, transported to Australia by Asian seafarers about 4000 years ago.*

# Dingo

## *Canis lupus dingo*

### Linnaeus, 1758

kah'-nis   lue-poos: 'wolf dog'

The Dingo is a primitive dog which evolved from the Indian Wolf, *Canis lupus pallipes*, about 6000 years ago and became widespread throughout southern Asia. Asian seafarers subsequently introduced it into Indonesia, Borneo, the Philippines, New Guinea and other islands, including Australia some 3500–4000 years ago. It eventually occupied all the Australian mainland from the tropical north through the hot central deserts to

*Canis lupus dingo*

**Size**
*Head and Body Length*
860–1220 mm
*Tail Length*
260–380 mm
*Shoulder Height*
440–630 mm
*Weight*
9.6–24.0 kg

**Identification**  Pelage typically ginger varying from sandy-yellow to red-ginger, occasionally black-and-tan or white. Most with white markings on the feet, tail tip and chest; some with black muzzle. Ears pricked, tail bushy. Distinguished from domestic dogs and hybrids in breeding once a year and having skull with narrower snout, larger auditory bullae and larger canine and carnassial teeth. Dingo–Dog hybrids of similar size and configuration are often distinguished by pelages with a dark dorsal strip or speckling in the white areas.

the cool southern forests and mountains and most islands except Tasmania. This colonisation was probably assisted by Aborigines, who had arrived in Australia many millennia earlier. Some Aboriginal groups used it to hunt game, especially macropods and possums and also as 'blankets' during cold nights. The Dingo probably contributed to the demise of the Thylacine and other native fauna on the mainland.

Most females become sexually mature at two years. Gestation is about 63 days and litters of one to ten pups (average five) are whelped during the winter, usually in an underground den. Pups become independent at three to four months or, if in a pack, at the commencement of the next breeding season.

Although often seen alone, many Dingos belong to a socially integrated pack, the members of which meet every few days or coalesce during the breeding season to mate and rear pups. At such times, howling and scent-marking is most pronounced. In remote and wilderness areas where Dingos are undisturbed by human control operations, discrete and stable packs of 3–12 occupy territories throughout the year. Within such packs, the dominant pair may be the only successful breeders, other pack members assisting in rearing the pups.

Territory size of a pack varies with the availability of prey and terrain but is not correlated with the size of a pack itself. Individual home

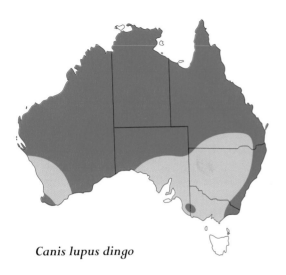

*Canis lupus dingo*

**Recent Synonyms** *Canis dingo*, *C. antarticus*, *C. familiaris dingo*.
**Other Common Names** Wild dog, Warrigal, Australian Native Dog.
**Status** Common to rare.
**Subspecies** The Australian population is *Canis lupus dingo*.
**References**
Corbett, L.K. and A.E. Newsome (1987). The feeding ecology of the Dingo III. Dietary relationships with widely fluctuating prey populations in arid Australia: An hypothesis of alternation of predation. *Oecologia* 74: 215–227.
Newsome, A.E. and L.K. Corbett (1985). The identity of the Dingo III. The incidence of Dingos, Dogs and hybrids and their coat colours in remote and settled regions of Australia. *Aust. J. Zool.* 33: 363–375.
Corbett, L.K. (1995). *The Dingo in Australia and Asia*. New South Wales University Press, Sydney.

ranges vary with age. The largest recorded territories (45–113 square kilometres) and home ranges (mean 77 square kilometres) are from the Fortescue River area of Western Australia. Mean home ranges recorded elsewhere are 33 square kilometres for arid central Australia, 39 square kilometres for tropical northern Australia and 10–21 square kilometres for forested regions of eastern Australia. Most remain in their birth area, but some individuals—especially young males—disperse: the longest recorded distance for a tagged individual is 250 kilometres over ten months in Central Australia.

Prey ranges from insects to Buffalo. However, within a particular region, a Dingo tends to specialise on the commonest available wildlife, changing group size and hunting strategy to maximise hunting success. For example, packs have greater success than solitary animals in hunting kangaroos, whereas individuals are better at hunting Rabbits. Staple prey are Magpie Geese, rodents and Agile Wallabies in the Top End (Kakadu); Rabbits, rodents, lizards and Red Kangaroos in central Australia; Euros and Red Kangaroos in the Fortescue River area; Rabbits in the Nullarbor Region; and wallabies and wombats in eastern Australia.

Since the early days of European settlement, the Dingo has harassed sheep and cattle. However, most attacks occur during periods when native prey are scarce (during droughts or as a result of human disturbance to habitats). When cattle die during drought, the Dingo scavenges their carcasses. Although the Dingo often assists in limiting numbers of Rabbits, Pigs and other pastoral pests, governments and landholders have attempted to control or eradicate it by offering scalp bonuses, hunting with trap and gun, poisoning and fencing. These attempts have been largely unsatisfactory, since most control measures merely harvest populations, or even promote increases in Dingo numbers by disrupting the social organisation of packs and allowing breeding rates to increase. Further, the widespread provision of subterranean watering points (bores and dams) for stock has allowed the Dingo to expand beyond the widely scattered natural waters, as has the provision of abundant non-native food sources: Rabbits in good seasons and cattle carrion during drought.

Today, the Dingo is under threat of extinction from another source. In the more settled coastal areas of Australia and increasingly in outback Australia, there is considerable interbreeding between the domestic Dog, *Canis familiaris*, and the Dingo, and the pure Dingo gene-pool is being swamped. Already in the south-eastern highlands about one-third of the animals are cross-breeds and the extinction of the subspecies seems to be inevitable.

L.K. CORBETT

*The introduced Fox, which has spread over much of the mainland, is a predator on many native animals.*

# Fox

## *Vulpes vulpes*

Linnaeus, 1758

vool'-payz: 'fox fox'

The European Red Fox was deliberately introduced into Australia in the 1860s and 1870s. First released in southern Victoria for sporting purposes, its subsequent spread was remarkable: by 1893 it had become a nuisance in north-eastern Victoria and by 1917 it had reached the region west of Kalgoorlie, Western Australia.

Next to the Dingo and feral Dog, the Fox is the largest terrestrial predator in mainland Australia, far larger than any carnivorous marsupial. It is found in habitats ranging from desert to urban, but is absent from tropical Australia and Tasmania. It does not appear to favour any particular habitat, its local distribution probably being determined by food supply and adequate refuge.

Although predominantly carnivorous, it is an opportunistic predator and scavenger. Diet varies with the season: wild fruits and insects being important foods in summer, small mammals

P. GERMAN

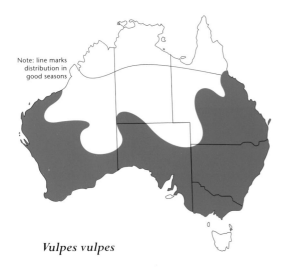

Note: line marks distribution in good seasons

*Vulpes vulpes*

(especially the Rabbit) and carrion in winter. Recent evidence confirms that the Fox is responsible for serious predation on some endangered or locally rare species such as rock-wallabies, *Petrogale lateralis* and *P. rothschildi*; Tammar Wallaby, *Macropus eugenii*; Brush-tailed Bettong, *Bettongia penicillata*; and Numbat, *Myrmecobius fasciatus*. Predation on farm livestock, particularly young lambs, can be serious and is possibly over-estimated.

The Fox is usually nocturnal and rests during the day in a den (usually an enlarged rabbit burrow), thicket, hollow log or leaning tree. In winter, when the range and availability of food is limited it may hunt and scavenge during the day. Faeces and urine are used to define territories by scent marking. Territory size averages 2–5 square kilometres for a family group.

Over a period of two to three weeks in early winter, females come into oestrus for two to three days. Males appear to be fertile throughout winter and early spring. Both sexes become sexually mature in their first year. Few adult vixens are barren and average litter size exceeds four young, factors which lead to a high reproductive success. Young are born in a den and first appear in late spring. Dispersal usually takes place in late summer or early autumn.

Population turnover appears to be high but mortality factors are poorly understood. Mange and distemper are probably important diseases and heavy hunting pressure in the 1980s (due to

## *Vulpes vulpes*

**Size**
*Head and Body Length*
610–740 mm (males)
570–670 mm (females)
*Tail Length*
360–450 mm (males)
380–430 mm (females)
*Weight*
4.7–8.3 kg (males)
4.0–6.8 kg (females)

**Identification** Reddish-brown above; whitish chin, throat, chest and belly. Distinctive tail tag, usually white but often black or dark red. Some variation in coat colour and pattern.
**Recent Synonyms** None.
**Other Common Names** Red Fox.
**Status** Abundant.
**Subspecies** The subspecies in Australia is probably *Vulpes vulpes vulpes*.
**References**
Coman, B.J., J. Robinson and C. Beaumont (1991). Home range, dispersal and density of Red Foxes (*Vulpes vulpes* L.) in Central Victoria. *Aust. Wildl. Res.* 18: 215–223.

Ryan, G.E. (1976). Observations on the reproduction and age structure of the fox, *Vulpes vulpes*, L. in New South Wales. *Aust. Wildl. Res.* 3: 11–20.

Kinnear, J.E., M.L. Onus and R.N. Bromilow (1988). Fox control and rock-wallaby population dynamics. *Aust. Wildl. Res.* 15: 435–450.

high skin prices) may have affected population characteristics. Nevertheless, there is no evidence to suggest that deliberate control measures have limited population size or distribution in the long-term. The main determinants of population size and distribution will probably continue to be food supply and refuge but the situation could be drastically altered in the event of accidental introduction of rabies to Australia.

B.J. COMAN

# FAMILY FELIDAE
## *Cat*

J. LOCHMAN

*Cats are major predators on native mammals and birds over most of Australia.*

With some three dozen species, native to all continents except Australia, this family has had questionable success. The larger species (tigers, lions, etc.) are endangered. Smaller species have fared somewhat better. The feral domestic Cat now occupies most of Australia.

## Cat

### *Felis catus*

### Linnaeus, 1758

fel'-is kat'-us: 'cat cat'

The origin of the feral Cat in Australia is uncertain. It quite possibly pre-dates European settlement, since many Dutch ships were blown ashore during the seventeenth century on the west coast of Australia while sailing to what is now Indonesia. Sailing ships usually carried Cats as pets and to keep down rodents and although the stark environment of the west coast was daunting to marooned seamen, it would not have been so for the Cat, which does not need to drink if feeding on fresh prey.

In Banks' journal of the voyage of the *Endeavour*, there is no mention of cats ashore at Endeavour River where the ship was beached for repairs, but 'catts' had been listed on board in October 1768 for the ceremony of crossing the line in the Atlantic and Captain Cook released 20 Cats on Tahiti in 1774. (It is said that Tathra, a township on the south coast of New South Wales, is an Aboriginal word meaning 'place of wild cats', referring to escapees from ships using the port.) The Cat now lives independently in all parts of Australia, even deserts, but some may revert to commensal life with humans in times of drought, as happened at Uluru in the mid-1970s when more than 200 were shot on the local rubbish dump.

Its diet is varied, including a range of small to medium-sized vertebrates, depending upon availability. Where the Rabbit exists, it is an important constituent, being caught by stealth, or by 'sit and wait' at warrens. Rabbits are few in temperate forests: in this environment, smaller mammals are more important in the diet of the Cat.

If the Cat arrived before European settlement, it may have caused substantial but undetected damage to the indigenous fauna. Certainly, the species can have a catastrophic impact on small mammals and sea-birds of offshore islands, and strenuous efforts are needed to eradicate it from these limited areas, as has been achieved in some New Zealand islands. Although few would doubt that the feral Cat poses a threat to Australian wildlife, its impact has never been measured. It is known to eat invertebrates, amphibians, reptiles, birds and mammals but, whereas the impact of the Fox has been assessed by removal studies, no such experimental work has yet been conducted on the Cat. On the other hand, one individual has been found with representatives of seven species of small reptiles in its stomach; and predation by the feral Cat is known to have hampered introductions of the Rufous Hare-wallaby, *Lagorchestes hirsutus*, in Central Australia.

*Felis catus*

### Felis catus

**Size**

*Head and Body Length*
448–617 mm (males)
380–555 mm (females)
*Tail Length*
235–335 mm (males)
230–316 mm (females)
*Weight*
3.8–6.2 kg (males)
2.5–4.4 kg (females)

**Identification**  No anatomical differences exist between feral and common domestic Cats. Identifiable (as a Cat) at a distance by rounded face and distinctive colouration. Most common coat colour is striped tabby (thin dark vertical stripes on lighter background), orange, black and, rarely, tortoiseshell (combination of orange and tabby or orange and black). Variable amount of white may be present, particularly on chest, belly or paws.
**Recent Synonyms**  None.
**Other Common Names**  None.
**Status**  Abundant.
**Subspecies**  None.

It is implicated among the difficulties attending reintroduction of other species. General studies of the diets of predators relative to prey abundance have shown that the main prey of a species may be eaten according to abundance and that scarce prey may be eaten by default in good times, but by intention at others. Thus, in times of drought in Australia, rare and endangered wildlife may come under severe pressure, especially if living in isolated pockets.

Any control program of the Cat in a particular area has to contend with movements of neighbouring individuals and their immigration. Studies of movements in Victoria indicate that females move over home ranges of 0.7–2.7 (mean 1.7) square kilometres; males, 3.3–9.9 (mean 6.2) square kilometres. Relative abundances were 0.7 per square kilometre in winter when the Rabbit

*The Cat obtains enough water from its prey to make drinking unnecessary: it is therefore able to live in arid environments.* C. ANDREW HENLEY/LARUS

was increasing and 2.4 per square kilometres in summer after breeding. Breeding may occur twice a year, spring and summer being the commonest periods. A study site of 6 square kilometres in central New South Wales contained five breeding females and two adult males, with kittens making a total of 28 (4.7 per square kilometre) in a year when the Rabbit was abundant. In autumn, prior to breeding, young males left their natal home-ranges: some moved 4–8 kilometres and settled in other favourable habitat already occupied by the species. Other immature animals moved into the study area and settled there. In the summer drought, there was a major dispersal, one Cat moving about 230 kilometres. In an earlier study at the same site, the Rabbit became very scarce in drought; several marked feral Cats deserted their home-ranges and preyed upon starving Rabbits coming to drink at a dripping tap at a nearby house in the heat of day. Subsequently, some of these individuals were reported 8–48 kilometres away, living commensally around a station home-stead and the nearest township, Mount Hope.

Such switching between feral and commensal existence is annual in the northern hemisphere, between summer and the winter snows. The switching is opportunistic in Australia but both cases demonstrate the Cat's adaptability. Its success in Australia is made possible in part by the climate. Young Rabbits are available as food in winter and spring; reptiles and insects in summer.

Feral Cats are difficult to control either by poisoning or by trapping. In New Zealand, the poison sodium fluoroacetate ('1080') has been used in fish-baits on islands; but persistent shooting, trapping and follow-up poisoning was necessary for its eradication. Biological control by infection with feline panleucopaenia ('cat-flu' or 'distemper') has been attempted on the sub-antarctic Marion Island, leading to a reduction in density from 6.7 to 2.7 animals per square kilo-metre. Thereafter, there was a change in diet, from sea-birds (54 per cent before control; 27 per cent afterwards) to the feral House Mouse, *Mus musculus* (increasing from 16 per cent to 41 per cent) possibly because the ratio of Mouse to Cat had increased. The distribution of Cats across the island did not alter with population decline.

Other prospects for biological control include feline herpes virus, which causes an upper respiratory tract infection. Reinfections may occur; and there is a vaccine to protect domestic cats. Being a herpes virus, it is amenable to genetic engineering.

A. NEWSOME

### References

Catling, P.C. (1988). Similarities and contrasts in the diets of foxes, *Vulpes vulpes* and cats, *Felis catus*, relative to fluctuating prey populations and droughts. *Aust. Wildl. Res.* 15: 307–317.

Turner D. and P. Bateson (eds) (1988) *The Domestic Cat: The Biology and its Behaviour.* Cambridge University Press.

Potter, C. (ed.) (1991) The Impact of Cats on Native Wildlife. Proceedings of workshop, 8–9 May, 1991. National Endangered Species Program, Aust. National Parks and Wildlife Service, Canberra.

# ORDER LAGOMORPHA
## *Rabbit and Hare*

Members of this group are characterised by possessing two pairs of chisel-like upper incisors that bite against one pair in the lower jaw—an arrangement comparable with that of the rodents and diprotodont marsupials. The teeth grow continuously.

## FAMILY LEPORIDAE
### *Rabbit and Hare*

The family Leporidae includes the Hares and Rabbits, each represented by one introduced species.

## Rabbit

### *Oryctolagus cuniculus*

### (Linnaeus, 1758)

oh-rik'-toh-lah'-gus  kue-nik'-yue-lus:
'rabbit digging-hare'

Introduced into south-eastern Australia from England in 1858, the European Rabbit spread during the following 60 years across the southern half of the continent to inhabit an area of 4 million square kilometres. Colonisation occurred despite the erection of thousands of kilometres of barrier fences and was marked by great population irruptions and crashes and by severe environmental disturbance. The Rabbit continues to probe Australian habitats and is presently making a slow advance northwards along the cool highlands in tropical Queensland and is becoming a permanent inhabitant around bores and other human-made watering points in arid, tropical environments. It is now firmly established as an integral component of the Australian fauna.

Populations are distributed in characteristic patterns within each major habitat type. In arid habitats, the species is closely associated with watercourses and swamps (except in wet years when it spreads widely into dunes and stony areas); in subalpine habitats it is confined to open grassy valleys; in agricultural areas distribution patterns are the result of complex interactions between food, shelter, predation and farming practices.

The Rabbit is a grazing animal which prefers green grass and herbage. It is capable of selecting the more nutritious components in a sward and of digging below the crowns of grasses to eat seeds and the roots, thus causing qualitative changes in the composition of plant communities. Under dry conditions it also feeds on bark, leaves and roots of shrubs, from which it obtains most of its water: it rarely visits water to drink if the moisture content of the food is high enough. An interesting feature in its digestive process is coprophagy: it reingests soft faecal pellets directly from the anus and swallows these whole. The pellet, formed by bacterial fermentation in the caecum, contain assimilable food and B-group vitamins.

During the breeding season, social groups of one to three males and one to seven females are formed, each group led by an aggressive male and female and protecting a territory. Groups tend to disperse when breeding ceases but dominant animals rarely vacate their warrens. Both sexes become sexually mature at the age of three to four months. The gestation period is 30 days and, a day or two before giving birth, a pregnant female constructs a short burrow with a nest of grass and fur into which the blind young are born. Their eyes open at the age of ten days and the young emerge from the burrow at 21 days. Litter size varies with age, season and nutrition but is usually four to seven. The number of litters produced in a year is one to two in the drier areas and up to five or more in more favourable environments. The mean number of young produced annually by a female varies from 11 in marginal areas to 25 or more in productive habitats.

Average longevity is very short. More than 80 per cent of animals die before reaching the age of three months but, once adulthood is reached, the mortality rate falls. Animals two to three years old are common in favourable areas and an age of six to seven years is not uncommon, especially in arid populations. The principal causes of mortality are myxomatosis and predation by the Fox and Cat. Myxomatosis is now endemic in wild popula-

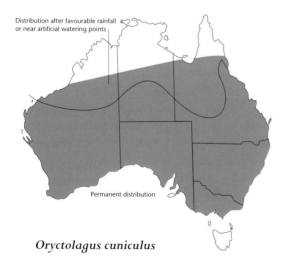

Distribution after favourable rainfall or near artificial watering points

Permanent distribution

*Oryctolagus cuniculus*

## *Oryctolagus cuniculus*

**Size**

*Head and Body Length*
356–424 (392) mm (males)
356–409 (385) mm (females)
*Weight**
0.98–2.21 (1.59 kg (males)
0.96–2.42 (1.57) kg (females)
* Recalculated from weights of animals with abdominal contents removed (so as to eliminate variation arising from the weight of foetuses).

**Identification** Burrowing and gregarious. Agouti-coloured coat; ears similar colour, without large black tips (as in Hare). Short front legs. Young born naked and blind in fur-lined underground nest.
**Recent Synonyms** None.
**Other Common Names** None.
**Status** Abundant.
**Subspecies** The Australian population is referable to *Oryctolagus cuniculus cuniculus*.
**References**

Myers, K., I. Parer, D. Wood and B.D. Cooke (1994). The rabbit in Australia (in) H.V. Thompson and C.M. King (eds) *The European Rabbit. The History and Biology of a Successful Coloniser.* Oxford Scientific Publications, London. pp. 108–157.

Stodart E. and I. Parer (1988). *Colonisation of Australia by the Rabbit.* Project Report No. 6. CSIRO Division of Wildlife and Ecology, Canberra.

tions and the initial case mortality rate of 99.9 per cent has fallen to less than 30 per cent.

Rabbits in Australia are infected by several species of ectoparasitic fleas and mites and by internal worms and coccidia. In wetter environments, internal parasites increase greatly and reproduction is adversely affected. Populations fluctuate differently in response to environments. In temperate areas of Mediterranean climate, when conditions are favourable, a combination of myxomatosis and predation suppresses population

*Introduced with difficulty, the Rabbit soon became abundant.* K. MYERS

*The Rabbit, which is a minor herbivore of Europe, became dominant when it was introduced to Australia.*

D. WATTS

numbers effectively. In such environments, a mortality rate of 85 per cent or more is required to match the annual ten-fold to twelve-fold net increase. In wetter areas and years, heavy parasite loads add a further burden on the breeding female. In arid areas, unpredictable rainfall and plant growth is followed by violent fluctuations in rabbit numbers, especially when predation and myxomatosis are out of phase with rabbit reproduction; plagues of rabbits reminiscent of pre-myxomatosis years are now recurring there at alarming intervals. In such areas, however, moderate mortalities of 50 per cent or less can send a population into decline when rains fail.

In Spain, the home of the stock from which Australian populations are derived, the Rabbit evolved in a highly predictable, seasonally harsh environment in which births and deaths are in balance. The main mortality factors in this region reside in food shortages and a complex array of more than 30 predators. In Australia, the environment is generally less predictable and kinder and, until myxomatosis was introduced, without major checks on population. Numbers are presently being held in check by an unstable alliance between Fox and Cat predation, myxomatosis and climatic variability.

Scientists in the Australian Animal Health Laboratories, Geelong, Victoria, are presently investigating a haemorrhagic disease as a new biological control agent in Australian Rabbits. This viral disease, first described in China in 1984, has steadily spread throughout the Rabbit-farming industry and wild Rabbit populations in Europe, causing considerable mortality.

A novel concept in Rabbit control has been launched by CSIRO scientists in Canberra, using genetic engineering, aimed at provoking an autoimmune reaction in the breeding Rabbit to specific proteins essential for successful reproduction, using a recombinant myxoma virus as a vector.

K. MYERS

# Brown Hare

## *Lepus capensis*

### Linnaeus, 1758

lep'-oos kay-pen'-sis:
'Cape (of Good Hope) hare'

The Brown Hare, a widely hunted animal in Western Europe, was brought to Australia by British settlers to provide similar sport. Although imported to Tasmania in the 1830s, it is not clear when it became established in the wild. On mainland Australia, it was introduced to Westernport Bay and Geelong between 1859 and 1865 and to Adelaide in 1869. It spread rapidly through naturally open habitats, pasture and croplands but avoided forests. In the sport of coursing, a captured hare was released in front of greyhounds, either being killed or escaping. Hares were captured and sold widely for coursing, accounting for part of the spread of the species. Isolated populations at Mackay, Ayr and Townsville in coastal Queensland almost certainly arose from hares that had escaped from coursing.

At the peak of its initial abundance, the Hare may have occurred further inland than at present. It has not become established in Western Australia (due to local opposition to their release), despite importation of small numbers from 1874 to the 1980s. Unlike the Rabbit, it has not been established on any offshore islands.

It did not spread far into inland eastern Australia, preferring the inland slopes, tablelands and coastal regions. The distribution in eastern Australia appears to be limited to areas that normally produce a growth of grasses and clovers in winter and spring. Preferring such conditions, the Hare would probably thrive if ever established in south-western Western Australia.

It spread extremely rapidly through sheep-rearing country in Victoria and New South Wales from the 1870s to 1890s, when sheep densities

*Lepus capensis*

**Size**
*Head and Body Length*
520–630 (590) mm (male)
520–640 (600) mm (female)
*Tail Length*
70–90 (80) mm (male)
70–90 (80) mm (female)
*Weight*
3.4–4.7 (4) kg (male)
3.2–5 (4) kg (female)

**Identification** Russet brown above, white on belly; clear russet on neck, chest and legs, but agouti on head, back and rump. Ears very long, with black tips. Distinguished from Rabbit by having longer ears and legs and black upper surface of the longer tail.
**Recent Synonyms** *Lepus europaeus*.
**Other Common Names** Hare, European Hare, European Brown Hare, Jackrabbit.
**Status** Locally common.
**Subspecies** In Australia, *Lepus capensis europaeus*.
**References**

Flux, J.E.C. (1965). Timing of the breeding season in the hare, *Lepus europaeus* and rabbit *Oryctolagus cuniculus* (L.). *Mammalia* 29: 557–562.

Rolls, E.C. (1969). *They All Ran Wild*. Angus and Robertson, Sydney.

Jarman, P.J. (1986). The brown hare: a herbivorous mammal in a new ecosystem (in) R.L. Kitching (ed.) *The Ecology of Exotic Animals and Plants: Some Australian Case Histories*. John Wiley and Sons, Brisbane. pp. 62–76.

were climbing towards their first historic peak. It preceded the Rabbit and Fox after Dingo densities had been reduced on pastoral land. In its first years in a district, Hare numbers rose exponentially, generally falling again within 10–20 years. At peak density, it was considered to be a major pest. In some districts in New South Wales, hundreds of thousands of bounties were paid per year and thousands could be shot on a single property in a day.

*Unlike the related Rabbit, the Hare is a solitary animal.*

In most districts, the fall in density coincided with the initial rise in Rabbit numbers. In some areas, its numbers may have risen again when myxomatosis (to which the Hare is not susceptible) reduced Rabbit populations in the 1950s, but the two species are reported to have fluctuated synchronously in Tasmania.

The Hare feeds on grasses and a variety of forbs; it also eats many crop species and will gnaw the bark of trees (usually exotics but also some native species) and vines and may thus damage orchards, plantations and vineyards. It is nocturnal, spending the day alone, crouched and concealed in a 'form' among vegetation or rocks. Unlike the Rabbit, it does not burrow. It escapes predators that approach closely by bursting forth and running at speeds as high as 50 kilometres per hour to another form.

Breeding begins in July, two to four weeks after the winter solstice and may continue until March or April. Females bear litters of one to five (usually two or three) leverets at six-week intervals. Litters tend to be largest in November to January. Occasionally, a second litter may be conceived before the first is born. Hares born in early litters may themselves be breeding by the end of the season. An adult female produces eight to ten young per annum, but is unlikely to survive more than two or three breeding seasons.

Leverets are born well-furred and open-eyed, being much more precocious than Rabbits. The female leaves the leverets (often splitting the litter) while she forages or crouches in her own form. She visits them only once or twice daily to suckle them. They are weaned at about four weeks.

The major predator in Australia is probably the Fox, although leverets would be vulnerable to the Cat and eagles. Hares rarely occur in country occupied by the Dingo. The species is now closely associated with modified pastures and croplands and generally absent from unmodified native plant associations.

P. J. JARMAN

**Lepus capensis**

# ORDER PERISSODACTYLA
## *Odd-toed Ungulates*

*P*erissodactyls are browsing or grazing mammals that support the body mainly on the elongated third digit of each foot. In members of the family Equidae (horses, donkeys, zebras) the feet are reduced to a single (third) toe, ending in a large hoof. Perissodactyls crop vegetation with their large upper and lower incisors and utilise microorganisms in the caecum and large intestine to reduce fibrous plant material to assimilable substances.

*Provided they have access to water, feral horses can survive in semi-arid regions.* G. STEER/TERRA AUSTRALIS

# FAMILY EQUIDAE
## *Horse and Donkey*

This small family (one genus, seven or fewer species) represents the culmination of a long evolutionary line of specialisation in grazing and running, but the success of the group has, to a large extent, led to its decline in the wild. Of the two species of Horse, *Equus caballus* has been domesticated for five millennia and is now unknown as a wild animal (although there are large secondarily feral herds). The wild horse, *Equus przewalskii*, survives only in small and precarious populations. Most of the world's donkeys are also domesticated, but isolated populations of the Asian Ass, *Equus hemionus*, and undomesticated Asian Ass or Donkey, *Equus asinus*, still persist. Zebras are the only wild equids surviving in substantial numbers.

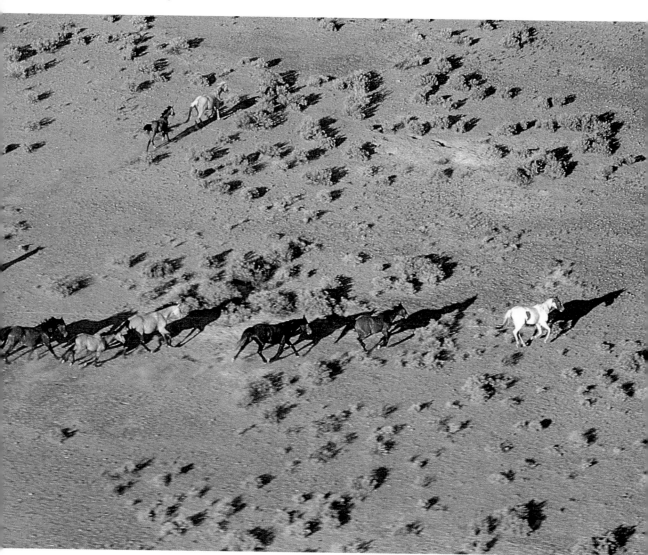

# Brumby

## *Equus caballus*

## Linnaeus 1758

ek'-wus  kah-bal'-us: 'horse horse'

Because the early development and pastoral occupation of Australian rangelands was achieved without fencing, the escape of Horses was a common occurrence. This led to the establishment of an abundant but variable population of feral animals. The feral Horse, or Brumby, now occurs over about half of Australia, being absent from much of the desert and from the intensively managed areas.

The preferred habitat is grassland and shrub steppe where pasture and drinking water are relatively abundant. In the more arid parts of its range it is found on Mitchell Grass plains, flood plains and on the dune fields of the Lake Eyre drainage basin which provide preferred grasses and ephemeral herbage. Herds of more than 100 individuals may be seen at watering or grazing places. These are made up of smaller distinct social groups which for much of the time move independently of each other. Harem groups generally contain a dominant male, one to three mares and their offspring. Young males, which are about two years old, are forced from their maternal harem

*Feral horses (Brumbies) occur in large herds in many parts of the mainland.* K. ATKINSON

group by their father and live in bachelor groups for approximately two years. They are then usually mature enough to acquire mares of their own.

Feral stocks have been derived from a variety of domestic breeds ranging from draught animals to Arabian and thoroughbred stock. Differences in origin of the founding population, geographical isolation, natural selection and culling by humans have led to the development of two recognisable types.

The inland or plains Brumby is said to display a 'coarseness' of head and body conformation but is renowned for its endurance and ability to survive under arid conditions. The alpine or mountain Brumby, isolated in the high plateaus of the Southern Tablelands of New South Wales and Victoria for almost a century, is generally smaller in stature but more strongly muscled and

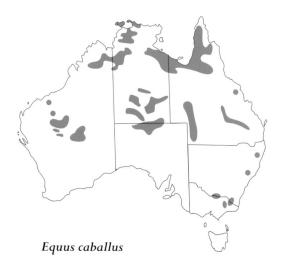

**Equus caballus**

is heavier-boned and noted for its agility and speed in rocky, mountainous country. Its conformation appears to be due partly to a substantial genetic contribution from draught horses which were once used in the mountain timber industry and partly to selection by local horseriders in favour of strongly muscled and boned mounts, referred to as 'clumpers'.

When present in large numbers, the Brumby can be a pastoral pest, destroying fences, fouling watering points and consuming pasture. Control operations are conducted to reduce this damage and avoid their suffering during drought when many die of starvation. Herds are regularly rounded up and a small proportion of the best animals are removed for domestication: the balance of those yarded are utilised for pet food or exported as meat for human consumption. Where they cannot be utilised the excess are invariably destroyed.

D. BERMAN

## Equus caballus

**Size**
*Height of Back (withers)*
up to 165 cm (plains type)
up to 150 cm (alpine type)
*Girth*
up to 180 cm (plains type)
up to 165 cm (alpine type)

**Identification** Distinguished from Donkey by shorter ears and absence of upright mane.
**Recent Synonyms** None.
**Status** Common.
**Subspecies** None.
**Reference**
Rolls, E.C. (1969). *They All Ran Wild*. Angus and Robertson, Sydney.

# Donkey

## *Equus asinus*

### Linnaeus, 1758

ah-seen'-us: 'donkey horse'

The Nubian wild ass, *E. a. africanus*, was domesticated in ancient Egypt around 4000 BC. The range of the wild form once extended across the African savanna zone from the northern reaches of Kenya to the Nubian Desert and Egypt. It now remains in a wild state only in isolated mountain areas of Somalia and Ethiopia.

The first Donkeys were brought to Australia in 1866. They were used extensively as pack animals and in haulage teams until the early 1900s, when they were superseded by motorised transport. Earliest reports of large feral populations come from the 1920s and 1930s and the species was officially declared to be vermin in parts of Western Australia in 1949. Use of the Donkey was particularly widespread in the Victoria River area of the Northern Territory and in the Kimberley pastoral district of Western Australia, where Kimberley horse disease or walkabout sickness (caused by an endemic toxic plant, *Crotalaria crispata*) restricted the use of the Horse. These regions currently harbour the highest den-

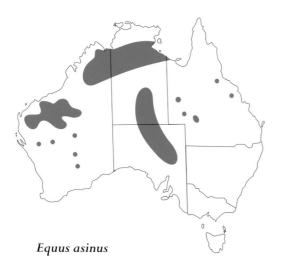

*Equus asinus*

---

*Equus asinus*

**Size**
*Head and Body Length*
250 cm
*Height*
180 cm
*Weight*
300–350 kg
**Identification**  Horse-like but smaller; with large ears; upright mane. Hair on tail forming a terminal tuft.
**Recent Synonyms**  None.
**Other Common Names**  Ass, Burro.
**Status**  Abundant.
**Subspecies**  None.
**References**
Choquenot, D. (1991). Density-dependent growth, body condition and demography in feral donkeys: testing the food hypothesis. *Ecology* 72: 805–813.
Freeland, W.J. and D. Choquenot (1990). Determinants of herbivore carrying capacity: plants, nutrients and *Equus asinus* in northern Australia. *Ecology* 71: 589–597.

---

sities of the feral Donkey in Australia, more than 10 per square kilometre being common in some areas. Densities decline into the semi-arid regions and deserts of central and western Australia, where populations become more susceptible to variation in seasonal conditions. Absolute numbers are difficult to estimate but there are certainly more than 2 million and perhaps as many as 5 million in Australia.

Virtually all research on the feral Donkey in Australia has been conducted in the northern third of its range. In these areas, it occupies tropical savannas, with regular cycles of grass growth and die-back associated with the annual monsoon. In these habitats, it continues to grow until three to four years of age. When fully grown, males are larger than females. Both reach sexual maturity at two to three years of age but females do not usually bear a foal until their fourth year. Females are

J. LOCHMAN

*Feral Donkeys can survive in a wide range of semi-arid and arid environments.*

highly fecund, more than 70 per cent breeding each year. This leads to annual population growth approaching 25 per cent under good seasonal conditions or when recovering from culling. As a population recovers and higher densities are attained, females continue to breed but their capacity to rear their offspring declines. This decline is related to increasing competition for favoured forage species; females ultimately being unable to procure a diet sufficiently nutritious to meet the demands of lactation. Mortality of juveniles eventually increases to more than 60 per cent in the first six months of life, with less dramatic increases in mortality of adults. At this stage, population growth stalls and an equilibrium with the food supply is established until disturbed by a shift in seasonal conditions or by culling.

Although the social organisation and behaviour of the feral Donkey has been studied extensively in semi-arid areas in North America, little work has been conducted in Australia. In America, dominant males maintain harems within loose territories. Harems consist of related females, their daughters and immature males. Younger males form associations but generally remain without established territories until successfully challenging for a harem.

The feral Donkey is considered to be an agricultural and environmental pest, probably competing with stock and native grazers for food and water. It is also blamed for extensive erosion in the rugged hills of the Kimberley and Victoria River regions. For these reasons, large-scale culling campaigns are regularly carried out in many parts of northern Australia. Donkeys are shot from helicopters with high-powered rifles. Between 1978 and 1987, 180 000 donkeys were shot in the Kimberley region; 83 000 were shot between 1981 and 1984 in the Victoria River area. The success of these campaigns in limiting the impact of donkeys has not been ascertained.

D. CHOQUENOT

# ORDER ARTIODACTYLA
## *Even-toed Ungulates*

embers of this large and successful group of browsing or grazing animals support the body mostly on the equally developed third and fourth toes of each foot, each ending in a hoof. The upper incisors are reduced or absent and vegetation is cropped by the lower incisors against a pad at the front of the palate. Fibrous plant material is fermented by micro-organisms in a multi-chambered stomach.

If there are processes on the head, these have a bony base at centre. When permanent, these are referred to as horns (as in cattle); when shed annually (as in deer) they are called antlers.

*The One-humped Camel has broad, padded feet and does much less damage to the soil than the hard-hoofed Horse and Donkey.* J. LOCHMAN

# FAMILY SUIDAE
## Pig

Although a relatively small family (five genera, eight species), the Suidae has species native to all continents except Australia. Pigs occur throughout The Old World; peccaries are confined to The New World.

Pigs are artiodactyls, but far less specialised than most other members of the suborder. The feet have four toes but most of the weight of the body is taken on the third and fourth toes, which end in small hooves. Unlike the more specialised artiodactyls, they are omnivorous; they retain upper incisors; and the stomach is not sacculated. The canine teeth develop as tusks, prominent in adult males.

Unlike other female artiodactyls, which usually produce only one young each year, a Pig may produce up to 14, suckled on up to 12 teats.

## Pig

### Sus scrofa

### Linnaeus, 1758

soos skroh'-fah: 'sow pig'

*Sus scrofa*

Descended from European domestic stock or possibly, in the north, from animals introduced from Timor or New Guinea, the feral Pig is now a very significant mammalian pest of agriculture and crown land throughout Australia, except in extremely arid areas. In many parts of the country, it is a reservoir of the viruses of Murray Valley encephalitis and Ross River fever. In north Queensland, it is commonly infected with meliodosis, brucellosis, leptospirosis and sparganosis, making it a health risk to humans during field-butchering and when poorly cooked meat is consumed. Great problems could occur in their role as an infection reservoir and/or transmitter of exotic disease if any

of these diseases entered Australia. Control of the Pig by Government agencies is of high priority throughout its range.

Present in most climatic regions of Australia, it is limited in its local distribution by the availability of appropriate food, access to water and reasonably undisturbed shelter. The Pig is an opportunistic omnivore with an alimentary tract similar to that of humans and (lacking a significant caecum) it cannot utilise gut microflora

## Sus scrofa

**Size**

*Head and Body Length*
110–165 cm (male)
110–140 cm (female)

*Tail Length*
15–30 cm (male)
15–30 cm (female)

*Weight* (5 years)
35–175 kg (males)
25–110 kg (female)

**Identification** Predominantly black (other colours indicate either recent crosses with domestic stock or expression of recessive alleles). Smaller and with narrower hindquarters than domestic breeds. Coarse mane erected when animal is stressed.

**Recent Synonyms** None.

**Other Common Names** None.

**Status** Abundant.

**Subspecies** Uncertain.

**References**

Giles, J.R. (1980). *The ecology of the feral pig in western New South Wales*. Ph.D. Thesis, University of Sydney.

Hone, J. (1980). Effect of pig rooting on introduced and native pasture in north-eastern NSW. *J. Aust. Inst. Agric. Sci.* 46: 130–132.

Pavlov, P.M. (1991). *Aspects of feral pig (*Sus scrofa*) ecology in semi-arid and tropical areas of eastern Australia*. Ph.D. Thesis, Monash University, Clayton, Vic.

G.A. HOYE

*The feral Pig eats a variety of plant material and is a predator on small vertebrates.*

effectively to digest the plant fibre which can comprise more than 30 per cent of its food intake. It competes with domestic livestock for pasture forbs during autumn and winter in southern regions and damages cereal crops and introduced pastures. During and after the wet season, it competes with domestic livestock for ephemeral swamp vegetation in tropical regions and it also does significant damage to sugar cane, peanuts and corn. Any available carrion is eaten and cannibalism also occurs. The Pig kills and eats lambs under two weeks of age and is a predator of other small mammals, amphibians, reptiles, ground-nesting birds, their eggs and offspring, as well as soil invertebrates. It has a negative impact on many plant species, their seeds and seedlings.

It is an aggressive competitor for food with native animals in tropical rainforests. It wallows in mud during hot weather, which helps cool the body by evaporation. If mud is unavailable, dust wallows are dug in shady areas to provide maximum contact with cooler soil. Group size varies with environment and degree of harassment. In savanna and flood plain areas of north Queensland, groups of 60 or more are common. In cloudy weather (with or without rain) it remains active throughout the day.

Males more than 12 months old are 10–12 kilograms heavier than females of the same age, developing a thickened cartilaginous shield over the shoulders and ribs, which protects them when fighting. Males are solitary after 18 months of age, joining groups only to breed.

Under favourable conditions breeding occurs throughout the year with a peak from May to

*The feral Pig causes considerable environmental damage by wallowing in waterholes.*

October. Females breed when about seven months old and can have two litters of four to six each year. A few days before giving birth, the female constructs a nest of grass, tunnelling into it to give birth. If nesting material is limited, the nest is simply a saucer-shaped heap. She remains in or near the nest with the young for approximately two weeks after which the piglets accompany her and the family will join a group (a protective advantage if the Dingo is present). The young are weaned when three to four months old and remain with her until the next litter is born. A female comes into oestrus before her piglets are weaned. Juveniles run together until the young sows mate. Males remain together until they are approximately 18 months old.

Populations fluctuate markedly, increasing rapidly in response to good conditions. During drought, on mainland Australia, sows may not reach the body weight (25–30 kilograms) or have a protein intake (more than 14 per cent) necessary for oestrus to be initiated. On islands off the east and north coast of Queensland, covered mainly by savanna vegetation, adult Pigs are up to 20 kilograms lighter than their mainland cousins of equivalent age and sows have been recorded to breed between at a weight of 13–15 kilograms.

The Dingo is a predator of piglets and adult Pigs in Queensland, the Northern Territory and Western Australia, but it is not known whether they limit populations. Large group-size reduces Dingo predation efficiency. In tropical areas, feral Pigs carry heavy burdens of helminth parasites from an early age, reducing the viability of some individuals during the annual dry season.

Deliberate release of piglets and juveniles by unscrupulous hunters in patches of scrub in rural areas and adjacent to settled areas has been a common occurrence in New South Wales and Queensland. This has led to a rapid increase in Pig distribution since the 1970s.

P.M. Pavlov

# FAMILY CAMELIDAE
## *Camel*

*The feral One-humped Camel is well adapted to life in arid regions of Australia.*

Camelids are ruminants and have only two toes (the third and fourth) on each foot. In these respects they resemble deer and cattle but, unlike these, they do not support themselves on the tips of their toes. The toes are directed more or less horizontally and rest on fleshy cushions: the hooves are small and serve only to protect the front of the foot.

Camelids originated in North America and the ancestors of modern camels did not reach the Old World until about two million years ago. Somewhat later, the ancestors of the llamas entered South America and camelids became extinct in their original home. Both camels and llamas have been domesticated.

# One-humped Camel

## *Camelus dromedarius*

### Linnaeus, 1758

kah-mel'-us drom'-ed-ar'-ee-us: 'dromedary camel'

The wild ancestor of the domestic One-humped Camel became extinct in prehistoric times. Only in Australia have Camels become feral in large numbers and only here do they live in natural populations. They are descendants of many thousands of animals introduced between 1840 and 1907, mostly from former British India. Camels were used as beasts of burden in desert explorations and construction of railway and telegraph lines, but primarily for supplying goods to remote settlements, stations and mines in the outback. The domestic population reached a peak of about 20 000 in the 1920s but, when motor vehicles became available, their numbers declined. Many unwanted animals were released and their descendants are feral. Current numbers are estimated at up to 100 000.

The only large, browsing herbivore in Australia, it is widely distributed over the arid and semi-arid regions of Central and western

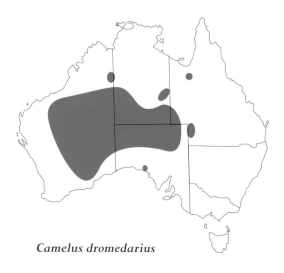

*Camelus dromedarius*

---

*Camelus dromedarius*

**Size**
*Head and Body Length*
225–345 (300) cm
*Shoulder Height*
180–225 (210) cm
*Weight*
600–1000 kg (including 20–200 kg of hump)

**Identification** Readily distinguished from all other Australian mammals. Distinguished from Bactrian Camel by a single hump.
**Recent Synonyms** None.
**Other Common Names** Dromedary.
**Status** Sparse.
**Subspecies** None, but two breeds are generally distinguished: a slender riding form (the dromedary proper) and a heavier pack-animal.

---

Australia. The unique dexterity of its split upper lip enables it to browse selectively: it can reach from ground level to a height of 3.5 metres. Camels in Australia feed up to 98 per cent on shrubs and forbs; grasses are of some importance only after rainfall, before forbs become available. The Camel eats more than 80 per cent of the plant species in its environment, preferring those of high moisture and salt content and tolerating high oxalate concentrations. It can utilise thorny, bitter and even toxic plants that are avoided by other herbivorous mammals.

Habitat preference is determined by climatic and nutritional conditions. The Camel is found in bushland and sand plains and, due to the abundance of shrubs, these habitats offer a regular and varied food supply throughout the year. In summer, sandplains are more attractive as succulent plants such as Parakeelia, *Calandrinia* spp., develop; desert oaks and other big trees provide shelter during the hotter hours of the day. Salt lakes and salt marshes are visited mainly in winter to feed on *Swainsonia* species, which are favoured foods.

The Camel is well adapted to desert environments. For example, its eyes are adapted to excessive light, protected by thick long eyelashes.

### References

Gauthier-Pilters, H. and A. Dagg (1981). *The Camel*. University of Chicago Press, Chicago and London.

McKnight, T.L. (1969). *The Camel in Australia*. Melb. Univ. Press, Melbourne.

Yagil, R. (1985). The Desert Camel. *Animal Nutrition* 5. Karger. Basel, Switzerland.

The eyelids are translucent, preserving some sense of direction in a sandstorm. The slit-like nostrils can be closed against driving sand. An ability to conserve water is its most important survival strategy: it has mechanisms to conserve water in urine, faeces and exhaled air. It has a low metabolic rate and a slow turnover of water and is able to endure extreme dehydration of up to 40 per cent of its body weight without serious effect. In hot weather it can sustain a body temperature of up to 41°C and does not need to sweat until the ambient temperature is above this level. To replace water losses, a dehydrated Camel may drink up to 200 litres within three minutes. In the Australian winter a Camel can go without water for several months, provided there is enough moisture in its food plants. In summer, if water is available, it may drink every other day, but it can live for weeks on succulent plants. The hump, consisting mostly of fat, is an energy store, not a water reservoir.

The Camel is a social animal, living in non-territorial groups. Three types of groups can be distinguished. Bull groups occur throughout the year and consist of males of all ages; elder bulls tend to live solitarily. Cow groups consist of females and their calves and occur only during summer, outside the breeding season. Breeding groups consist of one adult bull and several cows with their calves. These form in the breeding season, in winter. Group size ranges from 2 to 45 (average 11). In summer, groups may congregate, especially during droughts, forming large herds of hundreds of individuals.

Before parturition, cows segregate from their original group and give birth in seclusion. After having lived alone for up to three weeks they join

*The One-humped Camel is social, moving in groups of up to 45 (average 11) animals.* J. LOCHMAN

up with other young mothers, forming new groups. These core groups are subsequently stable until the young are weaned at one-and-a-half to two years. Other cows and some young bulls may attach themselves for varying periods.

During the rut, from May to October, a strong bull associates with a group of up to 24 cows and monopolises these against competitors. He chases away all young bulls, who then join other bachelors to form loose associations of 20 or more individuals. The young cows remain in their maternal group.

Rival bulls perform urination ceremonies, parallel walks and other displays and may engage in serious fights. Tenure of a group is three to six months and is terminated voluntarily or when a stronger bull takes over. During summer, groups of cows are escorted by younger and/or weaker bulls, or live on their own.

The gestation periods is 360–380 days. The single young weighs about 40 kilograms at birth and is suckled for one-and-a-half years. Birth intervals in feral Camels are 18–24 months; if the previous calf dies, birth intervals may decline to as little as 12 months.

The Camel does not seem to degrade the Australian desert environment. Its feeding habits are different from those of other ungulates: it usually spends little time in one locality, moving slowly and taking a bit here and there. Its independence from regular drinking enables it to use more remote areas and its soft foot-pads cause less soil erosion than the sharp hooves of other introduced herbivores.

The Camel has no enemies other than humans and, in contrast to other areas in the world, the Australian population is not affected very much by disease. Until a few years ago it was considered to be a pest and often shot on sight. Recently this attitude has changed and the species is now looked upon as a valuable asset.

B. DORGES AND J. HEUCKE

# FAMILY BOVIDAE
## *Horned ruminants*

*The feral Goat is well-adapted to arid environments but needs to drink.* C. ANDREW HENLEY/LARUS

Bovids are the most advanced and successful of the artiodactyls, being the most efficient eaters of grass. They comprise some 53 genera and 99 species but, like the deer, their adaptations to a variety of habitats—from tropical rainforests to tundra and deserts—have led to enormous subspeciation. More than 800 subspecies have been described and, even if half of these are dismissed as arising from the over-enthusiasm of hunter-naturalists, this is evidence of great adaptability. The subspecific diversity may, indeed, indicate that the group is still evolving rapidly.

Bovids have two hoofs on each foot, at the ends of the third and fourth toes, the other toes being absent or extremely reduced. They have a large and elaborately divided stomach for ruminant digestion. Females may or may not have horns, but these are always present in males.

Three closely related species of cattle—*Bos taurus* (European cattle), *Bos indicus* (zebu) and *Bos javanicus* (banteng)—have been domesticated. Generally, these are left free to find their own food and are rounded up periodically for branding, culling or medical treatment. Their condition, which is thus close to feral, may become so when, because of difficulty or loss of interest in their management, herds are left to themselves. The only predominantly feral species is the Water Buffalo.

Sheep receive more management than cattle and most of the breeds in Australia require attention from humans in order to remain in reasonable health. There are some small populations of feral sheep but these may not be permanent.

The Goat, which is well adapted to aridity, is able to browse and to graze on a wide variety of plants. It is a serious pest in many parts of Australia and, because of its agility in rocky terrain, it is difficult, if not impossible, to eradicate populations by shooting.

R. STRAHAN

# Swamp Buffalo

## *Bubalus bubalis*

### (Linnaeus, 1758)

bue-bah'-lus bue-bah'-lis: 'buffalo buffalo'

*The feral Swamp Buffalo is responsible for much soil erosion and the destruction of tropical wetlands.*

The original wild Asian Buffalo, *Bubalis arnee*, once widespread in mainland South-East Asia, is now endangered and confined to small areas in India (Assam, Madhya Pradesh) and Nepal. Two types of domesticated or feral Buffalo, *Bubalus bubalis*, are common throughout the

Old World: river-type buffalos, *B. b. bubalis*, which have curled horns and are native to the western half of Asia; and swamp-type buffaloes, *B. b. kerabau*, which have swept-back horns and are native to the eastern half of Asia from India to Taiwan. These are the stock from which animals were transported to virtually the rest of the world, including Australia; they are better referred to as Swamp Buffalos than Water Buffalos. Between 1825 and 1843 about 80 of the Swamp Buffalos were transported from Indonesia to the early settlements at Melville Island and Cobourg

### Bubalus bubalis

**Size**

*Head and Body Length*
up to 260 cm
*Shoulder Height*
up to 180 cm
*Weight*
450–1200 kg

**Identification** Large, long and wide swept-back horns which are triangular in cross-section (oval or round in *Bos*). Male horns are more massive and less curved than females and both have corrugations on the basal half. Pelage entirely slaty-grey, or sometimes with a white 'V' below the neck; off-white 'stockings' below the hocks and knees; and a dark band just above the hooves. Rarely albinoid or with depigmented areas. Body is sparsely haired. Ears relatively narrow and densely haired.

Peninsula. These settlements were abandoned by 1849 and so, too, were the Buffalos. They soon multiplied and colonised the major flood plain, woodland and sandstone escarpment habitats in the Top End of the Northern Territory where the annual rainfall is 1000–1600 millimetres and there are permanent and semi-permanent swamps and freshwater springs.

Size, growth and distribution of populations are dependent on the dry-season rainfall; low rainfall results in massive mortality and restricts the Buffalo to the plains around the northern rivers and associated swamps. The highest total population recorded in Australia (in 1985) was 340 000, representing more than half the world's free-ranging animals. Densities in preferred habitats, particularly the ecotone between forest and flood plain, were up to 34 per square kilometre.

The Swamp Buffalo is a ruminant which grazes selectively during the wet season (October–April) on sedges and aquatic grasses. It is unselective in the dry season and consumes most available grasses and herbage, as well as browsing on *Pandanus* and *Cathormium* leaves. An individual drinks about 20 litres of water per day and large bulls consume up to the equivalent of 30 kilograms of dry matter each day.

When food and water are abundant, it sleeps at night in a more or less permanent camp in wooded country, moving out at dawn to a feeding ground. Feeding ceases by mid-morning when it moves to water, first to drink and then to wallow. In mid-afternoon, it feeds again, returning to camp at dusk. If the dry season is extensive, it grazes at night and spends most of the day in wallows.

During the dry season, bulls and cows live apart. A maternal group, led by an aged female and consisting of other females and calves, mainly inhabits the heavy forested fringe of the blacksoil plains where water, shade and green food are plentiful. When adult females move onto the open plains to feed, the calves are left in a 'creche' at the edge of the plain with a 'nurse-maid'. Loose associations of adult males (bachelor groups) live more on the open plains, where there is little or no shade, or move further upslope in the open forests, where there is only dry feed.

At the onset of the wet season, males move into the areas occupied by females and associate with oestrus females. At this time, young males (two to three years old) are driven away from the maternal group whereas female calves remain with the groups and their mothers, possibly for life. Mating extends over eight months with a peak in March. One male may mate with several, but not all, cows in a group. The gestation period is 312–334 days and, in a lifetime of 20 years, a female may produce up to 12 young. The major natural predators are crocodiles and the Dingo (hunting in packs).

Both sexes, but particularly cows, are remarkably sedentary: groups occupy well defined home ranges of 200–1000 hectares for many years. Each range includes a grazing area, camp, wallow, drinking point and rubbing trees—all linked by well-defined trails. Groups of 50–250 may associate to form herds up to 500 animals. When food and water are scarce, individuals often die rather than move to another area; many die bogged in the viscous mud of shrunken waterholes. During an extended wet season, males may travel long distances: lone bulls have been recorded as far south as Tennant Creek, as far east as Townsville and as far west as Broome.

From 1880 until recently, the Swamp Buffalo has supported important industries in the Northern Territory: meat for human consumption (local and overseas), pet meat, hides, horns, animals for live export and trophies for big-game hunters.

On the debit side, the species has caused severe environmental damage, including accelerated soil erosion, channelling of floodwaters, saltwater intrusion into freshwater habitats, loss of vegetative cover, reduction in the diversity and abundance of wetland flora and fauna and disfigurement of landscapes by their wallows, trails and dung pats. The severest impact has been in the coastal blacksoil wetlands east of Darwin some of which is now incorporated into Kakadu National Park. In addition, the Swamp

**Recent Synonyms** *Bos bubalus*, *Bos buffelus*, *Buffelus indicus*.

**Other Common Names** Asian Water Buffalo, Asiatic Buffalo, Water Buffalo.

**Status** Sparse, locally common.

**Subspecies** *Bubalus bubalis kerabau* domesticated or feral, throughout eastern Asia, translocated to The New World and Australia.
*Bubalus bubalis bubalis* the River Buffalo, mostly domesticated, throughout western Asia and the Old World.

**References**

Freeland, W.J. and W.J. Boulton (1990). Feral Water Buffalo (*Bubalus bubalis*) in the major floodplains of the Top End, Northern Territory, Australia: population growth and the brucellosis and tuberculosis eradication campaign. *Aust. Wildl. Res.* 17: 411–420.

Tulloch, D.G. (1974). The feral Swamp Buffalos of Australia's Northern Territory (in) W.R. Cockrill, (ed.) *The Husbandry and Health of the Domestic Buffalo.* Food and Agricultural Organisation of the United Nations. pp. 493–505, 757–764.

Buffalo spreads cattle diseases, particularly tuberculosis (up to 25 per cent incidence in 1964). Consequently, both the Australian National Conservation Agency and the national Brucellosis and Tuberculosis Eradication Campaign (BTEC) require that feral Swamp Buffalo be eliminated and a massive shooting program (from helicopters) was commenced in the mid-1980s. As a result, the feral Swamp Buffalo has now been virtually eliminated from the northern wetlands. However, as interest in buffalo-farming grows, increasing numbers of redomesticated herds are being created, including one managed by the traditional Aboriginal owners of Kakadu National Park.

L.K. CORBETT

*Bubalus bubalis*

# Bali Banteng

## *Bos javanicus*

### D'Alton, 1823

bos jah-van'-ik-us: 'Java ox'

*The Bali Banteng has not spread far from its area of introduction on Cobourg Peninsula.*

This is the domesticated strain of the Banteng, a wild ox indigenous to South-East Asia, formerly occurring from India through Burma and Thailand to Indochina and south to Peninsular Malaysia and the islands of Borneo, Java and Bali. In January 1849, 20 domesticated Banteng were imported from Denpasar on the island of Bali (hence the name Bali Banteng or Bali Cattle) to the British settlement at Victoria on Cobourg Peninsula. The intention was to maintain the herd as a source of food but the settlement was abandoned in the same year and the cattle ran wild. Almost 100 years later, in 1948, the Bali Banteng was 'rediscovered' still confined to the Peninsula, which is now also known as Gurig National Park.

Population estimates from aerial and ground surveys from the 1960s to 1990s vary from 1500–3500. The species has been recorded in all habitats throughout the Peninsula but appears to be concentrated in the south-eastern and southern coastal areas. In 1985, the mean density was 2.5 per square kilometre. The preferred habitat is monsoon forest (which supported a density of 70 per square kilometre in 1988) and associated coastal plains (with freshwater lagoons).

The Bali Banteng feeds in open coastal areas, usually at night and crepuscular periods, retiring to monsoon forest and other densely forested areas during the middle of the day. The common sedge, *Fibristylis cymosa*, is a favoured food but it also browses on several species of trees and shrubs. It drinks fresh water every day and occasionally drinks sea water, possibly to acquire minerals.

In contrast to the Buffalo and Pig, it has not dispersed beyond its site of introduction. This may be because of the unique grassland–monsoon forest mosaic on Cobourg Peninsula which is partly maintained by Aboriginal burning practices. Wild Banteng in Asia typically live in forested areas but graze in glades and clearings. In Asia, this habitat mosaic is mostly maintained by slash-and-burn cultivation.

*Bos javanicus*

Herds vary in size and composition. Mixed herds consist of 2–60 animals, usually with one adult male per herd. Solitary males are common as well as male groups of up to nine animals. Solitary cows are rare but groups of cows and calves (up to 36 animals) are occasionally seen. In Asia, the size of mixed herds of wild Banteng varies from 2–25 or more, usually with one adult male per herd (although the herd is usually led by an old female). Wild male Bantengs may be solitary or live in small groups.

Growth is sexually dimorphic, males growing faster and attaining larger body size than females. On Cobourg Peninsula, females reach maximum size in three to four years while males continue growing until five to six years old. Males reach sexual maturity by five years of age while females mature at two to four years, producing their first calves at three to five years. On Cobourg Peninsula it breeds seasonally, mating activity peaking at the onset of the wet season during October and November. Most calves are born between June and September, but occasionally at other times. The timing of reproduction may serve to associate late lactation (the most nutritionally demanding phase of reproduction) with the flush of nutrient-rich pasture that accompanies early wet-season rains. Single births are the rule on Cobourg and the foetal sex ratio is 1:1. About one-quarter of the calves die before the age of six months: this may relate to nutritional deficiency at weaning or to predation by the Dingo.

The Bali Banteng is commonly believed to cause adverse effects on the coastal grasslands by overgrazing and trampling the burrows of small mammals. In dry years, there is a distinct browse-line in the monsoon forests and, in some dry years, many animals die from unconfirmed causes. However, a study in 1988 indicated that rooting by Pigs probably caused more soil damage than pugging by Bali Banteng and it is likely that the Buffalo also caused greater damage in earlier times when it was more abundant on Cobourg Peninsula.

There has been no consistent policy of management. In the 1960s, experimental herds were maintained near Darwin for commercial table use or interbreeding with Brahman-Shorthorn stock. These ventures were unsuccessful because of the highly temperamental nature of the Bali Banteng. In the 1970s, the Cobourg herd increased rapidly and a shooting program by rangers was instigated to reduce its adverse effects on natural ecosystems. This shooting ceased in 1978 and a fence

---

### Bos javanicus

**Size**

*Weight*
600–800 kg (wild, Asia)
up to 550 kg (feral males, estimated)
up to 400 kg (feral females, estimated)
mean 440 kg (captive males)
mean 300 kg (captive females)

**Identification** Similar to domestic cattle but smaller, more slender and easily distinguished by a sharply contrasting white oval patch on the rump and white 'stockings' from above the hocks and knees to the hooves. Young bulls and cows reddish to golden brown or fawn, whereas old bulls are black. Bulls have large, straight horns with the tips pointing out (wild Banteng bulls have angular horns with the tips turned forwards). Cows have smaller horns which usually point backwards.

# Goat

## *Capra hircus*

Linnaeus 1758

kap'-rah her'-kus: 'she-goat goat'

**Recent Synonyms** *Bos banteng, Bos sondaicus, Bibos javanicus, Bibos sondaicus.*

**Other Common Names** Bali cattle, Balinese cattle, Banteng.

**Status** Common, limited to Cobourg Peninsula.

**Subspecies** *Bos javanicus birmanicus,* mainland South-East Asia.

*B. j. lowi,* Borneo.

*B. j. javanicus,* Java to Timor and Australia.

**Extralimital Distribution** Java.

**References**

Bowman, D.M.J.S. and W.J. Panton (1991). Sign and Habitat Impact of Banteng (*Bos javanicus*) and Pig (*Sus scrofa*), Cobourg Peninsula, northern Australia. *Aust. J. Ecol.* 16: 15–17.

Choquenot, D. (1993). Growth, body condition and demography of wild banteng (*Bos javanicus*) on Coburg Peninsula, northern Australia. *J. Zool. Lond.* 231: 533–542.

was erected across the narrow neck of the Peninsula to limit cattle movements. Since 1981, the Bali Banteng has been an important economic resource to the traditional Aboriginal owners of Cobourg Peninsula. Clients pay safari operators up to $20,000 to shoot it and the operators receive about $2,500 for each trophy bull (about $400 for each cow). About 30 bulls and 80 cows are harvested annually and this appears to be an efficient management strategy.

A conservation quandary has recently arisen. The IUCN has designated the Banteng as vulnerable in Asia and it may therefore be that the Australian population represent a significant genetic resource for the species. There may be an argument for preservation of the Australian population, despite its exotic nature.

L.K. CORBETT

The Goat is hardy, thrives on rough grazing, reproduces rapidly and is a ready source of meat and milk—qualities that led to its introduction into Australia by the First Fleet in 1788 and on many subsequent occasions. John Macarthur's name is usually associated with the merino sheep that he introduced into Australia but, before this, he farmed goats. Special breeds were introduced later for their mohair and cashmere fibres and, recently, for their meat (boer goat).

Early pastoralists, townspeople, miners and railway construction workers took the Goat into inland Australia. Some were released or abandoned when commodity prices fell and others escaped but, even in areas where water was available, dense feral populations did not develop until the Dingo was controlled. Feral populations are most common in semi-arid to arid

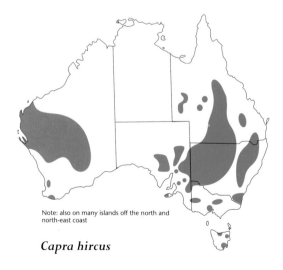

Note: also on many islands off the north and north-east coast

*Capra hircus*

pastoral areas used for sheep-grazing, where water is available and the Dingo is absent or uncommon. Smaller populations exist in conservation areas and other scrub patches of higher rainfall.

The Goat needs to drink during dry spells: individuals have been seen swimming between offshore islands to gain access to fresh water but, on oceanic islands, sea water may be drunk.

Numbers vary widely and reflect the intensity of previous control efforts. Although accurate figures are lacking, the total population, nearly

*Able to feed on virtually any foliage below a height of about 1.8 metres, uncontrolled feral Goats can cause immense environmental damage.*

all in pastoral areas, is possibly 2–3 million.

Most plants in pastoral areas are eaten to some extent. Trees and shrubs are usually favoured over grasses and ephemerals are eaten when available; there are clear preferences within each of these groups. In very high numbers the Goat is devastating: it removes virtually all foliage below 1.8 metres (or even higher if it can climb a plant or bend it down) and kills all plants within its reach. Even in low numbers, it has a noticeable impact on palatable species. Additionally, it makes the soil more vulnerable to erosion by removing protective vegetation and by breaking up the surface with its hooves.

Often in combination with the Rabbit, it has all but suppressed the regeneration of a number

J. McCANN

## *Capra hircus*

**Size**

*Head and Body Length*
120–162 (150) cm (males)
114–147 (133) cm (females)
*Tail Length*
12–17 (15) cm (males)
12–16 (14) cm (females)
*Weight*
16–79 (40) kg (males)
15–66 (27) kg (females)

**Identification**  Characteristic appearance.
**Recent Synonyms**  *Capra aegagrus hircus.*
**Other Common Names**  None.
**Status**  Abundant.
**Subspecies**  None.
**References**

Coblentz, B.E. (1978). The effects of feral goats (*Capra hircus*) on island ecosystems. *Biological Conservation* 13: 279–286.

Harrington, G.N. (ed.) (1982). The feral goat (in) P.J. Holst, (ed.) *Goats for meat and fibre in Australia.* SCA Technical Report Series No. 11. CSIRO, Canberra. pp. 1–73.

of palatable tree and shrub species in arid areas. Young plants are eaten as seedlings by the Rabbit and as growing juveniles by the Goat. Together, the two species have a far greater impact than either on its own and their depredations profoundly influence the composition of vegetation communities. Entire species have disappeared from some areas because mature plants which grow old and die are not replaced.

The Goat has been blamed for the decline of native mammals such as the Yellow-footed Rock-wallaby, *Petrogale xanthopus*, but this has not been proved.

The Goat is susceptible to many devastating animal diseases that occur overseas but not yet in Australia. Should these be introduced, they could be harboured and spread by feral populations.

Elimination of such reservoirs would be difficult because the feral Goat is hard to eradicate and because some of the measures employed could well disperse populations and spread the outbreak.

The high reproductive rate of the Goat frustrates many efforts at control and eradication. Breeding occurs through most of the year, albeit at a reduced level during early spring or in drought. Females breed when six months old and can produce kids every eight months or so. Gestation is five months. The first birth is usually single but, thereafter, twins and triplets become increasingly common. Females with a kid at foot are often pregnant. A population can increase by up to 75 per cent each year in the absence of control measures.

The feral Goat is very mobile: movements of up to 80 kilometres have been documented in New South Wales and South Australia. It moves freely through most fences in pastoral areas and readily reinvades areas from which it has been removed.

Populations are kept in check to some extent by mustering, trapping and shooting and numbers are usually well below the maximum that (albeit for only a short time) could be supported by the available vegetation. Should the species ever realise its full reproductive potential, it would attain densities that would transform most of its habitat to near desert. Any long-term approach to the species in Australia should aim at eradication rather than control.

Fortunately, the Goat is one of the few major mammal pests for which eradication can be seriously contemplated. Isolated colonies on islands and in scrub patches surrounded by farmland can be eradicated at reasonable cost. 'Judas' Goats carrying radio-transmitters and released into an area containing feral animals can be tracked with directional radio-receiving equipment, leading a hunter to within shooting range. Such individuals are far better than humans at finding feral Goats and their intensive use can lead to eradication of the last pockets of a population.

R. HENZELL

# FAMILY CERVIDAE
## *Deer*

A. YOUNG

*Female Rusa Deer and their young move in herds separate from males except in the breeding season.*

The cervids are a particularly successful group, distributed over Eurasia and the Americas in climates ranging from tropical to subpolar. Most species have a large number of subspecies, adapted to different environmental conditions.

Deer live in herds with complex social organisation, often involving considerable competition between males in the breeding season. Species of three genera—*Dama*, *Axis* and *Cervus*—are now established in Australia.

K. ATKINSON

# Fallow Deer

## *Dama dama*

## (Linnaeus, 1758)

dah'-mah  dah'-mah: 'fallow deer fallow deer'

This was the first of the cervids to become widely established in Australia. Being the common deer of the parks and large estates of England, it was comparatively easy to obtain and herds were thriving on and about the properties of several influential settlers by about 1850. Several consignments were shipped to the mainland from Tasmania—where it was first introduced and where some 20 000 animals now inhabit over 400 000 hectares of grazing country and fringe forest. It prefers grassland and forest edge to dense forest and has not colonised the native forest into which it retreated in the face of closer European settlement. Small populations survive in the vicinity of original points of liberation in all States except Western Australia.

The Fallow Deer is gregarious, occurring in groups which vary in size according to habitat and season. A group of three to four animals is usual

*The introduced Fallow Deer survives in a number of limited areas, where it appears to do little harm. It is shot for sport.*

in heavy cover; larger groups are common in undisturbed open country. Its senses are acute and it is extremely wary but when conditions permit, it remains in the open after feeding and may bed down in the morning sunshine before returning to cover. It normally feeds by day, but, in response to human disturbance, may become nocturnal. Predominantly a grazer, it moves from cover in the late afternoon to feed, at the forest edge and in clearings, on short grasses and

*Dama dama*

## *Dama dama*

### Size

*Head and Body Length*
134–165 (152) cm (males)
119–155 (148) cm (females)
*Tail Length*
to 23 cm (males)
to 19 cm (females)
*Shoulder Height*
77–98 (88) cm (males)
72–88 (83) cm (females)
*Weight*
50–97 (59) kg (males)
36–56 (38) kg (females)

**Identification** Often a group sighting. Similar to domestic goat in size. Wide range of colour, but common fawn and black forms predominate. Black and white marking of tail and buttocks conspicuous. Tail flicked continuously while feeding undisturbed; curled over back in alarm, displaying white underside. 'Adam's Apple' strikingly prominent in adult buck. Penis sheath well haired and prominent in all but very young fawns. Antlers multi-tined in adult males. Upper half of antlers palmated or flattened.

**Recent Synonyms** None.
**Other Common Names** None.
**Status** Abundant to common in Tasmania. Mainland populations mostly small and fluctuating.
**Subspecies** *Dama dama dama*, originally from Mediterranean region of southern Europe. Now widely distributed throughout Europe and introduced to Australia, New Zealand, Africa, North and South America.
*Dama dama mesopotamica*, Iran.

**References**
Bentley, A. (1978). *An Introduction to the Deer of Australia.* Forests Commission Victoria, Melbourne.
Chapman, D. and N. (1975). *Fallow Deer.* Terence Dalton, Lavenham, Suffolk.
Wapstra, J.E. (1973). *Fallow Deer in Tasmania.* Tasmanian National Parks and Wildlife Service, Hobart.

sedges. Acacias, blackberry and the tips of rushes and bracken are browsed. It is particularly attracted to improved pasture.

Males cast their antlers in October and regrowth is complete by mid-February. They begin to demonstrate mating behaviour in mid-March. A male takes possession of a territory or 'stand', marking the boundaries of it by thrashing bushes and saplings with its antlers and, at the same time, marking these with scent from the preorbital glands. The ground of the stand is marked by urinating into shallow scrapes made with the forefeet. Females in oestrus are attracted to the stand where the buck advertises its presence by a hoarse rattling call, very similar to that of a male Koala. The breeding season is usually from late March to the end of April but may vary slightly according to weather conditions and, perhaps, latitude. A single young is born in November or December: twins are very rare.

The Fallow Deer has numerous colour variants, four of which are found in Australia. In summer, the common form is reddish-brown with large, creamy white spots, with black markings on either side of the white-fringed tail; in winter it is greyish-brown and the spots are almost indiscernible. The black form, common in the eastern part of the Tasmanian range and in the vicinity of Lake George, New South Wales, has no white markings; it is black in summer and is tinged with brown on the flanks and shoulders in winter. The white form, not uncommon in Tasmania, has normal eye colour and is not an albino; it does not change colour with season. What is known as the menil form is paler in colour than the normal form, has brown markings and retains white spots throughout the year.

Destruction of cover in the preferred habitat has adversely affected several populations, particularly in south-eastern South Australia, but large numbers are now held on deer farms.

A. BENTLEY

C. ANDREW HENLEY/LARUS

# Red Deer

## *Cervus elaphus*

### Linnaeus, 1758

ser'-vus  el-ah'-fus: 'deer deer'

Made familiar to generations of British people by such paintings as Landseer's 'The Monarch of the Glen', the Red Deer is widely distributed over Eurasia. Those in Australia are descendants of English park deer and a few highland deer. In contrast to its success in New Zealand, the Red Deer has not flourished in Australia, the only major population being in the watersheds of the Brisbane and Mary Rivers where the well-watered, largely ringbarked habitat provides improved pasture and ample cover.

*The introduced Red Deer has not done well in Australia: the major population is in the watershed of the Brisbane and Mary Rivers, Queensland.*

| *Cervus elaphus* |
| --- |
| **Size** |
| *Head and Body Length* |
| 200 cm (males) |
| 190 cm (females) |
| *Tail Length* |
| 12–15 cm (males) |
| 12 cm (females) |
| *Shoulder height* |
| 114–122 cm (males) |
| 100 cm (females) |
| *Weight* |
| 136–158 kg (males) |
| 92 kg (females) |

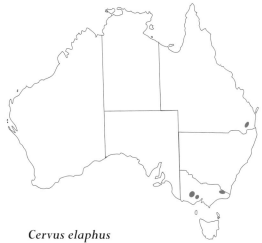

*Cervus elaphus*

is not uncommon over its range. Deer are protected in Queensland but, outside the hunting season, permits may be issued for their destruction where they cause serious damage to crops or pastures. Since 1980, some permits have been given to trap Red Deer to stock deer farms. Victoria has an open hunting season of one month each year but very few Red Deer are taken. Isolated by country developed for agriculture, the Grampians population cannot extend its range without specific management, but appears to have increased in recent years.

A. BENTLEY

A small and little-known herd, the result of a liberation near Aston, New South Wales, in about 1914, lives around the headwaters of the Snowy River and appears to be expanding slightly southward. Another is in the Grampians National Park, Victoria. Although much of the Grampians range is a sanctuary with abundant water and cover, habitat suitable to the Red Deer is limited. A remnant herd which lived in the softwood plantations and hardwood forest 30 kilometres south-west of Ballarat, Victoria and herds once established under a degree of human protection and management in South Australia and Western Australia, appear now to be extinct, or nearly so.

Breeding of the Red Deer is remarkably regular and six months out of phase with the Northern Hemisphere cycle: mating occurs from late March to April. Each mature male attempts to gather together a number of females and to hold these against the attentions of his rivals and it is at this time, usually in the morning and evening, that the male utters its lion-like aggressive roars. The young, initially spotted with white, are born from late November to December. Males cast their antlers in October or November and these are regrown by mid-February.

So far as is known, the Red Deer has no significant predator in Australia although the Dingo

**Identification**  Often group sighting of stags together or group of hinds and younger animals. Reddish colour in summer, greyish-brown in winter. Light coloured rump patch prominent. Tail not strikingly noticeable as in Sambar. Ears long and pointed. Age and abundance of food influence number of antler tines, adult stags being multi-tined: 6–8 tines common; 10–12 or more much less frequent. Antlers round in section; rarely exceeding 75–90 cm length in Australia.

**Recent Synonyms**  None.

**Other Common Names**  None.

**Status**  Queensland, common, limited; eastern Victoria, rare, scattered; western Victoria, sparse to common in limited habitat.

**Subspecies**  Some 11 subspecies. Australian population is *Cervus elaphus scoticus*, from Great Britain.

**References**

Whitehead, G.K. (1972). *Deer of the World*. Constable, London.

Darling, F.F. (1937). *A Herd of Red Deer*. Oxford Univ. Press, London.

# Rusa Deer

## *Cervus timorensis*

### de Blainville, 1822

tee'-mor-en'-sis: 'Timor deer'

The natural distribution of the Rusa Deer, which probably centres on Indonesia, has been extended by its transportation between the islands of Indonesia and Melanesia over a very long period by seafarers of different nations. Introduced into various Australian states in the latter half of the nineteenth century, it is also the most recent of the introduced deer. A herd of Javan Rusa in the Royal National Park, not far from Sydney, New South Wales, is descended from several animals

---

### *Cervus timorensis*

**Size**

*Head and Body Length*
152–185 (157) cm (males)
142–165 (149) cm (females)

*Tail Length*
circa 20 cm (males)
circa 20 cm (females)

*Shoulder Height*
93–109 (94) cm (males)
83–89 (86) cm (females)

*Weight*
58–115 (73) kg (males)
50–75 (53) kg (females)

**Identification** Greyish-brown pelage, often rough and coarse in appearance. Smaller and much less compact in body than Sambar. Tail not carried erect in alarm. Three tines to each antler in adult males. Antlers lyre-like, inner or posterior tine longer and forming continuation of main antler beam (70–75 cm long in Javan Rusa; 60–65 cm long in Moluccan Rusa).

---

obtained in 1907 from a shipment *en route* to New Zealand from New Caledonia, where it had earlier been introduced.

The Moluccan Rusa, from which the other herds developed, was brought from the Moluccan Islands in 1912 and liberated on Friday Island in Torres Strait. Deer swam from there to Prince of Wales Island where the main herd is now located. A number of these were liberated on Possession Island, 26 kilometres south-east of Thursday Island, in 1914. As recently as 1952, a male and three females were taken from Friday Island and released on an island off the northern tip of Groote Eylandt in the Gulf of Carpentaria.

Early liberations in southern Australia appear to have failed. Probably because cold winters limit the success of this thinly haired tropical species, the last sighting in Victoria was in 1940.

The preferred diet appears to be grass. In the Royal National Park south of Sydney, deer are attracted to improved pastures and on the Torres Strait islands Rusa Deer improve markedly in condition when grasses appear after the monsoonal rains. It is of interest that this population drinks sea water when fresh water is not available and appears to relish certain seaweeds.

Breeding occurs throughout the year, with a peak of mating activity from late June to August and a corresponding calving peak from about March to April. The Moluccan Rusa in Torres Strait mates in September and October, the

*Cervus timorensis*

C. ANDREW HENLEY/LARUS

calves being born in April and May. Calves are not spotted as in some other deer. During the mating season males carry masses of vegetation on their antlers, apparently as a threat gesture. Antlers are usually cast in January and February.

The Rusa Deer is gregarious but males form separate groups from the females and young except during the breeding season. An alarm bark and the shrill roaring of males during the mating season are the only known vocalisations.

It is not subject to any significant predators other than man. Among other measures taken to reduce numbers of Javan Rusa in the Royal National Park in New South Wales, permission has been given for them to be trapped for deer farms. A number of Moluccan Rusa have been taken from Prince of Wales Island for the same purpose.

A. BENTLEY

*Male Rusa Deer have lyre-like antlers.*

**Recent Synonyms** None.
**Other Common Names** Rusa.
**Status** Common, limited.
**Subspecies** Some eight subspecies. Australian population derived from *Cervus timorensis russa*, from Borneo and Java and *C. t. moluccensis* from the Moluccas.
**References**

Bentley, A. (1978). *An Introduction to the Deer of Australia*. Forests Commission of Victoria, Melbourne.

Whitehead, G.K. (1972). *Deer of the World*. Constable, London.

*Large, rounded ears are a distinguishing feature of the Sambar.*

# Sambar

## *Cervus unicolor*

### Kerr, 1792

yue'-nee-kol'-or: 'uniformly-coloured deer'

Since a Sambar invariably sees a human before it is itself seen, a bush encounter with this large deer is usually a surprise to both parties, often heralded by a startlingly explosive bark and followed by the impression of a very large dark brown animal with bat-like ears. As it turns away, the light colour of the underside of the tail and rump is most striking and the crash of the heavy body in the undergrowth confirms the identification.

Of the six species of deer established in Australia, the Sambar has proved to be the most successful. Factors contributing to its adaptability are its habit of living solitarily or in small family groups and its ability to use a wide range of the coarser vegetation abundant in the forests of south-eastern Australia.

Introduced from India, Sumatra and Sri Lanka into Victoria in the mid-nineteenth century, it is now widely distributed throughout Gippsland and has penetrated into New South Wales and the

### *Cervus unicolor*

**Size**

*Head and Body Length*
168–246 (216) cm (males)
162–241 (198) cm (females)
*Tail Length*
circa 30 cm (males)
circa 22 cm (females)
*Shoulder Height*
112–140 (127) cm (males)
102–125 (109) cm (females)
*Weight*
131–245 (192) kg (males)
109–182 (146) kg (females)

**Identification** Dark brown. Large, rounded ears about half length of head. In alarm, bushy tail raised over back, rump hair flared, displaying light coloured under-tail and inner sides of legs. Alarm often accompanied by extremely resonant bark. Only three tines to each antler in adult stags (brow tine and two tines on terminal fork). Length to about 75 cm. Front tine of terminal fork is seen as an extension of the main antler beam and may or may not be the longer tine of the fork (compare with Rusa).
**Recent Synonyms** None.
**Other Common Names** Ceylon Elk.
**Status** Sparse, possibly underestimated.

Australian Capital Territory. It was liberated on the Cobourg Peninsula in the Northern Territory in 1912, but its present range there is not known.

Although the preferred habitat is forested mountain country, it also inhabits open forest that includes suitable cover. A grazer and browser, it eats a wide variety of grasses, shrubs and tree foliage and finds the blackberry very attractive. It is not uncommon to find blackberry-infested gullies where the plants have been stripped to bare canes by the Sambar.

Much of its activity is nocturnal but, where habitat is not greatly disturbed by human activities, movement and feeding may begin in the early afternoon. It is extremely cautious and

**Subspecies** Some five subspecies. Australian population descended mainly from *Cervus unicolor unicolor* from Sri Lanka.

**References**

Downes, M.C. (1979). The Sambar Consultancy. *Australian Deer* 4: 9–20.

Downes, M.C. (1983). *The Forest Deer Project, 1982.* Aust. Deer Research Foundation, Melbourne.

Draisma, M. (1976). Data analysis—Sambar Deer. *Australian Deer* 1: 6–9.

moves into cover at first light to bed down and ruminate. After a hard frost or snow, it often lies in the sun on the edge of cover or on a more open northern slope.

Changing availability of preferred foods, severe weather, or disturbance by humans may induce local movement, but once the Sambar has colonised an area it rarely abandons it completely. Wider movement occurs in late spring and summer when animals move to the higher ground from the sheltered gullies at lower levels which are frequented in autumn and winter.

Calves, which are a uniform dark brown colour, may be born at any time of the year but there is a peak in May and June, corresponding breeding activity in September and October. Twins are very rare.

During the breeding season, long-standing wallows are opened up and the male defines its territory by rubbing its antlers against certain saplings and trees, perhaps transferring secretions from the prominent glands in front of the eyes. Most males cast their antlers in early summer but there is no regular cycle of casting and regrowth.

Eagles and the feral Dog have been known to take young calves occasionally but recreational hunting is responsible for most mortality. No detailed study has been made of populations but it has been conservatively estimated that there are about 8000 animals in Victoria.

Adaptable to a wide range of habitats, the Sambar appears to be the only deer in Australia capable of extending its range without human assistance. Its numbers appear to be controlled by the pressure of hunting.

A. BENTLEY

*Cervus unicolor*

# Chital
## *Axis axis*
### (Erxleben, 1777)

ax'-is ax'-is: 'deer (?) deer (?)'

The Chital was the first species of deer to be introduced into Australia. Close to the beginning of the nineteenth century, Dr John Harris brought a number from India to New South Wales and these produced a herd numbering about 400 animals prior to its dispersal around 1812. During the greater part of the nineteenth century, when acclimatisation societies were active and popular, the Chital was much favoured as a species for introduction and many liberations were made in various parts of Australia. All but one were

*The Chital, also known as the Spotted Deer, has not become well established. Only one small herd exists in the wild.*

unsuccessful and it is likely that the Chital's strong herding instinct and its preference for well-watered, open forest operated against its success.

Today, the only herd of Chital known to be established in the wild in Australia inhabits the country about the Maryvale Creek, some 130 kilometres north-west of Charters Towers, Queensland. The progenitors of this herd, two stags and two hinds, were liberated on Maryvale Station in, or shortly after, 1886. The open forest, watered by the Maryvale Creek and one or two small tributaries, provides ideal habitat for the now well-established herd.

Although the population is largely concentrated around the locality of the original liberation, there has been some extension of the range; availability of water being the most important limiting factor.

A very gregarious animal, the Chital is commonly seen in groups of 6–12, but concentrations of 50 to more than 100 animals are not a rare sight. Large groups often bed down in the shade of trees during the heat of the day. Feeding begins in the early afternoon and continues into the night. The diet comprises native and improved pasture grasses and other vegetation, including the leaves and fruit of the Chinese Apple which is common in some parts of the range.

The breeding season is not sharply defined. Young calves have been observed in April and May and from September to November. Twin calves have been reported more frequently than in the Hog Deer or Sambar. Calves are spotted white at birth.

The Dingo is common throughout the Chital's range and is known to prey on it, but the significance of this in the ecology of the herd is not known. Some years ago Chital populations in Queensland suffered high mortality, apparently from eating the noxious two-leafed stage of the Noogoora Burr which sprang up after drought-breaking rains.

Comparative isolation, the Queensland government's protective legislation and—even more importantly—the sympathetic concern of one or two station owners, have contributed to the

*Axis axis*

## *Axis axis*

**Size**

*Head and Body Length*

128–185 (164) cm (males)

? (females)

*Tail Length*

25 cm (males

? females

*Shoulder Height*

87–101 (93) cm (males)

80 cm (females)

*Weight*

68–113 (89) kg (males)

50 kg (females)

**Identification** Usually seen in a group. Dark band over the muzzle is a unique marking among the deer in Australia. The white throat-patch is a prominent feature of the strikingly bright markings. Three tines to each of the usually smooth, slender antlers. Brow tine often long and curved. Anterior or outer tine of the terminal fork invariably longer than the inner tine. Antler length of 70–75 cm not uncommon. Antler velvet reddish-brown during antler growth.

**Recent Synonyms** None.

**Other Common Names** Axis Deer, Spotted Deer, Indian Spotted Deer.

**Status** Common, limited. One known population.

**Subspecies** Two subspecies. Australian population derived from *Axis axis axis*, from peninsular India.

**References**

Deer Advisory Council of Victoria (1979). *Deer in Australia*. Fisheries and Wildlife Division of Ministry for Conservation, Melbourne.

Whitehead, G.K. (1972). *Deer of the World*. Constable, London.

Chital's survival. In 1979, trapping of Chital for commercial purposes and limited recreational hunting was instituted by the Queensland National Parks and Wildlife Service.

A. BENTLEY

# Hog Deer

## *Axis porcinus*

### (Zimmermann, 1780)

por-seen'-us: 'pig-like deer (?)'

Smallest of the deer in Australia (about the size of a sheep), the Hog Deer is distributed in isolated groups along the south-eastern coast of Victoria from the Tarwin River to the vicinity of Orbost and on several of the islands of the Noramunga Wildlife Reserve at Corner Inlet, Victoria. Descended from animals introduced from India and Sri Lanka in the mid-nineteenth century, the Australian population appears, now, to be the only viable wild population occurring outside its native range.

It is restricted to coastal scrublands and tea-tree swamps and has not penetrated into forest or highlands. It is primarily a grazer on native grasses, sedges and improved pasture grasses but browsing contributes to its diet. The Sallow Wattle appears to be an important food in parts of its range where a copper deficiency in the soil leads to a form of ataxia in domestic stock, which has also been reported in Fallow and Hog Deer. Feeding normally begins late in the afternoon and continues intermittently until early morning. There is some movement during the day but, where human

*Axis porcinus*

B. TRIGGS

*The Hog Deer is the smallest cervid to be introduced to Australia.*

disturbance is considerable, the Hog Deer becomes nocturnal.

Except in areas of low human disturbance where it may congregate in available feeding areas, the Hog Deer seldom occurs in groups of more than two or three. When alarmed, the group scatters, each animal seeking its own rather than a common escape route, a departure from herd behaviour which may be of survival value to an animal which is no larger than its predators and which does not occur in large numbers. The most commonly heard vocalisation is a sharp barking indicative of general alarm. A shrill call is given when an animal is startled into sudden flight.

Before Gippsland became closely settled there was considerable movement of deer between localities and even between islands and the mainland, but this is no longer apparent. There is evidence of strong territorial attachment, supported by instances of tagged individuals returning to their home ranges after being translocated almost 100 kilometres.

The Hog Deer breeds somewhat irregularly but many males are in mating condition in January and calves are most frequently seen between August and October after a gestation of about 230 days. Twins are rare.

Except perhaps in areas of limited habitat, the one-month annual hunting season, introduced in 1973, appears to have had no detrimental effect

## Axis porcinus

### Size
*Head and Body Length*
115–135 cm (males)
115–125 cm (females)
*Tail Length*
circa 14 cm (males)
circa 12 cm (females)
*Shoulder Height*
58–72 cm (males
53–58 cm (females)
*Weight*
45 kg (males)
25 kg (females)

**Identification** Yellowish to reddish-brown above, darker below; sometimes pale cream spotting in summer; dark brown in winter coat. Tail rather long, white below and at tip. Young calves spotted with white. Distinguished, within range, from other deer by small size. Antlers 30–35 cm long. One brow tine and terminal fork; additional tines not uncommon in older animals.
**Recent Synonyms** None.
**Other Common Names** Para.
**Status** Common to sparse.
**Subspecies** Two subspecies. Australian population is *Axis porcinus porcinus*, from continental India and Sri Lanka.
**References**
Presidente, P.J.A. and M. Draisma (1980). Hog Deer on Sunday Island, Victoria, Part 1. *Australian Deer* 5: 11–21.
Mayze, R. and G.I. Moore (1990). *The Hog Deer.* Aust. Deer Research Foundation, Melbourne.

on populations, nor has the slight mortality inflicted by the Dog and the Fox on young animals, or occasional poisoning by baits laid to control the Rabbit. Numbers appear to be limited mainly by the availability of suitable habitat, which is becoming less as a result of greater agricultural usage.

A. BENTLEY

# Glossary

*Allopatric* See *Sympatric*.

*Apical* At or near the tip of a structure (for example ear or tail). Opposite of *basal*.

*Arboreal* Living in a tree or trees. Contrasted with *terrestrial*, living on the ground; *aquatic*, living in water; *amphibious*, living on land and in the water.

*Arthropods* Animals such as insects, spiders, scorpions, millipedes, centipedes and crustaceans, which have jointed bodies, many legs and a hard, flexible covering or *exoskeleton*.

*Baculum* A bone present in the penis of some mammals. Its size and shape is sometimes used in identification (for example bats).

*Blastocyst* An early stage in development, when the embryo consists of a hollow ball of cells with no differentiation of organs.

*Browse, to* To feed on the leaves and twigs of trees and shrubs, *not* grasses. Hence, *browser*, an animal that browses. Compare with *graze*.

*Buccal* Pertaining to the interior of the mouth, or buccal cavity.

*Caecum* A blind branch of the alimentary canal, present in many mammals and best developed in certain herbivorous species.

*Carnivorous* Feeding on other animals. In practice, a distinction is often made between *carnivores* (flesh-eaters) and *insectivores* (eaters of insects and other arthropods). See also *Herbivorous*.

*Climax community* A stable plant community, which is the culmination of a succession of less stable states and, short of catastrophe, is unlikely to change further. In areas of high rainfall the climax community is usually a forest; in drier areas it may be savannah, shrubland or heath.

*Cline* A continuously graded variation in the characteristics of a species over distance. It is contrasted with the division of a species into subspecies (when variation is notably discontinuous).

*Cloaca* The terminal part of a gut which also receives the ducts of the urinary and female genital system.

*Convergent evolution, convergence* The evolution of similar characteristics, as adaptations to similar ways of life, by animals of quite different ancestry.

*Crepuscular* Pertaining to, or occurring around dawn or dusk. See also *Diurnal, Nocturnal*.

*Cryptic* Hidden. A cryptic species is one that hides effectively from zoologists.

*Diapause* Temporary cessation of embryonic or larval development. See *embryonic diapause*.

*Diastema* A considerable gap between the front (cutting) teeth and the rear (grinding) teeth, characteristic of many grazing and browsing animals.

*Diploid* The (normal) condition in whch each chromosome (except the sex chromosomes) in the nucleus of a cell has a similar partner. The total number of chromosomes (2N) is thus twice the number of different *kinds* of chromosome (N). The number of chromosomes characteristic of a species is usually expressed as the diploid number (for example 2N = 24).

*Diprotodont* Having the incisors of the lower jaw reduced to one functional pair, as in possums and kangaroos. (Small, nonfunctional incisors may also be present.)

*Distal* Towards or at the free (distant) end of a structure such as a limb or digit. Opposite of *proximal*.

*Distribution* The overall area in which a species is known to occur. It is not implied—and it is very rarely the case—that a species occurs in all parts of the area defined by its distribution. See also *Habitat, Home range, Range, Territory*.

*Diurnal* Pertaining to the day. An animal that is active by day is said to be diurnal. See also *Nocturnal, Crepuscular*.

*Drey* The nest of an aboreal mammal, usually more or less spherical, constructed in the branches of a tree.

*Embryonic diapause* The temporary cessation of development of an embryo. This phenomenon occurs in many bats, seals and macropods. Also referred to as delayed *implantation*.

*Epiphyte* A plant that grows upon another plant (usually a tree) and does not have roots reaching into the soil.

*Exoskeleton* The tough external covering of such invertebrate animals as *arthropods*.

*Extralimital* Outside the limits of the region under consideration. In considering the mammal fauna of Australia, we refer to the extralimital distribution of those species that also occur elsewhere.

*Exudate* A substance that has been exuded. Used here mainly to refer to the gums of certain trees and the nutritious excreta of some sap-sucking insects.

*Fluoracetates* A family of chemical compounds used as poisons, for example 1080 ('ten-eighty') against dingoes. Fluoracetates occur naturally in some plants, notably species of *Gastrolobium* from WA. Animals that have evolved in association with these plants can have considerable tolerance of the poison.

*Folivorous* Feeding on the leaves of trees.

*Forb* Any low, herbaceous plant other than a grass.

*Frugivorous* Fruit eating.

*Graze, to* To eat grass or herbs.

*Gular* Pertaining to the neck. Used here in reference to glands and pouches in the neck region.

*Habitat* An area providing the physical (rainfall, temperature, rock or soil structure, etc.) and biological (plants and animals) conditions required by a particular species. The habitat of a species is usually far less in extent than distribution indicated on a map. See also *Distribution, Home range, Range* and *Territory*.

*Hallux* First, inner or 'big' toe of hindfoot. Hence *hallucal*, pertaining to the hallux.

*Halophyte* A plant adapted to life in salty soil or salt water. Hence *halophytic*.

*Herbivorous* Feeding on plants. Herbivores may be further divided into *frugivores* (fruit-eaters), *folivores* (eaters of the leaves of trees) and into *browsers* and *grazers*.

*Home range* The area habitually traversed by an individual animal. It may be exclusive or overlap with the home ranges of other individuals of the same species. See also *Distribution, Habitat, Range, Territory*.

*Inguinal* Located in or near the groin.

*Insectivorous* Feeding on insects and other arthropods.

*Isohyet* A line on a map, linking places of equal annual rainfall.

*Karyotype* Appearance (number, size, shape) of the set of chromosomes characteristic of a species or subspecies.

*Lateritic* Pertaining to laterite, a reddish soil derived from the breakdown of underlying rocks rich in iron.

*Leguminous plant* One of a large group of plants of the family Leguminosae, including peas and wattles.

*Malvaceous plant* One of a large group of plants of the family Malvaceae, related to mallows and hibiscuses.

*Mid-successional* Here used to refer to a stage, in the course of regeneration of vegetation after a fire, between the earliest recolonisation and the establishment of a *climax community*.

*Mist net* A net made from extremely fine thread, used to capture small birds and bats.

*Myrtaceous plant* A member of the family Myrtaceae, of which the eucalypts are notable Australian examples.

*Nocturnal* An animal that is active by night is said to be nocturnal. See also *Diurnal, Crepuscular*.

*Oestrus* A state of sexual receptivity in a female (referred to in domestic animals as being 'on heat' or 'in season'). A female in oestrus is carrying a ripe ovum or ova, ready to be fertilised, and her behaviour stimulates a male, or males, to attempt copulation. Usually, copulation is not permitted until after a period of courtship on the part of the male.

The regularity with which non-pregnant females come into oestrus is referred to as the *oestrus cycle*: this may be described in terms of the underlying physiological events or merely expressed as the duration of the cycle—the time elapsed between successive oestruses. If the period of gestation is longer than the oestrous cycle, it interrupts that cycle.

A species in which the female has only one oestrus per year is said to be *monoestrous*; one that has several oestrous cycles per year is said to be *polyoestrous*. In some species, the female comes into oestrus shortly after giving birth: this is referred to as *post-partum oestrus*. Note that *oestrus* is a noun, *oestrous* an adjective.

*Olfactory* Pertaining to the sense of smell.

*Opportunistic* Used, in reference to diet, to denote the eating of any of a wide variety of foods, depending upon their availability. In respect of reproduction, it refers to a pattern of breeding that is linked with irregular favourable conditions (particularly unpredictable rainfall in arid areas) rather than to season.

*Patagium* An expanse of skin between the fore- and hindlimbs used in gliding (some marsupials) or flying (bats).

*Pinna* The protruding structure of skin and cartilage that surrounds the aperture of the outer ear. (What, in common usage, is referred to as the ear.)

*Pitfall trap* A smooth-sided container, sunk level with the ground surface and with lengths of netting or sheeting guiding small terrestrial animals towards the pitfall.

*Polyoestrous* See *Oestrus*.

*Polyprotodont* Having more than two well-developed lower incisor teeth (as in dasyuroids and bandicoots). See also *Diprotodont*.

*Pop-hole* One of a number of entrances to a complex burrow.

*Post-partum oestrus* See *Oestrus*.

*Precocial* Well furred at birth and able to move about independently.

*Prehensile* Able to grip. Used here mainly in respect of the tail of a mammal.

*Proximal* Towards the near or fixed end of a structure. Opposite of *distal*.

*Quiescent blastocyst* An early embryo (see *Blastocyst*) which has temporarily ceased development. See *Embryonic diapause*.

*Range* This term has the same meaning as *distribution*, which is a better term.

*Refugia* Plural of *refugium*, Latin for refuge. Areas to which remnant populations of a species (particularly an opportunistic breeder) may retire when conditions are unfavourable over much of its distribution.

*Relict* Surviving from the past. A relict species is one, or one of a few, surviving from a group that was once more numerous and/or more widespread.

*Rhinarium* A distinctive area of skin, often moist, around the nostrils of a mammal.

*Riparian* Pertaining to the banks of a river.

*Rumen* A compartment in the complex stomach of sheep, cattle and other ruminant animals where fibrous plant material is partially digested (fermented) by the action of micro-organisms.

*Sclerophyll* Pertaining to plants with tough leaves. Here used mainly to distinguish between two major types of eucalypt forest: *dry sclerophyll forest* which is open and *wet sclerophyll forest* which has a closed canopy. The two types intergrade.

*Sebaceous* Pertaining to oily or greasy substances. Here used in reference to *sebaceous glands* in the skin, which produce a secretion that lubricates hairs and renders them water-resistant.

*Sectorial* Cutting. Here used in reference to teeth, particularly premolars, that are adapted to cutting flesh or the bodies of invertebrate animals by a shearing action.

*Speciation* The evolutionary processes by which new species arise.

*Subspecies* An interbreeding population within a species, differing measurably from one or more other populations and usually geographically separate from these.

*Supratragus* A small flap, above the aperture of the outer ear, sometimes movable. See also *Tragus*.

*Sympatric* Living in the same area. Here used in respect of species of the same genus. Usually these occupy different areas and are then said to be *allopatric*.

*Syndactylous* Having the third and fourth toes of the hindfoot joined together except at the tip, where there are two claws (as in bandicoots and diprotodont marsupials). The condition of having the third and fourth toes separate (as in dasyuroid marsupials and most other mammals) is referred to as *didactylous*.

*Taxon* The scientific name of a category of animals. The most important is the name of a species, which consists of two words, the first being the genus name, the second the specific name. The names of subspecies, families, orders etc, are also taxa. The practice and study of naming organisms is known as *taxonomy*.

*Territory* An area occupied by one or more individuals and defended against other members of the species. A territory is usually centred on a more or less permanent nest, burrow, den or resting place. See also *Distribution, Habitat, Home range, Range*.

*Tragus* A flap, sometimes movable, situated at the front of the aperture of the outer ear. See also *Supratragus*.

*Trap-night* A measure of the intensity of trapping obtained by multiplying the number of traps used by the number of nights on which these were set.

*Type specimen* When a species is formally described for the first time, one of the specimens described is lodged in a museum to provide a permanent reference. It is known as the *type, holotype* or *type specimen*. Other specimens of the species, lodged at the same time as the holotype, are referred to as *paratypes*.

*Vibrissae* Long, stiff hairs, particularly on the face but also on the limbs, which extend the sense of touch of a mammal (for example the 'whiskers' of a cat or mouse).

*Volplane, to* To glide through the air.

# Index